THE NATURAL HISTORY OF TEXAS

INTEGRATIVE NATURAL HISTORY SERIES

Sponsored by Texas Research

Institute for Environmental Studies,

Sam Houston State University

William I. Lutterschmidt and

Brian R. Chapman, General Editors

BRIAN R. CHAPMAN

AND ERIC G. BOLEN

foreword by ANDREW SANSOM

THE NATURAL HISTORY OF TEXAS

TEXAS A&M UNIVERSITY PRESS

College Station

This paper meets the requirements of
ANSI/NISO Z39.48-1992 (Permanence of Paper).
Binding materials have been chosen for durability.
Manufactured in China through FCI Print Group
♾ ♻

Library of Congress Cataloging-in-Publication Data
Names: Chapman, Brian R., author. | Bolen, Eric G.,
author.
Title: The natural history of Texas / Brian R. Chapman
and Eric G. Bolen; foreword by Andrew Sansom.
Description: First edition. | College Station: Texas A&M
University Press, [2018] | Series: Integrative natural
history series | Includes bibliographical references and
index.
Identifiers: LCCN 2017014053 (print) | LCCN 2017018527
(ebook) | ISBN 9781623495732 (ebook) | ISBN
9781623495725 (printed case: alk. paper)
Subjects: LCSH: Natural history—Texas. | Ecology—Texas.
Classification: LCC QH105.T4 (ebook) | LCC QH105.T4 C485
2018 (print) | DDC 577.764—dc23
LC record available at https://lccn.loc.gov/2017014053

*Front cover photograph by Cookie Ballou and
Big Bend National Park, National Park Service.
Back cover photograph by Jude Smith.*

To naturalists everywhere,

past, present, and future,

but especially to

CLAUDIA "LADY BIRD" JOHNSON

(1912–2007)

A First Lady in conservation

as well as in the White House.

CONTENTS

FOREWORD

Not long ago, my wife, Nona, and I spent the weekend at Hershey Ranch near Stonewall in the Hill Country and finished off on Sunday at a farm near Manor on the east side of Austin in the Blackland Prairie. As we came down from the Balcones Escarpment and into the cornfields, I was reminded once again of the Lone Star State's sensational natural diversity and the striking contrast one often experiences when moving from one ecological region to another.

I was introduced to the wonderful diversity of the Texas landscape when I was a child growing up on the Gulf Coast, and my folks would take us each summer to an old family lodge on the Guadalupe River above Kerrville on the Edwards Plateau. There the contrast with our home ground was refreshing, particularly in the cool clear waters of one of the most iconic rivers of Texas. Later, as an undergraduate at Texas Tech University, I discovered the stark beauty of both the Texas High Plains and the stunning canyons, which are among the most beautiful in America.

As an adult, I had the great privilege of working for The Nature Conservancy in Texas, and in that capacity, I developed a full appreciation of the biogeographic provinces of Texas, which make our state one of the most biologically diverse in the country.

To date, no work has more elaborately described the remarkable biodiversity of Texas than this monumental volume by Brian R. Chapman, senior research scientist at the Texas Research Institute for Environmental Studies at Sam Houston State University, and Eric G. Bolen, professor emeritus of biology and marine biology at the University of North Carolina Wilmington.

Here in these pages, Chapman and Bolen carefully detail the biological constituents of each of the ecoregions of Texas and, in often eloquent prose, create a historical context for interpreting and understanding them and their place in the state's rich natural history. More importantly, this work is an essential guide for those who seek to protect the richness of the Texas landscape from the unprecedented population growth that will double the number of Texans in the next 50 years.

This explosive population expansion contributes to the greatest threat to the natural diversity of

Texas—the continued fragmentation of family lands across the state. The land in Texas is owned almost entirely by private citizens, and as our cities expand, as ancestral families experience generational change, and as highways, transmission lines, and other human infrastructure reach out into the countryside, the landscape along with the plant communities, the wildlife habitats, and the creatures it sustains continue to be broken up.

Another threat to the rich cornucopia of natural assets described here is the increasing incursion of exotic and invasive species of both plants and animals, which may in the long run pose an even greater threat to lands and waters, a threat unequaled anywhere else in the country in its variety and complexity.

And so, may *The Natural History of Texas* serve as both a guide to the rich natural history of Texas and a clarion call for its protection.

ANDREW SANSOM
Executive Director
The Meadows Center for Water
and the Environment
Texas State University
San Marcos, Texas

PREFACE

Texas famously evokes many images. A sea of oil rigs, for one, dusty rangelands packed with cattle, for another. And, of course, the Alamo, but most of all, an image of just plain BIG, whether it's state-sized counties, sprawling ranches, or expansive tales of just how much bigger and better things are in the Lone Star State—all true, of course. To take a ride from El Paso to Orange or from Texline to Brownsville, plan on two days of hard driving even on the best of roads. But on such a trip, look around and marvel at another of the state's assets: biological diversity. Indeed, few states, if any, can match Texas, not only for its mix of grasslands tall and short, its forests, deserts, and coastline, but also for its myriad species and ecological relationships in each of these landscapes. Geological features add to this varied realm—the weather-making Balcones Escarpment, dark bat-filled caves, fossil palms and mammoths, rocks embedded with the footprints of ancient monsters, and a Texas-sized sandbox replete with shifting dunes—all add to the wonders of Texas. For birders, Golden-cheeked and Colima Warblers and a winter flock of Whooping Cranes spice the rich avifauna of Texas. For others, stick-thin Texas Blind Salamanders, American Paddlefish, Nine-banded Armadillos, or American Burying Beetles might be of interest, or any of the more than 720 species of grasses, or the sites known for their "lost" pines and maples.

This, then, is the setting that stimulated—and certainly challenged—our wish to write about the natural history of Texas. The idea germinated as we worked on a new edition of a somewhat similar book with a continental view—*Ecology of North America*—whose scope, despite our wishes, precluded a closer look at Texas. True to its title, *The Natural History of Texas* partitions the state into well-known natural regions, each with its own distinctive flavor of plant and animal communities. Few Texans, if any, will fail to recognize regions such as the Piney Woods, Edwards Plateau, or Trans-Pecos, yet they may not be fully aware of the natural wonders of each. Others, visiting from distant states or provinces, should readily grasp the individual character of these regions with even a cursory review of our work. At least, that is our hope.

The first chapter provides an overview of natural history in Texas, and of how biologists in the past have dealt with defining and recognizing ecological areas within the state; a few brief biographies of the key figures in these efforts supplement the introduction. The remaining chapters each describe one of the 11 regions, first noting the dominant vegetation and other prominent characteristics, followed by a section entitled "Highlights" that provides a closer look at some of the region's more interesting features. In the High Plains chapter, for example, after a description of the shortgrass prairie and playa lakes, the "Highlights" include a look at the Dust Bowl and the ecology of Tumbleweeds, among other topics. These chapters close with a short overview concerning the conservation and management of each region's natural resources. At the end of the book are a glossary, a list of Latin names for the species mentioned in the text, a list of readings and references, and information about some of the plants and animals officially recognized as state icons.

"Infoboxes" appear throughout the text; these present biographies of special people (Geraldine Ellis Watson and Roy Bedichek), phenomena (squirrel migrations), historical events (extinction of the Passenger Pigeon and a war over salt), and cultural distinctions (the rise of "boogie woogie" in logging towns). With these, we gladly included Judge Roy Bean and his courtroom-saloon as well as Connie Hagar, the "Bird Lady of Rockport."

Each of us has had considerable experience in Texas; BRC is Texas born and bred and is still a resident, whereas Texas became the adopted state for EGB, who lived in the state for 26 years. Our professional association began nearly half a century ago at Texas Tech University and has continued in the years since, firmly cemented by our common training and research in field biology and concerns for a healthy environment. Our teaching roster included courses in vertebrate zoology, ecology, wildlife management, and conservation biology. This book thus seems an almost predictable outcome from our deep appreciation for Texas and its natural history.

BRIAN R. CHAPMAN, *Huntsville, Texas*
ERIC G. BOLEN, *Wilmington, North Carolina*

ACKNOWLEDGMENTS

Few books see daylight without the help of others, and we gratefully acknowledge those who have contributed to our effort, although we alone remain responsible for any errors that may mar the pages that follow. Our profound thanks go to the many individuals who provided their assistance, expert advice, and/or constructive critiques of various sections or chapters. These include Kathy Allen, Anna R. Armitage, Robert Bakker, Jeffery K. Barnes, Wylie C. Barrow, Jon A. Baskin, Laura Beloz, James F. Bergan, Brooke Best, Rena M. Bonem, Thadis W. Box, Alton Brown, Sherry L. Brown, Fred C. Bryant, Julia C. Buck, Horace R. Burke, Dava K. Butler, Thomas D. Byram, William Caire, Guy N. Cameron, Stanley D. Casto, Larry L. Choate, Karen H. Clary, Robyn A. Cobb, Leigh A. Cook, George M. Diggs Jr., Lou Dinsmore, D. Lynn Drawe, James A. Eidson, William R. Elliott, Steven D. Emslie, Anne Evans, Jonah W. Evans, C. Kirk Feuerbacher, Chris J. Flis, Dan L. Flores, Michael R. Forstner, Emmanuel Gabet, Kaitlen P. Gary, Ron R. George, Jonathan K. Gerland, Richard Gibbons, Selma N. Glasscock, Brice C. Glidewell, Andy G. Gluesenkamp, William Godwin, Walker Golder, James B. Grace, Joel Hambright, Chad W. Hargrave, Kevin M. Hartke, Peter J. Hatch, Stephen L. Hatch, Elizabeth H. Heise, Patricia Humbird, David M. Hurt, Ben Hutchens, Emily E. Hyatt, Diana Iriarte, Donald D. James, M. James, Robert L. Jarvis, Frank W. Judd, John P. Karges, Paul R. Katz, David L. LaVere, Jeffery G. Lewis, Barney L. Lipscomb, Ernest L. Lundelius Jr., William I. Lutterschmidt, Timothy W. Lyons, Michael H. MacRoberts, Joseph A. Marcus, Russell L. Martin Jr., Philip A. Matich, R. S. Matlack, Ann B. Mayo, Carmon McCain, R. Dale McCall, Danny L. McDonald, Christina Mild, Ashley R. Morgan, Michael E. Morrow, James M. Mueller, Paige A. Najvar, Lisa Neely, Daniel W. Noland, Lee C. Nordt, Mike O'Brien, James F. Parnell, Mary Parr, John B. Pascarella, James A. Plutino Jr., Maxwell B. Pons Jr., John H. Rappole, James R. Reddell, Sandra Rideout-Hanzak, Christopher W. Ringstaff, William E. Rogers, Francis L. Rose, David J. Rosen, M. Kent Rylander, J. C. Sagebiel, Roger W. Sanders, Fred S. Scharf, Norman J. Scott Jr., Kenneth D. Seyffert, Trent Shotwell, Evan Siemann, Jason Singhurst,

Melissa S. Sisson, A. Tucker Slack, Michael Smith, Autumn Smith-Herron, Rufus Stephens, Katie Swanson, Raymond C. Telfair II, Ellen Temple, Peter N. Thomas, Scott W. Tinker, Carmelo R. Tomas, John W. Tunnell Jr., Suzanne M. Tuttle, Doug D. Waid, Frank J. Weaver, D. Clark Wemecke, David B. Wester, M. White, Jennifer K. Wilson, Jeffery R. Wozniak, and Marianna T. Wright. The many people and agencies who graciously allowed use of their photographs or maps are acknowledged in the figure captions. Photographs lacking an attribution in the caption were taken by Brian R. Chapman.

We are especially grateful to the Botanical Research Institute of Texas in Fort Worth, the National Butterfly Center in Mission, the Lady Bird Johnson Wildflower Center in Austin, and The History Center in Diboll, Texas, for making their resources available to us. Special praise goes to the librarians at the University of North Carolina Wilmington for their deep-digging efforts to locate often obscure sources. We especially appreciate the kind thoughts of Andrew Sansom in the foreword.

The Texas Research Institute for Environmental Studies and the Office of Research and Sponsored Programs at Sam Houston State University (SHSU) provided financial resources that supported the development and production of this book. Richard Eglsaer, Jerry L. Cook, and William I. Lutterschmidt were instrumental in developing the agreement between SHSU and Texas A&M University Press to establish the Integrated Natural History Series. We thank William I. Lutterschmidt, coeditor of the Integrated Natural History Series, for providing valuable guidance during the preparation of this book, the first in the series. We appreciate the suggestions provided by Janice Bezanson, John P. Karges, and two anonymous reviewers. Shannon Davies, Stacy C. Eisenstark, and Katie Duelm offered much-needed editorial supervision of the book at Texas A&M University Press. The text of this book benefited from masterful copyediting by Laurel Anderton and we are grateful for her contribution. Final mention goes to our wives, Sandy and Elizabeth, for their loving patience and tolerance, without which the book would still be no more than an unfulfilled dream.

THE NATURAL HISTORY OF TEXAS

1

NATURAL HISTORY IN TEXAS

ESTABLISHING ECOLOGICAL BOUNDARIES

It may be the naturalists who save us in the end, by bringing us all back down to earth.

— ROBERT MICHAEL PYLE (2001)

A TEXTURED LANDSCAPE

Whether one travels north to south or east to west, Texas stretches for nearly 800 miles and sprawls across 267,000 square miles of the American Southwest. Besides bragging rights, such an expanse also endows Texas with what arguably represents the richest variety of ecological conditions of any state in the nation, including the "slightly" larger Alaska: rough-hewn desert, grasslands of several kinds, piney woods, coastal marshes, juniper-covered hills, and a semitropical floodplain represent a quick, yet still incomplete, look across the state's varied landscape (fig. 1.1). Missing from this overview are bat-filled caves, the shifting dunes at Monahans, the magnificent canyons along both the Rio Grande and the eastern escarpment of the Llano Estacado, the riparian swamps and hardwood bottoms, even a few mountains, and the coastal ribbons we call barrier islands, but also—well, you get the picture. Texas is textured by a landscape as varied as it is big.

EARLY NATURAL HISTORY IN TEXAS

Nineteenth-century naturalists recognized the uniqueness of Texas beginning with the expedition of Thomas Freeman and Peter Custis (1806) up the Red River. The latter separated Texas from the vast—and largely unknown—lands of the Louisiana Territory acquired by Thomas Jefferson. Jefferson, a president with more than a passing interest in natural history, had recently authorized the much heralded Lewis and Clark Expedition (1804–1806) and had similarly charged Freeman and Custis to explore the southern edge of the new addition to the United States. Regrettably, Freeman and Custis were turned back by the Spanish army just as their expedition entered Texas at present-day Bowie County. Had they been able to go on, the explorers might have continued upstream to the headwaters of the Red River and the majestic canyonlands on the eastern border of the High Plains—a discovery that awaited the trek of Army Captain Randolph Marcy (1812–1887) in 1852 (fig. 1.2). Like the Lewis and Clark Expedition, the Freeman and Custis

(*overleaf*) Figure 1.1. Extensive forests—habitats not always associated with Texas—in East Texas represent just one of its many-textured landscapes. Photograph by Jonathan K. Gerland.

Expedition included no naturalists, but also like their more famous predecessors, Freeman and Custis collected numerous specimens of interest to natural history, which included a then-unknown raptor, the Mississippi Kite.

In 1820, Major Stephen H. Long (1784–1864) led an expedition to explore the headwaters of the Platte, Arkansas, and Red Rivers. His troop included a botanist, Edwin James (1797–1861), a zoologist and entomologist, Thomas Say (1787–1834), and an artist-naturalist, Titian Peale (1799–1885). After reaching Colorado, the party turned south and split, one group descending the Arkansas River and the other (with James and Peale) striking even farther south in search of the Red River. Mistakenly, the latter instead followed the Canadian River and thus unknowingly turned their topographical error into the first scientific expedition to enter Texas.

Not long afterward, John James Audubon (1785–1851) visited Texas to continue his observations and paintings for his monumental *Birds of America*. Texas did not disappoint. He wrote, "The mass of observations that we gathered [in Texas] connected with the ornithology of our country has, I think, never been surpassed." Audubon's visit, made in the spring of 1837, included stops on Galveston Island and Buffalo Bayou. For those readers familiar with the current industrial development along the Houston Ship Channel and the lower reaches of Buffalo Bayou, note Audubon's description: "This bayou is . . . bordered on both sides with a strip of woods [where] I found the Ivory-billed Woodpecker in abundance." In keeping with his routine field methods, Audubon shot several specimens of what is today generally considered an extinct species in mainland North America. Lying just beyond the riparian forest bordering the bayou were extensive coastal prairies (chapter 11) where the artist-naturalist expressed surprise as to how well these grasslands provided a niche for Black-throated Buntings, today known as Dickcissels (fig. 1.3). Audubon and his son, John Woodhouse Audubon, concluded their 27-day tour by visiting with Sam Houston, then president of the newly established Republic of Texas.

Searches for a feasible route for a transcontinental railway across the vast North American interior spurred further exploration

Figure 1.2. Shaped by tributaries of the Red River, the red-hued Palo Duro Canyonlands remained uncharted until explored by Captain Marcy in 1852.

Figure 1.3. Extensive coastal prairies like this one in Austin County provided John James Audubon with opportunities to observe Dickcissels. Photograph by Sandra S. Chapman.

of the American West during the 1850s. One of these, led by Lieutenant Amiel Whipple (1813–1863) in 1853–1854, traversed the 35th parallel along the Canadian River, which might well have been selected as the route had Whipple's math been better (much of his pathway, however, later become famous as Route 66). Whipple's expedition included scientists associated with the Smithsonian Institution, and their reports, along with those from the other railroad surveys, filled 12 huge volumes with knowledge of the American West. With the marriage of the Union Pacific and Central Pacific Railroads in 1869, the way west opened and spurred the rise of competitive routes, notably that of the Southern Pacific, whose tracks crossed the Trans-Pecos. A network of branch lines soon spread outward from these trunk lines, providing access for a growing corps of intrepid botanists, zoologists, and geologists to an ever-shrinking American frontier destined to disappear by 1890.

Meanwhile, two European naturalists, Thomas Drummond from Scotland (1793–1835) and Swiss-born Jean Louis Berlandier (1805–1851), visited Texas. Both collected plants for the most part, but also birds and other vertebrates. Drummond made two trips to North America, the second of which, begun in 1831, included the Allegheny Mountains west to St. Louis and eventually New Orleans before he landed in Texas near present-day Freeport and headed inland. He collected about 750 species of plants and 150 birds between Galveston Island and the Edwards Plateau, especially along the Brazos, Colorado, and Guadalupe Rivers. Drummond commended two flowers, Tickseed and Firewheel, as "deserving of notice for their beauty." Both are now common roadside plants, with the copper tones of Firewheel inspiring the school colors for Texas State University in San Marcos (fig. 1.4). Drummond also attempted to collect insects but found few, which he attributed to the "custom of burning the prairies," a practice still widely employed today to maintain native grasslands in many areas in North America. After spending nearly two years in Texas, he left for Cuba, where he died shortly after arriving. Drummond's plant collections were widely distributed across Europe, where they generated heightened interest for additional botanical work in America (fig. 1.4).

Figure 1.4. Once called Firewheels (top), Indian Blankets caught the eye of Scottish naturalist Thomas Drummond, who sent plant specimens from South Texas to Europe for identification. English botanist Joseph Dalton Hooker named Drummond's Phlox (bottom) in his honor.

Berlandier began his fieldwork in Mexico, where he arrived in 1826, on behalf of Augustin de Candolle (1778–1841), at the time Europe's premier botanist. He served as a scientist on the Mexican Boundary Commission charged to settle the dispute with the United States over the western border of Louisiana. The expedition left Mexico City in November 1827 and entered Texas at Laredo— "a very desert place" in his words—in February of the following year. Berlandier collected along the entire route, which eventually included San Antonio, Gonzales, San Felipe, and Nacogdoches. His association with the commission effectively ended on the Trinity River in May 1828 after he was diagnosed with malaria. He was sent to Matamoros to recuperate and resided there the remainder of

Figure 1.5. Jean Louis Berlandier likely visited Presidio la Bahia at Goliad (Goliad County) in 1834.

his life except for a brief return to Texas in 1834 to collect near Goliad (fig. 1.5). Whereas Berlandier is remembered as the first botanist to collect extensively in Texas, his interests also included birds, fishes, reptiles, and insects, which he also collected and sometimes skillfully painted in watercolors. A tortoise—the smallest of its kind but with an unusually high-domed carapace—found in southern Texas and northeastern Mexico was named in his honor. Instead of burrowing like other species of tortoise, Texas (or Berlandier's) Tortoises more often clear resting places—known as pallets—under the protection of a bush or cactus; the pallets gradually deepen because of continued use for many years. The species has been protected in Texas since 1967.

Another immigrant, German-born Ferdinand J.

Lindheimer (1801–1879), regarded by some as the "Father of Texas Botany," fought in the Texas Revolution and subsequently collected plants for two prominent botanists, Asa Gray (1810–1888) at Harvard University and Georg Engelmann (1809–1884) of St. Louis. He collected extensively in the area around Fredericksburg and New Braunfels, but also in the bottomlands of the Brazos and Guadalupe Rivers, Galveston Island, and Matagorda Bay (fig. 1.6). Unlike others collectors, who were often members of larger parties, Lindheimer traveled alone in a two-wheeled Mexican cart pulled by a pony. His career as an active botanist and collector for Gray and Engelmann ended in 1852, when he began editing a German-language newspaper in New Braunfels. Nonetheless, many of Lindheimer's specimens, some of which

Figure 1.6. Ferdinand J. Lindheimer traveled by pony-drawn cart to collect plants in bottomland habitats like this near the Guadalupe River. Photograph by Paige A. Najvar.

represented new species, ended up in collections throughout the world and thereby established the richness of the Texas flora for botanists everywhere.

A large number of mostly amateur ornithologists dominate the cast of late-nineteenth-century naturalists in Texas—not surprising, given that its avifauna is the largest of any in the United States, leading California by about 80 species. The current list for Texas includes 639 species of birds and today boasts its own volume in the famed series of Peterson field guides (fig. 1.7). By the 1880s, at least 50 men and 2 women in Texas were actively engaged in some form of ornithology, often with a focus on collecting eggs, nests, or the birds themselves, but also compiling lists of species for counties or other specified areas. The interests of some included selling specimens to collectors elsewhere, but many specimens also ended up in museums in the United States and in foreign lands. Sixteen of the group were born outside the United States, and nine were native Texans. They resided in all regions of the state except for the High Plains and Trans-Pecos, and many added to the growing fund of ornithological knowledge with formal publication of their observations. Space limitations

permit mention of only a few of these pioneering naturalists. Among these are Harry Y. Benedict (1869–1937), an egg collector who later became president of the University of Texas; English-born Henry P. Attwater (1854–1931), collector of the type specimen for a subspecies of Greater Prairie-Chicken named in his honor; another Englishman, rancher Howard G. Lacey (1856–1929), who published a list of birds for the area near Kerrville; and a farmer from Corpus Christi, John M. Priour (1848–1931), who for four decades worked as a guide and collector for a number of prominent eastern ornithologists.

In terms of natural history, the second half of the nineteenth century corresponds to the Golden Age of Paleontology, in which Texas provided key sites for many discoveries. In 1877, Edward Drinker Cope (1840–1897), one side of the "Bone Wars" with Othniel C. Marsh (1831–1899) of Yale University, discovered a trove of fossils in the Permian Red Beds of north-central Texas (Baylor and surrounding counties). Cope, associated with the Philadelphia Academy of Natural Sciences, excavated thousands of fossils highlighting a critical point in evolution—the origins and dominance of

Figure 1.7. Populations of many species, including Whooping Cranes (above) and White-winged Doves (left), have declined significantly since the late 1880s. Photographs by Terri Tipping (above) and J. Byron Stone (left).

terrestrial quadrupeds that mark the beginnings of land-based ecosystems and the initiation of terrestrial food chains. *Dimetrodon,* a mammal-like reptile outfitted with a large finlike sail on its back, was the apex predator of its time and preyed on a community dominated by amphibians. Cope later discovered other deposits, notably the Clarendon Beds (Upper Miocene and Lower Pliocene in Donley County) and the Blanco Beds (late Pliocene in Blanco County), both rich in mammalian fossils. Marsh countered Cope with discoveries of his own fossils, some at the same locations, in what proved to be a competition employing shady methods, including sabotage, and attacks in their respective publications. In time, the reputations—and fortunes—of both men suffered, but their work nonetheless produced significant contributions, including huge collections of unexamined fossils that provided research opportunities for the next generation of paleontologists. *Copeia,* the journal of the American Society of Ichthyologists and Herpetologists, first published in 1913, honors Cope and his work. Texas has been the site of extensive paleontological work involving several other well-known scientists and important discoveries of both vertebrate and invertebrate animals. However, here it will suffice to mention only the discovery of dinosaur tracks in the Cretaceous strata in the bed of the Paluxy River near Glen Rose (Somervell County) early in the twentieth century, now the site of a state park (chapter 5). Since then, tracks have also been discovered at additional sites in the region (fig 1.8).

Figure 1.8. Two main types of dinosaur tracks appear in the rocky bed of the Paluxy River at Dinosaur Valley State Park. The stumpy feet of sauropods—huge herbivores that lumbered on four pillar-like legs—show little indication of toes. The three-toed tracks of theropods—fast-moving bipedal predators—also appear in the rock. Both types of dinosaurs followed the same muddy route; the footprints do not necessarily indicate a clash between predator and potential prey. Photograph by Glen Kuban.

Figure 1.9. Strecker's Chorus Frog was named to honor Texas naturalist John K. Strecker. Photograph by Suzanne L. Collins.

Final mention goes to John K. Strecker (1875–1933), whose life's work may justify the title "Father of Texas Natural History." By horse and buggy, he traveled widely in Texas, collecting animals of many kinds but especially reptiles, amphibians, birds, and mollusks, fulfilling his duties as the museum curator (and librarian) at Baylor University—accomplishments he earned with no more than a grade-school education. In 1940, the museum, now part of a larger museum complex at Baylor, was renamed in his honor. In all, he published 111 papers based on his fieldwork, plus many popular

articles. His studies of amphibians and reptiles established him as a recognized authority in herpetology. Similarly, a monograph dealing with freshwater mussels and another on aquatic and land snails likewise earned the commendation of malacologists; both papers remain widely cited. For fieldwork, he always dressed in a tie, believing that this formality convinced farmers and ranchers that he was worthy of access to their property. Strecker represented the last of the "old-time" naturalists in Texas—those who gained broad self-taught knowledge about wild things and their place in nature. Strecker's Chorus Frog, named in his honor, occurs in a broad band across eastern Texas, Oklahoma, and barely into southern Kansas; isolated populations also occur in Illinois, Missouri, and Arkansas (fig. 1.9). Unlike most other burrowing frogs and toads, this species digs using its stout front limbs and enters burrows headfirst.

The early era of natural history in Texas faded in the decade spanning the turn of the nineteenth century and the beginning of the twentieth, when a modern cadre of naturalists began fieldwork in the state. Notable among these were mammalogist Vernon O. Bailey (infobox 1.1) and ornithologist Harry C. Oberholser (infobox 1.2), who launched formal investigations of the state's biological resources on behalf of the Bureau of Biological Survey. Unlike the railroad surveys, in which natural history was a secondary interest and of relatively limited geographical coverage, fieldwork in the new era became the primary mission, whose sole purpose was cataloging plants and animals in *all* parts of Texas. This change in scope included

INFOBOX 1.1. VERNON O. BAILEY
Field Naturalist of the Old School (1864–1942)

Few biologists can match the field experience acquired by Vernon Bailey. Although born in Michigan, Vernon and his pioneer family settled in Minnesota on what was then still a largely untrammeled frontier. His parents provided his early schooling, followed by somewhat more formal training when the community eventually built a schoolhouse. Bailey farmed for a few years after finishing his rustic education but later briefly attended the University of Michigan and, still later, what is now George Washington University.

Meanwhile, in Washington, DC, a fledgling government agency was developing that would later become the Bureau of Biological Survey, the federal agency whose mission focused on what is now known as wildlife ecology and management. While in Minnesota, Bailey taught himself taxidermy and amassed a collection of specimens, some of which—particularly shrews—he could not identify. Accordingly, he contacted the agency's director, C. Hart Merriam, who quickly established an appreciation for the young man's skills and began buying Bailey's specimens for the national collection in Washington. Merriam also advised Bailey on ways to measure, label, and catalog his specimens. In 1887, Merriam hired Bailey as a field agent and ten years later appointed him as the agency's Chief Field Naturalist—the first and only person to hold that title—which he retained until his retirement from government service in 1933.

Bailey spent much of his career on fieldwork. The locations covered a broad swath of environments, from the northern forests and mountains of Glacier National Park in Montana to the dark recesses of Carlsbad Caverns in New Mexico. His fieldwork, which centered on mammals and was supplemented by observations of birds, reptiles, and plants, became the basis for 244 published works. His trapping efforts produced some 13,000 specimens of mammals for the bureau's collection; these also provided both Bailey and Merriam with material they used to designate a wealth of new species and subspecies. Both men developed a deep interest in taxonomy, and both were regarded as "splitters" (they often divided species into subspecies) when dealing with variations in pelage or skeletal measurements. Although several of their designations have not survived the rigor of modern scrutiny, a species of pocket mouse and a subspecies of Bobcat that Merriam named after Bailey remain unchanged.

In 1889, after earlier rejecting Merriam's wishes on the matter, Bailey made the first of seven trips to Texas. His initial fieldwork focused on the Trans-Pecos, but he later covered the entire state. Six of Bailey's larger works appeared in the North American Fauna series. One of these, published in 1905, bears the title *Biological Survey of Texas*. The monograph reports at length on the state's mammalian fauna, supplemented by lists of reptiles and birds he observed. He separated the plants and animals in the survey on the basis of "life zones"—ecological units proposed by Merriam based on biological communities that vary in relation to latitude (see main text). Among other observations, Bailey estimated that no fewer than 400 million Black-tailed Prairie Dogs occupied the huge colony stretching along the eastern edge of the Llano Estacado from San Angelo to Clarendon. He also noted when and where the last American Bison were killed in Texas. His survey remains a useful baseline for comparison with the current status of wildlife in Texas; reprinted editions of the bulletin are still published.

Bailey married Merriam's sister, Florence, herself an accomplished naturalist with a special interest in ornithology and the author of several books, including an acclaimed field guide. She accompanied her husband on his trip to South Texas, traveling by bumpy wagon throughout the brushy terrain between Corpus Christi and the border with Mexico.

In 1919, Bailey helped found the American Society of Mammalogists and later (1933–1935) served as its president. In the first issue of the society's journal, Bailey published a description of a new subspecies of beaver—*Castor canadensis missouriensis*—still recognized in mammalian taxonomy. He was also a member of ornithological societies and other professional groups concerned with the conservation of natural resources. Bailey remained active after retiring, including nature photography, lecturing, and even fieldwork. He also continued designing and perfecting live traps, particularly those developed to catch beaver alive and unhurt in order to restock areas where they had been extirpated—work recognized with prizes from the American Humane Association. At the time of his death, Bailey was planning another field trip to Texas in association with the US Fish and Wildlife Service. He died as he had lived: a field man at heart.

INFOBOX 1.2. *THE BIRD LIFE OF TEXAS*
The Life's Work of Harry C. Oberholser
(1870–1963)

Few biologists have devoted a career-long study to the birds of a state, especially when living elsewhere. Among these, Harry C. Oberholser comes close to standing alone in his dedication, in this case to the avifauna of Texas, which he began studying in 1900 and continued to study until his death 63 years later. Born in Brooklyn, New York, Oberholser briefly studied at Columbia University but left because of poor health. He then worked in his father's dry-goods store in Ohio, where he compiled a list of birds for Wayne County that was later published by the state's Agricultural Experiment Station—the first of nearly 900 publications, including a book about the birds of Louisiana. In 1895, he accepted a position in Washington, DC, as ornithological clerk for the Division of Economic Ornithology in the Department of Agriculture, precursor to the US Fish and Wildlife Service in the Department of the Interior. He resumed his education and in 1916 earned a PhD from George Washington University.

Oberholser undertook a variety of duties while in government service, but these always concerned birds. In 1900, the then 30-year-old Oberholser landed in Texas at Port Lavaca to begin fieldwork. Previously, field agents had collected specimens in parts of the state, but Oberholser was charged to expand the project statewide with his own fieldwork, supplemented by information gleaned from both local naturalists and published literature. This was the first of three trips, one of which (in 1901) included traveling with Vernon Bailey (infobox 1.1) and Louis Agassiz Fuertes (1874–1927), whose extraordinary artwork would appear decades later in pages of Oberholser's life's work—a manuscript initiated in his reports of these trips.

In 1928, Oberholser organized a census that continues today as the Winter Waterfowl Survey, an important tool in managing ducks, geese, and swans in each of North America's four flyways. He also successfully advocated that bird banding become a federal program coordinated and managed by the Bureau of Biological Survey. When time permitted, he served as a professor of zoology at both the Biltmore Forest summer school in North Carolina and the American University graduate school in Washington, DC. He was a member of 40 scientific and conservation societies, many in other countries.

Notably, he was a charter and honorary member of the Texas Ornithological Society.

In 1941, after a tenure of 46 years, he retired as Senior Biologist, having benefited from a presidential order that granted an additional year of service beyond the mandatory retirement age (70) to further his work on Texas birds. He then spent the next six years in Ohio as curator of ornithology at the Cleveland Museum of Natural History, retiring a second and final time in 1947.

Throughout his career, Oberholser maintained an interest in avian taxonomy and examined thousands of study skins for variations in plumage and morphology. Even the slightest differences between geographical populations of the same species caught his attention, and these often became the basis for declaring and naming subspecies. Oberholser soon earned a reputation as a "splitter," about which many of his colleagues did not always speak approvingly. In all, he named 11 new families and subfamilies, 99 genera and subgenera, and 560 species and subspecies of birds worldwide. He also identified thousands of birds sent to Washington by other agencies, museums, and private collectors. His detailed knowledge of feathers and avian anatomy also led to numerous appearances in court, where such evidence helped prosecute violators of federal wildlife laws.

Until his death in 1963, Oberholser continually added new information to a project—a comprehensive study of Texas birds—that had begun with a field trip across the state more than six decades earlier. In the end, the typed manuscript reached nearly 12,000 pages and three million words. A work of such length understandably overpowered potential publishers, and publication of his epic treatise indeed experienced several false starts. The original manuscript included four sections: the history of ornithological studies in Texas dating to 1828; species accounts, detailing the distribution of each species and subspecies; a gazetteer listing places where birds were observed or collected; and a 572-page bibliography. Much of this material was heavily edited, supplemented, or deleted in full (e.g., the gazetteer) in readying the original manuscript for publication.

Thus, only after his death and with considerable editorial work did Oberholser's epic finally appear (in 1974) in print, but even shortened, *The Bird Life of Texas* stands as a monument to a long and fruitful career in ornithology. For scholars, an unedited copy of Oberholser's original manuscript is available for study at the Briscoe Center for American History at the University of Texas at Austin.

similar work ongoing in other states and more or less coincided with efforts to develop systems to classify nature on a broad geographical scale.

ESTABLISHING ECOLOGICAL BOUNDARIES

Efforts designed to fit the landscapes of North America into discernible units were founded on taxonomic studies that, in turn, revealed distributional patterns. Plants and animals are not randomly distributed but instead show certain affinities, particularly those associated with climate and elevation and the vegetation they produce, but also with soils and the subtle variations in those conditions collectively known as "habitat." Some species, of course, have adapted to a wider range of conditions than others—White-tailed Deer, for example—yet others are far more restricted and become indicators of a relatively narrow set of ecological circumstances. Nonetheless, fine tuned or not, plants and animals sort out into ecological assemblages—the bricks of biogeography.

Biomes represent the largest units of biogeography in which similar vegetation, regardless of its location, is lumped together. Hence, the Great Plains of North America, the steppes of Russia, the pampas of Argentina, and the veldt of Africa are united in the Grassland Biome. Within North America itself, the grassland biome encompasses the Palouse Prairie in the northwest, the annual grasslands in California, the coastal prairies of Texas, and, of course, the Great Plains, each distinctive but nonetheless representing a single biome. Additional biomes in North America include Eastern Deciduous Forest, Desert, and Boreal Forest, among others.

In the late 1800s, C. Hart Merriam (1855–1942), then the head of what was soon to become the Bureau of Biological Survey and eventually today's US Fish and Wildlife Service, tweezed out smaller units he called life zones. His motives were primarily economic, as he hoped that the boundaries between life zones would guide agricultural development and lessen the need for costly attempts by trial and error. Merriam based his life-zone concept on the belts of vegetation, sorted by elevation, found in the San Francisco Mountains and surrounding areas near Flagstaff,

Arizona. He recognized six zones and, after adding two more to represent regions east of the Mississippi River, extrapolated and expanded these to cover all of North America. The Hudsonian Zone, for example, included the coniferous forests of northern Canada as well as those just below timberline on mountains farther south that were widely separated from each other by other zones. In so doing, Merriam relied almost entirely on temperature to explain the limits of plant and animal distributions—isotherms delineated the boundaries of each life zone in his system.

Whereas the life-zone concept initially gained general acceptance—it worked particularly well in the western regions—other biologists soon challenged Merriam's reliance on temperature as the primary factor controlling plant and animal distributions. Precipitation-evaporation ratios, for example, seemed as important as temperature in influencing the distribution of plants. In time, Merriam's life zones fell into disuse, in part because too many distinctive communities—among them sagebrush in the Great Basin and scrub oak chaparral in Arizona—were united within a single unit, in this case what he called the Upper Sonoran. For Texas, the same limitation—reliance on temperature at the expense of other ecological influences—united the semitropical Lower Rio Grande Valley with the eastern Panhandle and the arid Trans-Pecos, hardly a discriminating combination (fig. 1.10). Moreover, his isotherms considered only summer temperatures above 43°F, thereby ignoring the limitations posed by cold winters (in fact, because of a misunderstanding, the Weather Bureau provided Merriam with threshold values based on 32°F, thus invalidating the isotherms presented in his work). Still, Merriam deserves credit for attempting to explain the distribution of North American biota and for stimulating further interest in biogeography. But if his reliance on temperature made his calculations faulty, his descriptions of vegetation were not, and for many years they remained a useful resource for field biologists in the western United States.

In 1943, ecologist Lee Dice (1887–1977) presented descriptions and maps of 29 sizable areas called biotic provinces; he did not coin the term but refined the concept and applied it to all of North

Figure 1.10.
The serrated
beauty of the
Trans-Pecos
region reflects the
complex origins
of a habitat vastly
different from
that of the Texas
Panhandle or
Lower Rio Grande
Valley.

America. Each area is characterized not only by the dominant or climax vegetation but also by soil type, topographical features, biota, climate, and, when appropriate, unusual vegetation. The boundaries, however, do not reflect rigid geographic limits of species but instead portray the "distinctness and distributions of the various ecologic associations." Unlike the components forming a biome, which may occur in two or more widely separated regions, biotic provinces represent a single, continuous geographical entity unlike any other. Moreover, the demarcation between adjacent biotic provinces is certainly not a well-defined "line in the sand." Instead, one generally intergrades into another in zones known as ecotones that, as implied, blend adjacent ecological units with a mixture of species representing each. Interestingly, Dice often named his biotic provinces for prominent cultural entities. Hence, the Eskimoan Biotic Province coincides with the tundra of the far north, and Apachian designates the high grassy plains and mountains in a continuous area of southeastern Arizona, southeastern New Mexico, and parts of adjacent Mexico.

In 1950, biologist W. Frank Blair (a Dice student; see infobox 1.3) described the biotic provinces of Texas by associating the distribution of terrestrial vertebrates—excluding birds—with the dominant types of vegetation. In doing so, he started with the biotic provinces established by Dice, although he worked on a different scale—statewide rather than continental—and therefore benefited from a closer look at the plant and animal assemblages as well as soil types on which to map these units in Texas. Significantly, his analysis replaced Dice's Comanchian Biotic Province with a new and somewhat smaller unit, the Balconian Biotic Province, thereby highlighting the distinctive ecology of the Edwards Plateau. Both Dice and Blair recognized the Texas Biotic Province despite the highly varied vegetation within its boundaries, although Blair acknowledged it as an "unsatisfactory" designation. In Texas, this unit extends north to south from the Red River to the marshes on the Gulf Coast, a wide swath that includes the Cross Timbers and Blackland Prairies. The Texas Biotic Province indeed represents a transitional region—itself a large ecotone— between the humid forests of eastern Texas and the semiarid grasslands to the west. Later, closer looks at certain groups revealed further inconsistencies with the boundaries designated by both Dice and Blair. The amphibian and reptilian fauna in the Guadalupe Mountains, for example, matches better

INFOBOX 1.3. W. FRANK BLAIR
Herpetologist and Evolutionary Biologist (1912–1985)

A native Texan, W. Frank Blair earned a PhD in zoology in 1938 at the University of Michigan, where he remained for four years as a research associate. His studies during this period included those concerned with the home ranges of small mammals and the adaptations of pelage color to match the dark (lava) and light soils in the White Sands area of New Mexico. His doctoral work was supervised by Lee R. Dice (1887–1977), who championed the use of biotic provinces to designate ecological units in North America. Blair joined the Army Air Corps in 1942. After the war, he returned to Michigan but soon accepted an appointment to the University of Texas and began what was to be a distinguished career that continued until his retirement in 1982. Blair served as the first director of the university's Brackenridge Field Laboratory, which is dedicated to studies of biodiversity, natural history, and ecosystem changes.

Blair's academic life centered on herpetology. His particular focus on the phylogeny of toads established him as one of the world's authorities in the field of evolutionary biology. He organized summer field courses to study the biota in several regions of Texas, including Black Gap in the Big Bend area, the Glass Mountains in the Trans-Pecos, and the Canadian River Breaks in the Panhandle, among others. In keeping with his mentor's concepts of ecological classifications, Blair published *The Biotic Provinces of Texas* in 1950. Blair's early interest in mammals expanded to include the full range of vertebrates and led to publication (with coauthors) of *Vertebrates of the United States*, a landmark contribution and a standard reference for a generation of biologists. Another publication, *Evolution of the Genus* Bufo, met with wide acclaim as a seminal work of its genre.

After arriving in Austin, Blair and his wife moved into a home on land they maintained as a nature preserve. Affectionately known as their "ten acres," the site also provided Blair with an after-hours study area where he investigated the dynamics of the resident lizard population—his analysis produced another masterpiece, *The Rusty Lizard, a Population Study*. He bequeathed his beloved "ten acres" to the Travis Audubon Society, which operates the site—formally known as Blair Woods—as a natural preserve for ecological studies.

Blair believed that the mating calls of frogs were an important means of mate selection. To study these relationships, he and his students used tape recorders to capture the calls of frogs breeding in various areas, including those where the ranges of two related species overlapped. This research revealed that the calls varied geographically and played a role in maintaining reproductive isolation where similar species overlapped; the term "isolating mechanism" emerged from this and related work.

Blair's career is highlighted by service as president of the American Institute of Biological Sciences, the Ecological Society of America, the Society for the Study of Evolution, and the Texas Herpetological Society. He was a founder and guiding light, as well as president, of the Southwestern Association of Naturalists. He was active in the International Biological Program (IBP), which investigated the world's ecosystems, and served as the chair of the program's United States National Committee. Because of his contributions to IBP, Blair received the Joseph Priestley Award for 1977. More than 160 publications bear his name.

---

with the Chihuahuan than with the Navahonian Biotic Province. In all, Blair recognized seven biotic provinces in Texas, but only the Balconian lies exclusively in the state.

Nearly two decades later, botanist Frank W. Gould (infobox 1.4) noted that a less theoretical, more workable, and somewhat narrower system would better serve the needs of ecologists, including those concerned with practical matters such as range management. He thus recognized "vegetational areas" based on topographical, climatic, and edaphic (soil) factors and the similarities of plant communities they produced, with each area assigned commonly used local names (fig. 1.11). His designations were closely aligned with the "natural geographic divisions" delineated nearly 40 years earlier by William T. Carter and included, among others, units named Cross Timbers, Edwards Plateau, High Plains, and Trans-Pecos Mountains and Basins that gained wide acceptance among laypersons, amateur naturalists, and professional biologists alike.

Newer techniques, primarily aerial mensuration and reconnaissance, allowed considerable

Born in North Dakota, Frank W. Gould spent most of his youth elsewhere, notably in DeKalb, Illinois, where his father served as head of the geography department at Northern Illinois University. Gould stayed at home and earned his undergraduate degree at the same school. His collegiate career continued with an MS from the University of Wisconsin–Madison, where he surveyed the prairie communities remaining in Dane County. By 1941, he had finished a PhD in botany at the University of California, Berkeley, completing a taxonomic study of lilies in the genus *Camassia*. During this period, Gould also began what was to become a career focused on agrostology, the study of grasses.

In 1944, after two short teaching assignments in Utah and California, Gould assumed curatorial duties for the herbarium at the University of Arizona. Five years later, he left to become the curator of the Tracy Herbarium at Texas A&M University in College Station—his last position until his retirement in 1979. By that time—and because of Gould's diligence—the Tracy Herbarium had gained recognition as one of the most respected facilities of its kind in the United States. Throughout his career, Gould regularly acquired funding from the National Science Foundation to support his grass research and deftly used some of the money to further develop the herbarium. Today, the Tracy Herbarium houses more than 300,000 specimens, of which about 70,000 are grasses, including many that Gould collected and named (i.e., type specimens). He concurrently held a professorship

in the Department of Range Science and was later awarded the title of Distinguished Professor in the same department. In keeping with advancing technology, Gould supplemented the traditional methods of taxonomy with more sophisticated cellular techniques. His memberships included the American Society of Plant Taxonomists, the Botanical Society of America, and the Society for Range Management.

As the curator of a herbarium, Gould worked professionally with plants of many kinds, but grasses remained at the center of his personal interest in plant taxonomy. His work took him to lands as diverse as Costa Rica, Sri Lanka, Brazil, and England, among others. Gould produced some 80 publications, including monographs describing the grasses of, respectively, Texas, the southwestern United States, the Coastal Bend of Texas, and Baja California (the latter work published posthumously), as well as a critically acclaimed textbook, *Grass Systematics*. At the time of his death, he had completed much of the research and some of the writing for another major work, *The Grasses of Mexico*.

Of particular relevance, we highlight *Texas Plants: A Checklist and Ecological Summary*, a work Gould revised in 1969—a publication that provides the framework for our coverage of the state's major vegetational areas. His designations—Trans-Pecos and Piney Woods, for example—closely follow those widely used by naturalists of all stripes as well as by the lay public, all of whom we include in the target audience for this book. Gould, of course, did not coin these names, but his work did much to incorporate them into the lexicon of science.

refinement of Gould's approach. This system divided Texas into 12 large and 56 smaller ecoregions, which surely served many uses, but not necessarily those of natural history. Among other refinements, the newer system split off a small area of the Trans-Pecos to recognize the uniqueness of the Guadalupe Mountains, whereas many of the boundaries for the other ecoregions remained much as they were in Gould's work. Regrettably, at least from our standpoint, the refinements also included name changes that no longer reflected local usage or much originality. Thus, Coastal Prairies and Marshes morphed into Western Gulf Coastal Plains, and the wonderful Piney Woods lost its luster as the

South Central Plains. For us, and especially for our readers, we believe Gould's designations best serve the cause of natural history and therefore guide our presentation accordingly, albeit occasionally tempered with the additional clarity of newer work. But whatever the scale, units, and titles may be, one overarching fact remains—Texas presents a wonderland of biodiversity worth our notice.

THE STATE OF NATURAL HISTORY

Natural history, as often perceived by others, diminished into an unfashionable profession in the latter half of the last century. This slide was pushed by emerging demands for the products of chemistry

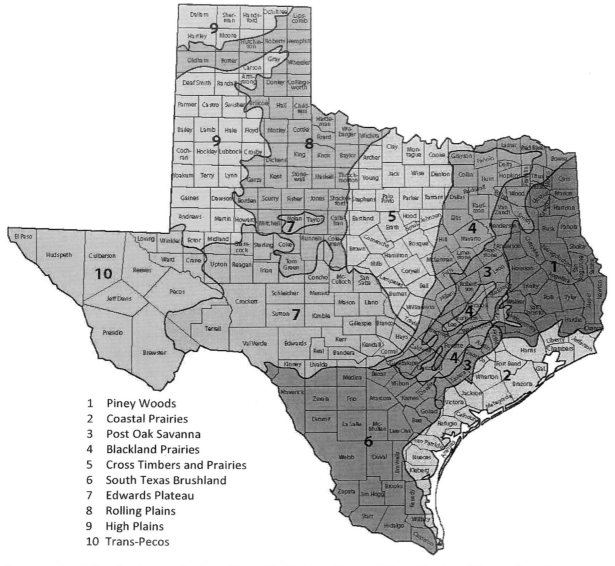

1 Piney Woods
2 Coastal Prairies
3 Post Oak Savanna
4 Blackland Prairies
5 Cross Timbers and Prairies
6 South Texas Brushland
7 Edwards Plateau
8 Rolling Plains
9 High Plains
10 Trans-Pecos

Figure 1.11. Map delineating the natural regions of Texas. Modified from *Illustrated Flora of East Texas* (Diggs et al. 2006); used with permission of the Botanical Research Institute of Texas.

and physics—the so-called hard sciences—associated with World War II and the Cold War that immediately followed. Sputnik, launched in 1957, added mathematics and microtechnology to the mix, spurring purely quantitative approaches to science, while at the same time casting an air of "soft" subjectivity on studies of wild nature. In this view, natural history represented an art, not a science, and a relic from the Victorian era of butterfly collections and cabinets filled with minerals.

Moreover, a world focused on microchips and molecular biology has largely captured the limelight

for students with an interest in science, fueled at universities by the lure of large research grants and steady promotions for its faculty who secure such funding. For budding biologists, the bandwagon shifted from outdoor to indoor studies—that is, into the cell and molecule and out of the marsh, prairie, and forest.

Meanwhile, cities and suburbs grew rapidly, severing large segments of the population from day-to-day ties with all but a well-paved landscape. Children now succumb to a world of video screens—electronic babysitters—as part of a lifestyle that often excludes treks through woods and fields or even to a park where some semblance of nature might stimulate their curiosity about the natural world. For many, any contact with nature may be solely by push button or plastic mouse.

An important component of natural history has been, and remains, the contributions of so-called amateurs—a term literally meaning "one who loves" but now corrupted to characterize someone with limited formal training in either botany or zoology. Such denigration is absurd. Amateurs, including Audubon, Strecker, and numerous physicians and army officers, who often were engineers, once formed the backbone of natural history. And they continue to make important observations. For more than a century, an annual Christmas census sponsored by the National Audubon Society and conducted largely by amateur birders has revealed trends in the distributions and populations of wild birds. Astronomy offers a useful analogy. Citizens armed with their own telescopes regularly discover a number of heavenly bodies—including comets and stray asteroids—which facilitates the work of professional astronomers tucked away in observatories. Thankfully, amateur naturalists remain actively engaged in outdoor activities, including physical as well as observational contributions, such as replanting storm-damaged dunes, serving as docents at nature centers, and assisting with prescribed burns. In contrast, however, the continued presence of an academic component in the modern corps of naturalists seems less assured.

Unfortunately, many universities have disposed of valuable plant and animal collections once used for both instruction and research. To illustrate, university collections often included bird eggs collected prior to World War II that enabled comparisons of the thickness of their shells with eggs laid during the DDT era that followed the war. These analyses documented that postwar eggshells were so thinned by pesticides that few could hatch—a discovery that provided the "smoking gun" leading to the banning of harmful agricultural chemicals. Not incidentally, many of these eggs were originally collected by dedicated amateurs known as oologists, including Texan Roy W. Quillin (1894–1974). Additionally, some universities no longer

support field trips, but even more harmful is that retiring professors representing disciplines directly related to natural history may not be replaced by those with similar academic interests. A laboratory specialist too often replaces a naturalist on the faculty. These events trigger a downward spiral: with fewer of these openings available, fewer students will prepare themselves for academic careers associated with natural history, which then slowly withers for want of both faculty and students.

Still, growing concerns about biodiversity and the health of ecosystems have somewhat checked this trend. The good news is the birth of conservation biology, a discipline dedicated to saving and restoring species and communities threatened—or worse, pushed to near extinction—by what are mostly anthropogenic agents. In response, state and federal agencies now seek college graduates—perhaps best thought of as "scientific naturalists"—equipped to deal with troubled populations of nongame species, including insects, mollusks, and amphibians and other groups often overlooked in the past, as replacements for their retirees. Moreover, new realms of life have been unveiled in environments as diverse as tree canopies and ocean vents, and these, too, require the attention of naturalists. Estimates suggest that less than a tenth of the world's species have been discovered and described—perhaps 100,000 to 400,000 in the United States alone—and the ecological role of many already discovered remains unknown.

Employment opportunities to meet these needs extend into a wide range of departments, including transportation, forestry, the military, and parks as well as wildlife management. Additionally, nongovernmental organizations—NGOs, in bureaucratic lingo—offer a growing job market for college-trained naturalists, among them the National Audubon Society, The Nature Conservancy, Sierra Club, National Wildlife Federation, and many others. All told, these opportunities require appropriate academic training, for which plant and animal taxonomy remains a common core—and a responsibility not to be shunned by universities enamored solely by molecules and test tubes.

2

PINEY WOODS
STEEPLE-HIGH TREES AND BOTTOMLAND SWAMPS

The clearest way into the Universe is

through a forest wilderness.

— JOHN MUIR

Piney Woods

Prior to settlement by Euro-Americans, a dense forest blanketed most of East Texas. While generally described as "Piney Woods," the cloak of vegetation also included large tracts of hardwoods. Parts of the southern forest were so impenetrable that settlers applied a more descriptive nickname still used today—the "Big Thicket." Steeple-high trees often exceeding 100 feet in height limited the light reaching the understory where Ocelots, Jaguars, Mountain Lions, Red Wolves, and Black Bears roamed in the shadows. By the mid-nineteenth century, however, settlers flocked to the region, eliminating most of the large predators and logging out the tall pines. Second growth and pine plantations eventually replaced the original forest, again creating the appearance of a pristine woodland (fig. 2.1). Sadly, travelers passing through other parts of Texas remain unaware that an extensive forest still blankets much of the eastern part of the state.

The Piney Woods represent the westernmost extension of the Eastern Deciduous Forest, a vast woodland once covering essentially all of the eastern United States. The belt of southern coniferous forest—one of several pine-dominated sections within the region of eastern woodlands—once encompassed approximately 15.8 million acres in 44 East Texas counties where soil and moisture favored the growth of pines and hardwoods. The Piney Woods bow westward from Louisiana into the northeastern corner of Texas and then arc southward along the eastern margin of the Post Oak Savanna (chapter 3) to the Coastal Prairies (chapter 11), giving out just north of Galveston Bay and Sabine Lake.

The gently rolling topography reaches elevations of 500 feet in the north and then gradually descends to almost sea level near the Coastal Prairies. The region experiences four distinct seasons, with average summer highs reaching 93°F and average winter lows falling to 36°F. Average annual rainfall varies from 46 inches in the north

to more than 54 inches in the south and feeds several large rivers, including the Angelina, Neches, Sabine, Trinity, and Sulphur, as well as numerous streams and bayous (fig. 2.2). As the rivers traverse the region, their courses are deflected eastward for considerable distances by the Kisatchie Wold, an ancient cuesta, or ridgeline, now represented by a line of asymmetrical hills stretching from west to east across the middle of the region. The forests developed on highly weathered soils—mostly deep, fertile sands and sandy loams. Historically, periodic fires played an important ecological role in shaping the composition of forest communities. Relatively cool understory fires ignited by both lightning and Native Americans occurred every five to ten years. Significantly, burns on higher, drier sites arrested the successional advance of hardwoods, thereby maintaining a fire-climax community dominated by pines, in turn restricting hardwood communities to wetter, low-lying areas where fires occurred less frequently.

MAJOR BIOTIC ASSOCIATIONS

Although widely used as a descriptive idiom, "Piney Woods" is somewhat misleading in ecological terms. Far from just pinelands, the region encompasses a rich variety of natural woodlands—remnants of once-extensive Longleaf Pines, galleries of Loblolly Pines and deciduous forests on mesic slopes, extensive stands of bottomland hardwoods bordering streams and rivers, and mysterious Baldcypress-Tupelo swamps. Moreover, shrubby bogs, natural prairies, desertlike dunes of gleaming white sand, and cleared agricultural lands punctuate the landscape, all distinguishing the Piney Woods as one of the most diverse regions in Texas. A description of the more widespread of these associations follows.

SHORTLEAF PINE FOREST

Because it grows well in many different soil and site conditions, Shortleaf Pine has the widest distributional range of any pine in the southern United States. Forests of this species occur in 22 states from New York to southeastern Oklahoma and northeastern Texas. In Texas, the Shortleaf Pine forest originally covered approximately 30,000 square miles in the northern half of the

(*overleaf*) **Figure 2.1. A dirt road enters the shadows cast by a tall stand of second-growth Loblolly Pines in Nacogdoches County.**

Figure 2.2. The Neches River pours across Rocky Shoals in Angelina National Forest near Rockland (Tyler County). Photograph by Jonathan K. Gerland.

Piney Woods. The forest extended south from the Red River to Angelina and Houston Counties and west from the Louisiana border to Hopkins County. North of the Sabine River, Shortleaf was the principal pine species, but farther south, Shortleaf Pines often integrated with Longleaf Pines and various hardwoods (fig. 2.3).

Shortleaf Pines usually occupy well-drained, infertile sites, but they also thrive in bottomlands and floodplains where the soils are deep, loamy, well drained, and acidic. Typically, Shortleaf Pines grow straight and tall—crowns reaching heights of 120 feet top trunks with basal diameters of 4 feet. Shortleafs may live for about 140 years, but most stands today represent second-growth forests that replaced virgin timber harvested a century ago, although logged-over sites on loamy soils were often converted to croplands. Most trees in the second-growth forests stand about 85 feet tall with trunk diameters of about 30 inches. Scattered Post Oaks, Southern Red Oaks, Black Hickories, Sandjack Oaks, and Flowering Dogwoods occur in these communities, often forming dense thickets after the pines were removed. Common understory species include Greenbrier, Yaupon, American

Figure 2.3. The smooth bark on Shortleaf Pines distinguishes them from Longleaf Pines where the two coexist.

Beautyberry, Beaked Panicgrass, and bluestem grasses.

Land-use changes, coupled with fire suppression, caused widespread declines in the abundance of Shortleaf Pines. Historically, Shortleaf Pines outnumbered Loblolly Pines, a species more vulnerable to fire. Where the two species coexist, Shortleaf Pine hybridizes naturally with Loblolly Pine—a characteristic foresters exploited to develop varieties with attributes such as increased resistance to fusiform rust and ice damage. Fire suppression and commercial plantings facilitated contact between the two species, thereby increasing the incidence of hybrids from 3 percent in the 1950s to 45 percent today. The loss of genetic integrity—the replacement of Shortleaf Pines with hybrids—in a species well adapted to fire and drought may prove unwise as the coniferous forests in Texas face the prospect of global warming and other environmental stressors.

LONGLEAF PINE FOREST

Longleaf Pine forests once blanketed approximately 90 million acres of the coastal plain from Virginia to Texas and were the region's most valuable resource, but the species now occupies just 3 percent of its original range. Before the first settlers ventured into Texas, Longleaf Pine savannas dominated an area of about 5,000 square miles in the south-central portion of the Piney Woods, extending eastward from the Louisiana border to the Trinity River and southward to the Coastal Prairies (chapter 11). The species thrives on many soil types, but most of the Longleaf Pine forest in Texas occupied elevated level sites with coarse, well-drained sandy soils. Settlers commented that a carriage could travel between the trees without much difficulty at sites known as "flatwoods." Fires maintained the uncluttered landscape and stimulated the growth of grasses and forbs while inhibiting invasions of shrubs and hardwoods (fig. 2.4).

Longleaf Pine represents a classic example of a pyrophyte, a fire-adapted species. Like the seeds of all conifers, its seeds develop in cones that, at maturity, dry and separate their bracts so the seeds fall away to be dispersed by the wind. Seeds landing on patches of bare soil germinate, but those falling on soils covered with organic litter remain dormant; hence fires that eliminate ground litter enhance the odds of sprouting (and also curtail a fungal disease harbored in the litter). The seedlings resemble tufts of grass and, instead of continuing to develop aboveground, spend the next 5–12 years extending their taproots to depths of 10–12 feet. During this stage of development, the cluster of needles surrounding the terminal bud insulates the seedlings from fires. After the central root fully develops, the saplings shoot upward while their taproots continue growing, eventually reaching depths of up to 20 feet. Mature trees develop straight trunks with diameters of up to 3 feet and heights exceeding 100 feet. Deep taproots allow Longleaf Pines to withstand severe windstorms and tolerate prolonged droughts better than most other southern pines (fig. 2.5).

Bluestem grasses once dominated the open understories in Longleaf Pine savannas—common species included Slender, Broomsedge, and Little Bluestem—but without periodic fires, dense thickets of Yaupon and Wax Myrtle replace grassy understories. With the prolonged absence of fire, however, Loblolly Pines, Southern Magnolias, American Beeches, and White Oaks eventually replace the Longleaf forest. Other oaks, Flowering Dogwood, American Sweetgum, and Red Maple may also invade areas that remain unburned after harvesting Longleaf Pines.

Because of its high quality, Longleaf Pine quickly became the most valuable and readily marketed timber resource in Texas. Early loggers easily maneuvered their mule-drawn wagons through the open stands to collect naval stores or transport logs to waiting railcars (infobox 2.1). Annual harvests commonly reached up to 750 million board feet during the early 1900s, but most stands of Longleaf Pines had disappeared by the 1920s. After logging, fire suppression limited the regeneration of new stands, and plantations of faster-growing species, such as Loblolly Pine, replaced many areas once sustaining Longleaf savannas.

LOBLOLLY PINE FOREST

Southwest of the Longleaf Pine region, a Loblolly Pine forest originally occupied approximately 7,000 square miles of hilly terrain. On low ridges of sandy loam, stands of Loblolly Pine produced

Figure 2.4.
Longleaf Pine
savannas still
occur on well-
managed forest
lands like these
near Boykin
Springs Lake in
Angelina National
Forest (Angelina
County).

Figure 2.5.
Various growth
stages of Longleaf
Pine, including
the grass stage
(green tufts,
lower left), can be
seen in this view.

INFOBOX 2.1. "TURPENTINERS" AND THE NAVAL STORES INDUSTRY

Many secondary but vitally important industries developed in association with logging and sawmills in East Texas between the 1880s and 1930s. To support these activities, temporary tent cities and makeshift shantytowns sprouted throughout the Piney Woods. Loggers, for example, necessarily lived near the cutting front, which frequently moved into increasingly remote areas, and mill towns along the railroad lines produced a variety of wood products.

Hardwoods were harvested and brought to camps where crews hewed railroad ties and barrel staves with broadaxes and foot adzes. Leftover slabs and chips fueled the fires in boilers and cookstoves. Specialized workers known as "turpentiners" collected resin from Longleaf Pines by "chipping" the trunk with a vertical series of V-shaped grooves—collectively called "catfaces" for their resemblance to cat whiskers—to remove bark, phloem, and the outer xylem. Metal gutters placed below the grooves directed the flow of resin into cups, not unlike collecting sap for maple syrup. When filled, the cups were emptied into barrels made on-site, which were then hauled to a turpentine camp where the resin was processed.

Most of the resin was sent to turpentine stills, but some was heated in kettles to vaporize the terpenes to produce rosin, a main ingredient of varnish. The stills, operated much like those that distill whiskey, generated spirits of turpentine used to manufacture medicines, paint, rubber, soap, and other products. Even the dead limbs of Longleaf Pine had value; these were collected and slowly heated in earthen kilns to extract resinous tar and pitch. These products—turpentine, rosin, tar, and pitch—were widely known as "naval stores" because of their applications on wooden ships to seal hulls and weatherproof ropes and sails; experienced sailors were thus known as "old tars."

Gathering and processing resin involved hard, hot, smelly, and sticky work. For their efforts, turpentiners received low pay (often only a penny per tree), which they had to spend for food and basic necessities at company-owned stores. Worse, exorbitant prices kept turpentiners in perpetual debt to the company, with prison or convict work gangs awaiting those who attempted to escape their "debt-slavery." Company owners provided booze—likewise at inflated prices—to break the monotony and hardships of the job. Predictably, turpentine camps became infamous for drunken brawls, gambling well spiced with cheating, and murderous knife fights, all of which earned the workers a reputation as the "meanest people who ever lived."

To reduce the violence, the companies imported musicians for entertainment. Instead of expensive bands, piano players, many of them African Americans, became the musicians of choice because they could produce both rhythm and melody on a single instrument. The inventive pianists, inspired by the sounds and moaning whistles of steam locomotives chugging through the pines, developed a style of music later known as "boogie-woogie." Always on the move, camp musicians introduced the catchy music throughout Texas and eventually beyond.

In 1915, the Texas turpentine industry peaked with about 25 distilleries concentrated in six East Texas counties—these produced about 21 percent of the nation's resin products before the industry began to fade. Most of the stills and camps had closed by the early 1930s, but the rhythmic music originating in the turpentine camps lived on. Carnegie Hall hosted several boogie-woogie concerts in 1938 and 1939, and elements of the distinctive style that once calmed rowdy turpentiners remain today embedded in blues, swing, jazz, and rock 'n' roll.

canopies so dense as to retard the growth of understory communities (fig. 2.6). Unlike Longleaf Pines, Loblolly Pines develop shallow root systems; hence the trees blow over easily in windstorms, thereby opening the canopy. Sunlight reaches the forest floor in these places, allowing Yaupon, Wax Myrtle, and tree saplings to initiate the early stages of forest succession. A deep layer of leaf mulch covers the thin topsoils typical of this association.

Several species characteristic of the eastern and Appalachian forests—Bloodroot, Jack-in-the-Pulpit, Trillium, Mayapple, and Trout Lily—reach their western limits in areas where hardwoods predominate. Damp leaf debris supports several eye-catching orchids, including the Southern Lady's Slipper, Crested Coral Root, and Three Birds Orchid.

The largest of the southern pines, Loblolly

Figure 2.6. The canopy formed by stands of mature Loblolly Pines may become dense enough to retard understory growth (left). William Lutterschmidt embraces a huge Loblolly Pine that has survived at Boggy Slough (Trinity County) for nearly 100 years (right). This tree, the Texas State Champion Loblolly Pine, is the largest of its kind in the state.

Pines normally reach heights of 98–115 feet, but exceptional specimens may top out at 160 feet (fig. 2.6). Loblolly Pines grow rapidly, and their resinous wood is harvested for lumber and wood chips. Unable to withstand fires, Loblolly Pines were once restricted to areas with damp soils where wildfires were rare. With the advent of fire suppression, however, Loblolly Pines invaded many areas in Texas once dominated by Longleaf or Shortleaf Pines. Their rapid growth makes Loblolly Pines the tree of choice for most commercial pine plantations, and the trees are now planted as row crops throughout East Texas.

PINE SAVANNA WETLANDS

Shallow, water-filled depressions known as flatland ponds dot the grassy, level plateaus covered with Longleaf Pines. How the depressions formed remains unclear, but many believe that wind deflation in the geologic past scoured away loose

sand at sites where the surface was dry and barren, leaving exposed hardpans that hold seepage and rainwater for most of the year. Consequently, deeper ponds, colorfully rimmed with Common Buttonbush and other shrubs, seldom dry up and so can support aquatic vegetation that includes Big Floating Heart and Swollen Bladderwort, a submersed carnivorous plant. Sedges, rushes, and aquatic grasses edge their shallow perimeters. Less permanent ponds accumulate decaying needles and leaves as they dry, creating conditions favorable for Prairie Sphagnum and other acid-loving plants. Others become choked with emergent sedges and rushes. In early spring, flatwood ponds erupt in a raucous chorus of anurans, among them Hurter's Spadefoots, Blanchard's Cricket Frogs, Spring Peepers, Gulf Coast Toads, Green Frogs, and Eastern Narrow-mouthed Toads. Bullfrogs add rhythmic bass notes to the nightly serenade while Marbled and Dwarf

Figure 2.7. Tannins from decaying leaves and pine needles add a tealike stain to the water in flatland ponds. In the summer, ferns, sedges, and other moisture-loving vegetation will encroach on the pond.

Salamanders quietly waddle through grassy pond margins.

Pine savanna wetlands also occur where lowlands collect the seepage and runoff from knolls topped by isolated stands of Longleaf Pines. Between wildfires, a thin layer of needles and other organic debris accumulates on the forest floor, retarding runoff and thus increasing the absorption of rainwater that can percolate downward through the sandy soils to the underlying clay hardpan. Unable to penetrate farther, the water travels laterally along the surface of the clay layer until it emerges and forms a wetland at a lower elevation. In some cases, the water oozes forth on hillside swales where it forms small, dead-end "hanging bogs" or originates rivulets known as spring branches. Streamlets slowly carve through the sodden soils in these wetlands, creating braided channels separated by low hummocks of mosses, ferns, and other acidophilic plants. Spring branches support a unique streamside plant community of Coastal Pepperbush, Hazel Alder, Harvestbells, and vibrant Joe-Pye Weed.

Slightly larger wetlands formed in this manner

commonly occur near the Coastal Prairies region. Here, shallow acid bogs remain soggy for most of the year at the base of terraces where spring branches drain onto flat floodplains (fig. 2.7). Various combinations of water depth and permanence at these locations produce distinctive acid-bog communities. For example, Sweet Bay, Black Highbush Blueberry, and Rose Pogonia prefer muddy borders, whereas Woolly Rose-Mallows, Yellow Fringed Orchids, and Pale Pitcher Plants flourish in wetter areas (fig. 2.8). Blackgum, one of the few trees that will tolerate the moist, acidic conditions, grows stunted and bushy in these bogs. Drained for real estate development, many of these wetlands no longer exist.

Much larger bogs—baygalls—develop where seepage and organic debris fill abandoned bends and channels of streams. Because of the dim light beneath the closed canopy of buttressed Sweet Bay and Large Gallberry, the tea-colored water in these wetlands appears almost black. An understory of Swamp Titi, Large Gallberry, Swamp Palmetto, and Poison Ivy forms impenetrable thickets that may foster worrisome imaginings

Figure 2.9. The nickname "Big Thicket" is best appreciated when peering into a baygall from just beyond the perimeter. Photograph by Sandra S. Chapman.

Figure 2.8. In the Piney Woods, carnivorous Pale Pitcher Plants (top) and Yellow-fringed Orchids (bottom) flourish at different sites in sunlit bogs. Photograph by Alan B. Cressler, Lady Bird Johnson Wildflower Center.

of eerie beings (the word "bogeyman" may have been derived from the word "bog"). Conversely, beautiful ferns counter the gloom in habitat where vines reminiscent of tropical lianas twist upward. Glossy Crayfish Snakes harmlessly probe crayfish chimneys, but Cottonmouths, with their namesake white mouths agape, challenge visitors crossing the dimly lit forest floor. In the thickets, Golden Mice scamper branch to branch, balancing themselves with semiprehensile tails. Unlike most mice, these arboreal rodents construct nests of leaves and vines perched well above the flood zone.

The dense undergrowth in baygalls gave rise to the name "Big Thicket" (fig. 2.9). The tangle of Swamp Titi originated another name for the habitat: "Tight Eye Thicket" resulted from "titi" being corrupted to "tight eye." Dense thickets remain, but these occur less commonly than in the past. After clear-cutting on slopes, silt in the runoff rapidly clogs spring branches and swales and flushes the acid bogs.

ARID SANDYLANDS

Ancient streams—apparently much larger than at present—deposited wide ribbons of sand at points where their courses turned sharply. Other sand ridges, often concealing buried troves of petrified palm wood (appendix A), originated as dunes formed on the coast of a prehistoric tropical sea. Such sandy areas are especially abundant on terraces of the Neches River and Village Creek watersheds in Hardin County, and, unlike the plant life in other parts of the Piney Woods, the vegetation in these locations remain sparse and desertlike. Although the average annual rainfall, which usually exceeds 50 inches, might appear to preclude aridity, rainwater in fact percolates rapidly through the coarse sand, carrying with it nutrient-bearing silts and leaving behind a dry, sterile surface.

The gleaming white-quartz sands support Eastern Prickly Pear, Gulf Coast Yucca, Texas Bullnettle, and other xerophytic vegetation more characteristic of deserts—each species specialized to survive dry conditions. A covering of hairs insulates the leaves of many broad-leaved plants against heat, while deeply indented pores on the leaves of others retard moisture loss. Despite the harsh environment, the sandy ridges bear the region's largest variety of wildflowers. Widely scattered, the dominant trees in the sandylands are Longleaf Pine, Bluejack Oak, Blackjack Oak, Post Oak, and Black Hickory.

BEECH-MAGNOLIA-LOBLOLLY ASSOCIATION

Seepage areas on slopes and upper floodplain terraces in the southeastern Piney Woods support woodlands dominated by American Beech, Southern Magnolia, and Loblolly Pine, along with White Oak, hickories, and Florida Maple. These forests originally formed belts squeezed between upland pines and floodplains—upper terraces where soil and mulch stayed too moist for periodic fires. With fire suppression, Loblolly Pines, fast-growing oaks, and American Sweetgum invaded cutover uplands and initiated an understory of shade-tolerant species such as Southern Magnolia and American Beech. Gums, oaks, and pines died out as beeches and magnolias matured into a shaded woodland of climax vegetation. Perhaps the most beautiful of all forests, these communities occur most often in Polk, Tyler, Jasper, Newton, and Sabine Counties. One of the best examples is Mill Creek Cove in Sabine National Forest (Sabine County).

Figure 2.10. Decomposition of logs and leaves eventually returns nutrients to the forest soil; some fungi add a touch of color in the process.

The forest is not without color. In the crisp days of early spring, the blooms of Flowering Dogwood, Eastern Redbud, and azaleas sparkle in the understory like stars on a dark night. In summer, Cranefly Orchids enliven the forest floor beneath the beeches, with each of their several flower stalks hoisting about 40 tiny blooms that look like craneflies dancing in the wind. Beechdrops, a parasite nourished by beech roots, push their delicate stalks bearing magenta blossoms above the leafy mulch, and the colorful sporophytes (spore-producing bodies) of fungi adorn decaying logs (fig. 2.10). After a blush of fall colors—reds, yellows, and oranges from hardwood leaves—patches of evergreen Christmas Fern and the yellow blooms of Witch Hazel brighten the drab hues of winter. The bright red clusters of berries on American Holly add a festive spirit to the darkest of days.

These forests support a diverse animal community. Texas Leafcutter Ants travel well-worn trails bearing a harvest of leaves on which they cultivate their diet of fungi in underground chambers—a food chain in miniature (fig. 2.11). Cavities in older trees and snags provide nesting sites for Barred Owls, Pileated Woodpeckers, and Evening Bats. Northern Cardinals, Blue Jays, Carolina Wrens, and Brown-headed Nuthatches nest in tree boughs while Gray and Fox Squirrels harvest and store acorns for future use. Periodic droughts sometimes limit the supply of mast, at times with interesting results (infobox 2.2).

Figure 2.11. Texas Leafcutter Ants follow a well-worn trail (above) with their burdens of leaf sections reaped from nearby Indian Paintbrush (left).

BOTTOMLAND HARDWOOD FORESTS

Confined mostly to rich alluvial soils on the floodplains of rivers and streams, bottomland hardwood forests occupy a transition zone between drier upland hardwood associations and the riparian forests lining the banks of rivers and streams. Vegetation in these communities survives periodic floods but cannot tolerate long periods underwater (as in a swamp). Because of species-specific tolerances to flooding, even small differences in soil and topography significantly affect the composition of these communities.

Bottomland hardwood forests typically develop in well-defined layers: an overstory of dominant trees, principally hardwoods; a recognizable and dense understory of trees and shrubs; and a ground cover of herbaceous plants (fig. 2.12). Forests of this type are especially well developed where the topography flattens toward the Coastal Prairies. Dense assortments of oaks, including Water, Willow, Cherrybark, Overcup, and Swamp Chestnut Oaks, intermingle with American Sweetgum and

INFOBOX 2.2. SQUIRRELS ON THE MOVE

Gray Squirrels initially occurred in the forested eastern third of Texas but, for various reasons, now occupy other parts of the state as well. They are one of the few mammals active during the day, as is well known by anyone walking a city street or hiking in prime woodland habitat—and certainly to those good souls with bird feeders in their backyards. Few of us, however, associate the phenomenon of migratory behavior with Gray Squirrels, but migrate they do, albeit not often—1857 marked the last record of a large-scale squirrel migration in Texas. Strictly speaking, "migration" is a term reserved for seasonal travels to and from a place of breeding and is not applied to one-way movements, but we will stick with it here nonetheless.

The year began with typical winter weather but suddenly warmed in February, enough so that farmers seeded their crops weeks earlier than usual. Then, in April, three successive waves of freezing temperatures crossed a swath of East Texas. The cold affected more than crops; oak and other timber looked as if a fire had swept through the forest. Acorn and other mast production failed accordingly, damaged further by a drought later in the season.

In contrast, growing conditions had been favorable the previous year and had produced a heavy yield of acorns and pecans. The surplus of mast likely triggered greater production of squirrels in both late 1856 and early spring 1857. However, the increased squirrel population could not sustain itself when the cold snap subsequently damaged the vegetation and curtailed the mast crop available for the remainder of 1857. Thus, the squirrels migrated en masse in search of adequate food. Residents vividly recall the migration. One family traveling by horse-drawn carriage in Hopkins County encountered a phalanx of squirrels moving through tall grass followed by a wave of even greater numbers. Most of the army passed around and under the carriage, but others climbed over the horses and through the carriage, much to the dismay of the passengers. Numbering in the thousands, the squirrels continued their march for about an hour—the cornfields were literally alive with squirrels, again suggesting that food shortages were responsible for the migration. Similar reports originated from cornfields in Dallas, Wood, and Marion Counties; in one case, cotton bolls were also damaged by the migrating squirrels.

Squirrel migrations have occurred elsewhere; the earliest of these events was reported by Meriwether Lewis when he traveled down the Ohio River to join William Clark in advance of their remarkable expedition to the Pacific. On September 11, 1803, the intrepid explorer encountered large numbers of squirrels swimming with "pretty good speed" across the river, heading south. Lewis checked the mast available on both sides of the river and found the nuts of walnut and hickory (but not beech) in "great abundance." Moreover, the animals were fat; hence Lewis attributed the migration to climate and not to a food shortage.

Migrations of squirrels were commonplace enough in the early nineteenth century that Audubon (1785–1851) and his coauthor John Bachman (1790–1874) recognized a separate species, *Sciurus migratorius*, for those populations with "wandering habits." Other migrations are far more recent. In 1946, a mixed horde of both Gray and Fox Squirrels migrated across a broad front in Wisconsin. The migration occurred in a year when the mast crop amounted to no more than 15 percent of normal following a year of abundant mast production. Some of the squirrels were hungry enough to gnaw at bark but otherwise appeared in good body condition. Earlier migrations in Wisconsin occurred in 1842, 1847, 1852, and 1857.

Another large-scale migration on record occurred in 1968 when squirrels moved en masse, but in random directions, throughout a large region extending from Vermont and New York to North Carolina and Tennessee. As in Texas, a poor mast year had followed a year (1967) of bumper acorn crops. This time, however, wildlife biologists were at hand, and their examinations revealed that the squirrels were in good condition, with full stomachs and normal loads of fleas and other ectoparasites.

Because starvation can be ruled out as the immediate cause of these movements, another theory emerged. Eastern Gray Squirrels are normally busy in early fall collecting and burying acorns and nuts as stores for the approaching winter. However, if these are not abundant, as in a year with a poor mast crop, the shortage of food provides the trigger for the migrations—a response enhanced when a bountiful crop in the preceding year paves the way for an enlarged population. Thus, the want of mast to bury, and not hunger itself, initiates the movements. While certainly plausible, the idea remains a hypothesis that will be difficult to test. Indeed, with ever-fewer mature forests and more squirrels residing in urban areas where food shortages are unlikely—remember those bird feeders—we may never again see movements akin to the Big Texas Squirrel Migration.

Figure 2.12. Note the diversity in this bottomland forest along White Oak Creek (Trinity County) near the Neches River. Photograph by Jonathan K. Gerland.

Blackgum in the overstory. Common Persimmon, Red Maple, American Hornbeam, Carolina Ash, and Possumhaw characterize the understory, which is shared with shrubs such as Dwarf Palmetto, Silver Bells, and American Beautyberry. As might be expected, fires rarely affect bottomland hardwood forests; when they occur (usually during droughts), the moist ground cover burns slowly and returns nutrients to the soil. However, with a prolonged absence of fires, accumulations of leaf litter can fuel high-intensity conflagrations that damage the canopy and alter the composition of the understory.

Known for their richness, bottomland hardwood forests may support up to five times more species of plants and animals than either pine or upland hardwood forests. Older stands with mature, hollow trees accommodate various cavity-nesting species, among them Wood Ducks, Screech Owls,

PINEY WOODS

Great Crested Flycatchers, Carolina Chickadees, and Brown-headed Nuthatches. The leafy canopy harbors nests of numerous songbirds, and drapes of Spanish Moss provide secure roosting sites for tree bats, such as Eastern Red, Hoary, and Seminole Bats. Raccoons, Striped Skunks, and American Mink patrol damp ground litter in search of Pickerel Frogs and Texas River Crayfish. Eastern Mud Turtles and Alligator Snapping Turtles occasionally leave the water and slog across muddy bottomlands.

BALDCYPRESS–WATER TUPELO SWAMPS
Semipermanent forested wetlands develop in sluggish areas where floodplains widen along riverbanks, sloughs, and oxbows. Some backwater areas—swamps—often remain deeply flooded throughout the year. Swamp water circulates slowly, often becomes anaerobic, and becomes stained dark brown to almost black from the tannins released from decaying leaves. Trees in the overstory feature root systems adapted to hydric soils lacking oxygen. Where the Piney Woods embrace the Coastal Prairies, the composition varies from mixed stands of Baldcypress and Water Tupelo to nearly pure stands of either species. The

trunks of both species are often buttressed (flared outward at the bottom), and the familiar woody "knees" arise from their roots (fig. 2.13). For many years, biologists believed that these structures absorbed oxygen, hence enabling submerged roots to survive. However, little evidence supports this idea, and scientists now think that buttressed trunks and knees simply help stabilize trees in soggy soil. Where soil accumulates around Baldcypress knees, large colonies of a showy orchid, Fragrant Ladies' Tresses, send up spiraling stalks with numerous large, sweet-scented blooms.

Whereas the roots of Baldcypress and Water Tupelo often remain submerged for long periods, the seeds of these species cannot germinate underwater, although they may remain viable when submerged for more than a year. The seeds of both species are examples of hydrochores—seeds that disperse by floating. Once the seeds are in the water, wind or floods carry them to new locations—some land on damp stream banks or other moist sites where they germinate into rapidly growing seedlings. Wind and water currents limit the distribution of the seeds, but bees distribute pollen between floodplains, facilitating genetic exchanges between otherwise isolated populations of these trees.

Understory and ground cover are generally absent in permanently flooded swamps, but aquatic plants, such as American White Water Lily, cattails, Duckweed, and invasive species such as Water Hyacinth and Giant Salvinia, often flourish in these locations. In seasonally flooded habitats, the understory includes Water Oak, Water Hickory, Swamp Blackgum, Eastern Swamp Privet, Water Elm, and Black Willow. Trumpet Creeper and Climbing Hempweed vines create tangles on the waterlogged ground where clumps of Common Beggar-Tick, Water Paspalum, Bog Hemp, and other species persist.

The most abundant fishes in swamps necessarily tolerate low oxygen levels—conditions, for example, that may force the Western Mosquitofish and Golden Topminnow to the surface for gulps of air. Other common species include Black Crappie, Warmouth, and Longnose Gar. Turtles abound in swamplands, and it is not uncommon to see up to seven species sunning on a single log. Northern River Otters, American Beavers, and Nutria remain in constant danger from the major swamp predator, the American Alligator. Nutria, which were introduced into Louisiana in 1938 and later invaded Texas, may reach densities large enough to cause "eat outs" of wetland vegetation. Sturdy branches of stately Baldcypress—the sentinels of the swamp—support nests of Bald Eagles, Swallow-tailed Kites, Ospreys, Great Blue Herons, and Great Egrets.

THE SAWDUST EMPIRE

The abundance of tall pines proved irresistible to fortune seekers who clearly saw the trees rather than the forest. One of the first sawmills in the Piney Woods appeared in 1829 on Carrizo Creek in Nacogdoches County, but the timber industry did not flourish until later in the nineteenth century. The demand for lumber began when settlers flooded into the area after Texas won its independence, and the availability of seemingly unlimited pine and hardwood forests soon produced charcoal, sawmills, and lumber towns. From 1883 to 1920, the state of Texas was the largest producer of charcoal—approximately 3,750 bushels of charcoal, or six railcar loads, once fueled the

prison furnace at Rusk. Convicts felled the trees, prepared the charcoal, and loaded the cars on the state-owned railroad that penetrated far into the forests. Initially, sawmills were located near streams and rivers; doughty river drivers floated the logs to millponds where the outflow was harnessed to power sawblades. The advent of steam-powered machinery—log loaders, narrow-gauge trains, and bandsaws—enabled the transport of larger loads of timber to more powerful and efficient sawmills (fig. 2.14).

Timber crews—usually 40 to 60 strong—selected, felled, scaled, and trimmed the downed timber to transportable size under the watchful eye of a crew boss known as the "Bull of the Woods." Access to transportation facilitated a harvest of about 146 million board feet of Shortleaf Pine in 1880. At the peak of the Sawdust Empire, more than 200 railroads chugged through the Piney Woods, hauling logs to more than 1,800 sawmills and some 3,700 other sites associated with the logging industry. About 100 of the sawmills could produce more than 80,000 board feet of lumber daily, enough to sustain large corporations and numerous company towns. Transported afar by trains and ships, shingles and lumber from the Piney Woods reached many western states, Central and South America, and the ports of Europe.

ENDANGERED SPECIES

Early logging companies seldom practiced sustainable forestry and therefore paid little attention to restocking cutover areas. Instead, logging operations simply moved to another location, leaving stump-covered sites to invasions of hardwoods and snarled thickets amid piles of debris. A few pine seedlings might appear, but cutover areas often remained unfavorable for regenerating new stands of pines. Because of these shortcomings, the pine forests, along with many company towns and most sawmills, disappeared by the early 1900s—an oil boom soon supported the regional economy. Pine plantations, ecologically sterile in diversity, later replaced woodlands, followed by fire suppression and hardwoods "deadened" by girdling for "timber stand improvement"—these conversions, along with other disturbances, permanently altered

Figure 2.14. Steam log loaders (top) and locomotives (bottom) sped the journey of timber from forest to sawmill and hastened the demise of the original East Texas forests. Photographs from 1907 courtesy of the History Center, Diboll, Texas (www.thehistorycenteronline.com).

Figure 2.15. The namesake patch of red feathers remains obscure on this Red-cockaded Woodpecker perched at a nest entrance, but note the resin seeping from the wells around the cavity (left). Restoration efforts for this endangered bird include inserting nest boxes into mature pines, such as this Shortleaf Pine (right). Photographs by John Maxwell, US Fish and Wildlife Service (left), and Brian R. Chapman (right).

many natural biotic communities. Some animals, adapted to ecological conditions no longer in place, persisted only in pockets of unaltered or marginal habitat, but other species disappeared completely from the Texas fauna.

A legendary species, the Ivory-billed Woodpecker, existed in East Texas as late as the 1930s but now seems extinct. With a 30-inch wingspan, it was one of the largest woodpeckers in the world. Ivorybills depended on large pine or Baldcypress snags for nesting sites, and the birds disappeared as the huge snags of these species fell or were chopped down. Because the trees take more than a half century to grow, die, and form snags suitable for nest cavities, the interval exceeded the life span of the few remaining birds. As Ivorybills began to disappear, commercial sellers and museums hurried to collect specimens, accelerating their decline. In the last two decades, several reports of living Ivorybills created hope that the species was not extinct, but the alleged sightings remain unsubstantiated.

In contrast to the large size of Ivorybills, the wingspan of Red-cockaded Woodpeckers (RCW) extends about 15 inches, about the same as a Northern Cardinal. Despite their name, red feathers appear as only a tiny streak on each side of the head—a cockade scarcely visible even with binoculars (fig. 2.15). RCWs, once abundant in Longleaf Pine savannas, began to disappear when the Sawdust Empire expanded in the 1800s. The territorial, nonmigratory species is the only North American woodpecker that excavates cavities exclusively in living trees, preferring Longleaf Pines but accepting any mature southern pine about 80 or more years old. Trees of this age often suffer from Red Heart Fungus, which softens the wood in the center of the trunk and therefore facilitates excavation.

Individual RCWs live in groups consisting of a mated pair and several male (rarely female) offspring from previous years—these serve as "helpers" for feeding nestlings of the current generation. The group maintains a territory consisting of up to 20 cavities in a cluster of trees dispersed across 3 to 60 acres. RCWs peck out small resin wells around the entrance of each actively used cavity—the flow of resin forms a sticky surface, presumably a defense against Texas Ratsnakes, Southern Flying Squirrels, and other arboreal predators. Ecologists regard RCWs as a keystone species in southern pine ecosystems because at least 27 species of vertebrates and numerous invertebrates are secondary users of RCW cavities. The specificity of their nesting requirements makes RCWs extremely vulnerable to habitat alterations—for example, most pines are cut long before they become infected with Red Heart Fungus. Fortunately, RCWs accept artificial nest cavities placed in second-growth pines, now an established management tool on national forests within the range of the species (fig. 2.15).

Sandy soils in parklike Longleaf flatwoods provided habitat for Louisiana Pinesnakes, which once occurred in 13 counties of East Texas and an area of similar size in Louisiana. Unlike ratsnakes and several other snakes in the region, Pinesnakes do not climb trees. Instead, they forage at ground level for small mammals, amphibians, turtle eggs, and birds. When fires or other dangers approach, Pinesnakes seek refuge in sandy burrows, especially those created by Baird's Pocket Gophers. Extensive harvests of Longleaf Pine, combined with the suppression of fire in the managed forests that followed, significantly increased the density of the understory vegetation. Because of these ecological changes, the range of Louisiana Pinesnake in East Texas gradually contracted to just four counties where periodic fires maintained stands of Longleaf Pine. The snake, however, has not been seen in these counties since 2008 and may be extirpated in Texas.

Red Wolf populations declined soon after settlers arrived and logging began. The much-altered habitat favored Coyotes, which soon expanded their range eastward, filling a void left as Red Wolves disappeared from the Piney Woods. By the 1960s, biologists realized that Red Wolves were a vanishing species hanging on in marginal habitat, and worse, many of the few remaining individuals were Red Wolf–Coyote hybrids. Because of a process known as introgression—or "genetic swamping"—Red Wolves were now losing their identity and the hybrids were more like Coyotes than Red Wolves. Fortunately, a small but genetically intact population survived along the coast in southwestern Louisiana and southeastern Texas (primarily Chambers County). To save the species, the last 17 animals in this population were captured, and those lacking any traits of Coyotes formed the core of a captive breeding program. After several failed attempts to reintroduce captive-bred offspring, four pairs released at Alligator River National Wildlife Refuge in North Carolina established a foothold. For now, the eerie wails of Red Wolves still echo across lowlands in North Carolina, but their serenades—long symbolic of wild America—are forever silent in Texas.

Timber harvests were not the only cause of alterations that affected wildlife in the Piney Woods. With settlement, demands arose to manage the rivers of East Texas for flood control and dependable water supplies, which spurred construction of dams. These obstacles further reduced the numbers of American Paddlefish, a species already suffering from pollution and overfishing (for meat and roe, a caviar). Essentially unchanged since the Late Cretaceous, American Paddlefish occur in deep, slow-moving rivers where they swim slowly, mouths open to scoop in plankton (fig. 2.16). A lengthy, broad, paddle-shaped rostrum (snout) extends from their head, a feature easily distinguishing the species from any other North American fish. Electroreceptors embedded in the rostrum and the anterior half of their body detect plankton and facilitate movement in dark waters.

Dams inundated hundreds of thousands of bottomland acres, altered natural flood-pulse cycles, and blocked the seasonal migrations of the paddlefish to their ancestral spawning grounds. In the past, spring floods stimulated their migration upriver, where the floodwaters temporarily inundated exposed gravel bars. Males and females spawn synchronously over these bars, where the externally fertilized eggs (zygotes) drop and adhere to the gravelly substrate. After hatching, the fry wash downstream and spend about a decade maturing in deep pools. Fortunately, American Paddlefish can be raised successfully in aquaculture farms, thus producing stock for replenishing diminished populations. For example, in 2014 the Texas Parks and Wildlife Department released radio-tagged American Paddlefish in Caddo Lake

Figure 2.16. The long rostrums of American Paddlefish detect concentrations of plankton that are scooped into their gaping mouth. Photograph courtesy of the US Fish and Wildlife Service.

and its tributaries; telemetry locations allow scientists as well as schoolchildren from 20 schools to track the movements of each fish. In addition, water releases from upstream dams are now timed to match natural flood-pulse cycles with the intent of mimicking natural spawning conditions. Continued management and protection offer hope for the survival of this ancient species.

HIGHLIGHTS

THE BIG THICKET

To folklorists and romanticists, the "Big Thicket" invokes dark legends veiled within a mysterious tangle where few dared venture. Yet for some biologists, the diversity of the area justifies its recognition as the "Biological Crossroads of North America." Somewhat analogous to beauty being in the eye of the beholder, the Big Thicket defies easy interpretation. Many citizens of East Texas assume that the Big Thicket includes all of the Piney Woods—an understandable misconception since the geographical extent of the region and its sprawling thickets has never been adequately delineated. Expansive thickets naturally occur in areas where the soils retain moisture—depressions, seepage slopes, and stream bottoms—but dense, unnatural undergrowth has also developed as the result of fire suppression. Soils covered with layers of moist humus, especially those containing leaves

of flame-retardant species such as American Beech, Southern Magnolia, and oaks, depress the spread of wildfire and encourage development of thick vegetation. On poorly drained soils, moisture- and acid-loving shrubs, such as Swamp Titi, Southern Magnolia, Large Gallberry, Wax Myrtle, Yaupon, Coastal Pepperbush, and azaleas, entwine in nearly impenetrable barriers.

The earliest reference using the term "Big Thicket" included all of Texas east of the Brazos River, but most ecologists more narrowly define the region as including all or parts of the southern half of the Piney Woods and accept at least one of the following descriptions. First, in the broadest view, the "Ecological Big Thicket" encompasses 3.5 million acres, including the entire basin between the Sabine and Trinity Rivers lying south of US Highway 190 between Livingston (Polk County) and Jasper (Jasper County) and continuing to I-10 between Liberty (Liberty County) and Beaumont (Jefferson County). Elevations in this region slope toward the Gulf, exposing a series of ancient sandy shorelines marking changing sea levels during the Pleistocene Epoch. The "Upper Thicket"—the more elevated terrain in the northern half—supports a mixed forest codominated by Loblolly Pine, White Oak, Southern Magnolia, and American Beech, except where nearly pure stands of Longleaf Pine cover the highest ridges (fig. 2.17). Fingers of the

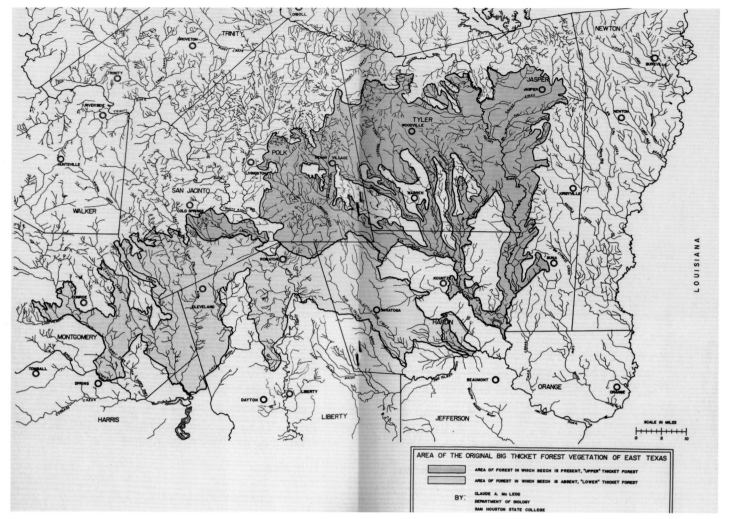

Figure 2.17. Delineation of the "Ecological Big Thicket" in East Texas by McLeod (1967). Reproduced with permission of Sam Houston State University.

mixed forest follow stream banks and penetrate the "Lower Thicket" in the flatlands farther south. The forest of the lower thicket differs only slightly—Chestnut Oak replaces American Beech as a codominant.

The second delineation assigns the "Traditional Big Thicket" to the watershed of Pine Island Bayou, an area of approximately 512,000 acres. Because of its impenetrable vegetation—essentially the "Lower Thicket"—this area remained untamed until long after other parts of the region had given way to settlement. For the same reason, it became known as the "Old Bear Hunter's Thicket." Hunters used dogs to chase Louisiana Black Bears (and Feral Pigs) into more open sites—one hunter boasted of bagging more than 300 bears in the Thicket. In another area, called the "Old Hurricane Section," a great storm flattened most of the large trees. Hot

mineral water bubbling to the surface near two salt domes, the Saratoga Dome and the Batson Dome, once attracted visitors to spas and luxury hotels, but the tourist industry ended with the development of nearby oil fields.

More narrowly, the third definition recognizes only the Big Thicket National Preserve, created in 1974 when Congress established and placed an archipelago of conservation units under the jurisdiction of the National Park Service. At the time, the action proved controversial; proponents faced serious threats yet bravely continued their efforts to protect this unique area (infobox 2.3). The preserve, later enlarged, today protects 109,085 acres of incomparable forest, although some sensitive areas remain at risk. A private organization, the Big Thicket Association, continues the work of saving these areas.

INFOBOX 2.3. GERALDINE ELLIS WATSON
Champion of the Big Thicket (1925–2012)

Unfortunately, unpleasant backstories often accompany the creation of a national park or preserve today enjoyed by thousands of visitors. Forgotten in the dustbin of history, these usually concern the trials and tribulations endured by those who bravely faced antagonists opposed to protecting an ecological treasure. We hope, however, that the contributions of Geraldine Ellis Watson will not go unnoticed by those visiting Big Thicket National Preserve—nine units of incomparable forest totaling more than 105,000 acres and six water corridors tucked away in the southeastern corner of Texas.

During her childhood in southeastern Texas, Geraldine's family moved from one sawmill town to another as her father sought work with timber companies and railroads. She learned forest vegetation in the company of her mother, who identified plants used for medicines and dyes, and from her father, who taught her to identify trees. Geraldine often visited undisturbed glades where she developed a deep affection for the splendor of the virgin forest and, for many years, kept extensive notes on the blooming schedules of forest plants. As might be expected, these experiences heightened her dismay when the beautiful savannas of Longleaf Pines fell to axe and saw and were then replaced with uniform, farm-like rows of Loblolly Pines. Likewise, she saw streams and waterfalls forced into channels or choked with brush and watched housing developments supplant flower-filled meadows.

Geraldine, realizing she lacked the scientific background necessary to document the biological diversity of the Big Thicket, enrolled at Lamar University. As an undergraduate, she collected plant specimens and created the university's herbarium. These efforts attracted the attention of prominent botanists, with whom she shared her knowledge of the Big Thicket and its flora. In recognition of her many achievements in education and conservation, volume 3 of *Useful Wild Plants* is dedicated to Watson.

Early in her career, Geraldine became an environmental activist dedicated to protecting the Big Thicket. She engaged the public with a weekly article appearing in the *Pine Needle*, a local newspaper in Hardin County—and often enough, she criticized the timber industry and other developers for their abuses of her beloved forests. For her efforts, she and her family endured constant harassment—some spat on her and publicly accused her of hobnobbing with communists, and

her teenage children were arrested and jailed for carrying her plant specimens, spuriously described by the local sheriff as "marijuana." Even the newspaper office was firebombed. Yet Geraldine persevered and became a founding member of the Big Thicket Association. As a member of the organization's board of directors, she traveled tirelessly across the state speaking about the values of biological diversity and the unique features of the Big Thicket. Her efforts to generate support included leading field trips to pristine areas of the forest, whenever possible taking along such dignitaries as then-Congressman George H. W. Bush (later President Bush), Senator Ralph Yarborough, and other officeholders. Her efforts produced results, and as public pressure intensified, Geraldine was among the authorities invited to testify before Congress, which eventually established the Big Thicket National Preserve in 1974.

For the next 15 years, Watson continued to document

Geraldine Ellis Watson expressed her love for the Piney Woods in both her books and her art. Her oil paintings depict the rich flora, including the Yellow Ladies' Slipper orchid shown here, and the distinctive culture of the region. Included with permission of the Natural History Museum, Sam Houston State University. Photograph by Steve Korevec.

the local flora while working at the preserve as a naturalist for the National Park Service. A talented artist, she also maintained a small gallery of her own illustrations of the regional biota as well as some features of the area's cultural history. After leaving the Park Service, she acquired and restored a property on Lake Hyatt, near Warren, where she developed her own refuge—the Watson Rare Native Plant Preserve. The area, open to the public, protects seven species of orchids, ten species of ferns, wild azaleas, blueberries, trilliums, and four of the five types of carnivorous plants native to North America.

Geraldine Watson wrote *Big Thicket Plant Ecology: An Introduction*, now in its third edition, which presents a definitive account of the region's plant communities and descriptions of each unit in the preserve. Her second book, *Reflections on the Neches: A Naturalist's Odyssey along the Big Thicket's Snowy River*, offers a peaceful memoir of a canoe trip—undertaken at the age of 63—along with insights about the river's natural history and encounters with the people she met. Both books project her love of the Big Thicket, a national wonderland that might not exist today were it not for her efforts.

CADDO LAKE

Most northern states boast hundreds of natural lakes gouged by the advance of Pleistocene glaciers. Although the Ice Age influenced plant and animal distributions in Texas, the glaciers never reached the state, which lists only one natural lake—all others were formed as impoundments. Caddo Lake, the largest natural body of freshwater in the South, covers 26,800 acres, half in Harrison County, Texas, half in adjacent Louisiana (fig. 2.18).

Two competing hypotheses address the formation of the lake, the first originating from the legends of the Caddo Indians. When a chief failed to obey the Great Spirit, he was punished by the New Madrid earthquakes (1811–1812), which caused the ground to subside and fill with water. Although these same earthquakes did create Reelfoot Lake in Tennessee, the presence of Baldcypress 400 to 600 years old—dated by tree-ring counts—at the lake site suggests that a swampy area existed there at least 200 years before the seismic shocks. Additionally, drill cores extracted from the lakebed and the surrounding area show no evidence of subsidence. Conversely, substantial evidence supports the second explanation—the "Great Raft Hypothesis." Spanish explorers in the 1500s and members of the Freeman-Custis Expedition in 1806 described a massive logjam on the Red River near present-day Shreveport. The huge raft of logs backed floodwaters into the river's tributaries, thereby creating a series of isolated impoundments, including the former Ferry and Soda Lakes, which

eventually coalesced into Caddo Lake and an interconnected wetland system.

Surrounded by bottomland hardwood forests, the "lake" more closely resembles an expansive swamp—open water in mysterious bayous and sloughs that weave through stands of stately Baldcypress draped with curtains of Spanish Moss (fig. 2.19). The lake received national notoriety in 1909 when the Great Caddo Lake Pearl Rush began. Fortune hunters streamed into the area after news circulated that pearls abounded in the lake's freshwater mussels. Most of these so-called pearl hogs went home empty handed, but a few got rich—in 1912, the lake yielded pearls valued at $99,200, equivalent to about $1.9 million today.

When the Great Raft existed, hundreds of steamboats traveled up the Mississippi to the Red River, up a small tributary south of the raft, and through Caddo Lake to Jefferson, the most inland city in East Texas accessible by navigable waterway. However, the lake drained and reverted to a shallow swamp after engineers removed the Great Raft in 1874. In 1914, a small dam constructed downstream restored the lake to levels that allowed barges and drilling equipment access to Caddo Lake for underwater oil exploration; the stabilized water levels also preserved the wetland vegetation.

In addition to more than 20 freshwater mussel species, at least 190 species of trees and shrubs, 93 different fishes, 22 amphibians, 46 reptiles, 220 species of birds, and 47 kinds of mammals occur in the ecosystem associated with Caddo Lake. This

Figure 2.18. Backwater from a log jam on the Red River formed Caddo Lake, the only natural lake in Texas; now impounded by a small dam, the lake extends into eastern Louisiana. Photograph by Kaitlen P. Gary.

census, however, does not include "Bigfoot"— an elusive apelike creature allegedly sighted in the area for decades. The 8,253-acre Caddo Lake State Park and Wildlife Management Area and the 8,493-acre Caddo Lake National Wildlife Refuge protect part of the region. Unfortunately, invasions of three noxious, exotic species—Giant Salvinia, Hydrilla, and Water Hyacinth—choke out the native vegetation and require ongoing treatment.

SAVAGE SAVANNAS— CARNIVOROUS PLANTS

Bogs collect water from seeps, but most of the moisture ponding in wet pine savannas originates as rainfall. Decomposing pine needles and hardwood leaves falling into these soggy depressions release acidic tannins that stain the water, and few inorganic nutrients cycle through these systems. The acidity slowly increases, with the result that few plants can survive in the moist, nutrient-impoverished soils in these small

Figure 2.19. Spanish Moss festoons Baldcypress trees at the swampy end of Caddo Lake. Photograph by Kaitlen P. Gary.

wetlands. However, one specialized group—carnivorous plants—cope by preying on small animals to compensate for the lack of certain soil nutrients, especially nitrogen. In short, carnivorous plants obtain chemicals, rather than calories, from their victims—as with other green plants, photosynthesis provides the energy needed for growth and development.

Carnivorous plants are relatively rare and are largely restricted to sunny wetlands with poor fertility. Carnivory developed independently in five different orders of flowering plants; these include more than 12 genera and 500 species distributed worldwide—a remarkable example of convergent evolution. Of the approximately 45 species of carnivorous plants in North America, only 14 occur in Texas. Ten of these are bladderworts, plants that flourish submerged in ponds, marshes, and other aquatic sites or on waterlogged soils. The other four species are emergent, thriving on soggy soils in grassy savannas or standing partially submerged in bogs. The carnivorous plants of the Piney Woods employ three different mechanisms for capturing prey.

The vegetative organs of bladderworts are not clearly differentiated into roots, stems, and leaves, but their bladder traps represent one of the most sophisticated trapping mechanisms in the plant kingdom. Each bean-shaped bladder consists of two convex halves connected by a hinge. After closing, water inside the bladder is pumped out, collapsing the walls inward and setting the trap. When a rotifer or other small animal brushes the tiny trigger hairs fringing a bladder, it snaps open, suddenly creating suction that pulls the prey inside. The trap closes instantly (in 1/460 of a second), after which digestive secretions dissolve the entrapped victim within a few hours. When larger prey, such as a mosquito larva, becomes partially ensnared, the digestive enzymes dissolve the animal in stages—the trap opens periodically to engulf more of the prey until it is completely consumed. Yellow or purple flowers rising above the water surface on thin stalks indicate the presence of bladderworts.

Two species of sundews and another of butterwort found in the Piney Woods capture animals in similar ways; the leaves of the three species form basal rosettes at or slightly above ground level. However, tiny erect tentacles on the leaves of the sundews secrete globular drops of sticky fluid (whereas a greasy mucilage coats the butterwort's leaves). Insects traveling across the leaves become mired in the glue-like secretions, after which digestion begins when the leaves (of all three species) slowly curl around the hapless animal. In places, sundews carpet the ground so densely that a single footstep may crush 10–20 plants (fig. 2.20).

The modified leaves of the Pale Pitcher Plant, the only representative of its kind in Texas, also capture insects. As the plant develops, the edges of a specialized leaf form an erect tube up to 26 inches in length. An umbrella-like flap extending over the tube's mouth prevents rainwater from diluting a nectar-like fluid in the tube that serves as both insect attractant and digestive fluid. Hairs pointing

Figure 2.20. Sweet gooey droplets sparkling at the ends of stalks on sundew leaves lure and entrap insects.

downward inside the tube thwart the prey's escape efforts, and victims eventually fall into the fluid, sealing their fate.

Ants represent nearly 75 percent of the prey captured in pitcher plants, but a variety of other insects and spiders also become victims. Surprisingly, the plants actually capture less than 1 percent of the animals that enter their tubular traps, but such a low capture efficiency may actually be beneficial. Whereas pitcher plants can satisfy their nutrient requirement with occasional captures, a higher rate of success might develop avoidance behaviors in the prey population and become an evolutionary backlash to the plants.

Not all organisms succumb to the digestive processes within pitcher plants. Indeed, the pools of liquid in their tubes provide unique habitats, known as phytotelmata, where organisms known collectively as inquilines—bacteria, fungi, protozoa, and the larvae of certain mosquitos, flies, and midges—flourish in miniature ecosystems. For example, the larvae of a flesh fly, the Pitcher Plant Sarcophagid, swim at the surface where they feed on newly drowned victims, whereas larvae of the Pitcher Plant Mosquito living at the bottom of the pool help digest the prey. Spiderwebs sometimes block access to the tubes, and Yellow Crab Spiders occasionally intercept prey just inside the lip. These examples illustrate two important ecological interactions: (1) mutualism—insect larvae consuming some nutrients while assisting with the digestive process; and (2) resource partitioning—spatial separation of the various predators, top to bottom, within the phytotelmata.

Most pitcher plants grow from rhizomes protected deep in the soil. Rhizomes allow the plants to survive catastrophic events such as hurricanes or the periodic fires that sweep through pine savannas. They flower in March and April and, somewhat ironically, depend on potential prey, especially bumblebees, for pollination. In addition to forming the leafy pitcher, each plant sends up a single stalk that supports one large, drooping flower. The plants die back at the end of the summer, releasing their seeds through slits in a round fruit.

CONSERVATION AND MANAGEMENT

During the Sawdust Empire era, which lasted from about 1850 to 1920, many large timber companies callously abandoned their logged-over lands. Inevitably, the last tree fell and some companies vanished in bankruptcy, but others moved west, where they continued "cut-and-run" logging in the Cascades. Left behind were thousands and thousands of stumps—grim reminders of a once-magnificent pine forest, not unlike gravestones scattered across a tattered battlefield of thickets—hardwood regrowth, and a few pine seedlings.

As early as 1898, it became obvious that a plan was needed to restore the forests in East Texas, including a strategy to harvest the timber on a sustainable basis. As part of the plan, the US Forestry Bureau, the precursor of the USDA Forest Service, initiated an agreement with several remaining Texas lumber companies and private individuals to responsibly conserve soil, water, and other natural resources (fig. 2.21). The plan bore fruit in 1933 when the Texas legislature authorized the USDA Forest Service to create national forests in the state. Guided by surveys made early in the planning phase, the Forest Service purchased land from 11 timber companies. These areas now constitute more than 90 percent of the four

national forests in East Texas, thus bringing a measure of protection to approximately 638,110 acres of woodlands. Now mature, these managed forests provide habitat for a diverse fauna and flora, including several endangered species. Controlled burns periodically reduce understory competitors and optimize conditions for Red-cockaded Woodpeckers and Louisiana Pinesnakes. As reminders of past abuses, historical sites in the national forests display the wasteful remains of the Sawdust Empire.

The battle to save the Big Thicket proved far more contentious than the earlier efforts to establish national forests in East Texas. At various times, up to 300,000 acres were proposed for a national preserve, but special-interest groups opposing federal appropriation of private lands succeeded in reducing the area to 84,550 acres in nine widely separated units and six stream corridors. Over the years, additional purchases by the federal government and private organizations added more than 24,500 acres to the original units, thereby protecting several exceptional habitats that otherwise would have been lost to development.

An effort by citizens' groups to establish permanently conserved places in the national forests resulted in the designation of five national wilderness areas in 1984—the only units of the National Wilderness Preservation System in Texas except for part of Guadalupe Mountains National Park in West Texas. These areas range in size from the 3,639-acre Big Slough Wilderness, which protects a relatively unaltered bottomland hardwood forest on the Neches River, to the 13,390-acre Upland Island Wilderness, which includes a variety of habitats from hillside bogs and Longleaf Pine forests to forested floodplains.

The enabling legislation establishing the Big Thicket National Preserve contained a compromise provision that allowed the recovery of minerals. Consequently, oil-drilling operations continue in some units of the Big Thicket, where the damage from roads and well pads is noticeable and regrettable. Perhaps of greater concern is the alteration of habitats immediately adjacent to the preserve. For example, a huge sanitary landfill was excavated next to the Lance Rosier Unit in the heart of the Traditional Big Thicket. Nonetheless, the

Figure 2.21. In 1967, Texas South-Eastern Locomotive No. 13 hauled logs to a sawmill in Diboll, Texas, evidence that some timber companies replanted and practiced sustainable timber harvests for decades after others abandoned their cutover lands. Photograph courtesy of the History Center, Diboll, Texas (www.thehistorycenteronline.com).

preserve protects some unparalleled habitats and a great diversity of species. Ironically, many former opponents of the preserve now proudly boast about "their" amazing forests.

Little Sandy National Wildlife Refuge (Wood County) offers a fine example of a joint conservation effort between the private sector and a federal wildlife agency. In 1906, a group of Dallas sportsmen purchased a tract of virgin bottomland hardwood forest and established the Little Sandy Hunting and Fishing Club, named after a creek that arises from springs flowing from outcrops of Carrizo Sands. The property includes oxbow lakes, shrubby swamps, and old-growth hardwood forests, which provide habitat for a rich community of both game and nongame wildlife—

and the largest Overcup Oak in Texas. When construction of a dam threatened to inundate the area in 1986, the sportsmen donated a conservation easement on 3,902 acres to the US Fish and Wildlife Service to create a federal refuge offering legal protection from the dam. Under the conditions of the conservation easement, the club is precluded from cutting timber, grazing, or using the land commercially in any way, while retaining their exclusive hunting and fishing rights. The habitat, one of the few remnants of its kind in Texas, thus will remain intact. While it is closed to the public, authorized field research is allowed, which has included studies of insects in relation to old timber, both upright and fallen.

3
POST OAK SAVANNA
LAND BETWEEN PINES AND GRASSLANDS

The country grew more attractive [in which] post-oak

forms a very prominent feature in Texas scenery and

impressions. It . . . appears wherever the soil is light

and sandy in a very regular open forest growth.

— FREDERICK LAW OLMSTED

As might be guessed, Frederick Law Olmsted's observations, made in the heartland of the Post Oak Savanna (Leon County), no longer reflect the status of the region's namesake woodlands, which today are better described as shrubby thickets. Gone, too, are the blankets of grasslands that once formed the understory beneath the trees. Thus little remains of the "very regular open forest growth" that captured the attention of early travelers (fig. 3.1). Fire suppression and the plow were the tools of change.

The region, which consists of two areas, lies between the Piney Woods (chapter 2) to the east and the Blackland Prairies (chapter 4) to the west and occupies about 20,600 square miles, or nearly 8 percent of Texas. The largest area extends as a band from the northeast corner of Texas (Bowie County) southwest to Guadalupe County. A smaller area parallels the southern third of the larger unit, from which it is separated by a narrow strip of Blackland Prairie. An isohyet of about 40 inches approximates the separation between the Piney Woods and Post Oak Savanna, whereas soil differences play a greater role in marking the boundaries with the adjacent Blackland Prairies. Overall, the region represents a large ecotone between the Eastern Deciduous Forest and the grasslands of the North American interior.

REGIONAL OVERVIEW

The Post Oak Savanna, true to its name, fulfills the physiognomy of a tree-studded grassland. In addition to Post Oak, Blackjack Oak contributes to the open woodlands superimposed on an understory dominated by Little Bluestem. This combination resembles the vegetation of the Cross Timbers and Prairies (chapter 5), and indeed, some authorities regard the two regions as parts of a single unit separated by the Blackland Prairies. Others, however, recognize the floristic complexity of the Post Oak Savanna as reason enough to justify separate recognition and cite a number of distinctive communities, including various types of bogs, in support of their decision. If anything, the floristics of the Post Oak Savanna are more

(*opposite*) **Figure 3.1. In the absence of fire, thickets of briers (shown here with white blossoms) and other shrubby vegetation often invade the grassy understory in the Post Oak Savanna.**

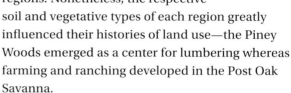

Post Oak Savanna

closely aligned with the Piney Woods (chapter 2) than with any other region of Texas, although many of the tree species that dominate Piney Woods forests are absent in the Post Oak Savanna. The abrupt increase in pines along the eastern edge of the Post Oak Savanna remains the most obvious feature separating the two regions. Nonetheless, the respective soil and vegetative types of each region greatly influenced their histories of land use—the Piney Woods emerged as a center for lumbering whereas farming and ranching developed in the Post Oak Savanna.

In places, remnant stands of old growth persist, in part because both Post and Blackjack Oaks lack commercial value; their poor growth form seldom attracted loggers, especially if other species were available. Studies of their tree rings indicate that trees 200–300 years old are not uncommon and belie their small stature. Indeed, the poor conditions—reduced moisture availability and infertile soils—that account for their slow growth and extended life span also saved the trees from the lumberyard. Moreover, as the trees age, the quality of their wood diminishes from an accumulation of defects, such as hollows and fire and lightning scars. Other characteristics of old-growth stands include canopies consisting of only a few bulky limbs, twisted and leaning trunks, exposed roots, and dead branches and tops.

Black Hickory, Yaupon, and Winged Elm add to the woodland component of the region, as does Eastern Redcedar. Some differences in vegetation occur between the northern and southern Post Oak Savanna; these result largely from the greater occurrence of sandy soils and the more irregular, dissected topography in the south. In the latter, pitcher-plant bogs and sandy exposures each support a distinctive flora. Mesquite is among the brushy vegetation invading the southern area of the region.

The characteristic vegetation developed on what are generally sandy soils that often overlie a dense layer of clay—some refer to the region as the Clay Pan Savanna. This relationship limits the downward flow of water and confines moisture

Figure 3.2.
The "Lost Pines,"
an isolated
Loblolly Pine
forest near
Bastrop, represent
the westernmost
limits of the
species.

available for plant growth to the upper horizons of the soil profile. Thus, given the rapidity with which moisture evaporates from sandy soils, aridity often prevails over much of the Post Oak Savanna. The region's sandy soils also likely influence the frequency of wildfires and the formation of fire-adapted communities. The flora at some sites includes endemic plants, but the region lacks endemic vertebrates.

SOME INTERESTING COMMUNITIES

LOST PINES

Stephen F. Austin (1793–1836), hailed as the "Father of Texas," knew opportunity when he saw it, in this case a forest of Loblolly Pines surrounded by grasslands. An isolated forest of Loblolly Pines amid the Blackland Prairies in what is now Bastrop County marks the westernmost limits of the range of this species, much of which extends along the coastal plains from Texas to Virginia (fig. 3.2). The "Lost Pines" lie in a belt 13 miles wide and more than 100 miles removed from the next-nearest location of the same species—the Piney Woods of East Texas. Along with others, Austin sought and

received a land grant from Mexico for a chunk of the forest, but the war with Mexico and his death in 1836 precluded Austin from reaping economic benefits from his holdings. Others did, however, and logging started in 1838 and reached its peak by 1880, although the area covered by the forest remained undiminished.

Two fictions persist regarding the Lost Pines. First, the trees at Lost Pines are uniquely associated with a favorable soil type that is unlike the soil in the surrounding area—ergo, an island of soil produced an island of pine forest in an otherwise grassland region. Although the pines do prefer sandy soils overlying clay, they are not dependent on such a profile and grow well on other water-permeable soils in the immediate area. Second, the genetic fingerprint differs between Loblolly Pines in the Piney Woods and those at Lost Pines. While there are indeed slight genetic differences, they are insufficient to characterize the trees at Lost Pines as either a variety or subspecies separate from those elsewhere in the range of the species. Nonetheless, the Lost Pines have adapted to local climatic conditions, notably in the ways they cope

Figure 3.3.
A devastating
crown fire in 2011
destroyed much
of the Lost Pines,
leaving snags and
a thick layer of
ash. Photograph
by Paige A. Najvar.

with drought. These include thicker coats of wax on their needles and fewer stomata, both of which reduce water loss, and a more fibrous root system that enhances water absorption in dry soils.

The origins of the Lost Pines remain unclear, but they likely represent a western refugium that remained after a once-larger area of contiguous forest retreated during the Pleistocene, leaving behind an island forest in Central Texas. Thus isolated, the population experienced a genetic "bottleneck" that eventually reduced diversity in the local gene pool; the result was an ecotype, an entity whose genetic makeup reflects adaptations to local conditions—droughts in this case—but is not different enough to warrant a separate taxonomic designation. Later, when favorable climatic conditions returned and the species reclaimed what is today the Piney Woods, prevailing winds carried pollen from the Lost Pines eastward,

thereby spreading genetic material to Loblolly Pines in East Texas and beyond. As a result, pines within the dispersal range of windblown pollen now reflect essentially the same genetic structure as those in the Lost Pine population. However, with more predictable rainfall in East Texas, the trees in the Piney Woods lack the selective pressure necessary to maintain the unique traits of the Lost Pines.

In 2011, a forest fire devastated the Lost Pines, burning more than 34,100 acres—about half of the relict forest—along with 1,600 homes and nearly all of Bastrop State Park (fig. 3.3). It remains the most destructive wildfire in Texas history. The disaster triggered an ecological quandary: What kind of Loblolly Pines should be replanted? Given the drought-tolerant properties peculiar to the Lost Pines, seeds from other sources would be poor stock to replant the burned area. Luckily, three

INFOBOX. HOUSTON TOADS AND THEIR TROUBLES

In 1970, the Houston Toad became the first amphibian listed as an endangered species—a distinction that at last recognized the plight of those species lacking the charisma of Whooping Cranes, but nonetheless needing protection. Houston Toads dwell exclusively in the woodlands in just a handful of counties in south-central Texas where stands of Loblolly Pine and Post Oak intermixed with Little Bluestem and other bunchgrasses characterize the vegetation. The sandy soils in these areas allow the toads to burrow in winter (hibernation) and again during peaks in summer heat (aestivation). Because of these requirements—sandy soils and forest cover—ecologists regard Houston Toads as habitat specialists.

For breeding, Houston Toads require slow-moving or still water that persists for at least 60 days. Their range in south-central Texas has steadily diminished because of urbanization, conversion of forest to agricultural lands, fire suppression, and the modification of wetlands in ways that preclude their suitability as breeding habitat. Habitat loss was particularly significant as the suburbs expanded around Houston, and today the toad populations in at least three of the surrounding counties (Harris, Fort Bend, and Liberty Counties) have faded away entirely. In other counties, the populations disappeared soon after they were discovered. Red Imported Fire Ants pose another threat, both by killing toadlets when they emerge from their natal wetlands and by eliminating much of the arthropod community on which the toads depend for food. Snakes and birds likewise prey on tadpoles and perhaps toad eggs as well. Hybridization represents a potential threat where Houston Toads contact two far more abundant species, Woodhouse's and Gulf Coast Toads. In the past, hybrids were uncommon, but the impact of hybridization increases as the numbers of Houston Toads diminish. The genetic integrity of the species lessens as hybrids form an ever-larger percentage of the population. Fortunately, the fungal disease that is killing amphibians worldwide, including in Texas, has not harmed the dwindling population already beset by other woes.

A phenomenon known as the "Allee effect" also

In the Lost Pines, ponds like this provide breeding habitat for the Houston Frog (inset). Photographs by Paige A. Najvar.

impacts Houston Toads, which rely on a large group of chorusing males to attract mates to a breeding habitat. Because breeding sites in the form of livestock ponds are overabundant within the toad's range and the toad population is reduced, only a few males chorus in each pond. Often, the magnitude of chorusing falls below the level necessary for successful breeding. As a result, populations decline and face extinction when they fall below a certain threshold.

Conservation efforts focus on habitat management, particularly in Bastrop County, the core area for the remaining population, and on maintenance of an assurance colony at the Houston Zoo. As a protection against extinction, the zoo colony preserves a "pure" gene pool and breeding stock. Initially, the zoo was stocked by a technique known as "head-starting," which involved taking Houston Toad eggs from ponds, hatching the eggs, rearing the young toads at the zoo, and returning the toads to their original habitat after they

matured. However, restoration efforts currently rely more on the release of eggs produced by captive breeding and less on head-starting. The releases occur primarily in Bastrop County at locations such as Bastrop State Park, which is actively managed for Houston Toads.

Unfortunately, the disastrous fire that destroyed much of the Lost Pines forest in Bastrop County eliminated a large area with a pine-oak overstory that many biologists regard as essential for adult Houston Toads. Meanwhile, the stocking program remains a viable method for restoring toad populations, but it requires patient monitoring for several years to fully assess its success. In 2015, Houston Zoo released some 450,000 eggs produced in its facilities and created a studbook to avoid inbreeding in the captive population producing the egg supply. Although still in progress, a Dogs for Conservation program intends to train dogs to locate wild Houston Toads; if successful, the technique will considerably improve monitoring efforts.

years before the fire, a farsighted employee of the Texas Forest Service had stored about 1,000 pounds of seeds from Lost Pine trees in an oversized freezer in a grocery store; from these, seedlings were grown at several nurseries in a program coordinated by the Lady Bird Johnson Wildflower Center (infobox 7.1). The availability of this stock enabled planting to begin immediately; additional seeds were later collected from trees spared by the fire. The replanting effort involves volunteers, including students from Texas A&M University, as well as the work of public agencies. The goal is to plant four million seedlings in five years. In addition to damaging a unique form of Loblolly Pine, the conflagration also destroyed much of the remaining native habitat for Houston Toads (see infobox above).

XERIC SANDYLANDS

Extremely dry conditions develop on outcrops of Carrizo, Queen City, and Sparta Sands, Eocene formations through which water drains rapidly. Also known as Oak-Farkleberry Sandylands, these sites often provide the hydrology that allows the formation of nearby wetlands (fig. 3.4). The deep sands act as a reservoir, holding water that flows

into adjacent seeps and springs and eventually into streams and rivers. In some topographic settings, the water moves laterally and emerges on the sides of slopes. The wetlands formed by the Xeric Sandylands contribute to the diversity of the Post Oak Savanna, including bogs with distinctive floras. Both the sandylands and bogs resemble their counterparts on deep sand hills in the Piney Woods (chapter 2).

Paper-thin coverings known as biological soil crusts form in arid regions where the surface remains undisturbed, producing cryptogamic crusts. These coatings typically develop in deserts but also occur in the Xeric Sandylands where local conditions produce extremely dry soils. They consist of fungi, algae, and lichens as well as cyanobacteria (also known as blue-green algae), whose sticky filaments bind soil particles and thereby lessen wind and water erosion. Moreover, the crusts cycle both carbon and nitrogen into the otherwise impoverished biological systems. Unfortunately, the delicate crusts are easily damaged—commonly by grazing and off-road vehicles—and may not recover for decades or even longer. Whereas botanists have identified foliose lichens in the crust as belonging to the genus

Figure 3.4.
Outcrops of
Carrizo Sands
provide habitat
for xeric-
adapted plants,
but rainwater
percolating
through these
soils feeds
nearby wetlands.
Photograph by
Matt White.

Cladonia, full analysis of the biological soil crusts in the Xeric Sandylands awaits additional study.

Trees characteristic of the Xeric Sandylands include Bluejack Oak, Sand Post Oak, and Black Hickory, in addition to Post Oak. Rattlesnake Flower, Reverchon's Palafox, Texas Sandmint, and the showy, endangered Large-fruited Sand-Verbena are among the herbaceous plants closely associated with these locations. The considerable number of species per unit area (richness) of the flora in the Xeric Sandylands seems related, at least in part, to the relatively small stature of the plants and their various forms of growth; many have only a brief life history aboveground. These conditions favor "species packing," hence encouraging a greater level of biodiversity than otherwise possible, which may be maintained by periodic fires and droughts. Together with the wetlands they foster, the Xeric Sandylands seemingly qualify as regional, if not national, ecological "hot spots" meriting proactive conservation.

BOGS

Enriching the flora of the Post Oak Savanna as well as in the Piney Woods, bogs are variously known as muck, peat, or more quaintly, "possum haw" bogs. In the Post Oak Savanna, three elements promote the formation of peat bogs: oxbow lakes isolated by the meanderings of the Trinity, Brazos, Colorado, and other major river systems; soils that remain slightly acidic in small wetland basins; and a topography that cuts into the Carrizo Sand aquifer, allowing water in the form of seeps and springs to reach the surface and emerge from hillsides. Peat bogs apparently develop on blocked-off stream meanders, although small streams pass through others. Peat accumulations reach depths of more than 16 feet at some sites, and pollen analyses indicate that bogs with high organic content have been present for thousands of years. A rich layer of herbaceous vegetation, including pitcher plants, covers the bogs, but thickets of shrubs may invade in the prolonged absence of fires. Other common plants include Sphagnum—the iconic peat moss of bog habitats—Cinnamon Fern, beak rushes, Yellow-eyed Grass, and numerous grasses. Bladderwort, also carnivorous, and Arrowhead characterize sites with standing water.

Bogs support unusual or rare plants at the limits of their distribution or with disjunct populations.

Figure 3.5.
The Dwarf
Palmettos
surrounding
bogs at Palmetto
State Park
represent disjunct
populations of a
species normally
found farther
east. Photograph
by Larry D. Moore
(from Wikimedia
Commons).

Pale Pitcher Plants offer such an example: the species reaches the western limits of its distribution in bogs of the Post Oak Savanna and also traps and digests insects and spiders in what represents a marvelous example of a microcosm (chapter 2). Sundew, another carnivorous plant, also occurs in these bogs; instead of tubular traps, however, sundews entrap their prey on stalks covered with a sweet sticky secretion that serves as both an attractant and snare for their hapless victims (fig. 2.20). In 1989, botanists discovered Twig Rush in bogs in Anderson and Henderson Counties—something of a revelation, as the range of the species otherwise lies far removed from Texas, in such places as Florida, Illinois, and South Carolina.

Vegetation at a bog is considered "immature" where the peat is not yet deeply embedded into the soil profile. In such bogs, shrubs and grasses intersperse with patches of pitcher plants and sundews. The latter occupy the more "boggy" locations where the gelatin-like surface shakes underfoot; these areas likely represent small ponds

that gradually filled with organic materials. Two species of orchids—Rose Pogonia and Fragrant Ladies' Tresses—along with pipeworts and Meadow Beauty also epitomize the bog flora at these sites.

Bogs also occur at Palmetto State Park (Gonzales County), named for an isolated population of Dwarf Palmettos, a species that is otherwise more abundant in the Piney Woods and along the Gulf Coast (fig. 3.5). The bogs form because of the site's unusual hydrology, which includes both seasonal overflows of the San Marcos River and seepage from water held in the Eocene sands. In addition to flooding, the changing bed of the river leaves behind oxbow lakes, which often become prime sites for bog formation. These circumstances enabled remnants of a once-widespread palmetto population to remain in place after the Pleistocene ended. The drier climate that followed reduced the distribution of the plants to wetter areas in the Piney Woods and farther east but left behind an "island" of palmettos in the Post Oak Savanna. Moreover, the park's unique environment provides habitat for more than 500 species of plants, some of which reach their respective eastern or western limits at this location. The dense vegetation no doubt contributed to tales of a mythical creature—a dark, hairy, but smaller version of "Bigfoot"—that allegedly haunts the area's swampy woodlands.

HIGHLIGHTS

SOILS AND FIRE FREQUENCY

When viewed broadly, an alternating pattern of tallgrass prairies, oak-dominated woodlands, and savannas emerges for the vegetation that historically characterized north-central Texas. These bands—the East and West Cross Timbers, Blackland Prairies, and Post Oak Savanna—coincide with the distribution of sandy (woodland) and clay (grassland) soils; the relationship was usually explained solely in terms of the soil moisture available, respectively, for each of the two types of vegetation. However, another idea may better explain the distinctive pattern—the soil-dependent fire frequency hypothesis.

This hypothesis suggests that because of their higher moisture content, clay soils produce large amounts of fuel in the form of grass, which heightens the likelihood and intensity of fires and, in turn, suppresses tree growth. Further, because grasses are fire adapted (chapter 8), they regrow rapidly and soon replenish the fuel supply for another fire. Conversely, sandy soils produce lesser amounts of fuel and thus experience fewer fires—a setting favorable for woodlands. Once established, trees further inhibit fuel accumulation and the occurrence of fires. However, given modern-day activities that suppress fires, the relationship between soil and fire frequency breaks down and woody vegetation can invade grasslands once "protected" by a shield of clay soil. Hence, the hypothesis that soil type mediates the frequency of fires offers a plausible explanation for the bands of Post Oak Savanna and other zones of distinctive vegetation in north-central Texas.

LENNOX WOODS, A TREASURE OF THE RED RIVER COUNTRY

The northeastern edge of the Post Oak Savanna touches a strip of riparian forest bordering the Red River that includes Lennox Woods, a unique, 1,335-acre preserve in Red River County protected by The Nature Conservancy. The forest marks the westernmost extent of the Austroriparian Biotic Province, the large region of pine and hardwood forests extending eastward to the Atlantic Coast (chapter 1). Notably, several features highlight the biological value of the site: an old-growth forest of large trees; "healthy" habitats never degraded by logging or other anthropogenic activities; a wealth of diversity—almost 300 vascular species—including populations of rare plants; and the near absence of invasive species. Lennox Woods lies entirely within the watershed of Pecan Bayou, and oxbow lakes, remnants of the stream's meandering course, add further to the ecological settings at the preserve.

The upland edges of Lennox Woods, which include centuries-old Post and Blackjack Oaks, also feature White and Southern Red Oaks, Shagbark and Mockernut Hickories, and Loblolly and Shortleaf Pines; some of these trees exceed 3 feet in diameter, and a count of growth rings revealed a Post Oak exceeding 300 years of age. The understory of smaller trees and shrubs includes Texas Mulberry, Flowering Dogwood, Farkleberry, and American Beautyberry.

Whereas the upland forest itself warrants due homage, it is the bottomland hardwood forest associated with Pecan Bayou that excels as a wonderland for visitors. Several species of oaks—Water, Willow, Bur, and Overcup—dominate the cathedral forest in the bayou's floodplain, and smaller species such as American Hornbeam and Winged Elm represent trees of the understory. A rich ground cover of herbaceous vegetation graces the bottomlands and includes several species rarely seen elsewhere in Texas, notably Hooked Buttercup, Southern Lady's Slipper, and Willdenow's Sedge. Occasionally flooded sites in the bottomlands host a population of Arkansas Meadow-rue, a wildflower rare enough to be listed as a threatened species.

Lennox Woods, as might be imagined, hosts scores of birds, both residents and those that appear seasonally. Several species of woodpeckers, including the spectacular Pileated Woodpecker, thrive in the old-growth forest. The mature trees also provide roosts for Rafinesque's Big-eared Bats, a species that prefers cavities in large trees near water—both readily available at Lennox Woods

Figure 3.6. Cavities in mature trees provide roosting sites for Rafinesque's Big-eared Bats. Photograph by James F. Parnell.

(fig. 3.6). Moths make up the bulk of their diet and are perhaps the reason that these bats become active only in full darkness rather than at dusk. Similarly, Lennox Woods provides safe harbor for an interesting insect.

AMERICAN BURYING BEETLES, NATURE'S UNDERTAKERS

Burying beetles locate, defend, and bury the carcasses of small vertebrates that become a food source for their offspring. One species, the largest and most distinctively marked of the 31 species occurring in North America, once ranged widely from the grasslands and meadows of the Great Plains to the oak-hickory forests of eastern North America. However, such a broad niche and wide distribution—normally safeguards

against extinction—failed to stem the wholesale disappearance of the American Burying Beetle, which now occupies just about 10 percent of its former range. Because of these circumstances, the American Burying Beetle was the first insect listed as endangered. Upland habitat at Lennox Woods is one of the two places in Texas where the beetles have recently been detected; the other is oak savanna in adjacent Lamar County. Another isolated population occurs on the opposite side of the Red River in Oklahoma as well as in a limited number of other states (Nebraska and Rhode Island). Soft soils suitable for digging seem more important than vegetative cover in determining the distribution of the American Burying Beetle.

In appearance, these beetles reach lengths of 1–1.4 inches, with irregularly shaped rusty-orange spots highlighting their shiny black bodies. Two spots occur on each wing and one on the head; the club-like tips of the antennae are also orange. An especially large orange splotch covering most of the midsection between the head and wings distinguishes the American Burying Beetle from related species (fig. 3.7).

American Burying Beetles are aptly known as "grave diggers," not just because they remove bodies but also because their reproduction is keyed to a buried source of food. Their odor-sensitive antennae can detect decomposing birds and small mammals as far away as 2 miles. After arriving at a carcass, beetles fight for possession until a dominant pair—usually the largest male and female—successfully claims the prize. After mating, the beetles burrow under a carcass until it settles belowground; they then cover it with soil, thereby lessening the chances of discovery by other scavengers. Once the carcass is buried, the beetles remove the hair or feathers, coat the remains with secretions that retard the growth of bacteria and fungi, and then work the body into a ball-shaped form. That done, they excavate a brood chamber next to the carcass and the female deposits eggs in small tunnels leading into the chamber.

Once the eggs hatch, the parents eat some of the larvae, lessening the number of mouths to feed in the days ahead. The adults remain at the site, continuing to tend the carcass on which they feed, regurgitating into the mouths of their larvae that beg for food. In most insects, including dung beetles (chapter 6), larvae forage for themselves, usually with their parents long gone; hence, the parental care shown by burying beetles is unusual. The larvae complete their metamorphosis in about 45 days and emerge as young beetles that will mature sexually in another 21 days. Some adults breed immediately, whereas most mate the following spring; in either case, they breed just once before dying.

NAVASOTA LADIES' TRESSES

An endangered species of orchid—Navasota Ladies' Tresses—occurs on the sandy loams bordering the intermittent tributaries of the Brazos and Navasota Rivers, especially in Brazos and Grimes Counties. The plants feature tall, slender stems 8–15 inches in height that arise from clusters of fleshy tubers, as do basal rosettes of slender leaves. During the annual cycle, the leaves disappear during late spring and summer and begin appearing again in early fall. Flowering occurs in the fall, with seed dispersal in December; when blooming, the small white flowers spiral counterclockwise on a spike at the tip of a leafless stalk (fig. 3.8). Thus, between late spring and early fall, no stems, leaves, or flowers appear aboveground, and the plants persist entirely underground. At times, the plants may not produce any aboveground structures for several

Figure 3.8. Several species of ladies' tresses orchids, including Navasota Ladies' Tresses (shown here), produce small white flowers that spiral counterclockwise around an erect stalk.

years, which obviously hinders monitoring their populations.

Navasota Ladies' Tresses require a lifelong association with at least one type of mycorrhizal fungus, but the specificity between the two organisms remains uncertain. Their fruits contain thousands of microscopic seeds that are difficult to cultivate, but collectors nonetheless illegally poach the plants for propagation, as methods now exist for transplanting whole plants. The plants often grow at the interface between patches of oaks and adjacent grasslands, typically near stream banks, but also along game trails, cow paths, fencerows, and power line rights-of-way, which may indicate their preference for either moderate disturbance or direct sunlight.

HEAVENLY CONTACT—OF A SORT

In 1980, a report described the discovery of a thin band of iridium in rocks, first in Italy and later elsewhere, indicating that it was the fallout from the impact of a huge meteor some 65 million years ago. Iridium, uncommon in rocks formed on Earth, often occurs with far greater abundance in rocks of extraterrestrial origin. Even more intriguing, the layer of iridium separated rocks marking the end of the Cretaceous Period and the beginning of the Tertiary Period—the so-called K-T boundary—a time highlighted by the extinction of dinosaurs. Hence, the band of iridium is sometimes known as the "Tombstone Layer." The same relationship turned up in the geological profiles of locations ranging from New Zealand to Denmark, including the banks of the Brazos River and a few of its tributaries. Eventually, geologists searching for oil discovered a deeply buried crater 112 miles wide on the northern rim of the Yucatán Peninsula in Mexico. Based on the immensity of the crater, an asteroid 6 miles in diameter hit with unimaginable force and subsequently deposited a worldwide layer of iridium.

As happens regularly in science, not everyone agrees that the impact of a huge asteroid produced an environment in which dinosaurs—and a good many other groups—could not adapt and forever perished. Extreme volcanic activity may have caused similar changes in climate and ecosystem structure. Whatever the causes, no one disputes the mass extinctions that followed—or that the impact of a huge meteor deposited a layer of iridium in rocks now exposed on the Brazos watershed in Falls County.

Another, much smaller meteorite hit Texas 58 million years ago in present-day Leon County. Originally thought to be a salt dome, the structure was named the Marquez Dome after the small community of that name near the site. Only later, after oil exploration revealed no underlying salt deposits, did geologists recognize the structure as the central point in a buried impact crater about 8 miles in diameter formed by a meteorite about 0.6 mile in diameter. The unusual nature of the site did not escape notice by settlers whose attempts to dig wells produced no results; the impact pulverized the underlying rocks to the point where the strata could not retain water. A phenomenon known as the rebound effect, caused by the impact, also brought deep rock formations to the surface, scattering blocks of Cretaceous limestone across the landscape and creating an "island" of black waxy soil in a "sea" of sandy or loam soils. Grains of shocked quartz provided further evidence of the impact.

CONSERVATION AND MANAGEMENT

Old-growth forests represent ecological treasures, and those few still remaining of Blackjack and Post Oak are no less so. In the Post Oak Savanna, stands occur in Milam, Guadalupe, Goliad, and Burleson Counties, as well as at Lennox Woods. With the exception of the latter, however, these scattered remnants are not in long-term conservation ownership and are thus at risk from human disturbances. Other overlooked stands that have not been authenticated by analyses of their tree-ring chronologies undoubtedly exist elsewhere in the region. Such circumstances warrant systematic searches and cataloging of these irreplaceable resources, pursued thereafter with formal agreements, including conservation easements, that afford their protection. These efforts, of course, will benefit from a campaign of public awareness regarding the living record and esthetic value provided by trees that predate European settlement of East Texas.

The Engeling Wildlife Management Area, owned and managed by the Texas Parks and Wildlife Department, consists of 10,958 acres of unusually diverse habitat, including forest, marshes, and bogs. The area represents the largest intact remnant of Post Oak Savanna held in state ownership. Andrew's Bog is the largest (257 acres) of three bogs at the Engeling area and features a cover of Southern Wild Rice along with carnivorous plants, pipeworts, Sphagnum Moss, and other bog-specific plants. A program of winter burning prevents woody plants from invading the bog vegetation and also maintains the integrity of other communities. Management efforts target a large population of Feral Pigs, including research designed to control their numbers. The area provides habitat for an isolated population of an uncommon butterfly, the Frosted Elfin, and a newly discovered species of dragonfly, *Cordulegaster sarracenia.*

The Engeling area encloses about half of the Catfish Creek Ecosystem, a spring-fed watershed of 4,460 acres designated as a National Natural Landmark. Because a significant part of this ecosystem remains in private ownership outside the boundaries of the Engeling area, the integrity of these lands remains potentially at risk. However, the landmark designation encourages the involvement of landowners, and indeed many have become active stewards who maintain the natural values of the Catfish Creek Ecosystem by participating in a voluntary nonbinding registration agreement.

The exact cause for the rapid decline of the American Burying Beetle remains elusive—a void that hinders halting further losses as well as restoring both their numbers and distribution. Likely no one cause is responsible. One idea suggests that forest fragmentation may be involved; in this scenario, the smaller units of forest, coupled with the extirpation of large predators, benefit larger populations of mesopredators—foxes, skunks, and Raccoons—that also feed on carcasses of small animals. Scavenging American Crows may be another source of competition. Correcting the effects of forest fragmentation, to the extent that it underlies the problem, seems unlikely, except in the case of publicly owned lands or those protected by private conservation organizations. Meanwhile, captive-breeding programs designed to restock the beetles in protected areas are currently underway in a few states, but these efforts have yet to include Texas.

Both federal and state agencies list Navasota Ladies' Tresses as endangered, a status directly related to the limited distribution of the species (13 counties in Texas), a complicated life history, habitat concerns, and poaching. However, most of this area is held in private ownership, which limits the effectiveness of legal protection. Habitat threats to the species are little different from those affecting other components of the regional biota, namely urbanization and the encroachment of thickets. Feral Pigs may present a more direct impact by uprooting the plant's fleshy tubers. Given the rapid development in the region, more needs to be learned about the effects of disturbance and habitat fragmentation on the welfare of the species, and ideally, conservation strategies that are compatible with the expansion of human populations need to be determined.

4

BLACKLAND PRAIRIES
GRASSLANDS AT THE EDGE

The rich black "waxland" soil of these prairies is

almost proof against burrowing rodents.

— VERNON BAILEY (1905)

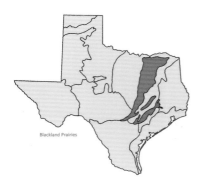

Blackland Prairies

Vernon Bailey's reference above concerned the "wide stretches" where pocket gophers were "entirely wanting" because of the barrier presented by the heavy clay soils that distinguish the Blackland Prairies. Instead, these common rodents abound elsewhere in the "sandiest and mellowest land" that favors their excavations. The 10.6 million acres of Blackland Prairies exceeds the size of Maryland and forms a wedge between the Post Oak Savanna (chapter 3) and the eastern edge of the Cross Timbers and Prairies (chapter 5). The broad upper end of the wedge-shaped triangle inscribes, west to east, the cities of Sherman and somewhat beyond Paris, and then tapers southward almost to San Antonio. Dallas lies in the upper third of the region. Two islands of similar habitat—the San Antonio and Fayette Prairies—lie isolated to the east of the much larger area of Blackland Prairie featured in this chapter.

Geologically, this region coincides with marine chalks, marls, and shales deposited late in the Cretaceous Period. Curiously, eons of weathering transformed and darkened these once lighter-colored deposits into the waxy black clays—gumbo—that now characterize the region. For millennia, these calcareous soils nourished lush grasslands, but these soon vanished when plows accomplished what the gophers could not (fig. 4.1).

VEGETATION AND SOILS

The rich flora of grasses in Texas—about 181 genera and 723 species—represents more than 60 percent of the genera and 40 percent of the species of grasses in the United States. More than 65 percent of these genera and 50 percent of these species occur on the fertile soils of the Blackland Prairies. Prior to cultivation, tallgrass prairie was the dominant vegetation of the Blackland Prairies, and for this reason, the area was similar to the True Prairie that once blanketed much of what is now the Corn Belt in Iowa, Illinois, and eastern Nebraska. Various combinations of Little Bluestem with two codominants—Big Bluestem and Indiangrass—distinguish the composition of the prominent tallgrass communities. These dominant species occur with other grasses or graminoids, commonly Sideoats Grama, Small-toothed Sedge, Tall Dropseed, and Texas Wintergrass, and forbs including asters, sunflowers, Prairie Bluet, Old Plainsman, Coneflower, Compassplant, Prairie-Bishop, and many others.

Two other, but relatively uncommon, types of prairie complete the overall view of the Blackland Prairies. The first of these—characterized by Eastern Gamagrass, Switchgrass, and Indiangrass—develops on poorly drained clays in the northern Blackland Prairies. Wild Blue Indigo, Coneflower, and Maximilian Sunflower represent the array of colorful forbs at locations such as Clymer Meadow in Hunt County (fig. 4.2). Eastern Gamagrass, Switchgrass, sedges, and rushes also colonize prairie openings in floodplains such as those at The Nature Conservancy's Cowleech Prairie (also in Hunt County). In the northeastern corner of the region, Silveus' Dropseed and Mead's Sedge form a community in areas with higher precipitation, where acidic sandy loams (alfisols) offset the influence of clay in the soil profile; at these sites Long-spike Tridens is a codominant with Silveus' Dropseed (a prime example is Tridens Prairie in Hunt County). Many of the forbs already noted occur here, as well as others with northern or eastern distributions. Mima mounds, a small but distinctive and somewhat mysterious topographical feature, commonly occur in these locations (see "Highlights" below). Boundaries between these communities remain well defined in keeping with changes in the dominant soil types.

Green Milkweed often grows at both Clymer and Tridens Prairies and, with two other species of milkweed, provides the Monarch Butterfly with habitat essential for the survival of its larvae. Unfortunately, milkweeds have become less abundant in recent years because of their increased exposure to herbicides in much of their range, a situation that now poses a somewhat greater threat to monarchs than the loss of critical winter habitat in Mexico.

(*overleaf*) Figure 4.1. Pale Prairie Coneflowers poke through the grasses at Clymer Meadow, a preserve of Blackland Prairie (Hunt County) managed by The Nature Conservancy. Photograph by David Bezanson.

Figure 4.2.
Maximilian
Sunflowers,
Prairie
Coneflowers, and
Prairie Thistle
enliven the
grassy expanse of
Clymer Meadow.
Photograph by
David Bezanson.

Black clays dominate the region, with a clayey component appearing in each horizon in the soil profile (fig. 4.3). They drain slowly and experience exceptional swelling and shrinkage in relation to changing moisture and temperature regimes. The predominant tallgrass communities, described above, developed on these soils—known as vertisols—whose unique properties also create small depressions called gilgai. Alfisols are less common but nonetheless influence both community and development; alfisols also produce the distinctive Mima mounds, mentioned above. Alfisols develop with a larger component of loams, some sands, and far less clay or calcium carbonate than vertisols. As is often the case, the presence of two rather dissimilar soil types represents a strong influence on the ecological settings where each occurs.

The Austin Chalk Formation offers another example of edaphic influences on the vegetation in the Blackland Prairies. This formation consists of countless billions of calcium carbonate platelets—coccoliths—that protected unicellular algae living in the deeper parts of the Cretaceous seas once inundating the North American interior (fig. 4.4). The calcified tests (shells) of microzooplankton, primarily foraminiferans, also contributed to this and other chalk formations, including the well-known White Cliffs of Dover on the western coast of the English Channel. Dark bands in the Austin Chalk indicate ash deposited during eruptions from a band of submarine volcanoes situated along what is today the Balcones Escarpment (chapter 7); the eruptions occurred 86 million years ago and coincided with the deposition of the chalk beds. Some elements of the Austin Chalk Formation include fossils of a diverse fauna, among them sleek-bodied marine reptiles and bony fishes, including a species apparently related to modern-day Skipjacks and other fishes common in coastal areas of subtropical regions.

Figure 4.3. The black clay gumbo that once supported lush prairies now nourishes hay, corn, and other agricultural crops.

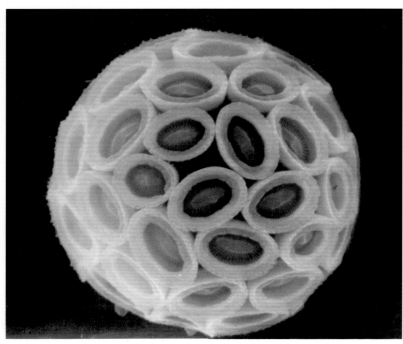

Figure 4.4. A coccolith, the "shell" that once encased one of the untold billions of microscopic unicellular algae deposited over thousands of years to create the Austin Chalk Formation, is shown in this microphotograph. Photograph by Carmelo Tomas.

Where outcrops of Austin Chalk are exposed (Grayson County), forbs such as Wild Blue-Indigo, Meadow Flax, Finger Poppy-Mallow, and Long-leaf Wild Buckwheat occur. In season, these and other forbs paint the outcrops in vivid color, but the same sites appear barren at other times. Near Dallas, Austin Chalk emerges as a bluff known as the White Rock Escarpment and extends upward for up to 328 feet above the surrounding terrain, producing a striking contrast—a wall of white chalk rising from a prairie of black soil (fig. 4.5). The escarpment and Austin Chalk outcrops in other counties support woody vegetation dominated by Texas Ash, Texas Red Oak, Cedar Elm, and Scrub Oak and associated species, with Texas Redbud representing the most important of the smaller trees in this flora. Seeps develop where water filters downward through the permeable chalk before encountering the underlying impermeable strata; it then flows laterally to emerge from the face of the escarpment. The flora developing at these localized sites further adds to the diversity of the escarpment's vegetation. As another example of a site-specific association, Mexican Buckeye prefers habitat near the rim of the escarpment. Farther south along the escarpment

Figure 4.5. The White Rock Escarpment, a bluff of Austin Chalk formed from the calcified shells of coccoliths and microzooplankton, rises high above White Rock Creek near Dallas (Dallas County). Photograph by Chip Clint.

(Bell County), the vegetation remains similar but includes characteristic species such as Sugar Hackberry, Texas Persimmon, Deciduous Holly, and a growing population of Eastern Redcedar. All told, the White Rock Escarpment and other sites where Austin Chalk breaks through the blanket of black clay provide microhabitats enriched by a large numbers of species. Species otherwise excluded from the surrounding area also occur in these sites. Regrettably, however, these same features—limited and distinctive in their occurrence—also make this environment vulnerable to human disturbances.

HIGHLIGHTS

GILGAI AND MIMA MOUNDS

Two interesting microhabitats dot the terrain of the Blackland Prairies. Gilgai develop exclusively on vertisols because of the extreme plasticity of these black clays as they alternately shrink and swell in

response to wide swings in soil moisture; these changes in volume produce significant movements within the soil profile. The term "gilgai" comes from an Aboriginal language and denotes a "small watering place," known locally in Texas as "hog wallows." These depressions vary from 10 to 19.5 feet in diameter, with depths of up to 20 inches. The vegetation and invertebrate fauna in gilgai vary greatly according to each depression's current hydrology, thereby creating a biological mosaic across the terrain. For example, at deeper sites where water persists longer, stands of Four-angle Sedge often develop, whereas Arkansas Dogshade characterizes comparatively shallow gilgai with shorter hydroperiods. Other species typical of the gilgai flora include Southern Dewberry, Eastern Gamagrass, and Sharpscale Spikerush; rhizomes of the latter typically form mats of vegetative cover where moist soils persist. During extended dry periods, the local flora gives way to elements typical of the surrounding prairie, but it returns when rainfall is again adequate to accumulate in the depressions.

Little is known about the invertebrates associated with gilgai, but these miniature wetlands surely provide rich habitat for various insects and crustaceans. Among the latter is the Parkhill Prairie Crayfish, named for its discovery in a 52-acre relict of tallgrass prairie in Collin County, but now known to occur in Hunt County as well. Fortunately, the prairie site is part of a preserve protected by the county government. Curiously, the species remained undiscovered until 1990, when specimens were caught in traps designed to capture snakes. Parkhill Prairie Crayfish prefer habitat in the open prairie instead of permanent wetlands, and they share a close taxonomic relationship with the similar and more widely distributed Regal Burrowing Crayfish. They tunnel almost vertically downward to the underlying water table and remain in their moist burrows during the day. At night, these crayfish perch at the entrance with their bright blue claws extending outward, as if on guard; on humid nights, they emerge and travel short distances in search of food or mates, but they plug their burrows during droughts.

Gilgai develop in a sequence of events. First, during extended dry periods, crumbled chunks of soil fall into the cracks formed when these soils shrink. The cracks may be sizable, some reaching depths of up to 20 inches and widths of 4 inches. Then, with a good rain, the chunks swell, exerting lateral pressure that pushes the surrounding soils outward and upward, creating a depression surrounded by a mounded rim. Each additional wet-dry cycle further exaggerates the process until the terrain is covered with depressions. This action, technically identified as argilliturbation but commonly described as "self-mulching," steadily churns and mixes the soil to the point where the first subsoil layer (B horizon) disappears, leaving only the topsoil (A horizon) overlying a deeper stratum, a layer of weathered, partially decomposed rock (C horizon), which is little affected by soil-forming processes. This process is also the basis for the name "vertisol," which is derived from the Latin *verto,* meaning "turn upward," and *sol,* for "soil."

Under pristine conditions, gilgai stored considerable amounts of surface water and, along with the dense root systems of prairie grasses, helped retard soil erosion. Some estimates indicate that gilgai might trap and hold the runoff from up to 6 inches of rainfall before the overflow washes across the surrounding terrain. Moreover, with thousands of gilgai once dotting the landscape, runoff and hence erosion were so reduced that streams in the region ran clear. Today, however, gilgai have largely disappeared under the plow, and erosion now strips the precious soil from the region's farmlands at rates among the highest in Texas.

Alfisols occur in some areas of the Blackland Prairies; these soils developed primarily on the edges of the clay-covered blacklands and lack the fertility of the latter but nonetheless produce their own distinctive topographical feature: Mima mounds. These features, also called "pimple mounds," develop as circular hills up to 46 feet in diameter and 39 inches high (fig. 4.6). Where they still occur, vegetation on the mounds appears zonal; the upper third is dominated by Big Bluestem and the midregion by Long-spike Tridens, and the lower zone largely resembles the surrounding prairie. Repeated mowing alters the composition, but the zonation remains apparent.

The origin of Mima mounds remains a mystery, but not for the want of ideas. These vary from the workings of pocket gophers to earthquakes, but after a century of speculation, no theory has gained scientific credibility. The presence of similar mounds in other parts of North America, such as the Mima Mounds Natural Area in Washington State, as well as in Kenya, China, and Australia, compounds the difficulty of finding a unifying explanation for their formation. Like gilgai, Mima mounds are no longer widespread in the Blackland Prairies, but examples persist at a few locations, including Tridens Prairie in Lamar County where the original prairie landscape has escaped the plow.

DOGWOOD CANYON—A WOODLAND TREASURE IN THE BLACKLAND PRAIRIES
Where it forms the western limits of the Austin Chalk Formation, the White Rock Escarpment rises above the Blackland Prairies. In southern

Figure 4.6.
Intact Mima
mounds remain
in only a few
protected sites
in the Blackland
Prairies.
Photograph by
Jason Singhurst.

Dallas County, a break in the escarpment forms Dogwood Canyon—a locale whose topography enables the mingling of species with, respectively, eastern and western affinities. Among these is the western extension of Flowering Dogwood—the canyon's namesake—typical of the Piney Woods (chapter 2), and the Black-chinned Hummingbird, a species more commonly associated with the Texas Hill Country (chapter 8) and western Texas. The nest of one in the branches of the other unites east and west. In addition to their attractiveness, the dogwoods produce lipid-rich drupes—fruits such as cherries with a central pit—whose pulp may contain of up to 35 percent fat, thus offering a fine source of energy available in the fall for migrating birds (fig. 4.7). The birds, in turn, disperse the indigestible seeds. Microbes attack—likewise for food—those drupes not consumed by birds, so the energy in this shortened food chain remains near the mother plant. Although the idea remains speculative, isolation of the dogwoods, along with their ability to thrive on the exposed "white rock" substrate, may have fostered a site-specific genetic signature that distinguishes this population from dogwoods elsewhere.

In places, the sides of the canyon extend upward for about 300 feet. Because of their relative aridity and protection from fire, the slopes serve as prime habitat for Ashe Junipers, which, when mature,

Figure 4.7.
In fall, the
red drupes
of Flowering
Dogwood provide
lipid-rich food for
migrating birds.

provide the shaggy bark necessary as nest material for Golden-cheeked Warblers (see chapter 7). The canyon's woodlands also meet the needs of nesting Black-capped Vireos, another endangered species (chapter 5). On the more mesic floor of the canyon, soils enriched by accumulating organic matter support a forest highlighted by Pecan, Cedar Elm, and Green Ash. Near the unnamed creek, sites prone to flooding include Black Willow, cottonwoods, and American Sycamore among other woody vegetation that tolerates saturated soils. The understory in these areas includes a rich flora of herbaceous plants that collectively provide most of the biodiversity in this and similar forests throughout North America. Among these, White Trout Lily is one of the more attractive

Figure 4.8. Colonies of White Trout Lilies sometimes develop where dappled sunlight reaches the forest floor. Photograph by Brenda K. Loveless, Lady Bird Johnson Wildflower Center.

species, flourishing on sites with deep soils; its white flowers—one per stalk—face downward, as if nodding, and form a graceful star-shaped whorl. Just 7–10 inches tall, each plant bears only two lance-shaped leaves, each dark green with a mottled pattern of purple blotches similar to the colors and markings of trout (hence the plant's name). White Trout Lilies, if left undisturbed, may develop in colonies on the forest floor (fig. 4.8), where they thrive on moist loams bathed with dappled sunlight, but the plants are slow growing and take about seven years to mature. Ants disperse the seeds of White Trout Lilies, lured into doing so by lipid-rich structures—known technically as elaiosomes—attached to the external surface of the seeds; the nutrient content in these structures far exceeds that of the seeds themselves. The ants carry the seeds to their nests, consume the elaiosomes or feed them to their larvae, and then discard the still-viable seeds in sites suitable for germination. Known as myrmecochory, this form of symbiosis—mutualism—seems associated with forest-dwelling herbs, perhaps because wind is less effective as a dispersal agent in forests than it is for similar vegetation in grasslands.

White Trout Lilies and other herbaceous understory plants stand at the core of the vernal dam hypothesis, an idea that concerns the flow and availability of nutrients in hardwood forests. This concept holds that understory plants—namely those initiating their growth early in the year—absorb and hold nutrients, particularly nitrogen and phosphorus, which would otherwise be lost in the runoff from spring rainfall or snowmelt. Later, when their leaves drop in concert with the closure of the forest canopy, the herbaceous understory plants release their store of nutrients. This benefits the vegetation just beginning its annual development. In other words, plants such as trout lilies accumulate nutrients at a time when they are rapidly adding biomass but before the nutrients are needed by later-growing species. In effect, they act as dams, or short-term nutrient sinks, that retard nutrient loss until nitrogen and phosphorus can be used by other species. Supporting data for this hypothesis remain inconclusive, but it seems clear that a full understanding of the role played by understory vegetation in the dynamics of forest communities awaits additional study. Indeed, the diminutive stature of these delicate plants may belie their much larger significance.

A group of coral root orchids in Dogwood Canyon and in other areas of the White Rock Escarpment likewise command the interest of naturalists (fig. 4.9). At least four species—three in Dogwood Canyon—are now known to occur in Dallas County, all of which are mycoheterotrophic in their lifestyle. Uniquely, these orchids depend on a fungus for their nutrition, not only as seedlings (as do many other orchids), but throughout their entire life span (in most other species, this relationship reverses when the orchids begin nourishing the fungi). In sum, this group of orchids, lacking chlorophyll and thus being nonphotosynthetic, therefore maintains a lifetime dependency on fungi. This relationship has likely developed into a high degree of specificity between orchid and fungus. Note, however, that the symbiotic relationship is not mutualism, as these orchids are obligate parasites that provide no benefits to the fungi.

The orchids in Dallas County are closely associated with soils of the Eddy-Brackett series that develop on hillsides (8–20 percent slopes) to depths of about 11 inches with a surface layer of gray-brown clay; these soils lie over a base of Austin Chalk. The association is strong, and soil maps are a useful tool for locating additional sites where

Figure 4.9. Spotted Coral Root represents one of the few species of nonphotosynthetic orchids with a lifelong dependency on fungi for their nutrition. Photograph by Doug Sherman, Lady Bird Johnson Wildflower Center.

orchids might occur. Oaks and junipers form the primary overstory community at these locations, but orchids are absent if the canopy cover exceeds 60 percent, perhaps because of its influence on air temperatures at ground level. Decaying leaf litter from the overstory, if undisturbed, provides organic matter necessary to support the fungi.

Interestingly, botanists only recently discovered the fourth species in this group in Dallas County—a range extension of nearly 430 miles. Previously, the species was known in Texas only from Pinyon Pine habitats at higher elevations in the Davis and Chisos Mountains (chapter 9). Nonetheless, in Dallas County the plants were found in the same habitat as the other three species. Remarkably, two species in the latter group of orchids similarly occur in isolated locations in the Trans-Pecos or elsewhere to the west as well as in Dallas County, all of which poses a knotty problem for biogeographers attempting to explain the disjunct distribution of these highly specialized species.

MAMMALS OF THE BLACKLAND PRAIRIES

Excluding domestic and exotic introductions, the mammalian fauna of the Blackland Prairies includes 78 species, compared to 139 for the entire state—only the Piney Woods (chapter 2) has fewer mammals. The homogeneous nature of the prairie habitat underlies this relationship—few niches to occupy—which became even more uniform as crops replaced the mix of native grasses. Additionally, the Blackland Prairies serve as an effective barrier against the incursion of eastern species as well as others from the west. As indicated earlier, the heavy black clays preclude the burrowing activities of pocket gophers. Likely for the same reason, prairie dogs are absent, except for a population originating from those introduced near Fort Worth. Red Foxes were also imported, in this case probably by hunters seeking game deemed worthy of their well-bred hounds; the slower Gray Foxes, native to the area, seemingly offered the dogs little challenge.

Several mammals, although never common, disappeared with settlement. As is usually the case, these included the larger species of predators, including Gray Wolves and Mountain Lions; even Jaguars occasionally visited the region but were soon extirpated. American Bison, too, vanished, along with Pronghorn. In the past, Raccoons, Coyotes, and other terrestrial furbearers gained economic importance in the Blackland Prairies. Changing fashions—bolstered by public distaste for trapping itself—curtailed fur prices, and the harvest of pelts was greatly reduced. Similarly, White-tailed Deer were once of economic importance. Some 75,000 deer skins were shipped from Waco in the mid-1800s, and as the network of railroads expanded across Texas, professional hunters supplied venison for the construction crews laying track.

On a finer scale, five species dominated the rodent community studied on a remnant of the Blackland Prairies known as Clymer Meadow, unplowed but with a history of haying, which may have mimicked periodic grazing by the bison herds that once foraged in the grasslands. In decreasing order of abundance, these were Hispid Cotton Rat, Fulvous Harvest Mouse, Eastern Harvest Mouse, White-footed Mouse, and Deer Mouse. Of these,

INFOBOX. OF SLOTHS AND SEEDS

For the giant herbivores of the Pleistocene—notably ground sloths and mammoths—the fruit of Osage Orange was not much of a mouthful, but it was something of a treat. The pulpy, seed-laden fruits, the size of softballs, grew on trees widely distributed across the plains where the huge animals flourished. In fact, for millennia the distributions of Osage Orange and the Pleistocene megafauna closely coincided across the vast grasslands of the North American interior. Sloths and mammoths unknowingly carried the seeds conveniently packaged in their own supply of fertilizer for deposit far and wide. Thus the distribution of Osage Orange became wedded to the fruit-eating giants of the Ice Age. Along the way, the fruits also became the sole source of food for caterpillars of the Hagen's Sphinx Moth.

The ground sloths were giants. Harlan's Ground Sloth, the largest of the species in North America, reached lengths of 10 feet and weights of 1.5 tons. The claw of the somewhat smaller, ox-sized Jefferson's Ground Sloth captured the attention of the soon-to-be president, who thought the fossil originated from a large lion. Based on fossil evidence, both occurred widely across North America.

Then, about 13,000 years ago came paleohunters whose stone-tipped weapons, according to the theory of Pleistocene Overkill, slaughtered the megafauna and, in doing so, ended the long-standing relationship with Osage Orange. Fruits selected by nature as food for large herbivores thereafter rotted uneaten beneath the trees, becoming an ecological anachronism akin to typewriters and carbon paper in modern offices. In short, a well-adapted tree would hardly waste energy evolving fruit that produced no biological payoff, which is exactly what happened when the megafauna expired. The once-widespread range of Osage Orange contracted, eventually reduced to a small area in north-central Texas and adjacent parts of Oklahoma and Arkansas. Moreover, the trees no longer grew on uplands but were now limited to the riparian woodlands on tributaries along a small part of the Red River watershed, at least one of which (in Grayson and Fannin Counties) bears the name Bois d'Arc Creek. In these settings, floods replaced the megafauna as the agent of seed dispersal—good enough for maintaining the local streamside populations but not for spreading the seeds across the higher ground lying beyond.

Human influences again intervened. Descendants of the paleohunters eventually discovered the strength and

Fallen fruits litter the ground beneath an Osage Orange hedgerow. The inset shows a mature "Hedge Apple."
Photographs by Barney L. Lipscomb.

flexibility of Osage Orange, which became the wood of choice for the bows of Native Americans—hence Bois d'Arc, "wood of the bow," a name bestowed by French traders and still widely used. However, the restricted distribution of Osage Orange precluded wide access to the trees, which provided the local Spiroans with a monopoly for trading the prized wood with other tribes both near and far.

Sharp spines protect the stems of Osage Orange, and their utilitarian value garnered notice from settlers (who, however, shunned the ill-tasting fruits). The spine-laden plants, when planted in rows and trimmed, soon produced thick hedges well suited for confining livestock. And it helped that Osage Orange responded to pruning with an outpouring of root suckers, further thickening the growth. By the mid-nineteenth century, Osage Orange was likely the most widely planted tree in North America, with some 60,000 miles of hedges established in 1868 alone. Even with the advent of barbed wire in 1874, many cash-starved homesteaders still relied on the spiny "Hedge Apples" for living fences aptly described as "horse high, bull strong, and hog tight." For those who could not wait for hedges to develop, Osage Orange provided rot- and insect-resistant wood for fence posts on which to stretch the newfangled wire. Thus, with settlement, and later, residential landscaping, the trees now occupy much of the United States. Still, the sloths represent vanished partners that left Osage Orange with adaptations that matched the lost world of the Pleistocene, but less so the modern world. As expressed by science writer Connie Barrow, the giant animals became the "ghosts of evolution."

both species of harvest mice preferred undisturbed tracts of prairie still dominated by Little Bluestem and Switchgrass. The occurrence of Eastern Harvest Mice on the Blackland Prairies represents a westward extension of the species' range in the southeastern United States. Eastern Woodrats— "packrats" of lore—constructed their stick nests at the base of multistemmed Osage Orange; also known as Bois d'Arc, these shrubby, spiny-twigged trees often develop in small clumps ("mottes"), scattered in grasslands as well as in riparian zones bordering small streams (see infobox). Hispid Pocket Mice, while occurring on sandy soils elsewhere in the immediate area, avoided sites with clay soils; this relationship illustrates how certain habitat features—soil types in this case—limit the ecological distribution of a species within its larger geographical distribution (fig. 4.10).

All told, the Blackland Prairies are not "hot spots" of animal diversity and lack any endemic species of mammals; indeed, the region lacks endemic vertebrates of any kind. Instead, the Blackland Prairies gain greater biogeographical significance as a transitional zone where the ranges of eastern species reach their western limits and, similarly, where the ranges of western species reach their eastern limits—albeit not a friendly place for pocket gophers.

Figure 4.10. Hispid Pocket Mice occur in the Blackland Prairies, but only in areas with sandy soils. Photograph by Troy L. Best.

MR. DEERE AND HIS PLOW— AN EARTH-TURNING EVENT

Initially, the North American prairies presented settlers with a puzzle: How could a farm possibly succeed where the soils seemed too poor to grow trees? Back east, farms were carved from the forest one tree at a time until a field was cleared, and the best sites for crops were where the trees had been both big and many. Indeed, the prairies represented a new environment for European immigrants—vast grasslands akin to the North American prairies simply did not exist back in the Old Country. In time, however, settlers realized the true fertility of the prairie soils, but then they faced two vexing problems: how to cut the thick sod of "wire-like fibrous roots interlaced and interwoven in every conceivable manner" and how to turn the heavy soils that quickly balled up and choked their iron or wooden plowshares. A team of horses might not be able to handle the work; oxen were better, but they too often stalled because the plows required frequent cleanings. The prairie was rich, but difficult.

Things changed in 1837 when a young blacksmith from Vermont fashioned a self-scouring plow from the hardened steel of a saw blade. John Deere (1804–1886) had settled in the frontier town of Grand Detour, Illinois, where his newfangled plow soon conquered the war on roots and sticky soil and, like the axe, forever changed an American landform. Now, with a smooth surface and sharp edge, Deere's plow turned over the soil without the toil of oxen, at once forever "breaking" both the sod and the spine of grassland communities. "Sod busting" had come of age, and with it, the tallgrass prairies of North America transformed from a complex, self-sustaining ecosystem into monocultures of annual crops fully dependent on yearly inputs of chemicals and energy-demanding tillage. Put another way, as an Indian chieftain of the era wistfully observed, "Grass no good upside down."

Compared to the Midwest, widespread cultivation was slow in coming to the Blackland Prairies in Texas. Homesteaders, to be sure, scratched out farms there early in the nineteenth century, but not until the 1870s and 1880s did the blackland tallgrasses give way to the reign of a cotton kingdom. An expanding network of railroads and Deere's plow, by then mass produced and complemented with other farm equipment, had triggered an agricultural revolution. Estimates vary, but all agree that almost all of the original prairie has long ago disappeared. Today, only a few relicts whose total area is barely larger than a single modern farm remain as reminders of the original Blackland Prairies.

CONSERVATION AND MANAGEMENT

As a viable entity, the Blackland Prairies were doomed with the discovery of their fertile soils and the development of machinery able to break the deep-rooted sod. Strong markets for cotton and grain fueled the exploitation of even marginal sites; hence soils exposed on the steeper slopes quickly eroded, leaving gullies and bare rock where tallgrasses once flourished. Indeed, field studies demonstrated that soil loss was 74 times greater after prairie was replaced with crops. Eventually, better farming practices such as contour plowing halted much of the soil loss, but the lesson of abuse was hard learned and long lasting. Some areas remained as unplowed hay meadows, and while these retained their prairie-like aspect, repeated mowings and herbicide applications altered the composition of the vegetation by reducing the incidence of forbs. Fire suppression encouraged the expansion of woody vegetation, notably Eastern Redcedar, into sites once dominated by prairie vegetation. Johnson Grass and King Ranch Bluestem are among the several species of nonnative grasses that have invaded the original flora. Johnson Grass is native to the Mediterranean region and was introduced into Texas in the 1880s; the forage value of this species may be lessened when certain growing conditions induce the production of toxins poisonous to livestock.

Although small, some remnants of the Blackland Prairies survive in the care of public or private agencies. Of these, two sites protected by The Nature Conservancy (TNC) deserve mention. Tridens Prairie, a remnant in Lamar County just 97 acres in area, features a grassland dominated by Silveus' Dropseed, Mead's Sedge, and its namesake, Long-spike Tridens. This community develops on alfisols and includes numerous Mima mounds. Regrettably, poachers at one time uprooted and removed large numbers of Blacksamson, a species of coneflower of alleged value as an herbal medicine, but the site otherwise remains undisturbed except for prescribed burning and limited grazing, the activities prescribed in the management plan. In contrast, vertisols characterize the soils at Clymer Meadow, where Eastern Gamagrass, Switchgrass, and Indiangrass characterize the prairie community. In addition to

Figure 4.11.
With continued
management,
including
prescribed
burning, the
prairie community
in this meadow
near Hallettsville
can be restored.
Photograph by
Paige A. Najvar.

the native vegetation, gilgai persist at this site, one of the few locations where these once-common structures still remain. TNC owns about 1,100 acres at Clymer Meadow. Another 300 acres of adjacent prairies are protected by conservation easements, which are legal agreements with landowners restricting future uses of the land to ways that protect its natural features. Because of its uniqueness, Clymer Meadow has served as a study area for several research projects.

Most estimates indicate that less than 1 percent of the original Blackland Prairies remains intact; hence locating and securing additional sites for preservation remains difficult. Except for a scattering of extremely small patches that escaped cultivation, locations historically managed as hay meadows represent the only remaining opportunity to protect sizable areas of unplowed ground, such as the privately owned Smiley-Woodfin Prairie in Lamar County. Whenever possible, these sites should become prime candidates for conservation easements or outright purchase. Mowing somewhat alters the flora of hay meadows, but these prairie communities can be restored rather quickly (fig. 4.11). Both TNC and the Native Prairies Association of Texas (NPAT) represent nongovernmental

organizations spearheading such efforts. The latter currently protects almost 2,800 acres of grasslands, including a small area of Blackland Prairie in Collin County and the 115-acre Mary Talbot Prairie Preserve in Bowie County. NPAT also publishes an online *Tallgrass Restoration Manual* and a management note for dealing with Johnson Grass.

Like many other parts of Texas, the Blackland Prairies are infested with Red Imported Fire Ants. At Parkhill Prairie, for example, fire ants reduced both the diversity (by 66 percent) and abundance (by 99 percent) of ants native to this site of virgin prairie. However, the impacts of fire ants on ground-nesting birds, small mammals, and other fauna, while studied elsewhere, are still unknown on the remaining fragments of Blackland Prairie. Fragmentation itself poses significant problems— among them increased nest predation—for Dickcissels, Grasshopper Sparrows, and other grassland birds. Nest predation, for example, is up to 24 percent greater on small fragments in comparison with nests lost on large fragments. Overall, these and other circumstances suggest that the current state of highly fragmented prairies contributes to the regional declines of grassland birds in much of North America and underscores another argument for restoring large blocks of Blackland and other prairies.

Crayfish and prairies generally do not captivate public attention the way iconic "charismatic species" such as American Bison or Coast Redwoods do. Consequently, these and other segments of the natural world too often get shelved despite their inherent uniqueness. Some 350 species of crayfish (of 500 worldwide) occur in the United States—including 40 taxa in Texas. While it is difficult to determine the status of their populations, some estimates suggest that half of the species in North America warrant some form of protection; currently, only four species are covered by the Endangered Species Act. In Texas, a clear link exists between Blackland Prairies—especially those where gilgai survive—and the Parkhill Prairie Crayfish, both particularly vulnerable because of their limited distribution and specialized niche. Whereas further discoveries may expand the currently known range of the species, the core population at a park in Collin County survives on just a small area of undisturbed habitat—a risky situation for any species. Field studies, including population surveys, of Parkhill Prairie Crayfish seem overdue as a component in protecting the species as fully as possible.

The Audubon Center at Dogwood Canyon (in southwestern Dallas County) highlights a larger complex of public lands consisting of Cedar Ridge Nature Preserve, Cedar Hill State Park, and Cedar Mountain Preserve; combined, these provide an oasis in a region otherwise occupied by urban sprawl. Established in 2011, the Audubon Center protects 205 acres of wooded habitat and sponsors school and family programs including workshops held in a spacious facility with a scenic overlook of the canyon. Trails offer birders and tourists access into the woodlands. The grounds feature a small area of restored prairie that includes milkweeds planted for the benefit of Monarch Butterflies.

5
CROSS TIMBERS
AND PRAIRIES
A CAST-IRON FOREST
IN A LAND OF GRASS

I shall not easily forget the mortal toil, and

the vexations of flesh and spirit . . . struggling

through forests of cast iron.

— WASHINGTON IRVING

Cross Timbers and Prairies

For Washington Irving (1783–1859), who is largely credited as the first American Man of Letters, the encounter with the Cross Timbers was indeed challenging. Irregular belts of forest, running north and south, abruptly interrupted the prairies and frontier cultures he had come to see. In places, the trees were well spaced, but in others they provided a trellis for tangles of vines and brush that challenged both horse and rider. Beyond lay more grassland, but for the moment, the forest hindered Irving's travels, about which he penned an image of a "cast iron forest" (fig. 5.1).

The broad belt of what is today known as the Cross Timbers extends southward from Kansas across Oklahoma and into Texas, where it splits into two segments, one running along the eastern edge, the other on the western edge of a grassland known as the Grand Prairie, thus becoming a Texas-sized sandwich of forest with a grassland filling. Just how the area was named remains uncertain, but one of the more common explanations suggests the landscape represented a landmark for pioneers heading west. Once they passed through these bands of woodlands—that is, crossed the timbers—only open country remained until the far-off Rocky Mountains rose from the plains.

Ecologists generally regard the region as a transitional zone, or ecotone, between the deciduous oak-hickory forests of the humid east and the interior grasslands of the semiarid west. Thus, both the flora and fauna of the Cross Timbers and Prairies are a mix of species with affinities to one or the other of the two climatic regimes and include few endemic species (infobox 5.1). Only five plants—among them two yuccas—qualify as endemic to the region.

To the west, the Cross Timbers and Prairies abruptly end when the soils change from the limestone and sands of the Cretaceous Period to the much older iron-rich Permian deposits marking the beginning of the Rolling Plains (chapter 6).

(*overleaf*) Figure 5.1. This dense line of trees jutting across a prairie in the Cross Timbers represents the seemingly impenetrable barrier described by Washington Irving.

Formed 92–96 million years ago (mya) from parent materials originating when the region lay under inland seas, the fossil-rich limestones represent seabed sediments and the sands indicate shoreline deposits. Bivalve mollusks, echinoderms, and ammonites represent the fossil fauna in the limestone deposits. Because the Cretaceous seas periodically advanced and receded, deposits of limestone and sands reflect variables such as water depth and distance from shore. Important aquifers underlie much of the Cross Timbers region. Rocks of the Pennsylvanian Period—the oldest in the region, dating from 286 to 320 mya—appear in the northwestern fringe of the Western Cross Timbers in rugged terrain known as "Palo Pinto Country."

Of the two units, the smaller Eastern Cross Timbers receives more annual precipitation than its western counterpart; the gradient runs from 40 inches in the east to 24 inches in the west, for a substantial difference of more than 15 inches. This, in addition to its deeper and somewhat more fertile soils, results in taller trees in the eastern unit (fig. 5.2). In places, the vegetation forms savannas instead of "pure" prairie or forest, so in reality, the Cross Timbers form a mosaic of cover types representing a textured landscape.

VEGETATION

THE FOREST

Two species dominate the Cross Timbers forest; together they form 90 percent of the canopy cover. The first is Post Oak, which belongs to a group known as the "white oaks," which, among other features, produce leaves with smooth-tipped lobes and "sweet" acorns; the latter are favored by squirrels, deer, turkeys, and other mast-eating wildlife (infobox 5.2). The leaves of a Post Oak are distinctively five lobed, and because the two large middle lobes end bluntly, each leaf resembles a cross. The branches are often twisted and gnarled, and instead of growing upward, they extend almost straight out from the trunk (fig. 5.3). According to some sources, the place-name Palo Pinto—Spanish for "painted wood" or "painted tree"—originated from the appearance of Post Oaks after their leaves have fallen. A hardy species, these trees survive droughts and may live for centuries, but surprisingly, instead of well-watered sites, the older

INFOBOX 5.1. LLOYD H. SHINNERS
Systematic Botanist (1918–1971)

Born into a Wisconsin family that had homesteaded in the Peace River country of northwestern Alberta, Shinners spent his early years in a frontier of unspoiled nature. When he was five, however, the family returned to Wisconsin, where the youngster started down a path of formal education that was to become a blaze of academic excellence. He graduated at the top of his high school class of 2,500 in Milwaukee and was elected to Phi Beta Kappa at the University of Wisconsin–Madison, where, after finishing his undergraduate studies in 1940, he remained to earn both MS and PhD degrees in the following three years.

Shinners initiated his career as a botanist with the Milwaukee County Park Board but after a year left for Dallas, where he began what was to become lifelong employment with Southern Methodist University (SMU), first as a research assistant but soon thereafter as the Director of the SMU Herbarium. Because of his tireless efforts, the plant collection expanded from about 20,000 to 340,000 specimens—one of the most complete herbaria available for a large region in the southwestern United States. Shinners personally prepared an average of 10,000 specimens per year during his 23-year tenure as director. Today, the SMU Herbarium forms the core collection associated with the Botanical Research Institute of Texas in Fort Worth. Likewise, the botanical library at SMU emerged as one the largest of its kind, with Shinners again personally responsible for adding many of the books.

Shinners penned 276 papers in which he contributed 558 new scientific designations. Much of his research concerned taxa in the huge family Compositae (now widely cited as Asteraceae). His 514-page book *Spring Flora of the Dallas–Fort Worth Area, Texas*, notable when published in 1958, remains a widely used reference for the region. Some of his taxonomic work proved controversial, enough so at times that orthodox journals rejected his papers. Undaunted, Shinners founded and edited a new journal in 1962 known as *Sida, Contributions to Botany*, after a genus of colorful plants in the mallow family. He actively sought and published manuscripts (including his own) that often "pushed the envelope" of systematic botany. Many years later, *Sida* was renamed the *Journal of the Botanical Research Institute of Texas* and remains an important venue for plant research. At the time of his death, Shinners was continuing work on a long-standing project, *Gulf Coast Flora*. He also helped establish (in 1953) the Southwestern Association of Naturalists and served as the first editor of its journal, *Southwestern Naturalist.*

By any standard, Shinners was a formidable intellect, not only as a prominent scientist but also as an accomplished composer of both music and poetry, a classicist, and a linguist with proficiency in seven languages; he often matched his musical compositions with his poems, as well as with those of others. Ornithology and entomology were among his other scientific interests. Shinners maintained a rigorous work ethic both in the field and in the SMU Herbarium, doing so in spite of recurring poor health associated with diabetes. Soon after arriving in Texas, he (with colleague Donovan Correll) undertook a field trip to the Guadalupe Mountains and alone decided to climb Signal Peak, the highest point in Texas and a challenge even for experienced climbers. During his ascent, Shinners fell about 80 feet and suffered multiple fractures. The pain from these injuries lasted for the remainder of his life, but he neither complained nor compromised his fieldwork, although he understandably did forgo further mountain climbing and instead focused his research and collecting on the flora of the Gulf Coast—a somewhat less precipitous landscape. Some colleagues said his detailed knowledge of plants and their distributions was gained intuitively, to which Shinners replied that they too could be "intuitive" if they would simply study plants in the field as well as inspecting lab specimens. Shinners in fact spent about five days per month in the field collecting plants and determining where they occurred. "Nature," he wrote, "gives away few secrets to the lazy, and none to the incompetent." His commitment to systematic botany did not exclude his concerns for what some disparagingly considered useless vegetation, about which he noted, "Conservation means wise use when the use is known. Conservation means preservation when no use has yet been found." By any standard, Lloyd Shinners, a man of small physical stature, emerged from a far-off frontier to become a giant of systematic botany both in Texas and well beyond.

Figure 5.2. In contrast with its western counterpart, the deeper soils and greater annual rainfall in the Eastern Cross Timbers allows the development of tall oaks.

Figure 5.3. Twisted and gnarled main branches typify the growth form of Post Oaks on dry upland sites.

trees are associated with rugged, rather arid upland locations.

Blackjack Oak belongs to the red oak group—oaks with bristle-tipped leaves and bitter acorns of lesser value as wildlife food. The large leaves, while variable, often develop with three shallow lobes producing an outline shaped like a cartoon of a dinosaur track; the upper leaf surface shines, but the underside resembles rough felt. The acorns of Blackjack Oaks commonly develop in pairs and take about 18 months to mature, at times producing crops of considerable abundance. Rough, dark, and "blocky" bark—somewhat like an alligator's skin—also helps identify Blackjack Oaks.

Both Post and Blackjack Oaks grow on leached, nutrient-poor soils and, in recent times, serve as indicators of worn-out, abandoned farmland—nature's signs of "buyer beware." Neither offers much commercial value other than for firewood, although the wood of Post Oak once provided a source for railroad ties and, as implied by the name, fence posts. In the Cross Timbers, mixed stands of both species are strongly associated with soils developing from sandstone. The oaks in the Western (or Upper) Cross Timbers commonly deal

INFOBOX 5.2. PASSENGER PIGEON
Gone Forever

Once the most numerous bird in North America—and perhaps on Earth—Passenger Pigeons exist no more, victims of what no species can long survive: unlimited exploitation, especially when breeding. In Kentucky, ornithologist Alexander Wilson (1766–1813) once watched for hours as an unbroken flight of what he estimated as two *billion* birds passed overhead. He, among others, also witnessed tree limbs breaking under the weight of perching birds, and shin-deep accumulations of their droppings killed the vegetation beneath frequently used roosts. About 136 million Passenger Pigeons occupied a nesting area in Wisconsin, and this was just one of many colonies in the birds' extensive breeding range. A single tree might hold as many as 90 nests. The abundance of these birds cannot be linked to fecundity—a clutch consisted of just one egg. Nor does abundance protect a species from extinction. Passenger Pigeons were gone forever about a century after Wilson marveled at their vast numbers.

Acorns and other mast formed the diet of Passenger Pigeons. In Texas, this caused some worries among the owners of free-roaming swine when the birds arrived; their hogs fattened on the same food. Newspapers of the day predicted dire consequences for swine production in several counties because of the competition between pigs and pigeons. In Leon County, an editor declared that the depletion of mast in the local forests by millions of pigeons would "cause a wail of despair to ascend from the throats of our beautiful Razorbacks." One can only wonder how well squirrels, deer, and Wild Turkeys fared as the mast supplies dwindled.

Because of their preference for oak forests, Passenger Pigeons favored northeastern Texas (the Cross Timbers) for their winter habitat, although at times large flocks ventured as far south as Austin and San Antonio and west onto the Edwards Plateau (chapter 7). Egg collections confirm that the birds at times nested in Texas, but the state served mostly as a wintering area. In general, the birds arrived in Texas in September and left for their northern breeding areas the following March. On one occasion, however, large numbers moved along the Colorado River to San Saba County, where they subsisted on the berrylike cones of junipers and remained to nest despite heavy hunting pressure. The effort failed when local residents burned the stands of junipers to prevent the birds from damaging their crops. Four areas in Texas bear names associated with the birds: two sites known as Pigeon Roost Prairie (one each in Hardin and Van Zandt Counties); Pigeon Roost Creek (Bandera County); and Pigeon Hill (Rusk County).

Like the Band-tailed Pigeon, an extant species of western North America, Passenger Pigeons visited mineral sites, one in Texas being the saline springs in Van Zandt County. Although apparently associated with breeding in Band-tailed Pigeons, the mineral supplements played an unknown role in the lives of Passenger Pigeons. Similarly, both species shared a common ectoparasite, *Columbicola extinctus,* a chewing louse named before it was discovered on Band-tailed Pigeons (biologists initially believed that the parasite and Passenger Pigeon concurrently met a common fate). This relationship reflects another biological relationship, namely that closely related birds often serve as hosts for closely related or even the same lice. Indeed, molecular evidence recently indicated that Band-tailed Pigeons are the closest living relatives of Passenger Pigeons, which suggests that the two species may have been eastern and western counterparts—forest-dwelling pigeons separated by the intervening plains.

Extinction was inevitable in light of the millions of Passenger Pigeons killed for food year round. The slaughter—netting, clubbing, and shooting—was especially merciless in nesting colonies because the tender nestlings (squab) were highly prized fare in fashionable big-city restaurants. At times, trees were felled to obtain large numbers of squab, thereby destroying nesting habitat in the process. As many as 1,000 professional pigeon hunters actively supplied pigeons for eastern markets, but large-scale commercial hunting did not develop in Texas, although some birds were sold locally—for 50 cents a dozen in one case. In a three-month period in 1878, more than 1.5 million birds were shipped to market from a colony in Michigan, where as many as 10 million were eventually killed. Railroads, aided by telegraph updates on where to find the birds, provided hunters with ready access to many colonies as well as a way to ship the harvest to market. Rail transportation provided one dealer in New York City with 18,000 birds a day. With no refrigeration, even more birds were killed to offset losses from spoilage during shipment. By March 1900, the last wild Passenger Pigeons were gone, and the few that remained were confined in zoos and private aviaries. Martha, presumed to be the last of her species, died alone in the Cincinnati Zoo in 1914.

with droughts and erratic rainfall and hence grow quite slowly, with some centuries-old trees not exceeding 30 feet in height; stands in Comanche and Throckmorton Counties represent some of the few examples of virgin forest still extant. Overall, the average height of both species decreases significantly from east to west.

Secondary species include Black Hickory, especially in the Eastern (or Lower) Cross Timbers, and Eastern Redcedar (actually a juniper). Black Hickory grows among the oaks on well-drained sites, where it may reach heights of 40 feet, and its nuts are a desirable wildlife food. Black Hickory is also the primary host for several species of magnificent moths, among them the Luna Moth, a striking swallow-tailed, pale green species with a 4.5-inch wingspan (fig. 5.4), and the Giant Regal Moth. The Giant Regal Moth's caterpillars are also spectacular. They attain lengths of nearly 6 inches and sport several large black-tipped horns, giving rise to the moniker "Hickory-horned Devils," despite being perfectly harmless. The caterpillars are bright green, but just before pupation, their color turns to turquoise.

Throughout the Cross Timbers, Eastern Redcedar has steadily encroached because of fire suppression in the decades following settlement. The cedars often close the canopy in areas that were once savannas. This expansion is facilitated by birds such as Cedar Waxwings that eat the cedars' fleshy "berries"—actually cones—but excrete the still-viable seeds within the berries when perched on isolated oaks. Large numbers of the seeds later germinate and develop into dense clusters of Redcedar immediately surrounding the center tree, in time filling the interspaces between the oaks, thus ending the structure of a savanna.

The woody understory varies by site but often includes Eastern Redbud, Chittamwood, Mexican Plum, Winged Elm, Coralberry, and Fox Grape. The herbaceous understory includes some of the prairie grasses, among them Little Bluestem and Tall Dropseed. The incidence of species associated with the aridity, among them Buffalograss and prickly pears, increases in the Western Cross Timbers. In the savannas, Sideoats Grama and Tall Dropseed are common in the understory on clayey soils, whereas Small Panicgrass becomes more abundant

Figure 5.4. Luna Moths rely on Black Hickory, a secondary species in the Cross Timbers. Photograph by Chris Jackson, DFW Urban Wildlife (www.dfwurbanwildlife.com).

at sandy sites. Drought is not uncommon in the Cross Timbers, and its occurrence temporarily replaces many of the more mesic plants in the prairie understory with those better adapted to drier communities. Buffalograss, a species common in the western plains and adapted to aridity, is several times more abundant in Palo Pinto County than elsewhere in the Western Cross Timbers.

Fire suppression, coupled with overgrazing, has modified the savannas, where widely spaced oaks once studded the grasslands and paved the way for invasions of brush, vines, and thickets of saplings. In the past, periodic fires—perhaps at intervals of about five years—kept out the invaders and, while seldom harming the mature trees, actually renewed the vigor and growth of the grasses. Almost nothing is known about the presettlement history of fire in the Cross Timbers. However, fire was a tool of Native Americans—as it is of modern conservationists—used to achieve specific objectives. But instead of grassland preservation, the goal of Native Americans was to attract bison, which sought the tender, protein-rich regrowth that soon poked through the charred terrain.

THE PRAIRIES
The Grand Prairie is the combined name for two adjacent ecological areas—both distinguished by

soils originating from limestone parent material—lying between the Eastern and Western Cross Timbers. The Fort Worth Prairie once featured grasses growing "belly high" to a horse. These included both tallgrasses and midgrasses, primarily stands of Little Bluestem—which formed some two-thirds of the original community—followed in prevalence by Sideoats Grama but also including lesser amounts of Big Bluestem, Indiangrass, Switchgrass, Tall Dropseed, and Hairy Grama. Today, more than a century of heavy grazing has altered the community structure of these grasses in some locations, where they have been replaced by Texas Wintergrass and Silver Bluestem. Surprisingly, extensive areas of high-quality prairie can still be found on ranches close to Fort Worth, but they are rapidly disappearing as development expands west from the Metroplex. Fortunately, the Bear Creek Ranch (Parker County), managed by the Dixon Water Foundation, provides an opportunity for visitors to see the native Fort Worth Prairie.

Like prairies elsewhere, the grasslands become flush with wildflowers in years with adequate rainfall. White Rosinweed is widespread in the Grand Prairie and owes its name to sticky secretions on its stems and leaves. Other prominent wildflowers include Pink Evening Primrose, Lemon Beebalm, Mexican Hat, and Texas Paintbrush. The latter species seems particularly adapted for pollination by Ruby-throated Hummingbirds, but production of its colorful blossoms may vary widely from year to year. The roots of Texas Paintbrush, although unusual for a wildflower, may penetrate the root systems of adjacent plants, commonly grasses, which they tap for additional nutrients (fig. 5.5a). Glen Rose Yucca is one of two endemic yucca species in the region; it was once known only from a sandy terrace on the Brazos River in Somervell County, but larger populations were later discovered in deep sands in Tarrant County (fig. 5.5b). The other endemic, Pale Yucca, prefers limestone outcrops in the region's prairies; true to its name, the leaves of this species are conspicuously whitish.

The Lyndon B. Johnson National Grassland (Wise County) harbors one of the few areas of protected native prairie remaining in the Cross Timbers. Hispid Cotton Rats dominate the rodent fauna in this area and increase in density in response to greater amounts of grass cover. Dense grasses provide food sources and habitat structure such as nesting material, cover from predators, and protection from weather. Populations of Hispid Cotton Rats experience surges and crashes in keeping with seasonal changes and weather conditions, reaching peaks in spring and fall, but decline rapidly in response to droughts.

A larger area, the Lampasas Cut Plain, forms the second component of the Grand Prairie. Here mesas, buttes, and intervening valleys provide a strikingly different image of a prairie landscape. Indeed, the region likely represents an extension of the Edwards Plateau (chapter 7), found to the southwest of the Cross Timbers region. The variety of topographical features is reflected in the flora, which includes some of the region's endemic species, such as Plateau Milkvine. Where Glen Rose limestone is exposed, the thin, rocky soils support a mix of shortgrass prairie grasses, oaks, and junipers. The junipers are particularly troublesome because of their rapid invasion of this region (and elsewhere) as a result of fire suppression and overgrazing (fig. 5.6). Similarly, mesquites have also expanded considerably in this area because of human activities.

RIPARIAN VEGETATION

Two major rivers—the Brazos and the headwaters of the Trinity—traverse the region, and their influence on soil moisture accordingly produces streamside forests. Moreover, the rivers provide corridors for the westward extension of eastern plants. In the Western Cross Timbers, species such as Little Walnut and Netleaf Hackberry prevail, whereas Cedar Elm and Pecan flourish in the Eastern Cross Timbers. The floodplains adjacent to the rivers in the Grand Prairie feature elms, Pecan, and hackberries.

A vegetative zone running parallel to the Red River includes the northern edges of, east to west, Lamar, Fannin, and Grayson Counties. Here microhabitats with special soil and moisture conditions foster vegetation often associated with the Piney Woods (chapter 2) and the Post Oak Savanna (chapter 3). These include well-developed stands of Southern Red Oak, American

Figure 5.5. Roots of Texas Paintbrush (bottom center), a common prairie wildflower, absorb some nutrients from the roots of adjacent plants (left). On a sandy Tarrant County outcrop, a rare Glen Rose Yucca is surrounded by aptly named Winecups (right). Photographs by David Bezanson.

Figure 5.6. Fire suppression in the Western Cross Timbers allowed Eastern Redcedars to invade limestone habitats above and below the escarpment of a mesa in the Lampasas Cut Plain.

Sweetgum, Loblolly Pine, Red Maple, and River Birch, and herbaceous vegetation such as Cinnamon Fern, Blood Milkwort, and Lanceleaf Loosestrife, among many others. In Fannin County, Southern Red Oak, Water Oak, Sassafras, and Blackgum reach their western limits. Likewise, in Grayson County, the westernmost in this area,

Black Oak, Pawpaw, Mayapple, Solomon's Seal, and Lindheimer's Beebalm represent examples of other species reaching their western limits. A fold in the sedimentary strata along the river in Grayson County brings older rocks to the surface, which contributes to the rugged terrain of deep-cut valleys in this area. These provide microhabitats for plants otherwise absent in other parts of the county, among them Largeflower Flameflower, Common Shooting Star, and Plains Nipple Cactus.

HIGHLIGHTS

CUESTAS

Layers of Cretaceous limestone lie for the most part unseen beneath the Grand Prairie, but edges of these strata become visible, predominantly in the Fort Worth Prairie. Here the layers of rock lie like shingles when viewed on edge, with the thick edge of the shingle forming a west-facing escarpment behind which a gentle slope dips eastward until it meets the foot of the next escarpment. In short, the layers overlap each other. These are cuestas— "sloping ground" in Spanish—whose geological ages decrease from west to east as an older "shingle" gives way to the escarpment of a younger, overlying shingle farther to the east. The exposed edges remain resistant to erosion, producing a prominent rocky outcropping rising above a less resistant surface and giving some relief to the otherwise gently sloping terrain.

Cuesta escarpments provide rocky habitat for several species of reptiles (fig. 5.7). Among these are the Eastern Collared Lizard and Prairie Lizard, and several snakes, including the Great Plains Ratsnake, Western Coachwhip, Flat-headed Snake, and Western Massasauga. The diminutive Flat-headed Snake belongs to a group of mostly black-headed species but commonly lacks this distinctive coloration itself. Both the coloration and size— 7 to 8 inches—of this shy species suggest a large earthworm; it often seeks refuge in the soft soil under rocks. The Western Massasauga belongs to a group of small rattlesnakes distinguished by heads covered with nine rather large scale plates, whereas a mix of small and large scales covers the heads of other rattlers. Massasaugas occur in moist areas on the prairie but also on cuesta outcrops, where they feed on rodents, lizards, and, when available, frogs. The potency of their hemotoxic venom exceeds that of many larger rattlesnakes, but because of their small size, less toxin is injected per bite; nonetheless, human victims should seek immediate treatment.

Figure 5.7. A male Eastern Collard Lizard surveys his surroundings from a vantage point on a cuesta.

AN ENDEMIC SNAKE ON THE BRAZOS

Harter's Watersnake, originally described in 1941, is one of two species of snakes endemic to Texas, and even within the state, its distribution is limited to the upper reaches of two disjunct river systems. This species is strongly associated with rocky shorelines and stream bottoms and rarely occurs where sandy or muddy soils replace the rocky strata.

The taxonomic status of this species remains unsettled—it may be one of two subspecies or a separate species. Our interest concerns the Brazos River Watersnake, which is found along the Brazos where it flows across the Cross Timbers. As an endemic taxon, it evolved in a region representing an ecotone for reptiles as much as for vegetation. To illustrate, the reptilian fauna in Palo Pinto County consists of 22 species (38 percent) with eastern affiliations and 24 species (42 percent) with western distributions, with the remainder falling into other categories.

The distribution of the Brazos River Watersnake is limited to just 180 miles of river corridor and the shores of two reservoirs, Possum Kingdom Lake (Palo Pinto County) and Lake Granbury (Hood County), one of the most localized ranges of all snakes in North America. Seldom found more than 10 feet from shore, these snakes avoid steep banks but will visit smaller tributaries bordered with riparian vegetation so long as the shoreline is rocky. In cool weather, adults bask on tree limbs hanging up to 6.5 feet above the water. Juveniles prefer shorelines near riffles, where they seek refuge under small, flat rocks less than 1.3 feet in diameter, whereas adults are more widespread

Figure 5.8. Shallow rocky areas in the upper Brazos River provide feeding and cover habitat for both juvenile and adult Brazos River Watersnakes (inset). Photographs by Dustin McBride.

and hide under larger rocks or in deep crevices (fig. 5.8). Presumably, the association between juveniles and shallow riffles reflects the increased availability and vulnerability of small fish as prey; minnows concentrate in shallow riffles to avoid large predatory fish.

Not all habitat along the upper Brazos River is suitable, especially for juveniles, and gaps develop in the snake's distribution where the river courses through extensive beds of clay or sand with few or no rocky riffles—areas that function as barriers and divide the snake population into five segments. Thus separated, each segment becomes vulnerable to extirpation at sites where, because of the barriers, recolonization would likely be difficult. Problems, real or potential, include dams that stabilize the seasonal floods needed to scour away sediments accumulating in rocky areas; takeovers of invasive species such as Saltcedar, which trap sediments that clog and bury rocky shorelines; and fish kills resulting from blooms of golden algae that might critically harm the food resources sustaining one or more of the isolated snake populations.

Because of its limited distribution, segmented population, and potential for habitat degradation, the Brazos River Watersnake is listed as threatened on the Texas list of endangered species.

BLACK-CAPPED VIREOS AND BROOD PARASITES

The Black-capped Vireo is one of several Neotropical migrants harmed by the unusual breeding biology of another species, the Brown-headed Cowbird. Because of this and other troubles—particularly habitat loss—the Black-capped Vireo was listed as endangered in 1987 (fig. 5.9). In Texas, the range of the Black-capped Vireo includes the Edwards Plateau as well as the Lampasas Cut Plain, where the mosaic of grasslands and woodlands provides desirable nesting habitat. A small, isolated population also persists in southern Oklahoma.

The vireos build small, open-cupped nests in the thick foliage of oak clumps, preferring those less than 10 feet tall. Periodic fires historically maintained the shrubby structure of these oak

Figure 5.9. After returning from its wintering grounds in Mexico, a male Black-capped Vireo first establishes a territory, helps construct a nest, and later brings food to the incubating female. Photograph courtesy of the US Fish and Wildlife Service.

clumps. Without these disturbances, the oak and juniper woodlands develop into full stature, which limits their suitability as nesting habitat. In contrast, Brown-headed Cowbirds prefer open country but seek the nests of other species along the margins of woodland habitat; hence the setting for the vireos matches well with the needs of the cowbirds.

Brown-headed Cowbirds evolved a form of reproduction known as nest or brood parasitism—they lay their eggs in the nests of other species and leave it to the hosts to incubate and raise their young. In short, the hosts become foster parents, often at the expense of their own nestlings. For cowbirds, brood parasitism likely developed when they followed bison to feed on the grassland insects kicked up by the roaming herds. Moving with the bison, the cowbirds could not remain in place long enough to construct their own nests, let alone incubate eggs or raise their young. Two features facilitate brood parasitism: a short incubation period and rapid development of nestlings (cowbird eggs hatch before those of the hosts, and their nestlings quickly become bullies that outcompete other nestlings for food). Indeed, the incubation period for Black-capped Vireos extends for 14 to 17 days, compared with 11 days for the cowbirds.

Black-capped Vireos in turn developed strategies to counter brood parasitism. They abandon parasitized nests at higher rates than unparasitized nests, which lessens the success of the former but

favors hatching in the latter. To compensate for abandonment and low nest success of parasitized nests, the vireos nest a second time, even after they successfully raise a brood. Hence, persistence helps overcome cowbird parasitism, but such parasitism coupled with diminished nesting habitat because of fire suppression means that the vireo population ultimately faces a marginal future, as witnessed by its listing as endangered.

A LAND TROD BY GIANTS

In 1908, a torrential flood raced through the watershed of the Paluxy River in Somervell County, scouring away the overburden to reveal the underlying bedrock known as Glen Rose Limestone. A year later, a roaming schoolboy discovered several large three-toed tracks embedded in the exposed bedrock, which he reported to the principal, who then organized a school field trip to the site. Word eventually reached Ellis Shuler, a paleontologist at Southern Methodist University, who formally described the tracks in 1917, although widespread recognition of the discovery did not occur until 1937, when Roland T. Bird of the American Museum of Natural History visited the site.

The three-toed, birdlike tracks belonged to a theropod, one in a group of carnivorous dinosaurs that includes the well-known *Tyrannosaurus rex*. In this case, a smaller version of *T. rex* in the genus *Acrocanthosaurus* likely left the prints in what was then a muddy area adjacent to an ancient sea that covered much of Texas during the Cretaceous Period. Like *T. rex, Acrocanthosaurus* chased its prey running upright on two large rear legs. In size, the predator varied from 20 to 30 feet in length and left tracks up to 24 inches long and 17 inches wide.

Bird soon found more tracks of the same kind and then discovered an entirely new sort of imprint—large, rounded footprints more than 3 feet long made by a huge, pillar-legged sauropod; somewhat smaller, horseshoe-shaped prints represented the forelimbs of the giant beast. These were the first footprints ever discovered of a sauropod, plant eaters that thundered about on four legs. Prior to discovery of these tracks, paleontologists generally agreed that sauropods, because of their great size, necessarily roamed in water deep enough to support their immense weight, but that belief was now dispelled—sauropods were terrestrial. The tracks also suggested these sauropods traveled in slow-moving herds, with adults protectively flanking the younger animals. In 1996, fossil bones of a sauropod discovered farther upstream on the Paluxy River in Hood County seemed to match the footprints. In 2007, an analysis of these bones indicated they represented a new species, *Paluxysaurus jonesi,* which was proclaimed the official dinosaur of Texas in 2009 (appendix A). *Paluxysaurus jonesi* was huge, weighing 15 to 20 tons, with a length of 60 to 70 feet, including its giraffe-like neck, which extended for about 26 feet.

Bird's fieldwork at Glen Rose uncovered still another first—a trackway with the clear footprints of both theropods and sauropods, hence a setting preserved in stone that may reflect a predator chasing its prey. This possibility was studied in detail, and although the circumstances remain unresolved, the tracks nonetheless present a scenario tempting to accept as a snapshot of a predator-prey relationship that took place about 112 million years ago. In 1940, interest in preserving the "chase sequence" resulted in removing a section of the trackway, parts of which were sent to the American Museum of Natural History in New York City and to the Texas Memorial Museum in Austin; another part was somehow lost or destroyed. Regrettably, the slab in Austin was not housed in a climate-controlled area and later deteriorated. However, with the photographs Bird had taken before the trackway was removed, the chase sequence was digitally reconstructed and thus offers a means for additional study of this remarkable look into the past. Other tracks remain intact at the original sites of their discovery.

To some observers, a few of the tracks seemed to resemble those of rather large humans and were once regarded as evidence that humans and dinosaurs coexisted some 6,000 years ago. Today, the "human" tracks no longer remain creditable—they are actually eroded theropod footprints—and the whole concept underlying this notion counters an immense body of solid paleontological and geological evidence. Dinosaur Valley State Park occupies 1,525 acres of the Paluxy River Valley,

where, in addition to seeing the ancient tracks, visitors may walk a nature trail and perhaps add Golden-cheeked Warblers and Black-capped Vireos to their life list of birds.

FOSSIL OYSTERS AND ANCIENT SEAS

Cretaceous-age soils and strata reflect a period when ancient seas covered much of the North American interior. Further evidence of the inland seas stems from the remnants of reefs formed during this long period of inundation. The reefs—massive beds of fossilized oysters—developed in association with the Walnut Clay Formation, a major deposition exposed in various sites in the Lampasas Cut Plain.

The reefs developed in an environment characterized by warm shallow seas whose salinity continually oscillated between typically marine and brackish conditions. Indeed, the reefs—today known as banks—developed in an environmental regime subject to the same conditions experienced by modern oyster beds. When circumstances were especially favorable for growth, the banks reached thicknesses of up to 30 feet; banks of this magnitude occur in Bosque and Coryell Counties. Some banks form prominent benches that extend across the landscape for several miles.

In places, several species of bivalves may be involved, but many banks consist almost entirely of an oyster popularly known as the Devil's Toenail. These occurred in immense numbers (up to 3,600 per cubic foot), which include tiny spat (an early stage of development) attached to old shells in a manner similar to modern oysters. The lower shell, or valve, is the larger of the two and, because of its gnarled appearance, was thought to resemble the wizened toenail of an unsavory character, at least to those with imaginative minds. This part of the oyster lay downward, embedded in the mud, whereas the flat smaller valve faced upward and opened like a lid on the hinge connecting the two shells. Another species, known as the Ram's Horn Oyster because of its curled shape, occurred in banks with more animal diversity, but these did not reach the thicknesses of the banks dominated by Devil's Toenails. The reclining posture of these species, as opposed to the vertical orientation of modern oysters, essentially disappeared as a way of

Figure 5.10. Illustrated in actual size, a 14-foot-tall Columbian Mammoth dwarfs visitors at Waco Mammoth National Monument. The fossil remains of an adult male lie below the platform, and those of a juvenile are exposed in the lower foreground.

life with the extinction of these and related forms at the end of the Cretaceous Period.

PLEISTOCENE MAMMOTHS

In 1978, a trove of mammoth fossils was discovered near the southern tip of the Eastern Cross Timbers, where the Lampasas Cut Plain converges with the Blackland Prairies (chapter 4) near Waco. The site lies close to the confluence of the Bosque and Brazos Rivers in McLennan County in what is today riparian forest, but the oldest fossils date to a time some 65,000 years ago when a grassland or savanna prevailed during a dry interval in the Pleistocene Epoch. The deposit represents at least two events, the first of which included 16 Columbian Mammoths—adult females and juveniles—killed when a flash flood trapped and drowned the entire herd (fig. 5.10). These fossils were thus identified as a nursery herd, the only deposit of its kind discovered to date. About 15,000 years later, still another flood drowned five more mammoths, four adult females and one bull. The site also revealed fossils of an extinct camel and a saber-toothed cat. The location, part of the Bosque River Basin, floods easily because of the area's impermeable bedrock and the influence of the adjacent Balcones Escarpment on rainfall—runoff from a storm dropping little more than 2 inches of precipitation

is enough to produce a 100-year flood. All of the fossils at the site precede the arrival of humans; hence the deaths of these animals were unrelated to anthropogenic causes, although elsewhere the remains of other mammoths from much later in the Pleistocene have been associated with Clovis cultures.

Several species of mammoths once roamed across North America, the Woolly Mammoth of the far north being perhaps the best known. At Waco, however, Columbian Mammoths dominated the Pleistocene grasslands adjacent to the Brazos drainage. One of the largest of their kind, they stood about 14 feet tall at the shoulder and weighed about ten tons, somewhat larger than either Woolly Mammoths or modern African Elephants. Their tusks were long and curved, and the largest extended 16 feet. The species became extinct about 12,800 years ago.

In 2015, President Barack Obama proclaimed the site as Waco Mammoth National Monument, thus incorporating it into the National Park System. Guided walking tours occur regularly and include the Dig Shelter, where visitors can stroll along a walkway overlooking the excavation. Columbian Mammoths are also featured in a display at the Mayborn Museum on the Baylor University campus.

Figure 5.11. A beard-like fringe of hairs on the ventral surface of their heads distinguishes Comanche Harvester Ants and enables them to carry several seeds at a time. Their inverted conical nests (inset) dot sandy soils near oak groves; the one shown here in the Fort Worth Nature Center and Refuge is about 10 inches in diameter. Photographs by Ann B. Mayo.

COMANCHE HARVESTER ANTS

From mammoths of the past, we now downsize to ants of the present, specifically a species closely associated with grasslands bordered by oak forests—a dead-on characterization of the Cross Timbers and Prairies region. Comanche Harvester Ants, named in honor of a well-known tribe of Native Americans, occur in Texas, Oklahoma, and Kansas, with a little spillover into somewhat similar habitat at sites in Arkansas and Louisiana. They are one of several species in a genus distinguished by a beard—a psammophore, for those interested in such matters—whose hairs are used to clean their legs and antennae (fig. 5.11). In the case of Comanche Harvester Ants, hairs may at times also be formed into a basketlike structure, thereby enabling an ant to carry more than one seed or a grain of sand. Comanche Harvester Ants— Comanche Pogos, for short—capture our interest based on their own merits but also because they represent a major item in the diet of Texas Horned Lizards (see appendix A) as well as of birds and other insectivorous predators in local food chains.

Comanche Pogos seek deep sandy soils for their nests—notably those in the Aquilla Formation— but sometimes settle for sites where a somewhat thinner horizon of sand lies over a deeper layer of clay. Sandy topsoil 1 to 2 feet in depth may suffice

in the latter circumstances, but unlike some related species, these ants clearly avoid clay topsoils. Thus, Comanche Pogos occur in the Cross Timbers region both in the oak woodlands where deep sandy soils predominate and at local sites in the adjacent prairies where shallow sands overlie another type of strata.

Some of the more familiar species of harvester ants clear large, flat disks around the entrances to their nests. Comanche Pogos, by contrast, construct craterlike mounds about 12 inches in diameter, with each nest supporting 5,000–6,000 adults. The mound lies at the center of a cleared area up to 39 inches in diameter. The ants forage for seeds and other foods, seeking whatever they can find and carrying the items back to their nest. There the food is stored until needed, and the seed husks and other debris are subsequently returned aboveground and dumped in middens along the outer edge of the mounds. Their diet, 80–90 percent seeds, is supplemented by flowers and other bits of vegetation, small arthropods, and grasshopper frass (excrement), so the debris that collects at the middens likely enriches adjacent soil. Comanche Pogos forage individually; hence a rich source of food may not be fully exploited, whereas other species of pogos often forage in streams of workers following established trails radiating from the nest. Some individuals extend their foraging activities into nearby woodlands where they collect elm seeds, but no evidence yet suggests that these are selected to fulfill specific dietary needs. All told, Comanche Pogos represent opportunistic feeders— seed-eating generalists—seemingly limited only by what they can grasp and carry in their mandibles.

At night, Comanche Pogos seal the entrances to their nests with soil. The next morning, a few workers that remained outside overnight reopen the entrance while others likewise remove the obstruction from inside the nest. The plan, of course, is to meet halfway, but at times the diggers lose their way and, for a few moments, fail to connect.

Like many other species of ants, Comanche Pogos move large quantities of soil when constructing their nests; networks of tunnels connect chambers dedicated to specific functions, such as food storage and housing for the queen.

In some cases, ant communities transport 280–540 pounds of soil per acre to the surface, thereby influencing the structure of the prairie soils.

In addition to Texas Horned Lizards, other predators attack Comanche Pogos, among them the Bee Assassin and spiders, including Black Widow Spiders. The Black Widows, which also prey on other ants, often build their webs near the entrance to the mound, where they lie in wait to attack. Bee Assassins—true bugs—may arrive at the mounds before the ants begin their daily activities and, like the spiders, wait in ambush. They attack with hollow, daggerlike beaks that first pierce and then extract body fluids from their victims. At times, however, the tables are turned and a would-be victim manages to kill its attacker.

In general, ants exert a positive influence on their environment and, at least in some cases, qualify as "environmental engineers"—organisms whose activities fulfill their own needs and concurrently affect the occurrence, abundance, and spatial patterns of other species. Tunneling activities also increase the distribution of water, as well as oxygen and carbon dioxide, through the soil profile. In places, these activities exceed those of earthworms. Moreover, ants recycle nutrients when they harvest and relocate their foods from one part of their habitat to another (for example, from forest to prairie), and their foraging may influence the abundance of those plants whose seeds the ants collect. Ant mounds may also create essential shelter for other species, thus mirroring the better-known role played by prairie dog burrows. Not all of these interactions have yet been determined for Comanche Pogos, so their full role as environmental engineers presently remains unclear—a situation no doubt not of any concern to a hungry horned lizard.

CONSERVATION AND MANAGEMENT

Much of the Cross Timbers and Prairies has experienced some anthropogenic impacts, and few areas can be regarded as untouched. Still, the Fort Worth Nature Center and Refuge offers a special opportunity for both visitors and researchers concerned with the natural history of prairies, oak woodlands, and riparian forests of the Western Cross Timbers. The site covers 3,621

acres and includes 240 acres of pasture for a bison herd of about 27 animals, all of which represent a genetically pure strain. In 1980, the facility was recognized as a National Natural Landmark. In Arlington, the Southwestern Nature Preserve protects a small, 58-acre relict of the Eastern Cross Timbers as part of the city's park program. Similarly, the city of Southlake, also in Tarrant County, offers environmental education programs and ecology hikes at the Bob Jones Nature Center and Preserve, a 758-acre tract of well-preserved habitat representative of the Eastern Cross Timbers. The Tarrant Regional Water District manages Eagle Mountain Park, a 400-acre ecotone between the Grand Prairie and Western Cross Timbers.

Fossil Hill Research Natural Area in Montague County, while privately owned, is accessible for authorized research activities. The 137-acre site features a remnant of old-growth Western Cross Timbers registered with the Ancient Cross Timbers Consortium based at the University of Arkansas (the University of Texas and the Texas A&M Forest Service are among the other members of the group). Two state parks occur in the region. Ray Roberts Lake State Park, located primarily in Denton County, includes sites representing the Grand Prairie, Eastern Cross Timbers, and Blackland Prairie. Still in development, Palo Pinto Mountains State Park, when opened, will allow visitors opportunities to visit more than 4,000 acres of rugged terrain in the Western Cross Timbers, including sites with old-growth oaks.

Owned and managed by the USDA Forest Service, the Lyndon B. Johnson National Grassland includes elements of both oak woodlands and prairie habitat on a 20,309-acre site in Wise and Montague Counties. The area, formerly known as Cross Timbers National Grassland, was renamed in 1974 to honor the 36th president of the United States. Like other national grasslands, this site is managed to restore the integrity of the original plant and animal communities. The Chisholm Trail, famed for its nineteenth-century cattle drives, passed through the area, which now includes about 75 miles of horse and hiking trails. Fall migrants pass through the area, thus offering good opportunities for birding.

True to its name, the mission of the Crosstimbers

Connection is to connect people with the natural history of the Cross Timbers and Prairies, with a primary focus on environmental education and conservation. Headquartered in Arlington, the organization sponsors field trips to the Big Thicket (chapter 2) as well as elsewhere in Texas and publishes an online journal and newsletter, *The Post Oak and Prairie Journal.*

Efforts to restore Black-capped Vireos include habitat improvement, primarily with a regime of prescribed burning that maintains suitable nesting conditions, such as multistemmed regrowth of oaks, while currently decreasing unsuitable vegetation, in particular, juniper invasions. Indeed, the birds often frequent sites recovering from recent fires. Overgrazing, especially by goats and other browsers, removes vegetation at exactly the same heights above the ground as those matching the needs of the vireos, and of course cattle attract cowbirds. Urban development in some areas reduces the habitat available to the birds. All these factors tend to isolate populations of Black-capped Vireos, which significantly increases the risk of extirpation.

However, habitat improvements alone do not necessarily limit the incidence of cowbird brood parasitism, which remains the primary target of management activities. At Fort Hood in Bell and Coryell Counties, all or much within the Lampasas Cut Plain and harboring the largest population of Black-capped Vireos in Texas under a single management authority, Brown-headed Cowbirds parasitized almost 91 percent of the vireo nests. As a countermeasure, managers trapped and humanely destroyed cowbirds, especially females. The effort soon produced positive results: parasitism dropped to just 12.6 percent six years after the program started, and the rate of vireo nest success increased from about 5 percent to more than 51 percent during the same period. Moreover, the number of vireo territories—an indirect measure of nesting activities—more than doubled in response to intensive cowbird control. However, these results may not yield similar benefits where small, isolated vireo populations are surrounded by large areas supporting large numbers of cowbirds. Additionally, removal programs are costly and must

be continued indefinitely, or else the gains will soon disappear. Despite these limitations, cowbird control remains a viable means of protecting the vireos from further population losses throughout much of their range.

Dams pose a serious threat for Brazos River Watersnakes; the obstructions slow the rate of flow and deepen the river, thereby increasing the distance between riffles and reducing the number of riffle areas where the juveniles forage on small fishes. The deeper water also covers shoreline rocks that otherwise provide the snakes with cover. Additionally, the slower rates of flow increase sedimentation that forms riverside habitat soon infested with thick stands of Saltcedar, which progressively entrap even more sediments that bury the rocky shorelines. These changes not only eliminate vital habitat for Brazos River Watersnakes but concurrently create conditions that likely favor competing species, such as Blotched Watersnakes.

Existing dams on the Brazos River no doubt will remain in place, but proposals for additional structures should be reviewed with caution, not only in regard to their effects on wildlife but also because of increased flooding caused by invasion of Saltcedar. One study revealed Saltcedar substantially increased both the flood stage and the area inundated; indeed, the vegetation had reduced the width of the river's channel by as much as 71 percent over a 38-year period (much of the reduction actually occurred in less than half of that period). The average width of the channel narrowed from 515 to 220 feet.

Captive-breeding programs are not a common tool for restoring endangered reptile populations, but some evidence, although inconclusive, suggests that such a program might work in the case of Brazos River Watersnakes. The species apparently breeds well in captivity, and 35 young were released at the site on the Brazos River where their parents had been captured in earlier years. The young were marked prior to release, but unfortunately no effort was later made to assess their survival. In the long run, however, a captive-breeding program makes sense only if the overarching problem of habitat suitability has been corrected.

6
ROLLING PLAINS
RANGELANDS AND RED-HUED CANYONS

Nary a bush to interfere with your vision, and you can see as far as your eyes are good up there on the Ballies—rolling country as open-faced as a Waterbury watch.
— "SHINE" PHILIPS

Rolling Plains

In a series of colorful stories set in the town of Big Spring, Earl Cleveland "Shine" Philips (1889–1968) provided brief glimpses into the hardscrabble life on the Rolling Plains during the early years of the twentieth century. Periodic droughts, severe storms, fires, and floods periodically wreaked short-term havoc on the landscape, making survival tenuous for the ranchers and farmers that Philips knew. Despite the harsh conditions, the reddish soils of the region were carpeted by prairies that had remained intact for thousands of years (fig. 6.1). American Bison and Pronghorn herds once roamed unfettered across the plains, and prairie dog towns stretched beyond the horizon.

Some authorities characterize the region as the Rolling Red Plains, as reflected by the dominant soil colors—a rich palette of reddish hues varying from delicate pinks to the dark reddish-brown tones of burnt sienna. The soils, which vary slightly from neutral to basic, formed from carmine-tinted shales, mudstones, and sandstones deposited in the western part of the region during the Triassic. Deeper layers, known as the Permian Red Beds, lie exposed farther east. Three large rivers, the Colorado (which means red), Concho, and Red, originate in the breaks of the Caprock Escarpment—these and their tributaries carry colorful loads of fine red particles eastward on their downstream journey. Nonetheless, the designation Rolling Plains well reflects the undulating terrain of this picturesque landscape.

The 24 million acres of Rolling Plains in Texas lie at the southern limits of the Great Plains, a region of more than 50,000 square miles extending from south-central Kansas through western Oklahoma into Texas. Gently rolling to moderately jagged landscapes of the Rolling Plains are bordered on the west by the Caprock Escarpment of the Llano Estacado (chapter 8) and to the south by the Edwards Plateau (chapter 7). The Cross Timbers and Prairies and the fragmented plateaus of the

Lampasas Cut Plain (chapter 5) define the eastern border. The elevation across the region varies from 800 to 3,000 feet, and many narrow valleys and sharp-edged canyons carry intermittent streams eastward and southeastward as they dissect the rolling topography. Average annual precipitation increases from 22 inches in the west to almost 30 inches at the eastern margin, but rainfall is unevenly distributed during the year. Typically, rainfall peaks in May and September—months that bracket a predictably dry summer marked by high temperatures and evaporation rates.

After ranchers and farmers "tamed" the land, fires were curtailed, thus removing an important ecological force, and large expanses of native prairie were either tilled or overgrazed. Wildlife populations slowly disappeared, decimated by both guns and the altered environment. Today, only scattered remnants of the original prairie are intact, and these are dominated by Sideoats Grama, Little Bluestem, and Blue Grama; most of the region has been widely invaded by "increaser" grasses as well as mesquite, Shinnery Oak, Tree Cholla, prickly pears, Redberry Juniper, and other brush species. Despite the disappearance of the original vegetation, the region retains its rugged charm and enchanting colors.

DISTINCTIVE SUBREGIONS

Many authorities consider the Canadian River Breaks (infobox 8.1) a component of the Rolling Plains. Cutting across the Texas Panhandle to separate the Canadian-Cimarron High Plains from the Llano Estacado, the broad Canadian River Valley shares many ecological features with the Rolling Plains to the southeast. Based on differences in soil, topography, and vegetation, four other subregions of the Rolling Plains can be distinguished: Caprock Canyons and Badlands, Fragmented Tablelands, Rolling Mesquite Plains, and Gray Limestone Plains.

CAPROCK CANYONS AND BADLANDS

Falling away in places as much as 1,100 feet to the prairie and canyons below, the escarpment of the Llano Estacado forms the western perimeter of the Rolling Plains. Redberry Juniper and a few disjunct stands of Rocky Mountain Juniper, a

(*overleaf*) Figure 6.1. Note the reddish soil typical of the Rolling Plains in this view of the Caprock Escarpment in Garza County.

Figure 6.2. After spending weeks crossing seemingly endless grasslands, the Spanish explorer Coronado welcomed the sight of trees below the rim of Palo Duro Canyon.

species native to the Rockies, establish footholds on the caliche rimrock and share escarpment slopes with Mountain Mahogany and small thickets of Skunkbush Sumac and Wild Plum. After struggling across a treeless plain for weeks, the Spanish explorer Francisco Vásquez de Coronado became the first European to see the deep canyonlands etched into the eastern edges of the Llano Estacado (chapter 8). Standing at the precipice of a huge abyss and apparently impressed with—at long last—the appearance of trees, Coronado is credited with naming the canyon *Palo Duro,* Spanish for "hard wood" (fig. 6.2).

The steep slopes of the Caprock Escarpment form the backstop for a network of canyons, isolated buttes, and rugged badlands that extend far to the southeast where tributaries forming the headwaters of three rivers—the Red, Brazos, and Colorado—steadily carved the second-longest canyon system in Texas (the canyon system along the Rio Grande is the longest). Ragged valleys and deep gorges remain separated by flat tablelands that resisted the erosional forces sculpting the adjacent landscape.

The best-known landmark in the system is Palo Duro Canyon, located in Armstrong and Randall Counties about 12 miles east of Canyon. The Prairie Dog Town Fork of the Red River, which eroded the canyon, is formed by the confluence of two smaller creeks, Tierra Blanca Creek and Palo Duro Creek, which originate on the Llano Estacado. Freshwater springs nourish the Prairie Dog Town Fork and other streams that flow off the Llano and cut similar canyons, such as Tule Canyon and Los Lingos Canyon.

A hike from the rim of the Caprock Escarpment to the streambed meandering through Palo Duro Canyon provides an instructive view of four geologic layers that together span 250 million years. The Ogallala Formation, a caliche layer known as the "caprock," forms the rim of the canyon walls (fig. 6.3). Often 20–40 feet thick, this formation in other locations may reach thicknesses of nearly 700 feet. Where chunks of caliche have broken off and tumbled downward, the Palo Duro Mouse, a subspecies of the Piñon Mouse endemic to these canyonlands, makes its home in the rocky debris below

Beneath the Ogallala is a thick layer of red to light yellow sandstone and mudstone—the Trujillo Formation—composed of "layered boulders" deposited about 181 mya. Farther below, bright

Figure 6.3.
This view of
the canyon rim
indicates the
thickness of
the Ogallala
Formation widely
known as the
"caprock."

bands of lavender, gold, maroon, orange, red, gray, and brown adorn the canyon walls. This colorful zone, the Tecovas Formation, is known locally as the "Spanish Skirts" for its obvious resemblance to the vibrant hues in the ceremonial dresses worn during Mexican celebrations (fig. 6.4). Layers of the Tecovas Formation contain fossils of phytosaurs, amphibians, and fishes—a fauna representing streams and swamps of the Triassic Period. At the bottom of the canyon lies the Quartermaster Formation, Permian rocks of the same age as the strata forming the rim of the Grand Canyon. Along the slopes framing the canyon floor, bands of satin spar gypsum and gray claystone lace the brick-red marine deposits of this formation. In addition to providing a chronicle of geologic history, the steep, multihued canyon walls offer a panorama peppered by splashes of green vegetation. The colors become especially vivid on those rare occasions when rainfall moistens the rocky layers of history.

More than 12,000 years ago, clans representing the Clovis and Folsom cultures became the first humans to venture into the Rolling Plains, where the canyons offered shelter from harsh winter winds. Numerous other cultures and tribes, including the Kiowa, Apache, and Comanche, left numerous pictographs and artifacts that still survive. In an all-out attempt to remove native tribes from the southern plains, the US Army engaged in what historians identify as the "Red River War of 1874." In a decisive action called the "Battle of Palo Duro Canyon," the Fourth Cavalry, led by Colonel Ranald S. Mackenzie (1840–1889), routed Native Americans camped in the upper reaches of the canyon at dawn on September 28, 1874 (fig. 6.5). The attack killed few Native Americans, but the loss of winter supplies and especially their large herd of ponies ended any further hostilities against white settlers on the Southern Plains.

Soon after the battle, Charles Goodnight (1836–1929; infobox 6.1), aware that Palo Duro's steep sandstone walls provided natural fencing for livestock, moved his herd of cattle into the canyon. His ranch, the first in the Texas Panhandle, became the harbinger of things to come—by the 1890s, ranches dotted canyonlands throughout the Rolling Plains. Few ranches later changed ownership, and

Figure 6.4. Colorful multiterraced formations in parts of Palo Duro Canyon suggest the swirling "Spanish Skirts" worn by Mexican dancers. The hues become even more vivid after a rain.

Figure 6.5. In 1864, the Fourth Calvary routed a Native American encampment near this location on the Prairie Dog Town Fork of the Red River in Palo Duro Canyon.

INFOBOX 6.1. CHARLES GOODNIGHT
"Father of the Texas Panhandle" (1836–1929)

If life experiences forge the nature of a man, then a lifetime filled with adventure and innovation surely distinguishes Charles Goodnight. All who knew him agreed he was a good and decent man who lacked any concept of his own importance. His pioneering spirit changed the Texas Panhandle and the Texas cattle industry, but more important, Goodnight's appreciation for the native plains bison preserved the iconic species for future generations.

Just five years after Goodnight's birth in Macoupin County, Illinois, his father passed away. After his mother remarried, the family moved in 1846 to the Brazos River bottomlands in Milan County, Texas. There, an old man named Caddo Jack taught young Charles to hunt and track. At 11, and after only six months of formal education, Goodnight left school without knowing how to read or write. He held a variety of jobs for the next two decades, gaining valuable experiences as a jockey, farmhand, supervisor of slave crews, ox-wagon freighter, cowboy, Texas Ranger, and Civil War soldier. These occupations prepared him for the future and introduced him to a variety of people and personalities. During this period, Goodnight met Oliver Loving (1812–1867), who became both a friend and business partner. Together, they rounded up cattle abandoned after the Civil War and drove the herd to market along a trail they blazed from Belknap, Texas, to Fort Sumner, New Mexico. Goodnight developed the first "chuckwagon" on subsequent drives along this route. Later, the Goodnight-Loving Trail became one of the most heavily used cattle trails in the American Southwest. Following Loving's death, Goodnight herded cattle for John S. Chisum (1824–1884) and extended the trail north from New Mexico to Colorado and Wyoming, all the while splitting his profits with the Loving family.

In 1876, while on a trip to Denver, Goodnight formed a partnership with financier John G. Adair (1823–1885) and established the first ranch in the Texas Panhandle. Known as the JA Ranch, the spread initially operated in Palo Duro Canyon where water and grass were abundant and steep canyon walls made fencing unnecessary. For the next 11 years, Goodnight improved the herd with new breeds of cattle and expanded the ranch to eventually encompass

1,325,000 acres stocked with more than 101,000 cattle. Where water was unavailable, Goodnight pioneered the development of artificial watering facilities.

When he moved his cattle into Palo Duro, Goodnight and his cowboys drove the resident American Bison herd down the canyon about 15 miles. In the process, cowboys delivered abandoned bison calves to the tender care of Charles's wife, Mary Ann (Molly) Goodnight (1839–1926), thus preserving animals with the distinctive genetic profile of the greatly decimated southern herd of American Bison. The JA Ranch later donated descendants of these captives as stock for the Texas State Bison Herd (appendix A), now located in Caprock Canyons State Park. In 1902, bison from the same herd were released in Yellowstone National Park; several zoos and ranches throughout the world also have bison herds whose lineages extend back to the stray calves in Palo Duro Canyon.

Because of a near-fatal illness, Goodnight sold his interest in the JA Ranch in 1888 and bought a small ranch in Armstrong County where he relocated his bison herd and built a spacious home. With Molly, he founded Goodnight College in the Rolling Plains town named in his honor. Having no children of their own, the Goodnights boarded college students and reared the son of their housekeeper. After Molly's death, Charles—at 91—married his 26-year-old nurse. Two years later, and still in ill health, Goodnight moved to Phoenix, Arizona, shortly before his death.

Goodnight's home, recently renovated and opened to the public, is located about 40 miles east of Amarillo. Listed in the National Register of Historic Places, the home forms the centerpiece of the Charles Goodnight Historical Center. He bequeathed his papers—rich with his accomplishments—to the Panhandle-Plains Historical Museum on the campus of West Texas A&M University in Canyon, where his friends also contributed artifacts from the JA Ranch. Along the Charles Goodnight Memorial Trail—the highway between Canyon and Palo Duro Canyon State Park—two state historical markers commemorate his life and the JA Ranch. Revered as an innovative cowman and a frontier legend, Goodnight was one of the first five inductees of the National Cowboy Hall of Fame. He is buried in a family cemetery in the little Texas town that bears his name, but in a larger sense, his legacy also lives on in a herd of shaggy American Bison—living reminders of a man who cared.

the founding families protected most pictographs and unique landforms. The "Lighthouse" in Palo Duro Canyon, the sandstone "Mitten" in Caprock Canyons State Park, and numerous colorful but spooky hoodoos, sometimes called "fairy chimneys," are appreciated today because of thoughtful land stewardship.

Severe storms often generate flash floods that roar without warning through the canyons (fig. 6.6). Natural riparian vegetation, including Plains Cottonwood, Black Willow, and Netleaf Hackberry bordering streams deep in the canyons, usually withstands the onslaught of floodwaters. However, some torrents denude the stream banks, enabling invasive species such as Saltcedar and Giant Reed to rapidly establish impenetrable thickets. Canyon flora also includes species common to other areas of the Rolling Plains, namely mesquites, prickly pears, Tree Cholla, and Fourwing Saltbush.

Salt springs and saline seeps welling up from underlying Permian strata enhance the surface flow of some rivers and creeks. The area's high evaporation rate produces salt crusts on the alluvium near these upwellings. Two endangered minnows, the Arkansas River Shiner and the Plains Minnow, inhabit salty, spring-fed tributaries of the Canadian River in the northwestern Rolling Plains. Both species depend on foods made available by disturbances to bottom sediments caused by the river's steady currents, but construction of Lake Meredith significantly reduced the river's flow and contributed to significant reductions in their respective populations.

Extensive zones of highly eroded, dry terrain—badlands—occur widely along the escarpment and river breaks of the Rolling Plains. An especially large, colorful area of exposed Triassic shale and mudstone badlands surrounds the North Fork of the Double Mountain Forks of the Brazos River north of Post. Badlands are characterized by perplexing, mazelike landscapes of low, steep-sided gullies separated by narrow-topped prominences. Scoured by wind and water, badlands continually erode, exposing fossils (infobox 6.2) and contributing heavy sediment loads to nearby waterways (fig. 6.7). Bones and teeth of plesiosaurs, fearsome aquatic reptiles that lived in shallow Triassic waters, collect in gullies and serve as

Figure 6.6. A severe thunderstorm over the headwaters of the Double Mountain Fork of the Brazos River (top) sent floodwaters racing downstream, changing a trickle (middle) into a torrent (bottom) in less than an hour.

INFOBOX 6.2. *SEYMOURIA*
A "Missing Link"?

In 1882, Charles H. Sternberg (1850–1943), an amateur but highly experienced fossil hunter, explored the Permian Red Beds on the Craddock Ranch in Baylor County near Seymour, Texas. During his searches, he discovered a cave well stocked with rattlesnakes and wisely looked elsewhere for fossils, soon stumbling upon a depression he described as "iridescent." The concavity contained numerous vertebrate fossils, which Sternberg collected, crated, and sent to the Museum of Comparative Zoology at Harvard University. These remained stored—apparently unopened—at Harvard until 1939, when Dr. T. E. White discovered and uncrated the trove of bones, but no record has survived of his analysis, if any, of the long-neglected fossils.

At about the same time that he sent the crates to Harvard, Sternberg described the fossils and their location in a letter to the curator of the Museum für Naturkunde in Berlin. Intrigued, the German curator, F. Broli, traveled to the United States to see the specimens firsthand, as well as to visit the location in Texas where Sternberg discovered the fossils. Realizing the unusual nature of at least one of the fossils in the bed, Broli collected additional specimens that he took with him on his return to Germany. There, in 1904 he described and identified what he believed was a new species of reptile—*Seymouria baylorensis*—named in honor of the town and county where the fossils were collected.

Broli's work indicated that *Seymouria baylorensis* was a small, semiterrestrial, swamp-dwelling animal no more than 32 inches in length. Although short and stubby legged, the awkward creature apparently moved slowly, undulating its backbone from side to side in a fishlike wiggling motion. The primitive animal, which lived about 280 million years ago, had a large skull that included well-developed jaws but enclosed a small brain. As adults, *Seymouria* may have been able to live far from water, but they most likely roamed through moist habitats in search of prey such as sedentary insects and small vertebrates, or the flesh of carrion. To date, no larval fossils of the species have been discovered, but those of closely related forms show external gills just as they occur in the larvae of modern amphibians, thus suggesting that *Seymouria* likewise relied on aquatic habitat for breeding.

Originally described as a primitive reptile, *Seymouria* actually displayed relatively few reptilian features and many more of those characteristic of amphibians. In response, the popular media of the day hailed the species as the evolutionary "missing link" between reptiles and amphibians. Indeed, the notion persisted in large measure because recognized authorities at times classified the curious creature in one or the other of these two groups of terrestrial vertebrates that define the Permian fauna. However, after the discovery of additional and somewhat older specimens, *Seymouria* was no longer touted as a "missing link." Today, bolstered by analyses of specimens obtained from other locations, paleontologists realize that *Seymouria* represents a highly evolved amphibian whose development, while advanced, does not indicate it was a direct ancestor of modern reptiles.

The lesson here is that so-called missing links are often creations of those who, for their own purposes, attempt to sensationalize scientific discoveries. Scientists usually avoid such dramatic terminology, especially in reference to fossils. In fact, paleontologists well know that evolutionary pathways are fraught with branches and dead ends (such as *Seymouria*) and rarely follow straight lines. The excitement thus lies not in heralding "missing links" but in expanding the library of fossil discoveries and fitting each stony remnant into a better understanding of how the biological world developed in the course of its long history.

---

reminders of an ancient subtropical sea once drowning the plains.

FRAGMENTED TABLELANDS

Surrounded by dissected topography, elevated flatlands of various sizes seem like boats floating adrift in a choppy sea. Composed primarily of sandy or silty loams lying atop Triassic or Permian Red Beds, the Fragmented Tablelands today grow cotton, sorghum, and wheat. Many farmers in the region, however, have replaced crops with grasses in voluntary compliance with the Conservation Reserve Program (CRP). The CRP provides farmers with monetary incentives to establish grassland habitat for wildlife on highly erodible soil. Unfortunately, the CRP in Texas requires seeding with Weeping Lovegrass, a fast-growing, nonnative species. Although drought resistant—a beneficial

**Figure 6.7.
Erosion in the
western Rolling
Plains exposes the
reddish deposits
of the Triassic
Era that contain
embedded
fossils—mainly
teeth, bone
fragments, and
coprolites—of
ancient aquatic
phytosaurs.**

characteristic for plants on the Rolling Plains—Weeping Lovegrass is an invasive species that crowds out native plants.

Scattered patches of native prairie habitat remain in areas inaccessible to cattle. These remnants are composed of Sideoats Grama, Little Bluestem, and other midgrasses characteristic of the southern Great Plains. Overgrazing usually eliminates these species, and they are usually replaced by shortgrasses such as Buffalograss, Purple Threeawn, and Blue Grama, which are better able to cope with aridity. Intensely overgrazed areas give way to invasions of mesquites and prickly pears and become more like savannas than prairies. The taller vegetation offers perches and cover for Western Meadowlarks, Cassin's Sparrows, Grasshopper Sparrows, and Dickcissels. Scissor-tailed Flycatchers nest in the tallest Honey Mesquite trees, beneath which Greater Roadrunners search for insects, lizards, and snakes.

MESQUITE PLAINS

To the east of the Fragmented Tablelands (fig. 6.8), the land flattens into a contiguous landscape of gently rolling plains. Herds of American Bison once roamed across this region, deftly avoiding the hazards of burrows dug by countless thousands of Black-tailed Prairie Dogs. Today, only a few scattered prairie dog towns remain—most were eliminated by ranchers who viewed the rodents as competitors for livestock forage. As the prairie dog towns disappeared, populations of many associated species also declined. Although Ferruginous Hawks and Swainson's Hawks, which depended on prairie dogs as a major food source, and Mountain Plovers and Texas Horned Lizards, which favored the closely cropped grassland habitat, are no longer as abundant as they once were, other species remain unaffected by the disappearance of prairie dogs. Collared Peccaries still root in arroyos and Coyote packs persistently serenade the moon—even on purple, moonless nights.

The subregion is named for the now abundant Honey Mesquite—a thorny shrub that sometimes becomes tall enough to be called a "tree." The distributional boundaries for the species have remained unchanged for centuries, but its density has increased significantly. Mesquites reproduce from seeds enclosed in sugar-rich pods (fig. 6.9). When mature, the pods fall to the ground, where

Figure 6.8. The gentle topography characterizing much of the Rolling Plains contrasts with this view of Fragmented Tablelands to the west.

they are often consumed by foraging animals that distribute the seeds in their excreta. Mesquite pods are especially favored as food during periods of drought or where overgrazing has diminished other forage, and expansion of the species apparently increases significantly in these situations.

Tributaries of the Concho, Brazos, and Colorado Rivers carve winding ravines as they flow eastward across the dusty Mesquite Plains. Oddly complex names—mouthfuls such as the North Fork of the Double Mountain Fork of the Brazos River or the Prairie Dog Town Fork of the Red River—likely reflect the difficulties early mapmakers experienced when tracing the intricate networks of wet-weather streams to a point where the flow justified naming a "river." Whereas many tributaries in the western part of the Mesquite Plains seem dry most of the year, water is usually percolating downstream less than 3.2 feet beneath the surface of a parched riverbed.

Rainfall is slightly greater on the Mesquite Plains than on the High Plains to the west. Where undisturbed sites still exist, the dominant grasses include Little Bluestem, Texas Wintergrass, White Tridens, and Sideoats Grama. With grazing or extended droughts, these are replaced by Curly Mesquite, Buffalograss, Hairy Tridens,

Figure 6.9. The sugar-rich pods of Honey Mesquite provide forage for many animals, especially during droughts, and the seeds in their droppings help spread the trees elsewhere.

and Tobosagrass, along with mesquites, Texas Prickly Pear, Tree Cholla, Tasajillo, Redberry Juniper, and numerous unpalatable forbs such as Broomweed and Purple Woolly Loco. On sandy soils, Broomweed and Purple Woolly Loco produce toxic compounds that poison livestock. Tobosagrass dominates the understory in some areas and reduces the quality of habitat for Northern Bobwhites and other ground-dwelling animals.

In dry canyons flanking intermittent streams, trees and shrubs struggle to survive, but where permanent water prevails, Pecan, American Elm, Black Willow, and Little Walnut dominate the riparian community. Deep sand banks along the upper Colorado River provide the only remaining habitat for the Texas Poppy-Mallow, an endangered species. Unlike most mallows, which typically hug the ground, this species grows to heights of more than 30 inches and produces striking deep red to merlot-colored flowers. Once the flowers are pollinated by bees, they never open again—an adaptation ensuring that pollen will be retained, thus guaranteeing seed production.

The Pease River and its tributaries dissect the clays and sandy soils of the northeastern Mesquite Plains. Toward the western edge of the Cross Timbers (chapter 5), the topography becomes irregular and densely blanketed by shrubby clumps of Honey Mesquite. Prior to human settlement, this part of the Mesquite Plains represented a transition zone between tallgrass and shortgrass prairies.

GRAY LIMESTONE PLAINS

North of the Callahan Divide (chapter 7), layers of gray or tan Permian-age limestone overlie Permian Red Beds. These thin, rocky soils are embedded with fossilized shells of brachiopods and clams. The topography is dissected and rough in those areas where erosion has carved and broken the mineral layers.

A prairie community of mixed grasses persists in low areas where the soil is thicker and the land has not been overgrazed. Widely scattered bushes of Honey Mesquite grow among the grasses, which consist primarily of Little Bluestem, Indiangrass, and Buffalograss. Near limestone outcrops and eroded areas, the soils become much thinner and hold less moisture after rains. Tree Cholla and

Mormon Tea serve as indicators of the rocky soils in such areas. Where the Limestone Plains approach the Edwards Plateau, scattered Ashe Juniper and Plateau Live Oak grow amid abundant Honey Mesquite.

ROLLING PLAINS WILDLIFE

NORTHERN BOBWHITE

Named for its distinctive two-note call, a whistled *bob-WHITE,* the Northern Bobwhite is one of the best-known and most well-studied birds in Texas. Bobwhites are especially abundant on the Rolling Plains where clumps of Lotebush, mesquites, and prickly pears provide loafing and nesting cover, protection from predators, and refuge from the region's extreme heat.

During most of the year, bobwhites associate in social groups (coveys) consisting of 10–15 birds. Coveys move as a group when feeding. A large part of the bobwhite winter diet (95 percent) consists of dry seeds, especially those of Texas Doveweed, Ragweed, and panic grasses, but tender greens, such as newly sprouted seeds or new growth on perennials, are consumed in the spring. Vitamin A, contained in green vegetation, was once considered essential for reproduction, but now its importance is less certain. Insects are eaten throughout the year but become more important in summer.

Pairs form after coveys break up at the onset of the breeding season (fig. 6.10). Each pair selects a nest site and jointly builds a ground nest of dead vegetation in protective cover, typically beneath grasses or shrubs more than 18 inches tall. In overgrazed areas, bobwhites often place their nests beneath clumps of mesquites, prickly pears, Lotebush, or Tree Cholla. After the females lay a clutch of 12–15 eggs, the mates share incubation duties until the eggs hatch about 23 days later. The eggs hatch synchronously—within a single day—and, except when following their parents in search of food, the downy young remain closely brooded for about two weeks.

Populations of Northern Bobwhites in northern Texas have historically experienced years of boom or bust. Lower than expected numbers usually result from adverse weather conditions, especially drought, during the breeding season; conversely, booms generally coincide with years of abundant

Figure 6.10. Black facial markings distinguish male Northern Bobwhites from females. Photograph by Steve Maslowski and courtesy of the US Fish and Wildlife Service.

rainfall. In the spring of 2010, higher than average rainfall greened the prairie and increased both habitat quality and food supplies, leading biologists to expect a bumper crop of quail on the Rolling Plains. For a short period, bobwhite populations indeed swelled, but a dramatic crash followed. Apparently the same spring rains that promised a booming quail population also produced an explosion of insects, especially crickets. Bobwhites became infected with Parasitic Eyeworms when they ate the crickets, an intermediate host of eyeworms. Although these parasites are tiny, they cause inflammation and swelling, which impair a bird's vision and lessen its ability to locate food and detect predators. Parasitic Eyeworms likely contributed to the general decline currently experienced by bobwhite populations, and regrettably, these parasites may also infect Lesser Prairie-Chickens, producing similar results.

An unlikely predator, the Feral Pig, entered the Rolling Plains several decades ago and quickly adapted to the dry, mixed-grass brushland. Some evidence suggests that these opportunistic omnivores accounted for more than 33 percent of the depredation on simulated quail nests placed under mesquites and Redberry Juniper trees. Hence, Feral Pigs may create additional concerns for the future of bobwhites on the Rolling Plains.

Northern Bobwhites represent an important source of income from hunting leases for ranch owners on the Rolling Plains. Nearly 100,000 Texans hunt bobwhites each fall and winter, so most landowners manage their property to provide the best possible habitat conditions for the birds. Concurrently, biologists seek better methods for managing Tobosagrass and brushy vegetation in ways that produce optimal conditions for both livestock and Northern Bobwhites. Unfortunately, overgrazing, modern clean farming methods, and the conversion of ranchland to small-acreage ranchettes seem to be increasing on the Rolling Plains, further fueling concerns for the future of bobwhites—one of the state's most aesthetically appealing and economically valuable species.

TEXAS BROWN TARANTULA

When Captain Randolph B. Marcy (infobox 8.2) traversed the Rolling Plains to locate the source of the Red River, members of his exploratory party collected and preserved numerous specimens of the region's fauna. To the party's amazement, one of their discoveries was the largest spider they had ever seen—the Texas Brown Tarantula—the first species of tarantula described in North America.

The leg span of the males can exceed 4 inches—females are slightly larger. Their dark chocolate-brown body, black abdomen, and legs are covered with short, fuzzy hairs (fig. 6.11). When disturbed, Texas Brown Tarantulas face potential attackers head on while raising their heads and forelegs in an intimidating manner—repeated flexing of their forelegs enhances the menace. If this posture fails to dissuade, tarantulas lunge forward, kicking coarse, barbed hairs from their abdomens toward the enemy. The hairs only mildly irritate the eyes, mouth, and nasal tissues of mammalian predators but usually discourage further harassment.

Most Texas Brown Tarantulas dig and reside in vertical burrows. Some, however, adopt burrows abandoned by other animals or find protective cover in natural cavities under rocks, dried prickly pear pads, or mesquite logs. They line their hideaways with silk webs but extend a few lines of silk outward that serve as trip wires to detect passing prey. Tarantulas remain deep in their burrows during the day and then move to the entrance at night to wait for a potential meal. Occasionally, Texas Brown Tarantulas completely

Figure 6.11. Mysteriously, Texas Brown Tarantulas migrate en masse during the summer. Photograph by Chris Jackson, DFW Urban Wildlife (www.dfwurbanwildlife.com), and courtesy of the US Fish and Wildlife Service.

emerge from their burrows and wander up to 8 feet away in search of large insects and caterpillars—their typical prey. With a target at hand, the spider leaps forward—head elevated—and inserts a pair of fang-like mouthparts into its quarry. Venom injected by the fangs immobilizes the victim, which the spider carries back to the burrow.

The venom of Texas Brown Tarantulas does not affect mammalian predators, and Striped Skunks and Raccoons eat the spiders with full immunity. Their most effective predator, however, is the Tarantula Hawk—a large spider-hunting wasp. The two most common Tarantula Hawks in Texas reach lengths of up to 2 inches; both are easily recognized by their striking color pattern—a glossy blue-black body and bright rust-orange wings. Because they can deliver an intensely painful sting—one that is considered the second most painful insect sting in the world—the wasps' colorful appearance likely represents an example of aposematism—a visual warning to stay away.

Adult Tarantula Hawks feed on nectar, typically visiting flowers during the cooler periods of a summer day. When hunting, female Tarantula Hawks occasionally drop to the ground to inspect locations where a spider might loiter. They locate tarantulas by smell and when one is detected, the wasps expel the spider from its lair before deftly stinging its side or abdomen. Within seconds, the spider becomes paralyzed—alive but incapable of movement. Taking advantage of the opportunity, the wasps often drink fluids oozing from the fresh wound. The wasps then

drag their victim back to a burrow, lay an egg on the tarantula's abdomen, and seal the entrance. After hatching, the tiny grub ingests the spider's fluids. After further development, the larva tears open the spider's abdomen and feeds voraciously, selectively avoiding vital organs into order to keep the paralyzed spider alive for as long as possible. In time, the larva transforms into a pupa, from which an adult later emerges that digs out of the sealed chamber in search of nectar and, for females, eight-legged prey.

An amazing (and rather intimidating) event occurs somewhat unpredictably during the midsummer months on the Rolling Plains. Presumably cued by weather conditions, which remain unknown to biologists, hundreds of male Texas Brown Tarantulas simultaneously leave their burrows and migrate en masse across the plains. The spiders all appear to move in the same direction and resist any effort to redirect their path. The mission underlying these movements remains a mystery, but many assume the males are searching for mates. If so, such movements likely reduce inbreeding, thus ensuring genetic mixing and population viability. Despite the potential benefits to the spiders, even seasoned naturalists become apprehensive when afoot in a landscape alive with large, hairy spiders.

INTERIOR LEAST TERN

Those small white birds hovering and diving headlong into shallow headwaters of the rivers flowing across the Rolling Plains are Interior Least Terns. They are a subspecies of the smallest of the North American terns, measuring only 10 inches in length and 20 inches from wingtip to wingtip.

After spending the winter months on the coasts of Central and northern South America, Interior Least Terns typically arrive on the Rolling Plains in early April. Soon after arriving, the birds begin a two- to three-week period of courtship activities. Pair formation is sealed when a female accepts a fish offered by a courting male. The pair remain together for the balance of the breeding season, sharing duties associated with nest construction, incubation, and nest defense.

Nests, nothing more than shallow scrapes in barren sand or gravel, are located on midriver

Figure 6.12. Interior Least Terns lay their eggs in shallow scrape nests on river sandbars. Photograph by Chris Jackson, DFW Urban Wildlife (www.dfwurbanwildlife.com), and courtesy of the US Fish and Wildlife Service.

sandbars or along barren shorelines (fig. 6.12). Colonies usually consist of 5 to 20 nests situated on the highest elevations available, away from the river's edge and near areas of shallow water teeming with small fish. Colony sites change year to year in response to the unending shifts of river channels and sandbars.

All members of a colony participate in its defense, which involves vigorous diving attacks on intruders and spirited vocalizations. Lying in shallow depressions, the cryptically colored eggs, which are pale tan to olive buff with dark streaks and spots, can be difficult for a predator to find. Small stones, twigs, chunks of wood, and other flotsam near the nest add visual clutter that further improves nest concealment. Although capable of flight three weeks after hatching, the young continue to be fed by their parents until all begin their migratory journey to the south in August.

Nesting success varies greatly from year to year. Untimely storms occasionally flood the colonies, wiping out nests, eggs, and flightless chicks in mere moments, but droughts take a different toll. Falling water levels during the nesting season can reduce the food supply and allow predators access to formerly isolated sandbars. Nesting habits of the Interior Least Tern evolved in concert with the normal summer regime of river flow, but discharges from reservoirs add an unpredictable element to nesting success—releases of impounded water may flood colonies during the nesting season. Where historical flood cycles once scoured vegetation from riverbanks and created sandbars, impoundments now allow encroaching vegetation

to choke the riverbed downstream, while sediments trapped upstream behind dams starve sandbars, an essential nesting habitat. Water pollution and human disturbances, including camping and fishing in nesting areas, also pose significant threats.

As a consequence of widespread population declines and altered riverine systems, the Interior Least Tern was listed as an endangered species in 1985. Since then, state and federal agencies have censused the birds annually, dissuaded human disturbances at nesting locations, and developed breeding habitat. Riprap structures designed and placed to divert river flows in ways that create sandbars suitable for nesting colonies represent one example of efforts to improve the future of this lively small tern.

MOUNTAIN BOOMERS

How the Eastern Collared Lizard earned the nickname "Mountain Boomer" remains a mystery, but it was not because it makes a deep-throated noise—it produces no vocal sounds at all. Some suggest settlers may have heard a frog call or a gust of wind rush through a canyon just when they first observed a brightly colored lizard sunning on a rock. The name "Collared Lizard," however, is appropriate—the lizard's distinctive coloration includes bands of black that resemble a collar encircling the neck and shoulders.

After spending the winter deep in burrows under large rocks, Mountain Boomers emerge in April and remain active until late September in the canyons and mesas of the Rolling Plains. The lizards sun themselves atop large rocks in the early morning before foraging on insects, spiders, small mice, and other tiny creatures found in rocky ledges and sparse vegetation. When threatened, Eastern Collared Lizards rely on speed to escape— some have been clocked at 16 miles per hour. Like many lizards, Mountain Boomers often run on their hind legs with their long tail acting as a counterbalance—and perhaps as a steering mechanism.

Eastern Collared Lizards reach up to 14 inches in length, and although they are large, they would probably go unnoticed if it were not for their distinctive coloration. Adult males are typically green with white spots distributed across the back; females are tan. During the breeding season, a male's head becomes yellowish, and after mating, a gravid female develops bright orange spots on her neck and sides—a signal announcing her condition and thwarting further advances from males. Sexual dichromatism in Mountain Boomers represents an example of sexual selection that evolved when females selected males with vivid colors and markings. The genes of those males falling short in the eyes of females simply pass out of existence, with each succeeding generation reinforcing the power of the selection process. Females, on the other hand, benefit from cryptic coloration that improves their ability to blend into the background, with predators eliminating those that fail the test of natural selection.

Both males and females engage in territorial defenses against intruders of the same sex, but conspicuous coloration and use of high, open perches exposes males to a greater threat of predation from hawks, Loggerhead Shrikes, Greater Roadrunners, and a variety of snakes. Where predation becomes excessive, males maintain constant vigilance and rely on their extraordinary speed and agility to escape.

BLACK-TAILED JACKRABBIT

Common names of plants and animals can often be misleading, and such is the case for the Black-tailed Jackrabbit. To be sure, its tail is black, but the animal is not a rabbit—it is a hare. The distinction is significant because hares are larger than rabbits, and most have much longer ears and hind legs. Names aside, Black-tailed Jackrabbits are enigmatic denizens of open grasslands and desert scrub habitats of the Rolling Plains and ecologically similar regions of western North America.

Despite their large size, Black-tailed Jackrabbits are easily overlooked in grassy habitats. They begin foraging at twilight and stay active all night, relying on their acute visual and aural senses to avoid predators. With eyes on the sides of their heads, jackrabbits gain a wide panoramic view of their surroundings, and their large ears constantly twitch as they scan for sounds of danger. During the day, jackrabbits find shallow, shady depressions under a bush or tree where they crouch motionless, ears

lowered. Their buffy pelage, densely sprinkled with small black streaks, blends perfectly with their surroundings. If disturbed, however, Black-tailed Jackrabbits race away at speeds of up to 30 miles per hour and avoid predators with quick zigzag turns and jumps of 20 feet or more.

Summer diets include prickly pears and a wide variety of green grasses and forbs, including agricultural crops. In winter, jackrabbits subsist on dried vegetation and the leaves of woody plants. During the digestive process, jackrabbits absorb enough water to meet their needs without drinking. Ingested foods collect in a chamber between the small and large intestines where symbiotic bacteria assist in the digestion of cellulose and production of B complex vitamins. Black-tailed Jackrabbits produce two types of droppings. One, a soft pellet that contains vitamins and proteins, remains intact after the initial digestive passage. These pellets are consumed—a form of coprophagy more precisely termed "reingestion" because it differs from eating a true fecal pellet. After passing through the digestive tract a second time, soft pellets remain intact in a special part of the stomach where, in a double-digestion process similar to rumination by cattle, bacteria again digest cellulose and extract nutrients. When this process is completed, a dry fecal pellet is excreted—and abandoned.

Females produce as many as six litters, each containing one to six young, during the year-long breeding season. Although a legion of predators, including hawks and owls as well as Golden Eagles, Coyotes, foxes, Bobcats, and American Badgers, exert constant pressure on Black-tailed Jackrabbit numbers, their fecundity rate often results in dense populations. Approximately 128 Black-tailed Jackrabbits can devour as much range vegetation in a year as seven sheep or one cow. In some areas of the Rolling Plains, densities reach 150 jackrabbits per 0.4 square mile. Such abundance often occurs where livestock have already overgrazed the vegetation, thereby creating a sparse rangeland favored by Black-tailed Jackrabbits. Somewhat incongruously, an overabundance of jackrabbits becomes a warning signal to ranchers of a badly overstocked range.

Figure 6.13.
Overgrazing
often promotes
invasions of
prickly pear.

HIGHLIGHTS

PRICKLY PLANTS AND PACKRATS

Captain Marcy's 1854 description of the Rolling Plains indicated that much of the area consisted of grasslands scattered with large mesquites. To some extent, periodic range fires likely kept the prickly plants in check before settlers arrived with their herds of cattle, sheep, and goats. As uncontrolled grazing degraded the original grassland communities, the vegetation that fueled these wildfires also diminished, resulting in fewer and far less intensive prairie fires as well as further expansion of mesquites, prickly pears, Tree Cholla, and Tasajillo. Besides bearing sharp spines that impede browsing, mesquites and cacti possess several other attributes that enhance their ability to spread.

Prickly pears grow into dense, jumbled structures composed of flat, rounded stem joints called cladophylls, or simply "pads." Of the seven species occurring on the Rolling Plains, the Texas Prickly Pear—the largest of the group—becomes the most abundant on overgrazed lands (fig. 6.13). Well adapted to life in arid and semiarid regions, this cactus stores water and can photosynthesize year round. When drought reduces the water content of the cladophylls, their stomata no longer

open to absorb the carbon dioxide necessary for photosynthesis. Instead, the process continues by using respiratory carbon dioxide and stored water.

Each mature pad produces from one to eight yellow flowers (some plants produce orange or red flowers) located above the ovary. After fertilization, the ovaries develop large fruits ("tunas"), each containing up to 300 seeds. Succulent, sugar-rich tunas become desirable forage for many animals, but cactus seeds are protected by a hard coat. Ingested seeds undergo mechanical scarification and acid leaching as they pass through an animal's digestive tract, which increases their ability to germinate. Livestock and wildlife, especially Black-tailed Jackrabbits, are important agents in seed dispersal.

Prickly pears also reproduce vegetatively. Each pad, or part of a pad, broken from its parent can potentially develop into a new plant. When the pad comes in contact with soil, roots rapidly emerge from auxiliary buds and soon supply the developing cactus with water and nutrients. The combination of vegetative reproduction and long-term seed viability poses difficulties for controlling prickly pears. Currently, moderate to dense stands of prickly pears occupy about 29 percent of the Mesquite Plains.

Figure 6.14. Tree Chollas develop large purple blossoms at the tips of their multibranched stems (left). The jointed stems of Tasajillo, like those of cholla but thinner (right), break off when their spines attach to passing animals; the joints later fall off and frequently produce new plants.

With large purple flowers rivaling the beauty of roses (fig. 6.14), Tree Chollas develop as multibranched shrubs reaching heights of 5 feet. The plant's cylindrical stem-like structures somewhat resemble strings of sausages. Green photosynthetic tissue overlies a woody skeleton, and flowers develop on the ends of specialized terminal joints. When the fruit ripens, the ovaries turn yellow and remain available to herbivores throughout the winter. In addition to relying on seeds, Tree Chollas also reproduce vegetatively. Each joint bears clusters of up to ten reddish spines, each about 1.2 inches long. Whereas the long, barbed spines are difficult to remove, the outermost joints dislodge easily and attach to animals feeding on fruit. Eventually, the joints drop free in new locations, where some take root in areas of exposed soil. After droughts and overgrazing expose large areas of barren soil, Tree Chollas undergo what is often described as a "wave of invasion."

Tasajillo, a much thinner version of Tree Cholla, forms dense shrubby tangles less than 5 feet high.

On the Rolling Plains, the plant typically grows in protected areas under mesquites or within clusters of prickly pears or Tree Cholla (fig. 6.14). Their cylindrical stems, each no more than 0.5 inch in diameter, consist of many small joints that bear small inconspicuous leaves and, from July to August, tiny greenish flowers. White-tailed Deer, Wild Turkeys, and other birds relish the bright red fruits, dispersing seeds in the process of consuming and excreting them. Like Tree Cholla, Tasajillo also reproduces vegetatively when animals dislodge and carry its joints afar.

Stiff spines arm mesquites and cacti, but cacti additionally produce dense clusters of slender and extremely sharp glochids (fig. 6.15). Almost invisible, these tiny barbed hairs surround the base of spines on pads, joints, and fruits and dislodge with the slightest touch. White-tailed Deer and livestock occasionally suffer from a condition known as "pearmouth," which results when glochid injuries foster bacterial infections on the victim's lips, tongue, and palate.

Figure 6.15. Close-up photography reveals the tiny, sharp glochids protecting prickly pear cladophylls and tunas.

Spines and glochids deter many animals, but Eastern White-throated Woodrats and Southern Plains Woodrats seem immune to both defenses. These large rodents construct shelters, or "middens," at the base of a cactus or mesquite from piles of sticks, cactus pads, and other plant materials, including dried dung. As decoration, they often add shiny bits of trash and snail shells—a behavior that earned woodrats the nickname "packrat." Many species, including beetles, moths, spiders, lizards, snakes, frogs, toads, mice, shrews, and rabbits, also find lodging in woodrat middens, which offer both protection and escape from summer heat.

Western Diamond-backed Rattlesnakes also commonly seek shelter in woodrat middens. For many years, biologists wondered how packrats and rattlesnakes coexisted until Allan H. Chaney (1923–2009) observed that woodrats survived rattlesnake strikes. Investigators also discovered a somewhat unexpected reversal of roles: some woodrats kill rattlesnakes visiting their middens. Further study revealed that woodrats could neutralize the toxic properties of rattlesnake venom. After succumbing to rattlesnake attacks for thousands of generations, enough packrats with some measure of resistance to the venom survived and passed on the advantageous trait to their offspring, in time producing generations of woodrats well adapted to cope with rattlesnakes. Thus, prickly plants and a genetic defense provide woodrats with a good measure of safety inside their spine-studded castles.

TUMBLEBUGS: NATURE'S CLEANUP CREW

Prior to European settlement, millions of American Bison wandered across various kinds of grasslands in North America. Because the herds moved continuously, plains grasslands were not overgrazed and soon recovered after the animals moved on. Still, the shaggy herds left two reminders of their passage, the first being "buffalo wallows," shallow depressions that formed where the animals repeatedly rolled about to take a dust bath (chapter 8). The second, however, left no lasting physical evidence—every day each animal deposited about 56 pounds of fresh dung. Each fresh patty—a "buffalo chip" in western lingo—temporarily provided a moist, nutrient-rich habitat for a thousand or more coprophilous mites and insects. No one knows how many species once thrived on bison dung, but a hint emerges from the more than 450 species of insects so far discovered feeding on cattle dung. Most of the species in North America almost certainly represent descendants of the insect fauna once associated with the long-gone bison droppings.

The behavior of dung beetles varies: some species prefer fresh manure while others visit only older, drier patties; some are active at night but others search only during the day. Dung beetles also show species-specific differences in the ways they deal with manure. Because of their behavior, Tumblebugs represent a specialized group of scarab beetles known colloquially as "rollers" (fig. 6.16). Other dung-feeding scarab beetles—"tunnelers"—bury the patties for safekeeping belowground, whereas others, the "dwellers," simply move inside a patty to conduct their housekeeping activities. In all, Texas is home to about 58 species of dung beetles, all of which possess unusually shaped antennae. The end of each antenna resembles a platelike oval club adorned with three to seven expandable leaves. The large surface area of these leaves effectively detects odors—an obvious advantage for locating dung.

Tumblebugs, common on the Rolling Plains, fly up to 10 miles to locate a fresh patty, from which they wad a clump into a ball with their hind legs and roll it away. Most often, a male faces backward and pushes the ball with his rear legs while a female follows closely behind. The procession follows a

INFOBOX 6.3. RATTLESNAKE ROUNDUPS
Inglorious Festivals

In Texas, spring is a time for outings and yearly traditions, not the least of which are leisurely trips along roadways to see the annual kaleidoscope of wildflowers. Another, held each year in March, attracts some 35,000 spectators to a controversial event in Sweetwater, a small town on the Rolling Plains of Texas. Billed as the "World's Largest Rattlesnake Roundup," the attraction features a huge pit containing more than 3,000 coiled vipers—the centerpiece for snake handlers, rattlesnake dinners, cold beer, and vendors peddling rattlesnake-themed curios. Similar events at 20 other Texas towns likewise attract large crowds; in 1993, a roundup in Freer drew a crowd exceeding 70,000.

Rattlesnake roundups originated as efforts to reduce the risk of snakebites to both people and livestock. Initially advertised as service-oriented educational events, the annual roundups soon evolved into carnival-like festivals with strong commercial overtones. Although the roundups provide communities with social and economic benefits—sponsors reinvest most of the profits into community projects—the highly touted educational component often sends an unclear message. Information about the natural history and ecological value of rattlesnakes is frequently inaccurate, superficial, or even omitted from educational presentations, often replaced with well-garnished tales of dangerous encounters and the risks of snakebites. Such presentations commonly occur in a sideshow atmosphere where daredevils perform their stunts while handlers harass snakes until they rattle and then decapitate and skin the bodies.

With a fossil history dating back about five million years; rattlesnakes evolved into one of nature's most highly specialized serpents. A pair of heat-sensitive pit organs located on either side of the head between the eyes and nostrils detects infrared radiation, thereby enabling rattlesnakes (and other pit vipers) to locate and target warm-blooded prey on the darkest of nights. Two fangs, folded back until needed, become erect when the mouth opens during a strike. The venom soon immobilizes the victims—rodents and other small mammals—and initiates internal digestion of their tissues. Most species, including Western Diamondbacks, produce hemotoxic venoms, but a few (e.g., Mojave Rattlesnakes) inject potent neurotoxins as well.

For humans, snakebites lie at the core of a common phobia but in truth present little risk—fewer than 15 fatalities occur out of about 8,000 humans bitten each year by venomous snakes in the United States. According to the Centers for Disease Control and Prevention, more than five times as many people perish from lightning as die from snakebites, although such statistics likely do little to assuage those who fear snakes. Recent research indicates that rattlesnakes are not "dumb animals," as many believe. They possess a sense of self-identity and at times live in social groups; they also recognize spatial arrangements of objects in their environment and remain with their young for extended periods. All told, rattlesnakes are complex creatures that play significant roles in their ecosystems, most notably as links in food chains.

Rattlesnake roundups offer trophies and prizes to contestants who capture the most snakes (based on weight) and the heaviest or longest individuals, along with special recognition for new state records. Sponsors purchase and resell the captured snakes to dealers of exotic skins or curios. In the field, contestants often pour or spray gasoline, ammonia, or other toxic substances into dens to flush out aestivating snakes, with little regard for Eastern Box Turtles and other animals sharing the den that become "collateral damage." Many gas-soaked dens remain useless for years afterward. (The Texas Parks and Wildlife Commission is considering new rules regarding rattlesnake gassing.)

It is difficult to determine the point at which rattlesnake roundups reduce local populations beyond the point of recovery. Unlike many serpents, female Western Diamond-backed Rattlesnakes bear live young. To raise their body temperature—and that of their developing embryos—pregnant females aggregate at communal dens where they periodically emerge to bask in early spring. As a consequence, pregnant females remain vulnerable for capture, and the loss of this component poses serious threats to the stability of local populations. Indeed, to ensure success, roundup sponsors at times import and release rattlesnakes caught elsewhere to supplement populations depleted by years of exploitation.

Animal rights activists decry the brutal and inhumane treatment of rattlesnakes at these events, and they are not alone. A position paper published by the American Society of Ichthyologists and Herpetologists voices similar concerns about rattlesnake roundups. In fact, all scientists adhere to strict requirements for treating animals humanely, which prompts the expectation that the participants in roundups treat rattlesnakes with equal consideration.

Figure 6.16.
Tumblebugs roll
a dung ball to a
burial site in soft
soil. Photograph
by Howard Garrett
(www.DirtDoctor
.com).

remarkably straight path to an area with soft soil where the ball is buried. The pair then mates and divides the dung ball into small portions, forming each of these into a hollow sphere—a brood ball into which the female lays an egg before burying each orb deeper in a separate tunnel. Each of these brood chambers thus protects an egg and later provides the larva with food. Young beetles eventually emerge from the remains of their brood balls, dig a new tunnel to the surface, and fly away to renew the cycle at a fresh patty. Some parent Tumblebugs remain nearby to protect their offspring until the larvae mature and emerge as adults.

The slaughter of bison on the plains of North America coincided with the introduction of domestic cattle. Because of differences in their respective diets, the exchange of cattle for bison perhaps caused the extinction of some species of coprophilous insects, but the theory seems impossible to test. Fortunately, most species of dung beetles adapted to the somewhat different composition of cattle manure and survived the transition. In recent years, introductions of several species from other continents expanded the list of dung beetles in North America.

Despite their unsavory food habits, dung beetles provide ecological benefits to grassland communities. Their tunneling activities increase the capacity of the soil to absorb and hold water, and buried dung enhances nutrient cycling. In contrast, a fresh patty remaining in place loses up to 80 percent of its nitrogen content by means of volatilization. Because larvae consume only

about 50 percent of a brood ball before pupating, bacteria and fungi convert the remainder into nutrient-rich humus. In addition to enriching the soil, Tumblebugs increase grazing conditions when they remove dung piles that ruminants otherwise avoid, thereby leaving up to 10 percent of the grass untouched. Removal of manure patties also reduces populations of Horn Flies and Face Flies— two biting insect pests that lay their eggs in fresh manure and, as adults, feed on the blood of wildlife and cattle.

Nocturnal Tumblebugs in Africa guide their linear movements to a brood site using polarized moonlight and the alignment of stars in the Milky Way—an amazing feat for such humble organisms. Most Tumblebugs in Texas, however, move about during daylight hours and are easily observed, but their means of orientation awaits discovery.

CLIFF DWELLERS

Biologists have long studied complex ecological relationships in virtually all of the many environments in Texas and elsewhere. Until recently, however, cliffs remained almost completely overlooked as sites for ecological research, perhaps because of the difficulties of sampling vertical surfaces and the view of such places as hostile and lifeless. In other words, people looked elsewhere for more accessible and appealing terrain to study. Nonetheless, cliffs occur widely throughout the world, and those in the Caprock Canyonlands of the Rolling Plains are as abundant as they are attractive. Some sheer cliffs on the sides of narrow, high-walled canyons in this region drop more than 700 feet—awe-inspiring formations aptly described by historian-naturalist Dan Flores as "wilderness cathedrals." Thanks to eons of erosion, seemingly endless banks of cliffs in the Canyonlands also offer a colorful and majestic laboratory for adventuresome ecologists.

A cliff is usually defined as a high, steep, or overhanging face of rock descending downward from a plateau to a more or less horizontal base. What constitutes a cliff, however, remains relative to scale. For humans, a cliff usually extends upward for more than 10–13 feet—a distance clearly exceeding our own height. From the perspective of an insect, however, the height of a "cliff" is far less, although for an ant, vertical distances of any size seldom seem intimidating. No hard-and-fast angle differentiates a cliff from a slope, but a practical approach assumes that rocks falling from a cliff drop in a free fall before hitting solid ground below. In contrast, rocks bounce and tumble as they travel down a slope.

Many cliffs develop with overhangs (or undercuts) and thereby become even more diversified as habitat than vertical surfaces alone. Sheltered sites beneath overhangs, for example, offer shaded microhabitats where spiders hang webs and Pipe Organ Mud Daubers sculpt hundreds of tiny mud pellets into a series of tubular chambers that indeed resemble organ pipes. Such sites become arenas for the dynamics of predator-prey relationships. The mud daubers capture and immobilize the spiders, which are packed into the chambers, and after laying their eggs, they seal the tubes. In parallel with Tarantula Hawks, the mud tubes serve as an incubation compartment where mud dauber larvae feed on the comatose spiders. Overhangs also provide shelter where Eastern Phoebes construct their cup-shaped nests crafted from an adobe-like mix of mud pellets and dry grass (fig. 6.17). In just such a place in Pennsylvania, John James Audubon (1785–1851) placed a silver wire on the leg of a nestling phoebe that returned a year later to nest at the same site—a phenomenon known today as homing behavior.

Unique plant communities often develop in the shelter of overhangs beneath waterfalls or adjacent to permanent pools and semipermanent tinajas. Mosses and lichens carpet splash zones in such places, and the green wash of algae coats moistened rocks lying in shade for most of the day. Here, too, luxuriant cloaks of Maidenhair Ferns cover cliff walls where they protect Plains Leopard Frogs and Blanchard's Cricket Frogs from desiccation.

A few ledges projecting shelflike from cliff faces collect enough soil and rubble to give plants a tenuous foothold. In some places, twisted and gnarled junipers blasted by wind and sand somehow endure for centuries on their precarious roosts. American Kestrels, Great-horned Owls, and a number of hawks perch on ledges from which they launch hunting sallies and

Figure 6.17. A small overhang on a sheer cliff provides shelter for the cup nest of an Eastern Phoebe (top), whereas Cliff Swallows construct gourd-shaped nests in large colonies under the eaves of larger overhangs (bottom).

return to eat their prey. Golden Eagles and Red-tailed Hawks construct nests on isolated shelves high on southwest-facing cliffs—sites where southerly exposures moderate cold temperatures during incubation and brooding activities in the early spring. Exposed to the sun, south-facing escarpments absorb heat from solar radiation during the day and then radiate warmth at night as the rocks gradually cool in the night air. Telltale whitewash staining the cliffs, visible from afar, alerts visitors to nesting sites favored by raptors.

Vertical cliff faces protected from winds and situated near reliable sources of mud and water attract Cliff Swallows, whose colonies may contain

100 to 2,000 tightly clustered mud nests. To construct their nests, each member of a mated pair scoops up a pellet of wet soil from a nearby mud bank. Mixing the mud with sticky components in their saliva, the birds deftly daub the pellets onto the cliff face—each pellet is one of hundreds added over several weeks to form a gourd-shaped structure with an opening in the neck (6.17). When completed, the nests house clutches of four to five spotted eggs laid on a thin cushion of grass; after hatching, the featherless and helpless (altricial) young remain in the nest until they develop feathers and motor skills and fledge.

The birds often return to the same colony for consecutive breeding seasons, thus expending less energy repairing existing nests than would be required to start anew. However, the energetic benefit may be offset by the buildup of ectoparasites at colonies occupied for consecutive years. Two species of ticks, a swallow bug, and a community of mites emerge at night from nearby cracks and crevices to feed on adults and their nestlings. Even if the swallows abandon a colony for several years, most of the ectoparasites survive until the birds return, sometimes gaining their nourishment from bats roosting in the empty nests. Swallow Ticks endure these absences deep in cracks where, in a state of torpor, they survive on water absorbed from the surrounding sandstone. In laboratory simulations, these ticks remain alive without feeding for up to seven years— presumably a reflection of their persistence under field conditions. When the Cliff Swallows return, their calls activate the ticks, which then renew their nightly quests for meals of blood.

The cracks and recesses in cliff faces provide habitat where Common Side-blotched Lizards, Texas Spiny Lizards, and Texas Greater Earless Lizards forage for insects. Rock Wrens and Canyon Wrens nest in holes and crevices as well as in the rocky rubble accumulating on ledges and below. Unlike Rock Wrens, Canyon Wrens build cup-shaped nests attached to the rocky walls of crevices or large recesses. Because Canyon Wrens and Rock Wrens coexist in the same habitat and feed on similar prey, the two species provide a good example of competitive exclusion—they forage and nest in different locations on the same cliffs.

Rocks of various sizes accumulate at the foot of most cliffs, where Annual Sunflower and Skunkbush Sumac invade the older piles of rocks, often forming dense thickets. Hairy Rattleweed and Buffalobur Nightshade, however, thrive in sandy soils adjacent to debris in talus accumulations. After drying on the plant, disturbed rattleweed pods imitate the sound of rattlesnakes, much to the concern of anyone nearby. Nocturnal predators, including Raccoons, Ringtails, White-backed Hog-nosed Skunks, Western Diamond-backed Rattlesnakes, and Bullsnakes regularly patrol cliff bases looking for rodents, fallen nestlings, and other prey.

Only a few species can survive the stressful conditions imposed by cliffs. Animals dwelling in escarpments constantly battle the forces of gravity and relentless winds. Plants face even tougher conditions. They also deal with wind and gravity, but plants additionally tolerate thin soils that lack space for roots and hold little water, receive virtually no direct precipitation, and, on south-facing precipices, withstand full exposure to direct solar radiation. As a result of these harsh conditions, cliffs present unique environments with a limited number of microhabitats.

SHIMMERING LAKES AND DUST DEVILS
On a blistering summer day, an exhausted traveler stumbling across the Rolling Plains might believe that a cool, blue lake shimmers invitingly in the distance. But as the wanderer advances, hoping to slake his thirst, the gleaming shoreline evaporates only to reappear farther afield. Seemingly a hallucination, the vision is actually a mirage—an optical phenomenon commonly observed when heat dances across arid plains.

The optical illusion, properly described as an "inferior mirage," develops when the sky appears as if on the ground. The deception occurs only on days when intense sunlight scorches sandy surfaces, warming the air immediately above the ground more intensely than the air overhead. Because cool air is denser than warm air and therefore has a higher refractive index, light rays traveling downward from the blue sky at a shallow angle—the angle at which humans view distant objects—are refracted upward near the ground. The observer's brain, however, incorrectly interprets the view as sky reflecting off water. Mirages receding in the distance on hot highway surfaces form in the same way. Inferior mirages, regardless of how much they shimmer, indicate extremely high ground temperatures—and harsh conditions for plants and animals.

Calm, clear summer days often create another common heat-related phenomenon on the Rolling Plains and in similar semiarid environments. Whirlwinds, known locally as "dust devils," form above flat, sandy surfaces as air heated at ground level ascends rapidly into the cooler air above. In the process, complex thermal interactions cause the air to rotate as it rises. As more hot air rushes in, the spinning effect intensifies and, as the vortex becomes self-sustaining, results in momentum that drives the whirlwind toward new sources of hot air nearby. Dancing across the landscape, fully formed dust devils become visible when sand and loose debris collect in the swirling maelstrom (fig. 6.18). Although visible for great distances, most dust devils remain relatively small and rarely exceed 200 feet in height. However, a few scour the surfaces of rocks or loose soil and propel microbes, tiny animals, seeds, and other propagules upward as much as 1,000 feet. There, winds aloft may disperse these materials, depending on their mass and shape, for great distances.

Disappearing lakes and dust devils, although ephemeral, are as much a part of the environment—and culture—on the Rolling Plains as any living organism. Native Americans believed dust devils were the wandering spirits of their revered ancestors, and they carefully watched the swirls for omens. A dust devil spinning clockwise indicated a good spirit and future bounty, whereas swirls rotating counterclockwise heralded the coming of bad medicine and hard times. Today on the Rolling Plains, as in the past, dust devils still swirl across the landscape, often near imaginary lakes shimmering on the distant horizon. Watch, and interpret these as you wish.

CONSERVATION AND MANAGEMENT

For most of the last century, the Rolling Plains remained under the control of a few large ranches. As a result, urbanization and fragmentation

Figure 6.18. Formed from whirlwinds of heated air, dust devils enliven the plains, transport seeds and small organisms, and, for some, presage omens. Photograph by Wesley Burgett.

developed slowly in comparison with many other regions in Texas. Fire suppression accompanied settlement, before which lightning-caused fires probably scorched large areas every 5 to 20 years— often enough to check the spread of mesquites, cacti, and Redberry Juniper. Also in the past, abundant vegetation and accumulations of mulch fueled intense, so-called hot fires. Now, after years of overgrazing, bare ground and clumps of thorny bushes have replaced the grasses and forbs, lessening the availability of adequate fuel. Soil erosion accompanied these changes, further diminishing the potential for fires to burn intensely enough to destroy most species. Thus, fires today, whether natural or prescribed, may initially reduce unwanted vegetation, but many species recover rapidly thereafter.

Modern fire regimes, in particular, allowed invasions of Redberry Juniper. Originally restricted to buttes and escarpments, Redberry Juniper soon expanded into grasslands, increasing its distribution by 61 percent between 1948 and 1982. Tobosagrass forms clumps of dense sod in many places, but its coarse tissues offer limited value as forage, and infestations of either species significantly reduce the carrying capacity of rangelands for livestock and wildlife. Properly conducted, controlled burns help control both Tobosagrass and Redberry Juniper as well as decrease the overstory of mesquites, but herbicides prove more effective for controlling cacti.

The combined impacts of overgrazing, droughts, and invasions of Saltcedar have reduced or eliminated Plains Cottonwoods from many riparian woodlands on the Rolling Plains. In particular, the loss of the older trees creates situations somewhat analogous to a trophic cascade—the collapse of a food chain when a significant species is eliminated. In this case, populations of Wild Turkeys decline when large cottonwoods disappear. The first settlers wrote of huge flocks of Wild Turkeys leaving the open prairie at dusk to roost well above the ground in mature cottonwoods. Whereas turkeys might otherwise roost on lower branches of shorter trees, such locations increase predation beyond what the flocks can withstand. Where monocultures of Saltcedar today choke stream banks, broken skeletons of dead cottonwoods stand as grim monuments of times past.

Humans attracted to landscapes with extreme topographical features often exploit cliffs for recreation. Rock climbing and rappelling each involve direct contact with cliff faces, whereas hiking, cycling, hang gliding, and the fun of riding horses or all-terrain vehicles are among the many other activities people enjoy in rugged terrain. Still, these activities often scar the landscape and scrape algae and lichens from rocky surfaces. Trampling easily damages young trees and shrubs already facing a tenuous hold on life. Unfortunately, the impacts of human activities on cliffs remain essentially unstudied, but on the brighter side, the isolation of cliffs and canyons on the Rolling Plains has so far buffered most locations from needless abuse.

Much of the Rolling Plains is held in private ownership, which limits access to most areas. Fortunately, several highways enable travelers to gain impressions of the vivid terrain, and three state parks offer opportunities for hiking and camping in a remarkable landscape. Recognizing the magnificent geological features of Palo Duro Canyon, the state of Texas acquired a large area at the site for a state park in 1933. (As early as 1908, the location was proposed for a national park, but political infighting in Washington eventually killed the initiative.) Palo Duro Canyon State Park opened to the public a year later, and the Civilian Conservation Corps spent the next five years building access roads and cabins. Located east of the city of Canyon, Palo Duro Canyon State Park, the second-largest state park in Texas, currently provides access to 28,000 acres of public land. Another facility, Caprock Canyons State Park, opened in 1982, protects 15,313 acres of awe-inspiring terrain. In addition to its majestic geological formations, the park, located north of Quitaque (Briscoe County), is home for the Texas State Bison Herd (see appendix A). Farther east, gray-green bands of raw copper add even more color to the red-hued canyons, arroyos, and small mesas at Copper Breaks State Park. This park, covering 1,900 acres on the Pease River between Quanah and Crowell, provides grazing for part of the state's official herd of Longhorn Cattle.

7
EDWARDS PLATEAU
THE TEXAS HILL COUNTRY

*Since well over a century ago, the region has
been a sort of reference point for natives of other
parts of the state, and mention of it usually brings
smiles and nods.*

— JOHN GRAVES (2003)

Edwards Plateau

Rolling highlands separating broad valleys characterize much of the Edwards Plateau, a slightly elevated region covering approximately 24 million acres in the west-central portion of Texas. The boundaries of the somewhat oblong plateau are defined primarily by the underlying geology—horizontal layers of Cretaceous-age limestone covered by shallow calcareous soils. Deep beneath the strata lies the Edwards Aquifer, an enormous underground reservoir that feeds many crystal-clear springs (fig. 7.1). The plateau slopes gently from about 600 feet above sea level on the eastern edge to about 3,000 feet in the central and western regions.

The Pecos River Canyon delineates the western margin of the Edwards Plateau, but the northeastern and northwestern portions intergrade without noticeable geological demarcation into biotic associations typical of the Cross Timbers and Prairies (chapter 5) and Southern High Plains (chapter 8), respectively. Only the western areas of the Edwards Plateau, where broad expanses of relatively flat uplands are dissected by shallow, gently sloping valleys, represent a true plateau. A small outlier—designated the "Callahan Divide"—is separated from the central plateau by river valleys and wide lowlands to occupy a northern fragment extending eastward from Coke and Nolan Counties to Callahan County in the middle of the state.

Sweeping in a wide arc for nearly 200 miles, a distinct change in topography defines the southern and southeastern edges of the Edwards Plateau. Along this margin, the Balcones Escarpment follows the Balcones Fault Zone, an older geologic feature rising abruptly more than 300 feet higher than the South Texas Brushland (chapter 10) to the south and the Blackland Prairies (chapter 4) to the east. The southernmost expanse of the plateau adjacent to the escarpment—affectionately known as the "Texas Hill Country"—is incised by numerous picturesque river canyons and steep ascents creating an undulating, wooded landscape

(*overleaf*) Figure 7.1. Cold, clear water flows from Dolan Springs (Val Verde County). Photograph by Paige A. Najvar.

composed of hills or low mountains instead of a tableland. An older geologic province, known variously as the Llano Uplift, Granitic Central Basin, or Central Mineral Region, is exposed in a comparatively small area of about 2,300 square miles along the northern margin of the Edwards Plateau. Surrounded by layers of younger Cretaceous limestones, Llano Uplift bedrock consists mostly of granitic and metamorphic rocks more than a billion years old.

Although much of the plateau is relatively flat, the region is most well known for its scenic network of crystal-clear, spring-fed streams and rivers traversing canyons cut through the layers of limestone and exposures of granite. In undisturbed areas, scattered clumps of low trees interrupt the grassy savanna. After spring rains, bluebonnets, Texas Paintbrushes, Indian Blankets, and other colorful wildflowers decorate winding Hill Country roadsides with beautiful vistas unlike any found elsewhere in the state (infobox 7.1).

STRUCTURE AND CLIMATE
GEOCHRONOLOGY AND STRUCTURE

The oldest rocks on the Edwards Plateau occur in the Llano Uplift area. Originally intruded as molten magma deep in the Earth's crust around 1.1 billion years ago, the well-known pink granites—known as Town Mountain Granite—in the Llano Uplift solidified underneath layers of rock. In the late Cambrian, seas inundated large parts of the continent, resulting in deposition of marine sedimentary strata—sandstones, shales, and limestones—layered directly atop the ancient rocks. Consequently, the granite reposed underground for 200 million years.

During the Permian Period, a tectonic event uplifted the region and subsequent erosion exposed the granite. When shallow continental seas periodically covered exposed surfaces of the ancient rocks beginning about 100 mya, widespread sediment deposition produced the shale and limestone layers that now characterize the entire Edwards Plateau. Evidence from across the plateau shows that the limestones were deposited in shallow tropical waters—bivalve, gastropod, and echinoderm fossils abound in some areas, reef deposits occur in others, and dinosaur footprints

INFOBOX 7.1. TEXAS WILDFLOWERS
Lady Bird's Legacy

In early spring, roadsides in the Hill Country and elsewhere in Texas present an eye-popping tapestry alive with a palette of vibrant colors—wildflowers have again renewed their timeless cycle across the landscape. Texas Paintbrushes and Texas Bluebonnets are among the first to weave color into the waysides browned by winter, but these soon give way to a dazzling yellow coverlet of Coreopsis. Still later, the warm and softer hues of Indian Blanket interlaced with the cool blue aura of Cornflowers and Spiderworts herald the end of another growing season. The blooms, coupled with the interplay of light and topography, easily enthrall even the most jaded traveler. The display represents a lasting legacy of a remarkable advocate for nature and the beautification of the nation's highways—Claudia Alta "Lady Bird" Taylor Johnson (1912–2007).

Shortly after her birth in Karnack, Texas, baby Claudia attracted the attention of a nurse who famously noted she was as "purty as a ladybird," thus giving rise to the nickname that would forever identify her and her good works. A shy girl during her childhood, Lady Bird spent much of her time fishing, swimming, and wandering in the forests and bayous near her home in East Texas, where she developed a deep appreciation for nature. After graduating from the University of Texas, Lady Bird met and married Lyndon Baines Johnson—a union that changed the course of history for both the state and the nation. Lady Bird used funds from an inheritance to launch her husband's first political campaign, thus initiating a career that would eventually lead both to the White House. She had a keen nose for business, and another of her investments eventually made the Johnsons millionaires.

Soon after that fateful day in 1963 when Lyndon Johnson became the 36th President of the United States, Lady Bird created the First Lady's Committee for a More Beautiful Capital. After the election in 1964, she expanded her vision to include conservation and beautification at the national level, which incorporated a focus on cleaning up the nation's highways. Her efforts produced the Beautification Act of 1965, popularly known as "Lady Bird's Bill," and the Surface Transportation and Uniform Relocation Assistance Act of 1987—funding for the latter requires that highway landscaping projects include planting native flowers, shrubs, and trees and places limits on billboards at these locations. For more than 20 years, Lady Bird personally presented awards to those highway districts in Texas that planted native vegetation to beautify roadsides and adopted mowing schedules that permitted these plants to reseed the rights-of-way. She continued these efforts until her death, and today the native plants that color roadsides, urban parks, and trails across the state and nation mark her enduring legacy.

On her 70th birthday, Lady Bird Johnson founded the National Wildflower Research Center using her funds, 60 acres of land she purchased and donated, and a major financial contribution from her friend, actress Helen Hayes (1900–1993). The facility later moved to a larger site located on 279 acres in the Texas Hill Country (10 miles from downtown Austin) and, on her 85th birthday (1989), was renamed the Lady Bird Johnson Wildflower Center. The Wildflower Center, which is now part of the University of Texas at Austin, supports research and education and serves as a nationwide clearinghouse for information about wildflowers and native plants. Each year, more than 100,000 visitors enjoy the facility's 4.5-acre wildflower garden, the 16-acre arboretum, and a display of some 700 species of native trees. By the Internet, the center's online database fields queries concerning the horticultural and ecological characteristics for more than 7,200 species of plants. The center recently initiated the Millennium Seed Bank Project, which collects seeds from plants native to Texas, some of which helped reestablish dune vegetation destroyed when Hurricane Ike devastated Galveston Island. The center also played a major role in replanting the Lost Pines after a catastrophic fire burned much of this island forest in 2011 (see chapter 4).

A grove named in Lady Bird's honor was dedicated at Redwood National Park, and among many other recognitions, she posthumously received the Rachel Carson Award from the Audubon Women in Conservation. Her portrait appeared on a postage stamp, and she served on several boards, including those of the National Park Service and National Geographic Society.

Her daughter, Lucy Baines Johnson, provided perhaps the most fitting tribute to her mother's far-reaching impact when she said, "I think there is no legacy she would more treasure than to have helped people recognize the value in preserving and promoting our native land." The Wildflower Center bearing her name and the carpet of native wildflowers gracing roadsides each year testify to Lady Bird's contributions. In her own words, "where flowers bloom, so does hope"—so pause a moment when enjoying a trip through Texas to remember her gift to us all.

Figure 7.2. This flat stretch of limestone in the Guadalupe River near Hunt (Kerr County) once served as the westbound lane of Farm Road 1340; the eastbound lane ran along the riverbank. Some curbing remains in the riverbed.

and evaporite deposits (e.g., gypsum) mark the locations of ancient shorelines.

Beginning in the Cretaceous Period and continuing into the Miocene Epoch (27–12 mya), the Balcones Fault Zone moved vertically, elevating the Edwards Plateau and Hill Country, while the land to the south and east subsided. Interestingly, the limestone strata were not folded or contorted during the uplift, and the flat-lying layers are visible today in riverbeds, road cuts, and canyon walls throughout the region (fig. 7.2).

During the millions of years since the region was uplifted, runoff eroded the tableland, creating the familiar topography of the Edwards Plateau. The central plateau is less rugged and broken than the celebrated Hill Country, but erosion unearths rocky slopes and limestone formations in many areas. Between the Devils and Pecos Rivers in the extreme southwestern portion of the Edwards Plateau, the topography shifts to a highly dissected region featuring steep slopes flanking mesa-like highlands.

CLIMATIC CONDITIONS

The Edwards Plateau—especially the Hill Country—experiences both wet years and extended droughts; rainfall amounts usually range between 25 and 35 inches per year, but extremes of 11 inches and 41 inches have been recorded in successive years at the same location. Rainfall patterns are best described as "erratic" on the eastern margin and "undependable" farther westward. In a relatively normal year, however, rainfall remains low from November through April and peaks in May and June, and a second period of rain usually occurs in September and October. The normal pattern may be interrupted in some years by the remnants of hurricanes that move inland from the Gulf of Mexico. On such occasions, a storm may cause severe flooding, especially when a torrential downpour stalls over an area for several days.

Torrential rainfall and heavy flooding along the Balcones Escarpment may be greater than anywhere else in the world because local conditions—location and physiography—generate conditions that spawn intense storms. The escarpment lies in a transition zone between two climatic regimes—a humid subtropical climate to the south and east, and a semiarid climate to the west. When cold fronts move into the region from the north, unstable air behind these fronts often collides over the escarpment with moisture-laden air arriving from the Gulf of Mexico (fig. 7.3). The change in elevation along the escarpment forces the moist air mass upward just as it meets the cold, dry air mass, often producing what local meteorologists call "rain bombs"—sudden, intense thunderstorms and flash flooding. Some of the highest rainfall events ever recorded in the world

Figure 7.3. Intensive thunderstorms build near the eastern edge of the Balcones Escarpment and often produce heavy flooding in the Pedernales River and other rivers in the region. Photograph by Jonathan Gerland.

were generated in this region. As examples, in 1921 a storm near Thrall (Williamson County) produced 36.4 inches of rain in just 18 hours—a world record that still stands—and another near D'Hanis (Medina County) in 1935 dropped 22 inches in less than 3 hours. In combination, the frequency of thunderstorms and physiographic factors favoring rapid runoff make the Balcones Escarpment one of the most flood-prone areas worldwide.

Whereas summer visitors from the Llano Estacado might complain about the region's humidity, residents of Houston or Corpus Christi might find the Hill Country satisfyingly dry. Despite these differences, most would agree that the region is hot—not much of a surprise given that Austin, at the eastern edge of the Edwards Plateau, lies at the same latitude as Cairo, Egypt. The daily high temperature in August averages nearly 97°F, but thermometers in Austin occasionally register 101°F or higher for several days.

Winter temperatures vary considerably from day to day and year to year on the Edwards Plateau. Winter storms—known locally as "cold snaps" or "northers"—sometimes drop temperatures as much as 50°F in a single day. Whimsical cowboys once claimed that northers froze coffee fresh from a campfire so fast that the ice stayed warm for hours. Freezing temperatures and snows, though rare, are somewhat more common in the western areas than in the southern or eastern regions of the plateau. When they occur, such conditions rarely persist for more than a few days—never long enough to freeze the ground. Nevertheless, freezing temperatures occur with enough frequency to keep out tropical and some subtropical vegetation.

BIOPHYSIOGRAPHIC ASSOCIATIONS

Prior to the arrival of Euro-Americans, the Edwards Plateau formed a grassy savanna studded by groves of mesquites, oaks, and junipers (fig. 7.4). Additionally, the landscape included a mosaic of plant communities maintained by periodic, naturally occurring fires. Records from the early 1800s indicate that grass cover was meager in places, likely because of unpredictable and highly variable rainfall; trees and woody brush were confined chiefly to steep slopes and canyons where fires were less common. Scattered vestiges of these communities remain, but fire suppression, agricultural development, and, at some locations,

Figure 7.4.
The presettlement
landscape of
the Edwards
Plateau may have
resembled this
oak-grassland
savanna, a
restored habitat
on a well-
managed ranch.

extensive urbanization subsequently modified the original vegetation.

Botanists catalog approximately 2,300 species of native vascular plants on the Edwards Plateau. More than 200 additional nonnative species also occupy "wild" habitats (those not directly associated with human influences). Endemic species represent only about 10 percent of the native flora of the Edwards Plateau, but more than 100 of the 400 species endemic to Texas occur only in this region—some are considered endangered because of their limited distribution. For example, Texas Snowbells, small trees producing clusters of small, bell-shaped white flowers in April and May, persist in only a few canyons. The only remaining wild population of another endangered endemic, Texas Wild Rice, grows in a few patches along a limited stretch of the San Marcos River.

Many authorities recognize four biotic associations on the Edwards Plateau, whose respective characteristics provide links with the plateau's major topographic features. Our tour across the region will take us from south to north before veering westward toward the Devils and Pecos Rivers.

BALCONES CANYONLANDS

Where the Balcones Escarpment juts abruptly upward from the South Texas Plains, the plateau features a highly dissected topography embracing numerous springs, streams, and rivers and steep-sided canyons (fig. 7.5). Although many Texans regard the entire Edwards Plateau as "Hill Country," the margin of the escarpment, more than anywhere else, portrays the enduring vision of purple ridgelines, clear creeks, and, of course, iconic hills.

Within this rumpled expanse, which some ecologists describe as the most distinctive biotic region in Texas, an oak savanna covers the uplands between drainages and extends downward onto exposed midslopes and into some wide alluvial valleys. The assemblage of trees on the highlands includes Plateau Live Oak, Texas Red Oak, Ashe Juniper, Cedar Elm, and Escarpment Black Cherry. Although open areas dominated by native grasses such as Little Bluestem, Indiangrass, and grama grasses were once widespread, fire suppression in recent decades has allowed woodland vegetation to invade the ridgetops. Grasslands remain in heavily grazed areas and pastures where ranchers remove

Figure 7.5. Steep canyon walls etched into level limestone strata enclose the Frio River and many other rivers in the Hill Country.

Figure 7.6. A narrow riparian forest lies sheltered by a cliff along the Frio River in Uvalde County (top). Massive Baldcypress trees flank the Guadalupe River and many other rivers and streams flowing through the Edwards Plateau (bottom).

junipers to increase forage production for their livestock.

Where the climate becomes more arid on the western rim of the escarpment, shrubby vegetation replaces oak savanna, but even these species may eventually succumb to prolonged drought. High, flat expanses and south- and west-facing slopes support scattered clumps of Ashe Juniper, oaks, Texas Persimmon, Sotol, Cenizo, and mesquites. The north- and east-facing slopes typically support woodlands dominated by Live Oak.

Many canyons of the Balcones Escarpment orient in a southeasterly direction and are deeply incised enough to provide limited protection from searing heat and sun. Rich in endemic shrubs and small trees such as Texas Persimmon, Texas

Mountain Laurel, Mexican Buckeye, and Agarita, sheltered canyons also afford habitat for many woody species more typical of forests in East Texas. Narrow strips of deciduous trees occur on north-facing slopes just below the limestone caprock (fig. 7.6). The borders of cool, clear watercourses are often lined with ranks of stately Baldcypress, American Sycamore, and Black Willow—typically eastern species that may have been stranded in Central Texas at the end of the glacial Pleistocene (fig. 7.6). Similarly, Swamp Rabbits also reach their westernmost limits in the river canyons of the Balcones Escarpment.

Many animals whose life cycles are wholly or partially associated with conditions such as soft soils (fossorial species), the hydrologic effects of faulting (the fauna of aquifers and springs), or specific plant-soil relationships may be limited to either the eastern or western sides of the Balcones Escarpment. For example, Spring Ladies' Tresses, an orchid bearing tall spiral spikes with up to 50 white flowers, inhabits open Blackland Prairie sites lying east of the Balcones Escarpment, whereas regularly moistened limestone outcrops along the escarpment mark the limits of its western congener, Giant Ladies' Tresses. Mexican Ground Squirrels and Prairie Skinks occupy nonrocky, grassland-dominated soils east of the escarpment, whereas Rock Squirrels and Great Plains Skinks occur in the canyons and rocky uplands to the west. All told, about 70 percent of all Texas reptiles and amphibians and high percentages of other vertebrates occurring in Central Texas are limited (either eastward or westward) by the Balcones Escarpment.

EDWARDS PLATEAU WOODLANDS

The elevated, central part of the Edwards Plateau features a gentle terrain of low, flat-topped hills descending into wide, flat valleys. Scattered groves of Plateau Live Oak, Texas Red Oak, and Ashe Juniper once punctuated grassy sites covering the hills and valleys—a savanna maintained by recurring fires started by lightning or Native Americans. Moreover, the fires curtailed the expansion of Ashe Juniper, which remained at low densities on hillsides. Oaks, in contrast, send up viable root sprouts after fires that grow into dense, shrubby thickets known as "oak mottes" and become a common feature of the central plateau. The mottes—then as now—provide a shaded nursery for many shrubs, including Agarita and Texas Persimmon, and provide key nesting habitat for the endangered Black-capped Vireo (chapter 5). The region is renowned for the colorful wildflowers that blanket the landscape after spring rains. The vivid sparkle of Texas Bluebonnets and Golden Waves in pastures and on roadsides dazzles in contrast with the softer hues of Winecups and Texas Paintbrushes.

The deeper upland soils of the Edwards Plateau probably always supported a savanna-like mosaic of grasslands interspersed with woodlands. Intensive

Figure 7.7. In wet years Ashe Juniper berries provide abundant food for many animals that in turn facilitate seed dispersal.

grazing decreased some prairie grasses, especially Little Bluestem, Indiangrass, and Sideoats Grama. In heavily grazed areas, these once-common species often give way to Curly Mesquite and Purple Threeawn as well as brush species like prickly pears, junipers, Texas Persimmon, and mesquites. Large areas of rangeland are now dominated by an invasive Asian grass, King Ranch Bluestem, which government agencies recommended in the mid-twentieth century for soil stabilization.

Early accounts describe closed-canopy juniper thickets, but these were confined to canyon slopes, especially along the Balcones Escarpment. During the past two centuries, however, two species of juniper—Ashe Juniper throughout the region and Redberry Juniper on the far western plateau—invaded the uplands. "Cedar brakes"—dense stands of juniper—now blanket highlands and slopes in many areas where deeply eroded limestone formations are exposed. The invasion and expansion of cedar brakes are often attributed to periodic droughts, introductions of grazing (and overgrazing) livestock, and elimination of recurring, natural wildfires.

Both junipers are shrubby, multistemmed evergreens rarely exceeding 19 feet in height at maturity. Abundant crops of berries are produced in wet years and provide desirable forage for both mammals and birds (fig. 7.7). Of these, American Robins may be the most effective agents for

dispersing juniper seeds; perched at overgrazed sites, the birds excrete seeds where reduced competition favors the survival and development of juniper seedlings. In the absence of dispersal agents, the heavy berries simply drop onto the underlying layer of rich soil where they remain protected by decomposing leaves. Upon the death or mechanical removal of the tree, clusters of small seedlings spring up. Few herbaceous plants are capable of germinating or surviving beneath the canopy, but cedar duff is thought to foster germination of Texas Madrone seeds. A showy member of the mint family—Cedar Sage—produces tubular red flowers on spikes arising from a basal rosette of heart-shaped leaves, and it occurs almost exclusively in association with Ashe Juniper.

Juniper trees consume large amounts of groundwater and their dense foliage intercepts rainfall, reducing the amount reaching the ground. Ranchers in the Texas Hill Country generally consider junipers undesirable because they suppress grasses, thereby reducing the carrying capacity for livestock. Consequently, many landowners regularly thin or remove cedar stands to improve both water supplies and livestock forage. The cost of removing juniper is often defrayed by the sale of wood products gathered from cleared lands. Two regional industries harvest Ashe Junipers for raw materials, but Redberry Junipers lack the dense, oil-bearing heartwood necessary for commercial uses. Straight trunks of mature Ashe Junipers are harvested as "cedar" fence posts (fig. 7.8). These posts resist decay both above and below the ground. Gnarled Ashe Juniper trunks unsuitable for posts, along with stumps, roots, and branches—often the trimmings from fence-post production—provide small distillation facilities with materials for extracting oils. Aromatic compounds in the oils are blended into soaps, household sprays, and germicidal bathroom cleaners, many of which use the word "pine" in the product name. Obviously, the botanical distinctions between juniper, cedar, and pine, though real, blur into the charm of local cultures.

Years of hunting and trapping as well as agricultural and urban development have reduced or eliminated wildlife once common on the central plateau. Although occasional reports of Black Bears, Mountain Lions, and Red Wolves still surface, all were extirpated from the area long ago. A few Black Bears and Mountain Lions have moved back onto the central plateau, likely after crossing the Rio Grande near Del Rio, but they remain scattered and thinly populated. In contrast, populations of American Badger, Coyote, and White-tailed Deer have increased in recent decades. The increase

Figure 7.9. Enchanted Rock, a huge dome of pink Town Mountain Granite, is the largest and most well-known feature of the Llano Uplift. Photograph by Jonathan Gerland.

of woodlands has also favored greater numbers of North American Porcupines and expanded the breeding range of White-winged Doves farther northward.

LLANO UPLIFT

Despite the implication of its name, the Llano Uplift is actually a basin. The formation, which covers 1,673 square miles, is sometimes called the Central Mineral Region in recognition of the gold, copper, iron, tourmaline, smoky quartz, Texas blue topaz—the State Stone (appendix A)—and other valuable deposits in the area's Precambrian rocks. Numerous rose-colored domes of mostly barren granitic rocks punctuate the flat to rolling terrain within the depressed area. Visible from afar, Enchanted Rock, with its summit 425 feet above the local ground surface, is the largest of several domes rising prominently above the surrounding plain (fig. 7.9). It was from this height in 1841 that Texas Ranger Captain Jack Hayes, as part of a survey party

out of Fredericksburg, held off a band of Comanche with his two newly patented Colt revolvers. This well-known landmark and the smaller domes in the area represent the exposed and visible parts of a vast underground granite shield covering more than 62 square miles.

The granite domes appear solid and durable, but in reality they are composed of onion-like layers that continually erode and slough off. Massive sections periodically undergo exfoliation— a process whereby the outer layers break into smaller pieces along lines of parallel sheet fractures and slide downslope, fashioning a domelike structure. As a result, their surfaces present a mosaic of barren granite, jumbled rocks, and shallow soils where grasses, cacti, shrubs, and stunted trees somehow gain a foothold. Cracks and protected areas between boulders often shelter unique communities of small plants such as Spikemoss, Purple Cliffbrake Fern, and Nuttall's Stonecrop. For decades, a lone Live Oak—now a memory—occupied a prominent spot high on Enchanted Rock; offspring of the old tree now compete for space among the boulders below.

Climbers struggling to the peak of Enchanted Rock usually pause for breath along the route. While resting, they may notice splotches of color—yellows, oranges, grayish greens, and grays—dotting the rocks (fig. 7.10). Not paint drops left by a clumsy artist, they are lichens—unique, plantlike organisms that are often the only indication of life on the otherwise barren surfaces. It is tempting, but inaccurate, to call them "plants." Instead, lichens are "composite organisms"—a combination of two, or sometimes three, species of organisms that mutually benefit from a shared structure. The crusty growth combines a photosynthetic organism—either algae or cyanobacteria—with a fungus. The fungus provides its partner with water and minerals mined from the substrate or absorbed from the air and, in return, receives organic food molecules and oxygen from its live-in companion. To further aid the partnership, the fungus also produces colored pigments that shield the delicate chemical processes associated with photosynthesis. Lichens grow slowly and should not be touched—a grayish lichen only 4 inches in diameter may well be over a thousand years old!

After rains, water collects in low spots—weathering pits—where chemical reactions disintegrate the granite and winds remove residual mineral particles when the depressions dry. The pits eventually deepen into vernal pools, which may retain water for several weeks and thereby provide microhabitats in the unforgiving landscape. Water-filled vernal pools often harbor a unique suite of plants and animals. For example, after hatching from unseen eggs, tiny, translucent Fairy Shrimp swim upside down while devouring algae and plankton that spring to life from spores deposited when the pools last held water. Rock Quillwort, an endangered nonflowering plant endemic to vernal pools, also germinates from spores when water fills the depressions. The thin, pale green leaves of this diminutive grasslike plant tower as much as 2 inches above the water's surface (fig. 7.10). When the pools again dry, the foliage withers and blows away, leaving behind spores to join those of algae and the eggs of Fairy Shrimp as a dustlike

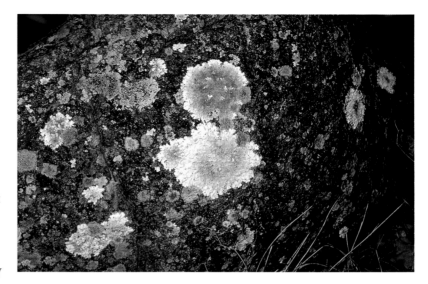

Figure 7.10. Colorful lichens (top), some more than 1,000 years old, often adorn the granitic surfaces on Enchanted Rock, where vernal pools (bottom) also offer habitat for mats of Rock Quillwort and other diminutive plants. Photographs by Brian R. Chapman and William R. Carr, Lady Bird Johnson Wildflower Center.

layer lining the depression. A process known as cryptobiosis allows these eggs and spores to lie dormant—sometimes for years—in the dry, hot concavities while awaiting the return of moisture conducive to regeneration.

Slopes descending from the plateau surrounding the Llano Uplift support stands of Ashe Juniper and Texas Red Oak, but these species do not occur on the granitic soils within the basin. Blackjack Oak, Catclaw Mimosa, Soaptree Yucca, Honey Mesquite, Little Bluestem, and various dryland grasses cover

the land between the domes. Deep, sandy soils within the basin support Basin Bellflower, a species endemic to the region, and stands of Cedar Elm, Black Hickory, and various oaks. Wild Turkeys and Mourning Doves roost in scattered Post Oaks and Black Hickories at night while Ringtails reconnoiter the rocks below.

For more than 11,000 years, Native American tribes explored the large dome and camped at its base. Many of their legends described eerie creaking and groaning sounds emanating from the rock—likely noises resulting from contraction as the granite cooled after a hot summer day. Early Euro-American settlers named the dome Enchanted Rock based on legends about its supernatural powers. After many decades of private ownership and restricted access, The Nature Conservancy purchased Enchanted Rock and then deeded the property to the State of Texas. Enchanted Rock State Natural Area was opened to visitors in 1984, the same year the site was added to the National Register of Historic Places.

SEMIARID EDWARDS PLATEAU
The western third of the Edwards Plateau represents a transition zone between the arid Trans-Pecos region to the west, wooded savannas to the east, grasslands to the north, and brushlands to the south. Annual precipitation is too low to support the dense cover of woody vegetation found eastward on the plateau.

Although it is hard to imagine today, grasslands once covered much of this area. Beginning in the late nineteenth century, shrublands gradually replaced the shortgrass prairie, resulting in the extirpation of many grassland-dependent species such as Montezuma Quail. The shift to scrubby vegetation—often associated with desertification—began with gradual erosion of the shallow soils resulting from year-round overgrazing by large herds of domestic livestock and control of once-frequent grass fires (fig. 7.11). The process accelerated during the twentieth century as water tables declined and surface water disappeared for long periods between rains. Today, groves of Ashe and Redberry Juniper, oaks, mesquites, and Papershell Pinyon grow on the rocky mesas and mountains and in the canyons. Curve-billed Thrashers abound in residual patches of grassland dotted with scattered mesquites. Toward the Pecos River, shrubs typical of the Chihuahuan Desert become more abundant, indicating a much drier environment.

Numerous rugged, steep-sided canyons lead to either of the two major tributaries—the Devils and Pecos Rivers, whose rock-strewn channels carry water only after heavy rains. Heat waves bounce from exposed rock surfaces during the day, but deep within shadowy canyon bottoms, small pockets of moisture persist and become havens for Maidenhair Ferns and Red-spotted Toads. Prairie Lizards dart between rocks, peering out as if monitoring their habitat from afar, and Rock Squirrels and White-ankled Mice scramble among the boulders deftly avoiding Black-tailed and Western Diamond-backed Rattlesnakes lurking in the rubble. Black Bears and Mountain Lions occasionally cross the Mexican border to wander unfettered in the canyons. In all, the sun-baked canyons are both scenic and, in their own way, vibrant with life.

Remote and largely unspoiled, the Devils River bisects the semiarid Edwards Plateau for all of its 94-mile length, but only the southern half reliably lives up to the name "river." The river originates where six normally dry tributary canyons converge near Sonora, but the upper part of the river flows only after heavy rains flood one or more of these dusty draws. After flowing underground for stretches of up to 20 miles (fig.7.12), the water emerges again and flows clear after thorough cleansing by sand, gravel, and limestone during the long subsurface journey. Springs along the lower half of the river gradually increase the volume of permanent flow, and even in the absence of flooding, the current often runs swiftly. Freshwater sponges adhere to submerged rocks and cobble where the river drops over Dolan Falls and splashes through a series of Class IV rapids (infobox 7.2). Listed as a threatened species, the Devils River Minnow occupies stretches of fast-moving water, but much of this habitat was lost when the Amistad Dam on the Rio Grande inundated the last 12 miles of the Devils River with impounded lake water.

KARST, SINKHOLES, AND CAVES

As raindrops fall, they absorb carbon dioxide in the atmosphere and become mildly acidic before contacting the ground. Consequently, minute layers of soft rocks, such as Edwards Plateau limestone, imperceptibly—but steadily—dissolve each time it rains. Centuries of irregular rainfall slowly sculpt the softer areas of surface limestone,

Figure 7.12. Water in the "Dry Devils River" flows underground for long distances during the dry season (top). With added inflows from several springs along its route, the Devils River flows swiftly for the last half of its course to Amistad Reservoir on the Rio Grande (bottom).

producing a landscape pitted with solution pans, rain pits, and rills (fig. 7.13). Rainwater also drains into rock fractures to continue the dissolution process in deeper layers, where it develops karst—a limestone-based topography characterized by sinkholes, caves, and underground drainage systems. On the Edwards Plateau, the development of karst created extensive underground networks that include large aquifers and caves large enough for tour groups to enjoy. The formation of aquifers and caves happened long enough ago in geological time to allow extensive speciation, making the plateau one of the richest areas in the world for karst-endemic invertebrate and fish species.

INFOBOX 7.2. LIVING SPONGES IN TEXAS RIVERS?

Most people associate sponges with oceans, unaware that a few species also occur in the freshwater systems of Texas, including the Edwards Plateau. Freshwater sponges share many attributes with their marine relatives but rarely grow as large. Easily overlooked or mistaken for algae or "slime" encrusted on underwater rocks, sponges are multicellular organisms lacking either tissues or organs. Instead of such structures, specialized cells fulfill their complex life cycle and basic biological functions.

Water movements influence the growth form of a freshwater sponge—some look like crusts, whereas others of the same species might form spheres or fingerlike projections. Numerous holes of varying sizes perforate the external surfaces of sponges. Water enters through these pores into a large internal cavity and from there continues through a network of incurrent canals to chambers lined with flagella and microvilli. The whiplike action of the flagella pushes the water farther along the canal system, while the microvilli absorb dissolved nutrients and engulf particulate matter in a process known as phagocytosis. The food is then passed along to other cells for digestion. Thus cleared of food, the water is expelled through excurrent canals that lead to a large exit pore.

Freshwater sponges are often bright green because of the extensive populations of algae living within sponge cells. In a symbiotic relationship—mutualism—somewhat similar to that between algae and fungi in lichens, the algae's photosynthetic activities produce oxygen and carbon for the host, whereas the sponge provides nutrients, such as nitrogen and phosphorus, required by the algae. Some sponges remain totally dependent on this relationship and decline in the absence of their mutualistic partner.

The body of a freshwater sponge consists of a mineral skeleton formed by filamentous spicules of silicon bound together by a rigid framework of collagen. Sponges have few predators, likely because the sharp spicules in their tissues function as pincushions that would injure the mouth of any attacker. However, the larvae of insects commonly known as spongeflies parasitize freshwater sponges. The cycle begins when adult spongeflies lay their eggs, protected under a web, on leaves overhanging a river. After hatching, the larvae drop into the water to begin life as ectoparasites on the outer surface of freshwater sponges; using specialized mouthparts, they pierce the cells of their hosts to withdraw nourishment.

After further development, the larvae end their parasitic ways and leave the water in search of cover under rocks or behind tree bark; thus hidden, they spin cocoons and pupate until emerging as adults. Far from being parasites, the omnivorous adults feed on a diverse menu ranging from pollen to animal carcasses.

Individual sponges function as either a male or female during a reproductive season, but gender is not fixed and can vary from year to year. Specialized cells produce either sperm or eggs, depending on the current sex of the individual. Gamete development occurs for only a short period during which a local population reproduces in synchrony. In response to unknown environmental cues, the males in each population simultaneously release their sperm, which drift until they enter the canal systems of female sponges or die. Although the mechanism remains unclear, water circulation in the females likely conveys the sperm to the egg cells for fertilization. Freshwater sponges are viviparous; the larvae are nourished by nurse cells while undergoing extensive development within a female's body. After acquiring a covering of flagella, the larvae are released in a free-swimming form that eventually settle on a suitable substrate and metamorphose into adults.

The flexible nature of sponge cells also permits a form of asexual reproduction known as fragmentation. In this process, small pieces of sponge torn away by currents or other disruptions develop into fully functional sponges if they land on a suitable substrate. Another form of asexual reproduction, however, is much more complex. This occurs when, in response to normal environmental stresses, some freshwater sponges form one or more masses of cells surrounded by a resistant coating. These structures, called gemmules, contain yolk in specialized cells but maintain low metabolic activity and resist adverse conditions, including anoxia. When the stresses wane, the gemmules are released and thereafter absorb their yolk, which increases their metabolic rate and initiates their development into active sponges.

At least ten species of freshwater sponges occur in Texas, but recent discoveries of *Spongilla cenota* in the Llano and Devils Rivers underscore the great diversity of this group in unpolluted waters of the Edwards Plateau, where rock or cobble surfaces at depths of 1.5 to 8 feet provide favorable habitat. In addition to their interesting natural history, sponges also serve as useful bioindicators of healthy freshwater ecosystems.

Figure 7.13. Rainfall or flooding often forms pits by dissolving softer deposits in exposed limestone surfaces, as shown here near the Pedernales River (Blanco County). Photograph by Jonathan K. Gerland.

SINKHOLES

The sudden collapse of apparently solid ground often makes frightening headlines, especially when a sinkhole occurs in a densely populated urban area. Although sinkholes occasionally develop from human activities, as when a water main breaks or old sewer pipes disintegrate, they occur more commonly in karstic regions. Natural processes create most sinkholes—also known as dolines— when mildly acidic water flows through pores or cracks and into the underlying limestone. As the subsurface limestone dissolves over time, a void enlarges until there is no longer any support for the land above. The sudden collapse forms what is known as a "cover-collapse sinkhole." Solution sinkholes, an alternate form of doline, are also common in the region. These develop where surface water collects in natural depressions and the dissolution of soluble surface layers creates a slope-sided pit that steadily enlarges in width and depth (fig. 7.14).

Devil's Sinkhole, the largest and most well-known doline on the Edwards Plateau, formed when an underground cavity collapsed and left a void 351 feet deep. Native Americans once camped

Figure 7.14. The entrance to Longhorn Cavern (Burnet County) provides a fine example of a cover-collapse sinkhole, a solution sinkhole, and solution cave. Photograph by Jonathan K. Gerland.

Figure 7.15. Mosses line the walls of Devil's Sinkhole. Note the climbing ropes (center) and streaks from falling water droplets. Photograph by Dr. Jean K. Krejca, Zara Environmental, LLC (www.zaraenviromental.com).

around the gaping 40- by 60-foot hole, which appears abruptly on an otherwise flat expanse of sparsely vegetated limestone. The wives of early settlers gave the sinkhole its current name in preference to the less genteel "helluva hole" used by their husbands. The lip of the abyss is undercut, beneath which a circular shaft more than 82 feet in diameter drops nearly straight down before gradually widening into an immense, oval room measuring 250 feet wide by 450 feet long. A central dome composed of fallen rock at the bottom of the pit rises more than 66 feet above the floor (fig. 7.15). Pools of water extending outward into underground "lakes" on three sides of the room are constantly refreshed from seeps on the shaft's walls. Mosses and lichens color the damp walls wherever light pierces the enveloping gloom.

More than just a geological curiosity, Devil's Sinkhole also pulses with life. In the twilight zone in the upper levels of the pit, Cave Swallows and Eastern Phoebes raise their broods in cup-like nests constructed from mud pellets laboriously attached to rock slabs. Nearby, Honey Bees buzz about pendulous combs dangling beneath an overhanging limestone eave. A dimly lit grotto at the bottom provides seasonal roosting habitat for one to three million Brazilian Free-tailed Bats. Each summer evening, the bats emerge from Devil's Sinkhole and other grottoes in smokelike

columns that quickly dissipate as they disperse in search of moths and other aerial insects. An equally spectacular show awaits in the half light of early morning when the bats return, at first circling several times above the entrance and then swooping headlong down to their roosts below. Thick layers of guano collect on the floor beneath the roosts, providing both habitat and food for a specialized community of invertebrates—guanophiles—adapted to life in a darkened, fetid environment. Two endemic members of this diverse fauna, including a troglobitic isopod, *Cirolanides texensis,* and a troglobitic amphipod, *Stygobromus hadenoecus,* live in the three "lake rooms" at the bottom of the sinkhole, nourished by the abundant organic matter falling into the water. In times past, the thick layer of guano—a rich source of nitrogen—at Devil's Sinkhole was mined for either fertilizer or the production of gunpowder (infobox 7.3).

Protected within 1,859 acres of state-owned land near Rocksprings (Edwards County), the Devil's Sinkhole State Natural Area is managed by the Texas Parks and Wildlife Department. In partnership with the state agency, a nonprofit organization—the Devil's Sinkhole Society—offers public tours for viewing the wispy "bat tornadoes" at dawn and dusk.

CAVE FORMATIONS

Most of the more than 4,000 known caves in Texas are located in limestone formations on the Edwards Plateau, and many more undoubtedly await discovery. Most limestone caves form in a manner similar to that described for sinkholes, and indeed, many sinkholes collapse into deep caverns, creating new pathways to hidden realms. The Caverns of Sonora, however, formed 1.5 to 5 mya when gases rose upward through a fault zone and mixed with water in an aquifer. The extremely acidic solution dissolved limestone layers, which enlarged the aquifer until the water drained off and left the cavernous space.

As defined by the Texas Speleological Survey, a cave is a naturally occurring cavity with a minimum of 15.5 feet of traverse length and an entrance whose dimensions do not exceed either the length or depth of the cavity. With these criteria, Bexar

INFOBOX 7.3. GUANO, GUNPOWDER, AND BAT BOMBS

In the warm glow of each tranquil summer evening, thousands of people gather at observation decks near the mouths of sinkholes and caves to thrill at the emergence of smokelike columns. Drifting upward and away into the fading light, the murky wisps are actually thousands of hungry Brazilian Free-tailed Bats departing from subterranean grottoes, popularly known as "bat caves," to feed on flying insects. The spectacle can be observed at several Edwards Plateau caverns and the Devil's Sinkhole, which harbor vast populations of roosting bats during the summer and fall. One large bridge, the Congress Avenue Bridge over Lady Bird Lake in downtown Austin, famously offers a similar evening experience.

Brazilian Free-tailed Bats typically roost within the same zone in a cave year after year and commonly pack themselves densely into clusters of more than 100 bats per square foot. Caves and sinkholes regularly used by large numbers of bats also become known as "guano caves," because guano accumulates beneath bat roosts in deep piles that sometimes extend for great distances. Guano deposits can be quite unsettling, especially when the bats are nervously fluttering above, and visits are usually not permitted and certainly not recommended. Guano, bat urine, and live mites drop from the ceiling in a constant rain. Small gnats fill the fetid air and alight repeatedly on the eyes, nose, and mouth. Ammonia levels in some cave galleries reach levels high enough to bleach hair, but most disconcerting, the guano-covered cave floor appears to be in constant motion—seemingly alive with the undertakings of untold millions of beetles, fleas, mites, ticks, spiders, pseudoscorpions, daddy longlegs, and other troubling creatures. Guano thus serves as both habitat and food source, supporting an incredibly complex community of cave invertebrates that rely on the material as the first link in their food chains. Flatworms, blind crayfish, and other invertebrates are also nourished by guano that falls into permanent pools.

In addition to supporting a unique faunal community, guano caves also played significant roles in historical events. Joseph LeConte (1823–1901), an ardent secessionist and

The exodus of Brazilian Free-tailed Bats from roosts beneath Austin's Congress Avenue Bridge draws crowds each summer evening. Photograph contributed by the Austin Convention and Visitors Bureau.

professor of chemistry and geology at what is now the University of South Carolina, developed for the Confederate Army a method for producing an essential component of gunpowder from guano. After extraction from a cave, guano was dried and mixed with wood ashes before being roasted in a kiln to produce saltpeter, the colloquial name for potassium nitrate. About 2,500 pounds of guano had to be shoveled out of the caves each day to produce just 100 pounds of pure saltpeter crystals. To facilitate removal, mule-drawn rail carts were used to remove the raw material from the deposit in Frio Cave (near Concan). The Confederate Army operated gunpowder kilns at New Braunfels and near Frio Cave to process guano from several caves—the remains of the kilns can still be seen at the Concan location. A simple mixture—seven parts saltpeter mixed with five parts charcoal and five parts sulfur—provided firepower for Confederate guns until the end of the Civil War.

The bats of Frio Cave were again enlisted during the Second World War to support a strange but ambitious effort named Project X-Ray. After watching clouds of bats exiting Carlsbad Caverns in New Mexico, Dr. Lytle S. Adams (1883–1970), a dentist, was driving home on December 7, 1941, when he heard the announcement on his car radio that Pearl Harbor had been attacked. Incensed, he devised a plan for American retaliation. His idea involved tying small, time-fused incendiary bombs to bats, which would be released from airplanes over Japan. He reasoned that the bats would seek shelter in buildings constructed of rice paper and wood on the islands, and when the fuse detonated, the structures would ignite and cause widespread devastation. After the concept received approval by President Franklin D. Roosevelt (1882–1945), traps were placed at Frio Cave and several other area caves to "recruit" the living messengers of destruction. Testing of the "bat bomb" was conducted in December 1943 on a remote area of Dugway Proving Grounds, Utah, where a mock Japanese village had been constructed. When the bats were dropped, they bypassed the target structures, heading instead to newly constructed buildings on a nearby military airfield, which were set ablaze. Although the "bat bomb" test proved effective—albeit misdirected—the project was brought to a halt by a far more effective weapon—the atomic bomb.

County alone has 575 cavities and more than 1,000 caves that occur within 40 miles of San Antonio. At least 137 caves in Texas exceed 990 feet in length, and 132 reach depths of at least 99 feet.

Amazing speleothems—distinctive hanging, upright, or columnar structures—add splendor to some Edwards Plateau caves. The most common speleothems are stalactites, which form when mineralized water solutions drip from cave ceilings. As mildly acidic water percolates through limestone, calcium carbonate in the rock is transformed into a solution of calcium bicarbonate. As each drop falls, a minute ring of calcite remains behind and gradually elongates to form a fragile, hollow tube known as a "soda straw." Growth of these formations depends on many factors, but the average stalactite grows about 0.12 inch per year. Some soda straws become long and delicate, but most often, debris plugs the central tube, causing minerals to accumulate on the outer surface and form a conical structure (fig. 7.16). When drops fall to the cave floor, the deposits gradually build cones upward, forming stalagmites. Stalactites and stalagmites sometimes meet to form a ceiling-to-floor column, which steadily increases in diameter through the ages. Natural Bridge Caverns, a commercial cave near New Braunfels, features spectacular examples of stalactites and columns.

Calcite speleothems can take several other forms. Flowstones in the form of thin, linear "draperies" develop when mineral-bearing water streams across a sloping surface before dripping downward. Usually translucent, or nearly so, some draperies are called "cave bacon" because brownish, meat-like streaks separate narrow whitish layers. The most delicate speleothems, helictites, defy explanation; they grow outward, not downward or upward, and develop in countless shapes. To imaginative minds, various helictites resemble ribbons, flowers, saws, curly fries, rods, worms, clumps of worms, fishtails, or hands. Perhaps the "butterfly" formation in the Caverns of Sonora (Sutton County) represents the best-known helictite in Texas and remains gorgeous even after part of one wing broke off in 2006. No theory has yet explained what forces shape helictites, but

Figure 7.16. The droplet about to fall from this soda straw will leave behind a tiny mineral deposit that imperceptibly lengthens the delicate tube (left). Over time, millions of such droplets build stalactites, stalagmites, and columns in karst caves (above).

many believe that wind or capillary action may be involved.

Seven caves in Texas are open to the public. One, Caverns of Sonora, is widely considered the most beautiful cave in the world because of its abundant and distinctive speleothems. All of the "show caves" contain a variety of interesting formations and fossils and allow a glimpse into the mysterious underworld. Nine state-owned "wild caves"—those lacking trails or electric lights—are also open for guided tours.

CAVE FAUNA

At show caves, visitors enjoy an assortment of formations, and most experience a brief period exposed, perhaps for the first time, to a stark environment totally devoid of light. Most tours, however, steer visitors well away from the biological curiosities endemic to caves—the animals adapted to an eccentric life in a dark wilderness.

Descent into a cavern begins in the dimly lit "twilight zone" just inside the entrance. Within this short section, temperature and light regimes vary daily and seasonally even though light penetrating the zone is limited. Ferns, mosses, and a few types of flowering plants—species able to continue photosynthetic activities in reduced light—persist here. Some animals that seek food or shelter in a cavern's threshold venture into the twilight zone but rarely proceed farther.

Beyond the twilight zone, environmental conditions gradually cease to be influenced by season, time of day (or night), or surface weather patterns. The number, size, and orientation of cave entrances may influence airflow, but even so, conditions in the interior remain relatively stable throughout a day or year. Temperature deep within a cave generally mirrors the annual average for the region above.

Few animals venture deep into caves, and fewer still are adapted for survival within a cavern's foreboding recesses. Nevertheless, a specialized branch of biology—biospeleology—is dedicated to the study of cave organisms and their adaptations. Of special interest are those cave-adapted creatures that seemingly "evolved backward." Indeed, fascinating studies await those wishing to discover the genetic mechanisms by which some species lose eyes, pigments, and other structures that remain functional in their aboveground counterparts. To facilitate such research, biospeleologists developed a classification system based on the degree to which these species have adapted to cave existence.

Troglophiles complete their life cycles with equal success both in caves and on the surface. This group includes certain species of insects, spiders, scorpions, crayfish, and salamanders.

Green plants form the base of almost all aboveground food chains, and in their absence cave-dwelling animals necessarily depend on energy from organic matter imported from the outside world. The role of "energy importers" is played by trogloxenes—species that periodically depend on caves for roosting, reproduction, or hibernation but regularly exit, usually to feed. Bats represent the foremost organisms in this group. Bats emerge nightly to feed on moths, beetles, and other insects and return at dawn to congregate in tight clusters on cave walls and ceilings. Beneath these roosts falls a steady rain of fecal droppings and urine, producing deep deposits of nutrient-rich guano on the cave floor (infobox 7.3).

Other trogloxene species include Cave Crickets and Cave Swallows. Cave Swallows, the only North American birds that require caves for nesting, construct their nests in the twilight zone of caves with large entrances. Unlike bats, these swallows feed during the day and spend nights in their nests or roosting on narrow shelves. Porcupines, skunks, Ringtails, Raccoons, and Bobcats frequently enter caves in Texas, but only Raccoons explore the dark zone. Raccoon droppings provide food for some cave-dwelling invertebrates, but more significantly, the excreta nourish a fungus that serves as an important food for springtails and other organisms.

Troglobites are so adapted to cave conditions that they can no longer cope in the outside world and never venture forth. Typical adaptations include the loss of skin pigmentation and vision (fig. 7.17). For animals in total darkness, it is no longer beneficial to expend energy to produce pigments for camouflage, protection from the sun, or sexual recognition, and functional eyes are useless. As these species lost their vision, they developed other sensory organs. Eyeless cave fish,

Figure 7.17. A cave-obligate millipede, *Speodesmus echinourus* (top), in a Travis County cave and the endangered Texas Blind Salamander (bottom), a denizen of the Edwards Aquifer, exhibit the classic features of troglobites—blindness and loss of pigmentation. Photographs by William R. Elliott and Joe N. Fries, US Fish and Wildlife Service.

for example, locate prey and avoid obstacles by detecting minute changes in water pressure with lateral lines or elongated barbels.

The specialized adaptations of troglobites limit their dispersal ability, and consequently many remain confined to either a single cave system or a few caverns within a small area. As a result, many troglobites, such as the Texas Cave Shrimp, are extremely vulnerable to extinction. Species in areas undergoing rapid urbanization are the most at risk. Just a year after the Endangered Species Act of 1966 became law, the Texas Blind Salamander became the first troglobite placed on the endangered species list by the US Fish and Wildlife Service. The salamander, a "poster child" troglobite (fig. 7.17), faces threats from declining water levels in the aquifer as well as pollution from the city of San Marcos, which lies directly above the aquifer. The species is endemic to the San Marcos Pool of

Figure 7.18. Barton Springs, among others fed by the Edwards Aquifer, provides recreation but also habitat for the endangered Barton Springs Salamander (inset). Photographs by Paige A. Najvar and Lisa O'Donnell (inset).

the Edwards Aquifer, but individuals occasionally emerge from springs and survive in deep stream pools.

Although the caves, sinkholes, and aquifers throughout the Edwards Plateau contain a rich cave-dwelling fauna, much remains unknown. Many taxa, especially mites, centipedes, and miscellaneous insects, remain unidentified and undescribed, and many cave systems have yet to be thoroughly explored. Virtually every region of the Edwards Plateau contains one or more species endemic only to cave systems in that same area. Nineteen invertebrates—eight beetles, six spiders, three harvestmen, an amphipod, and a pseudoscorpion—are protected by the Endangered Species Act. Several others are "candidate species" (the status and vulnerability of their populations are under review for federal protection). As might be expected, listing these "cave bugs" generated public controversy in the Austin and San Antonio areas, but most of the development projects temporarily stalled by concerns for these species were eventually approved after provisions for protecting the caves they inhabit were guaranteed.

HIGHLIGHTS

LAND OF 1,100 SPRINGS

The Pearl Brewing Company began brewing beer in 1883 using pure spring water issuing from the Edwards Aquifer in San Antonio. Just over a century later, the company developed an advertising campaign using its most well-known slogan, "From the country of 1100 springs." Although the company's estimate of springs in the Texas Hill Country might not be accurate, the slogan recognizes the region's many sources of crystal-clear water bubbling up through limestone layers perforated like Swiss cheese (fig. 7.18).

Recognized as one of the most prolific artesian aquifers in the world, the Edwards Aquifer, located beneath the southern border of the Edwards Plateau, discharges about 900,000 acre-feet of water annually. Water-bearing rock extends northeast in a 180-mile arc, but the entire aquifer complex is considerably larger and consists of three distinct zones: the contributing zone, the recharge zone, and the artesian zone.

Prior to 1896, most geologists believed that the underground water in the central part of Texas originated from the Rocky Mountains. In actuality, the contributing zone encompasses about 5,400 square miles in the central Hill Country where water collects, often sinking into the ground to be discharged from springs into streams and rivers. Rivers and creeks in the contributing zone flow toward the recharge zone, which runs in a narrow band along the Balcones Escarpment. Outcrops of highly fractured and faulted limestone on the 1,250-square-mile surface of the recharge zone act as a sieve and allow water to trickle into the aquifer. Although some recharge comes from rainwater, approximately 75–80 percent of the water entering the aquifer is supplied by streams and rivers crossing the recharge zone. After heavy rains, some streams on the Edwards Plateau overflow into sinkholes, which transmit the water directly to the aquifer. For example, the large Valdina Farms Sinkhole (Medina County) swallows 1,770 gallons per second during a flood.

Recharge of the Edwards Aquifer is also augmented by contributions from another large water deposit—the Trinity Aquifer—underlying northern portions of the Edwards Aquifer. Where the two water deposits are juxtaposed, upwellings from the Trinity Aquifer add at least 59,000 acre-feet of recharge annually to the Edwards Aquifer. Because of this connection, the aquifers are often referenced as the Edwards-Trinity Aquifer.

Water percolates downward by gravity from the recharge zone through porous limestone to the artesian zone, where the liquid is trapped between impermeable layers. As new water flows into the aquifer, its weight creates tremendous hydraulic pressure on the liquid imprisoned deeper in the formation. The resulting tension forces water upward through faults and wells to burst forth at the surface. When the first municipal well was drilled to supply water for San Antonio, for example, the water gushed upward 26 feet.

Springs gushing from underground aquifers afford a luxuriant microhabitat supporting a distinctive flora and fauna (fig. 7.19). Colonies of Chatterbox Orchids cling to dripping ledges and thrust their showy orange blossoms above verdant stands of Southern Shield Ferns, Maidenhair Ferns, and horsetails congregated near the water source. A little farther away, dense clumps of Seep Muhly and Little Bluestem protect Western Slimy

Figure 7.19. Verdant microhabitats often surround the clear, cool springs that erupt from aquifers beneath the Edwards Plateau. Photograph by Paige A. Najvar.

Salamanders and Red-striped Ribbonsnakes from predators.

Although the Pearl Brewing Company no longer exists as a corporate entity and its brewery in San Antonio has closed, the original advertising slogan remains a viable descriptor for the Edwards Plateau. The many springs giving birth to clear rivers flowing through verdant canyons still provide the biological diversity and scenic beauty for which the Texas Hill Country is so justly admired.

HILL COUNTRY RIVERS

Many Texas rivers and their tributaries with headwaters on the plateau owe their origins to springs and, as a result, typically flow year round (fig. 7.20). The Guadalupe River originates from two spring-fed branches in Kerr County and runs

Figure 7.20. Clear water from a nearby spring cascades over a small waterfall on Spring Creek (Burnet County), a tributary of the Colorado River. Photograph by Jonathan K. Gerland.

clear and swift beneath high limestone bluffs and ancient Baldcypress trees. Recurring floods along the middle and lower parts of the river prompted construction of a dam several miles upstream from New Braunfels. Completed in 1964, the dam impounded Canyon Lake, a reservoir with approximately 8,230 acres of surface area and 80 miles of shoreline. Although there are no natural lakes in the Texas Hill Country, reservoirs designed for flood control, water storage, and recreation occur on most of the rivers on the Edwards Plateau.

Two rivers, the Colorado and Concho, originate outside the region before crossing the Edwards Plateau. Riffle systems in the Concho and middle Colorado Rivers host the Texas Map Turtle, a species endemic to the Edwards Plateau. Baldcypress, Pecan, American Sycamore, and Black Willow dominate narrow riparian strips, but somewhat larger gallery forests of oaks, elms, hackberries, Black Walnuts, and Eastern Cottonwoods develop on wider floodplains, which occasionally experience catastrophic flooding.

Deep channels and vegetated shallows in several rivers provide habitat for native Guadalupe Bass (see appendix A) and introduced Smallmouth Bass; tiny, spectacularly colorful Orangethroat Darters rush from rock to rock in fast-running riffles.

Streamsides are busy places day and night—crevices in canyon walls offer refuge for Texas Alligator Lizards during the day and calling perches for Cliff Chirping Frogs at night. The diminutive frogs call from damp recesses, but their voices are almost drowned by the cacophony from Blanchard's Cricket Frogs and Spotted Chorus Frogs favoring backwater pools left when the rivers change their course. Armadillos shuffle noisily through the floodplain, searching leafy debris for grubs and insects while Painted Buntings, Summer Tanagers, and White-eyed Vireos advertise their territories from perches high in the canopy.

BRIGHT-FACED SONGSTERS

The clear, buzzing songs of Golden-cheeked Warblers—the only bird species nesting exclusively

Figure 7.21. The endangered Golden-cheeked Warbler (top) requires thin strips of bark exclusively from Ashe Juniper trunks (bottom) for nest construction. Photographs by Steve Maslowski, US Fish and Wildlife Service (top) and Brian R. Chapman (bottom).

in Texas—herald the beginning of spring in the cedar brakes (fig. 7.21). Upon returning to the Edwards Plateau after overwintering in southern Mexico and Central America, the small, colorful males establish breeding territories typically located in closed-canopy communities near streams or on canyon slopes. After selecting a mate, females construct nests woven from thin strips of bark collected exclusively from mature Ashe Junipers—no other source suffices—but the nests themselves may be located in any suitable tree regardless of species. As a result, the loss of closed-canopy, old-growth stands of Ashe Juniper poses the greatest threat to Golden-cheeked Warblers. Efforts are underway to preserve 76,320 acres of Golden-cheeked Warbler and Black-capped Vireo habitat in both the Balcones Canyonlands National Wildlife Refuge (in Burnet, Travis, and Williamson Counties) and a private initiative in Travis County, the Balcones Canyonlands Preserve System. Several other federal, state, and private entities manage habitat on the Edwards Plateau for the protection of these two endangered birds.

NATURE'S WATER FILTERS

Perhaps the least obvious inhabitants of the Edwards Plateau are the freshwater mussels living largely unnoticed on the bottoms of ponds, streams, and rivers. Of the more than 50 species in Texas, approximately 21 occur in the Texas Hill Country. As filter feeders, freshwater mussels require a bountiful supply of diatoms, desmids, and algae. Because of their diet, mussels rarely occur in headwater pools and streams on the Edwards Plateau, where the cool, clear water flowing directly from aquifers lacks sufficient food. Some species require moderately to swiftly flowing water, but others tolerate a variety of conditions where oxygen levels are high enough and pollution levels are low enough for their survival. The Concho River was named for its abundance of freshwater mussels, especially the Tampico Pearlymussel. In contrast, stretches of the Medina, Guadalupe, and Llano Rivers flowing north of the Balcones Fault Zone generally lack mussel populations because of scouring that disrupts their beds during severe floods.

Because their soft bodies are enclosed by two

shells—or valves—hinged with ligament-like tissue, freshwater mussels are known as bivalves (as are clams and oysters). Adult freshwater mussels spend their lives entirely or partially buried in the bottoms of permanent bodies of water. A muscular "foot" can be extended ventrally from the mantle that encases their inner body, and along with movements of the two valves, it can facilitate both vertical and horizontal travel in the substrate. During warm seasons, mussels reside near the surface of the substrate, where, by slightly opening their valves, they expose two tubelike siphons, which constantly move water containing oxygen and food across the gills and out again. Food particles suspended in the water become entrapped in the gills and then move by ciliary action to the mouth. Mussels can change locations but seldom move as long as the water bears enough essential nutrients for their needs.

Freshwater mussels have a unique but complex reproductive cycle that varies somewhat by species. Typically, males release their sperm, which the females take in through their incurrent siphons. At about the same time, eggs released by females become fertilized on their gills, parts of which serve as brood pouches collectively known as marsupia. Within each marsupium, fertilized eggs develop into tiny parasitic larvae called glochidia; to continue their development, the larvae become parasites that must attach themselves to the gills or skin of a host fish. The glochidia of some species of mussels parasitize only one kind of fish—a classic example of host specificity—whereas others can survive on fishes of several species.

Remarkably, freshwater mussels employ wily ruses to entice fish close enough to transfer their larvae. For example, the Texas Fatmucket—a mussel endemic to the Guadalupe–San Antonio and Colorado drainages of the Edwards Plateau—accomplishes this task with a mantle modified to resemble a minnow complete with eyespot, tail, and waving fins (fig. 7.22). When a passing fish attempts to dine on the "minnow" protruding seductively from the substrate, the mussel suddenly engulfs the unsuspecting predator with a blast of glochidia. Other mussels acquire hosts by releasing glochidia in gelatinous packets that resemble small minnows or tiny insect larvae. After attaching to a host with

Figure 7.22. The mantle of the female Broken-rays Mussel resembles a minnow that lures predatory fish near enough to be infected with the mussel's parasitic larvae. Photograph by M. Chris Barnhart.

minute hooks, glochidia embed themselves in the fish's tissues, where they change little in size but develop most of their adult characteristics, all without much harm to the host. When development is complete, a miniature version of an adult mussel breaks free of the host and falls to the bottom. Once there, the now independent mussel becomes a lifelong filter, quietly helping maintain the sparkling waters of the Hill Country.

LOST MAPLES

Relicts of the past—fragments of rock fences, sagging farmsteads, and rusted hulks of farm tractors overgrown by the very vegetation they once subjugated—lie scattered across the Edwards Plateau. Less obvious, however, are the biological remnants of bygone ages in which the environment differed significantly from conditions today. Some biological relicts may be the few remaining descendants of a once-large assemblage of related species (taxonomic relicts), or the enduring individuals of a more widespread population (biogeographic relicts).

High canyon walls bordering the headwaters of the upper Sabinal River and adjacent drainages in northwestern Bandera County provide a sheltered enclave for a relict population of Bigtooth Maples (fig. 7.23). Far removed from any other maple forest today, these isolated trees descended from maples that were once widespread in Texas about 10,000 years ago. Although Texas remained ice-free

Figure 7.23. Lost Maples, a relict population of Bigtooth Maples, lies sheltered by high canyon walls along the upper Sabinal River.

during the Pleistocene Epoch, the frosty conditions imposed by northern glaciers nevertheless pushed many species of plants and animals southward. The ancestors of Sugar Maples thus migrated to the cool, wet periglacial environments that developed in the southern half of North America, where, separated geographically, they evolved into eastern and western forms.

With the retreat of the glaciers and a warming climate, the two new forms—the Sugar Maple in the east and Bigtooth Maple in the west—moved northward as conditions gradually became too warm and dry for continued survival in the south. A few isolated populations of Bigtooth Maples, however, persisted in protected southern environments such as the Sabinal River Canyon, enabled by the deep, rich soil, dependable water, and shelter from drying wind and searing sun afforded by the dissected canyons of Bandera County.

Lost Maples State Natural Area, near Vanderpool, protects 2,900 acres surrounding the canyons that host the relict population of Bigtooth Maples and rare endemic wildflowers, including Canyon Mock Orange, Big Red Sage, and three recently described new species. In autumn, the leaves transform from green to golden yellow and then turn red before falling—visitors to Lost Maples experience a colorful palette worthy of a calendar and one of the most vivid displays of fall foliage in Texas. Rewarding scenery and a variety of birds await visitors at other seasons, as well. During the spring and summer, for example, energetic trills of Canyon Wrens echo off the sheer limestone walls, and Ruby-throated Hummingbirds clash in aerial combat near Mexican Flame bushes. Above the stream, jewel-like Green Kingfishers flash perch to perch in search of vulnerable prey. The network of sun-dappled canyons at Lost Maples, in addition to ecological importance as a natural refuge for thousands of years, also offers stunning scenery that some describe as the "Yosemite of Texas."

A WORLD WITH OXYGEN

Deep in the heart of Texas—Mason County, to be precise—limestone cliffs along the Llano River date to a time when life consisted predominantly of single-celled organisms so delicate as to barely persist as fossils. These included cyanobacteria—once called blue-green algae—of the late Precambrian Era, a block of time beginning with the Earth's formation 4.6 billion years ago and lasting until about 541 million years ago (about 88 percent of geological history). Cyanobacteria survived in an atmosphere of methane, carbon oxides, and other gases, but it contained essentially no oxygen. Toward the end of the Precambrian, some multicelled, soft-bodied forms shared this environment with the cyanobacteria, but life

Figure 7.24. A kayak piloted by Tonya Huston floating near a limestone boulder in the Llano River demonstrates the size of some fossilized stromatolite columns. Photograph by Tony Plutino.

nonetheless remained sparse and uncomplicated for many millions of years.

The Llano's limestone rocks, some of which fell into the riverbed (fig. 7.24), bear fossils known as stromatolites, laminated structures of fossilized cyanobacteria that evolved more than three billion years ago. When living, they grew as stumpy columns in shallow lagoons of ancient seas. The sticky cap of living cyanobacteria at the top of the columns captured fine sediments with each tidal cycle, thus adding a tissue-thin mineral layer on which developed yet another colony of cyanobacteria—a process that added about 2 inches of height per century.

As time passed, the stromatolites, supplemented by mats of cyanobacteria, nonetheless slowly— very slowly—began changing the physical world. Cyanobacteria lack nuclei, but they are endowed with a marvel of nature's engineering— chloroplasts—and thereby produce photosynthetic oxygen in the same way as do modern green plants. For millions of years, however, dissolved iron in the seawater chemically captured the oxygen, as shown by layers of oxidized iron, primarily hematite,

forming dark bands in Precambrian sedimentary rocks. This period of "mass rusting" ended when the iron became fully saturated, after which a surplus of oxygen accumulated in what is known as the Great Oxygen Event. The Precambrian thus ended, still with a simple biota, but primed by an oxygenated atmosphere.

About 542 million years ago, at a time when oxygen likely reached current levels, life underwent a burst of speciation known as the Cambrian Explosion. Ironically, the arrival of these new forms of life curtailed the distribution and abundance of stromatolites, which thereafter survived only where extremely high water temperatures or salinities excluded herbivores and competitors. Some still persist, notably those protected at Shark Bay World Heritage Site in Western Australia.

The relatively sudden appearance of such a full range of animals—swimmers, crawlers, and burrowers—puzzled Charles Darwin, as no ancestors for these creatures were apparent on which to establish their evolutionary lineage. What is clear, however, is the role played by cyanobacteria, often in the form of stromatolites, in producing an oxygen-rich atmosphere in a developing world—and those in Mason County were part of the action.

EXOTIC SPECIES: THE WORLD'S WILDLIFE IN TEXAS

Travelers crossing the Edwards Plateau often spot animals native to Africa or other semiarid regions of the world. For most, an unexpected glimpse of a Plains Zebra or a Reticulated Giraffe when rounding a highway curve can be surprising—and quite distracting.

By definition, any plant or animal from a foreign area that is transported either intentionally or accidently to another location represents an exotic species. In Texas, however, the term "exotic" is commonly applied to large, nonindigenous mammals and birds, especially those introduced for sport hunting. In addition to two large flightless birds—Emu and Greater Rhea—as many as 40 species of exotic mammals from Africa, Eurasia, and South America roam some well-known large ranches on the Edwards Plateau.

Nilgai—an Asian antelope introduced on the King Ranch in South Texas in 1930—was the first exotic species released in Texas. For many years thereafter, the Texas Parks and Wildlife Department resisted requests to release more exotics, but the agency eventually developed import regulations and relaxed the restrictions. Some landowners maintain exotic animals simply for their novelty or because breeding the captive animals helps conserve a species endangered in its native land, but income from hunting stands foremost as the reason for managing a herd of exotics.

Leasing private land to hunters began in Texas in the early 1920s as a way ranchers might gain a reliable source of income in a region where an unreliable environment too often governed their economic success. Because exotics were not subject to "closed seasons," nonindigenous game animals provided hunters with other options besides the annual season for White-tailed Deer. Indeed, some ranchers obtained steady income from year-round hunts of exotic game. There are, however, some major concerns about "exotic game ranching," including the introduction and spread of diseases and parasites, competition with native wildlife, range degradation, and the uncontrolled expansion of exotic populations.

Importation of exotic species requires quarantines designed to reduce the threat from foreign diseases and parasites, but "mistakes" nonetheless happen. For example, after fulfilling the quarantine, Black Rhinos released on a Texas ranch still carried Tropical Bont Ticks, which serve as vectors for heartwater disease in Africa. In another case, Axis Deer released on the Edwards Plateau suffered an outbreak of malignant catarrhal fever. Underlying these concerns is the worry that native animals may be more vulnerable to foreign diseases than their exotic hosts because they lack long exposure to these pathogens in their evolutionary history.

When exotic animals are introduced, competition often begins between native and domestic animals for the available resources— food, water, and cover. To illustrate, the preferred foods of White-tailed Deer—green, succulent forbs and browse—also dominate the diets of Axis Deer, Sika Deer, Fallow Deer, and Blackbuck Antelope. However, when forbs or browse become scarce, the exotics easily shift to grasses, whereas White-tailed Deer soon face starvation. Because of their adaptability, Sika Deer and other exotics have been likened to "sheep in deer clothing." Given that only so much natural food exists per unit area, exotic ungulates easily outcompete whitetails, especially when both share high-fenced pastures.

Many landowners on the Edwards Plateau and elsewhere in Texas protect their investments in native and exotic wildlife with high (average height 8 feet) "game fences." The fences are intended to control the movements of managed wildlife, reduce immigration of other animals, and deter poaching. Game fences sometimes cause serious biological and ecological issues—vegetation degraded by overbrowsing, for example—but with proper care they often work well if the confined populations remain slightly below carrying capacity. Still, even well-constructed fences cannot prevent escapes, and consequently, Axis Deer and other exotics now range freely on the Edwards Plateau. Among other issues, the growing number of vehicular collisions with Axis Deer raises new legal questions about liability.

More than 70 exotic species now reside in Texas, and most of these are held on the Edwards Plateau. Whereas bluebonnets form an indelible element of the regional landscape, so today do exotic animals. For better or worse, a springtime trip through the Hill Country now features sightings of Emus or Blackbuck Antelope as well as wildflowers.

CONSERVATION AND MANAGEMENT

The ready availability of water from abundant seeps, springs, and rivers on the Edwards Plateau surely influenced the region's long history of human presence. Caves in the region, for example, bear evidence of occupation at least 10,000 years ago, and a prominent hypothesis suggests that overhunting by Paleo-Indians caused the extinction of mammoths and other large mammals of the Ice Age. However, the most lasting environmental impacts coincided with the arrival of settlers. At first, the goats, sheep, and cattle introduced by Spanish settlers simply provided subsistence, and the small herds posed little threat to the character of the vegetation. By the late 1850s, however, ranches with overstocked ranges began depleting much of the original grassland vegetation; palatable species such as Big Bluestem all but vanished under intensive grazing pressure. Shortgrasses, among them Buffalograss, Curly Mesquite, and threeawn grasses, as well as King Ranch Bluestem, replaced the tallgrass. So great was the damage that some evidence suggests that the original grassland composition would not return even if grazing miraculously ended.

Invasive species continue to enter the region from every direction, but the harmful effects, if any, of many of these remain unknown. Red Imported Fire Ants, however, have reduced populations of Eastern Cottontails and both White-footed and Northern Pygmy Mice throughout the Edwards Plateau. These troublesome ants have also invaded hundreds of caves in the region, where they pose a major threat to cave fauna, including many species already listed as endangered or under consideration for listing. Unfortunately, Red Imported Fire Ants are difficult to control, but some toxic baits have at least temporarily eliminated individual colonies.

The region's major environmental threat—urbanization—has accelerated in recent years. Rapid expansions of the human population, subdivisions, and 1- to 10-acre "ranchettes" together degrade native vegetation and threaten the integrity of both surface and underground water systems. Land developed for residential and light industrial uses, especially near the larger cities, invariably changes the dynamics of recharge zones for watersheds and aquifers, as well as lessens water quality itself, commonly because of contaminated runoff from paved surfaces. Increasing demands for water by metropolitan, agricultural, and industrial entities have already reduced the flow of Comal Spring, San Marcos Spring, and many other springs. The survival of many aquatic organisms, including species federally listed as endangered, depends on the continued health of the beleaguered Edwards Aquifer (infobox 7.4).

Freshwater mussels quickly succumb to pollution and thus serve as sensitive indicators of water quality. To be useful in this regard, however, regular monitoring of their populations and distributions is required, already a necessity for the False Spike, a critically endangered species found only in a short stretch of the Guadalupe River, and five Edwards Plateau species considered candidates for listing as threatened or endangered. Dams or other structures that alter stream flow also threaten freshwater mussels, first because increased siltation upstream suffocates their beds, and second because the diminished flow reduces fertilization in some species or eliminates the habitat requirements of fishes serving as hosts for larvae in others. Obviously, these threats can be identified only if thorough population surveys are conducted before new impoundments interrupt the natural flow of streams and rivers.

According to some authorities, the cascading loss of biotic and genetic diversity in the Edwards Plateau region can be checked only by establishing several large biological reserves, each selected on the basis of its resemblance to pristine conditions. Efforts continue to preserve large tracts of undisturbed habitat within the Balcones Canyonlands National Wildlife Refuge and the Balcones Canyonlands Preserve System. Several other federal, state, and private entities seek sites for managing endangered species. For example,

INFOBOX 7.4. CLARK HUBBS
Ichthyologist and Eminent Naturalist (1921–2008)

As a boy, Clark Hubbs accompanied his father, Carl—a celebrated ichthyologist and naturalist—on collecting trips, lured in part by rewards of one dollar for discovering a new species and five dollars for a new genus—good money for a lad in the 1930s. Thus began a career of field biology highlighted by a robust work ethic and a love of discovery that would remain a lifelong trademark. Indeed, Clark—at 87—was busy collecting fish and field data just a few weeks before his death.

After receiving a BS in zoology from the University of Michigan in 1942, Hubbs served in the US Army during World War II in the Pacific Theater, notably at Okinawa. When the war ended, he enrolled in Stanford University, where he earned a PhD in 1951; during this period, he was associated with both the Hopkins Marine Station and the Scripps Institution of Oceanography. Before completing his doctorate, Hubbs joined the faculty at the University of Texas at Austin in 1949 as an instructor—the humble beginning of an academic career that lasted for nearly 60 years and produced more than 300 publications. He went on to serve the university in many capacities, including as chair of the Division of Biological Sciences. His distinguished service led to an appointment named in his honor—the Clark Hubbs Regents Professor—which, following his retirement in 1991, continued as the Clark Hubbs Regents Professor Emeritus.

Hubbs sampled more rivers, streams, and springs and collected more fish specimens in Texas than anyone else, but special interests always centered on the spring-fed streams of the Trans-Pecos and Edwards Plateau. Whereas his collections contributed significantly to the conservation of the state's fish fauna, he also acted firsthand in furthering the protection of fishes. He testified as an expert witness in a lawsuit that established the Edwards Aquifer Authority, a state agency whose mission in part ensured that enough spring water remained available for the survival of the San Marcos Gambusia and Fountain Darter. When Hubbs realized that dredging had destroyed the original habitat for the endangered Comanche Springs Pupfish and Pecos Gambusia, which thereafter persisted only in a highly altered canal at Balmorhea State Park, he enlisted the aid of Texas Parks and Wildlife, volunteers, and students from several universities. Their efforts eventually succeeded in diverting water from the canal to re-create the natural habitat for the fish—a desert ciénaga.

Hubbs developed the first taxonomic key for identifying the state's freshwater fish, which may have triggered his last major contribution to the natural history of Texas—a checklist for the fishes of Texas. This project in turn evolved into the Fishes of Texas Online, aided by two of his former students who provided additional information and technical expertise. The mass of data provides the sampling site for every preserved fish ever collected in Texas, including tens of thousands of those contributed by Hubbs himself, plus the taxonomy, characteristics, habitat, and biology for each species. Fishes of Texas Online serves as an invaluable resource for research and conservation of the state's fish fauna as well as a monument to Hubbs and his contributions to ichthyology.

During his career, Hubbs actively participated in several professional organizations. His colleagues elected him president of the American Society of Ichthyologists and Herpetologists, the American Institute of Fishery Research Biologists, the Texas Organization for Endangered Species, and the Texas Academy of Science. He edited *Copeia* and the *Transactions of the American Fisheries Society* and helped found the Southwestern Association of Naturalists, for which he served as editor of its journal, as president, and as a long-term member of the Board of Governors. The association honored him with the W. Frank Blair Eminent Naturalist Award in 1991 and with the George Miksch Sutton Award in Conservation Research in 1997. Still, his greatest and most enduring legacies remain in the students he inspired, the survival of endangered species he championed in his research and conservation efforts, and the threatened aquatic habitats he identified and helped protect. By any reckoning, Clark Hubbs will long be linked with ichthyology in Texas.

in 1985 Travis County purchased and protected 232 acres surrounding Hamilton Pool, a large swimming hole in a grotto formed by a collapsed sinkhole. In addition to the uplands, where stands of mature Ashe Juniper provide nesting material for Golden-cheeked Warblers, the shaded recess near the waterfall tumbling into Hamilton Pool harbors Chatterbox Orchids, Red Bay, Maidenhair Ferns, and a variety of mosses, as well as a nesting site for Cliff Swallows.

INFOBOX 7.5. THE NATURE OF ROY BEDICHEK (1878–1959)

Adventures with a Texas Naturalist thrust Roy Bedichek into the hearts and minds of readers with a love of nature. Bedichek crafted the book, published in 1947 when he was 69, from three decades of notes accumulated as he wandered across Texas visiting sites from prairie to forest. Usually traveling alone, he often camped out, cooking over an open fire. Then, at the urging of two influential friends, famed historian Walter Prescott Webb and literary giant J. Frank Dobie, Bedichek secluded himself to write about his experiences and observations. For want of an isolated pond and rustic cabin, he instead lived and wrote Walden-like for a year in a large room in a stone building on Webb's ranch near Austin. There, with the companionship of a curious Canyon Wren that popped in and out of a hole in the ceiling, Bedichek wrote at an oak table once in the service of gamblers and confiscated by Texas Rangers. He stocked this hideaway with 1,200 reference books, cooked with iron pots in the fireplace, and for inspiration, gazed across the landscape of the Texas Hill Country.

By any measure, Bedichek had sampled life. He had picked berries in New Jersey, waited tables in Montreal, homesteaded in Oklahoma, slaughtered hogs in Chicago, and labored on river boats on the Ohio. He had also tried his hand with the newspaper business in Fort Worth and San Antonio, taught high school English in San Angelo, and, with a friend, hiked across Europe. This done, Bedichek pedaled a bicycle from Falls County in north-central Texas to Deming, New Mexico, where he edited a local newspaper, served as secretary for the chamber of commerce, owned a small ranch—and married. He had traveled much—and often—since his birth in Illinois and move to Texas with his family six years later. In 1903, Bedichek graduated from the University of Texas, where, in 1925, he would also earn a master's degree.

His job hopping over, Bedichek started work in 1917 with the Austin-based University Interscholastic League, then a part of the University of Texas Bureau of Extension, where he remained until retiring in 1948. His duties required frequent travel, which further enabled his visits to natural areas across the state. On one such trip, he camped among the dunes near Monahans, where he presaged their preservation as a state park—an event realized in 1957 (chapter 8). Marveling at the acorn production from the miniature forests of Shinnery Oak among the dunes, he ventured that the fruit-to-wood ratio might be like no other in the world, which remains unproved but is likely accurate. In the Hill Country, he likened the areas cleared of "cedar" to a landscape diseased by leprous spots, where future studies will reveal the effects of this practice on wildlife, perhaps beneficial in some cases, but certainly harmful for Golden-cheeked Warblers (chapter 7).

Fences, Golden Eagles, and especially Northern Mockingbirds held a special place in Bedichek's world. Fences interfere with the healthy circulation of natural life in ways not unlike hardened arteries in the human circulatory system. Still, fences along rights-of-way maintain roadside corridors where native vegetation can persist free from ruinous grazing. Eagles, shot from aircraft for their love of lambs, deserve better, and Bedichek opined that improved range conditions might restore natural prey to levels where the birds would no longer hunt livestock. As for mockingbirds, Bedichek admired their character and mannerisms even more than their song and espoused at length the heresy that they did not at all mock other birds (see appendix A).

Readers can quickly see that *Adventures* is no mere account of nature's splendor, as the book reflects Bedichek's bent for blending scientific objectivity with poetry and philosophy. The text cites top-ranked scientists of the day along with references to the likes of Browning and Whitman—all testimony to his perception of the fullness of the natural world in which we dwell. Not least among these is Tennyson's celebrated "Nature red in tooth and claw," which Bedichek notes is but a partial view of a whole to be seen not as a still, but as a motion picture of the limitless drama of nature. He deplores how technology and mechanization "have broken the rhythm of life," including the redistribution of our population into "huge clots, called cities," and advances Huxley's *Brave New World* as a case in point. Bedichek thus laments, "We have been expelled from an environment in which we were part and parcel in the other life about us," reminding us that "Though inland far we be, Our souls have sight of that immortal sea, Which brought us hither."

At Barton Springs in Austin, a sculpture identifies Philosopher's Rock, named for the site where three titans in the intellectual history of Texas regularly conversed, and one of the bronze likenesses portrays a genteel naturalist sometimes heralded as the state's "Most Civilized Soul."

8
HIGH PLAINS
PLAYAS, SHORTGRASS, AND SHIN OAK

They complain that the Plains are flat and bare.

Naturally—that is the beauty of them.

The scarcity of landmarks, the scale and sweep

of the High Plains are what make their sublimity.

— STANLEY VESTAL, *SHORT GRASS COUNTRY* (1941)

In Texas, the High Plains landform is represented by two geographic entities, each with similar characteristics but physically separated by the Canadian River Breaks (infobox 8.1). The northern area forms the tip of the Texas Panhandle and part of a huge plateau extending over a vast area of Oklahoma, Kansas, Colorado, Nebraska, and South Dakota. Some systems designate the part in Texas as the Canadian-Cimarron High Plains, today an area devoted largely to rangeland (fig. 8.1). In the past, however, the land was extensively tilled and emerged as the epicenter of the infamous Dust Bowl in the 1930s. The destructive force of this prolonged drought—and the erosion that followed—initiated the Soil Conservation Service (now the Natural Resources Conservation Service) and land-management practices, among them shelterbelts, contour plowing, and sites purchased expressly to restore grassland cover on severely damaged soils.

South of the Canadian River Breaks lies a tableland widely known as the Llano Estacado. The area was isolated—high and dry—during an uplift that enabled the Pecos and Canadian Rivers to capture and divert the ancient rivers flowing southeast from the Rockies. A rocky layer of whitish calcium carbonate developed as water in the mineral-rich upper soil horizons evaporated, leaving a cement-like layer that was later covered with a veneer of wind-deposited topsoil. Where erosion has removed the topsoil, as at the edges of the Llano, this hardened stratum emerges as a shelflike formation widely known as "caprock" (fig. 8.2).

THE LLANO ESTACADO

The thirst for gold is a powerful force, but Francisco Vásquez de Coronado (1510–1554) may have had some second thoughts as he crossed—day after day—a land that stretched flat and featureless to the horizon. The Spaniard and his troop were trekking across a grassland that seemed as endless as it was lacking in gold—a prairie cropped by

(*opposite*) **Figure 8.1. Windmills, flat terrain, and Sandhill Cranes highlight the features of the High Plains. Photograph by Jude Smith.**

herds of American Bison that were so numerous that not a day passed in which Coronado lost sight of the shaggy "cattle." Historians have reconstructed several versions of the expedition's eastward route (in 1541), one being a sweeping loop eastward from the modern town of Hereford to Canadian. Coronado later returned empty handed but with the distinction of being the first European to encounter the High Plains of Texas.

Coronado is credited with naming this vast region the Llano Estacado, usually translated as "Staked Plains" for any one of at least three reasons. The first attributes the "stakes" to the tall stalks of yucca, a plant still commonly seen on the plains today. Another supposes the Spaniards placed stakes as markers along their route as a means of retracing their steps on their return trip, although the source of wood for these stakes remains unexplained, nor did Coronado in fact take the same route when returning to Mexico. Another suggests the translation should be "palisaded plains" because of the fort-like façade of the escarpment confronting Coronado as he approached the vast tableland from the west (fig. 8.2). An altogether different explanation has nothing to do with stakes. Some believe *estacado* is a corruption of *estancado,* meaning "stagnant water," a reference to the thousands of playa lakes dotting the plains. Coronado in fact complains that the necessity of marching his troop around the plentiful lakes slowed his advance.

Two communities dominate the native vegetation of the Llano Estacado. Before settlement, a shortgrass prairie occupied most of the area and formed the featureless terrain heralded in the journals of explorers (fig. 8.3). Indeed, the shortgrass prairie seemed desolate, monotonous, and endless to Captain Randolph B. Marcy (infobox 8.2). In his words, the landscape appeared as "a vast illimitable expanse of desert prairie [and] a desolate waste [that] always has been, and must continue, uninhabited forever." Buffalograss and Blue Grama characterize the shortgrass prairie. Both form thick sods of densely packed roots; up to 70 percent of

INFOBOX 8.1. CANADIAN RIVER BREAKS

The vast High Plains of the North American interior are subdivided by the Canadian River, which cuts west to east across the northern end of the Texas Panhandle. North of the river, the High Plains continue unbroken into Oklahoma, Kansas, and far beyond, whereas the famed Llano Estacado lies isolated south of the river, mostly in Texas but also spilling over into eastern New Mexico. Over time, the river's wear carved a valley with red sandstone walls replete with gullies and ravines—rough terrain collectively known as the Canadian River Breaks. The walls form steep-sided canyons in New Mexico, but these become less abrupt and form a broad valley in Texas, where the riverbed lies some 500–800 feet below the plains. The slopes of the valley are conspicuously studded with clumps of "cedars," the regional name for those tall shrubs that botanists primly note are really junipers. After leaving Texas, the river crosses Oklahoma, where it merges with the Arkansas River just west of Fort Smith.

For some birds, the Canadian River Breaks serve as a conduit that facilitates penetration of the High Plains by such species as House Wrens, Northern Flickers, and Eastern Bluebirds. Similarly, Red-headed Woodpeckers likely invaded New Mexico along the pathway formed by the riparian habitat bordering the river.

The Canadian's swath across the High Plains provided explorers, beginning with the conquistadors of the sixteenth century, with a well-watered route across a landscape famed for its aridity. As the river cut downward through the strata, it sliced through the water table lying beneath the High Plains, which in turn created outpourings of springs and seeps that added to the Canadian's flow. Native Americans, notably those that anthropologists today identify as the Antelope Creek Phase of the Pueblo Culture, had encamped in the valley long before the Spaniards tramped through in search of golden cities that never were. People of this culture occupied the region between 1200 and 1500 AD and left behind ample evidence of their hunting, gathering,

and horticultural way of life. Even earlier cultures—Clovis, Folsom, and Plainview—mined the red bluffs of the Canadian River Breaks for multicolored flints whose excellent edge-holding properties made them ideal for arrow points, knives, scrapers, and other tools. In one case, Alibates flints were associated with a mammoth kill site on Blackwater Draw in eastern New Mexico. Technically a type of Permian-age dolomite, the flints were highly prized and widely traded and hence often ended up in Montana, central Mexico, and east of the Mississippi River. More than 700 pits, once up to 25 feet wide and 7 feet deep, remained in continuous use for some 12,000 years; these sites, including piles of chipped flakes, now form the centerpiece of Alibates Flint Quarries National Monument, where, with advance reservations, ranger-led tours can be scheduled.

Given the river's location, "Canadian" seemingly represents a hefty case of geographical awareness run amok. The present-day descriptor apparently stems from the French Canadian trappers who once traded along the river as early as the 1740s, which prompted its designation as the "River of the Canadians." The trappers were real enough, but another scenario for the origin of the river's name seems more likely, namely that something went awry in translation. Indeed, deletion of a Spanish tilde from *cañada*, meaning "sheepwalk," and its amplified form *cañadon*, which translates as a really large and deeply cut *cañada*, offers a far more plausible explanation. Thus, when Anglicized in print and speech—and lacking the squiggle—"canadon" easily drifted into "Canadian." The misnomer may have become forever cemented when the formal reports of US Army explorers such as Lieutenant James W. Abert in 1845 (whose manuscripts included the tilde) were printed as government documents without the important tilde. In any case, *cañadon* reflects an accurate image of the valley's appearance— a natural corridor of canyons regularly used by traders and herders—and highlights a strong connection between language and landscape.

the root biomass develops in the top 6 inches of the soil profile, and the remainder extends about 5 feet deeper. Most of the sod-forming roots are less than 0.04 inch in diameter, thus maximizing the surface area available to absorb moisture. These features developed as adaptations to the region's

sparse rainfall—some 18 inches per year on average, but at times varying from 12 inches to as much as 45 inches—and to the grazing pressure of bison. Because of these extensive root systems, some ecologists quip that these grasses are "short only aboveground." A few species of midgrasses, such as

Figure 8.2. A steep escarpment forms the northwestern rim of the vast Llano Estacado. Photograph by Larry L. Choate.

Figure 8.3. Shortgrass prairie stretching to the horizon in the western Llano Estacado seems desolate and endless. Photograph by Larry L. Choate.

Sideoats Grama and even the taller Little Bluestem, also mingle in the composition of the shortgrass prairie, as do scattered plants of yucca and small cacti.

Distinctive fingers of sandy hills and dunes poke eastward from New Mexico into Texas, differing from the flat prairie landscape on either side. One such area, known as the Muleshoe Sandhills, extends for about 58 miles into Bailey, Hale, and Lamb Counties. A newly acquired state wildlife management area in nearby Cochran, Yoakum, and Terry Counties protects another area of dunes, benefiting Lesser Prairie-Chickens. Geologically, these formations likely developed when windblown sands accumulated in relict channels of an ancient river. The deep sands offer a moisture regime that supports taller grasses—Sand Bluestem, Little Bluestem, and Sideoats Grama, for example— and other vegetation dissimilar from that in the surrounding areas of shortgrass prairie. In all, some six communities can be identified within the sandhills, of which two—Sand Sage–Midgrass and Sand Sage–Ragweed–Yucca—occur most commonly. This unique area attracted Paleo-Indians and later cultures, some of whom occupied large campsites.

Periodic fires maintain most grassland systems, including the shortgrass prairies, against the

Figure 8.4.
Fire suppression
enables junipers
and other brushy
vegetation to
invade the High
Plains from
its bordering
escarpments.
Note the thick
"caprock" of
calcium carbonate.
Photographs by
Larry L. Choate.

invasion of woody plants. Three advantages give grasses the upper hand in this relationship. First, when grasses burn, only the current year's growth is lost to the flames, whereas woody plants lose several years of growth that take many years or even decades to replace—if they survive at all. Second, and related to the first, grasses can replace their seed-bearing abilities rather quickly after a fire, commonly within weeks. In comparison, most woody plants require many years to reach sexual maturity. Finally, the growing points (buds) of most grasses lie protected in the middle of the plant at or just below ground level, whereas the buds of woody plants develop at the tips of branches where they are fully exposed to fire damage. Because of these features, fires clearly inhibit woody vegetation and favor the development of grasslands. Unfortunately, the reverse is also true: when fires are suppressed, grasslands give way to communities of trees and shrubs. Hence, mesquites have invaded some areas of the shortgrass prairie, often enhanced by overgrazing. Junipers creep out of the adjacent Rolling Plains and onto the flat plains along the eastern edge of the Llano, providing another indication of the changes wrought by fire suppression (fig. 8.4).

The second community develops where sandy soils dominate the western edge of the Llano, and here Shinnery Oak characterizes the vegetation, along with lesser amounts of Sand Sage. Shinnery Oaks typically form tangles less than 3 feet tall, although some individuals may reach twice that size. Their leaves, although distinctively lobed like those of most oaks, are irregularly shaped, and their waxy and leathery surfaces help retain water. During droughts, Shinnery Oaks often drop their leaves and delay the regrowth of new foliage.

The root systems of Shinnery Oak, like those of shortgrasses, are immense when compared to their aboveground biomass, again reflecting an adaptation to the rigors of drought. They extend downward 15–20 feet, and some are as thick as a human thigh. The underground system of Shinnery Oak also includes thick rhizomes—bud-bearing underground stems concentrated in the upper 24 inches of the soil profile (fig.8.5). Like roots, rhizomes store water but also produce shoots that initiate new growth from the parent plants. Nonetheless, Shinnery Oaks grow slowly and persist in old stands that only gradually colonize vacant areas nearby. Whereas the aboveground biomass usually lives for less than 15 years, the underground

Figure 8.5. When exposed, the root and rhizome networks of Shinnery Oak (top) reveal a biomass several times greater than the aboveground biomass of the same plants (bottom). Photographs by Russell D. Pettit.

systems persist much longer—clones resulting from asexual reproduction represent centuries of continued growth.

Forbs and grasses fill in the spaces in the low overstory. "Shinnery country," as it is known locally, provides the remaining core of habitat for Lesser Prairie-Chickens, but shrubby oaks also provide food and cover for a number of other species, including Northern Bobwhites, Pronghorns, and several small mammals.

A site along the southwestern base of the Mescalero Escarpment features a landform known as the Monahans Sandhills, the namesake of a

nearby town, where a dune field of striking white sands extends northwestward for about 200 miles (well into New Mexico) and the largest dunes reach heights of 70 feet. The origins of this formation remain obscure, but one idea proposes that erosion of mountains in New Mexico provided the sands, which accumulated in river systems flowing between the mountains and the escarpment. Later, when the Pleistocene ended and the rivers dried, the sands blew eastward until they piled against the base of the Llano's southwestern escarpment. The result was a Texas-sized sandbox associated with the southern flank of the Llano Estacado, presenting, in the words of the intrepid Captain Marcy, "a most singular and anomalous feature in the geology of the prairies" (fig. 8.6).

The sandhills continually shift as winds move the dunes in an ever-changing landscape, but parts of the formation are relatively stable, anchored primarily by the rugged root systems of Shinnery Oak. In most years, Shinnery Oak provides a bountiful source of acorns, which offer a key link in the rather meager food web of the sandhills. Whereas other plants in the sandhills lack the stability of Shinnery Oak, all are psammophilous, an ecological mouthful meaning "sand loving," and certainly apt for the area's environment.

Despite their aridity and shifting dunes, the Monahans Sandhills do not represent a true desert; the area receives an annual average of just over 12 inches of rainfall, which often ponds between some of the dunes. Where the water table lies near the surface, water accumulates in semipermanent seeps that may develop isolated patches of vegetation including willows, bulrushes, and cattails. This surprising abundance of water enabled Captain Marcy and his troop to cross the seemingly parched sandhills on their return trek to Fort Smith, Arkansas. His mission, completed in 1849, was to locate a route for settlers heading west, in part because of the gold rush to California. Years later, the Texas and Pacific Railroad, now part of the Union Pacific system, selected Monahans as a site to refill its locomotives steaming between the Pecos River and Big Spring.

Today, this unique area includes Monahans Sandhills State Park, a 3,840-acre facility located in parts of Winkler and Ward Counties. Even

Figure 8.6. Winds constantly shift and sculpt the Monahans Sandhills, enveloping willows and other vegetation growing in the slacks between the dunes. Low seeps in some slacks hold small ponds that attract thirsty wildlife and support wetland vegetation during years with abundant rainfall.

at this size, the park covers only a small part of the extensive dune field, which extends into southeastern New Mexico. The park features active dunes that shift in keeping with the prevailing winds and offer visitors opportunities for sandboarding and other outdoor experiences not readily available elsewhere in Texas.

PRAIRIE DOGS AND BISON: GRASSLAND ICONS

Two icons of prairie ecosystems are, like the grasslands themselves, largely memories of a bygone era. True, some Black-tailed Prairie Dog towns remain, but free-roaming bison are gone from the Llano Estacado, where both species once numbered in the millions. In Texas, a small herd of bison persists in the sanctuary of Caprock Canyons State Park, but these scarcely reflect the vast herds that once foraged on the resilient shortgrasses.

The end of what was known as the "southern herd" succumbed to hide hunters in 1870, when the last of some seven million of Coronado's shaggy cattle faded into history. Similar destruction, although not as permanent, befell the immense Black-tailed Prairie Dog colonies in Texas. Field biologist Vernon Bailey (infobox 1.1) estimated that 400 million prairie dogs occupied a colony some 100 miles wide and 250 miles long lying just east of the Llano Estacado. For all of Texas, Bailey proposed a total population of 800 million. Today, after decades of government-sanctioned poisoning, small colonies remain on farms and ranches, where they are not always welcome, but other colonies find protection at Mackenzie Park in Lubbock and in the Muleshoe National Wildlife Refuge (infobox 8.3).

INFOBOX 8.3. MULESHOE NATIONAL WILDLIFE REFUGE

Established in 1935 in Bailey County, Muleshoe is the oldest national wildlife refuge in Texas. The refuge now consists of 5,809 acres dedicated primarily to the management and protection of migratory waterfowl and Sandhill Cranes. A thriving town of Black-tailed Prairie Dogs and a host of songbirds are among the other species of wildlife benefiting from the three saline lakes and more than 5,000 acres of shortgrass prairie within the refuge boundaries. For the most part, the lakes depend on the runoff from precipitation, but springs supplement the water filling Paul's Lake. When full, the lakes together provide about 600 acres of surface water.

In the past, as many as 250,000 Sandhill Cranes spent all or part of the winter at Muleshoe, but 100,000 is more typical of the winter population at the refuge. Similarly, ducks and geese at times numbered in the tens of thousands, but these too have dwindled in recent years. Nonetheless, Muleshoe remains a viable link in the Central Flyway, one of four migratory routes running north and south across North America. Three saline lakes, each divided into two management units, provide the wetland habitat for these birds. Additionally, vigilant birders may see eight or more species of raptors, including Bald and Golden Eagles, in the course of one day. Coyotes and American Badgers are among the larger mammals on the refuge.

In 1944, the large number of ducks then concentrated at Muleshoe apparently triggered the first-known epizootic of Fowl Cholera in North America. The bacterial disease, while deadly for birds, is unrelated to the dysenteric cholera that infects humans. Some 60,000 ducks at Muleshoe died from the disease in the winter of 1956–1957. The disease remains a risk for waterfowl throughout the region.

As elsewhere on the High Plains, Blue Grama and Buffalograss are the prominent species, but Alkali Sacaton predominates on the lowlands where high salinities prevail in the soil profile. Unchecked, this salt-tolerant grass forms large clumps; the interspersed patches of bare ground expose the soil to erosion. Accordingly, the refuge manages the Alkali Sacaton community with prescribed burning to refresh the growth of the species in a manner that better protects the soil. The area of shortgrass prairie protected at the refuge is one of the few large blocks of this habitat in Texas that has never been turned by a plow. Limited grazing, under refuge supervision, is allowed as a means to simulate the influence once exerted by bison.

PRAIRIE DOGS AND ASSOCIATES

Of the five species of prairie dogs—a group endemic to North America—Black-tailed Prairie Dogs were (and are) the most abundant and widely distributed (fig. 8.7). Their range extended broadly across the western plains from Texas north to Montana and into Canada. Lewis and Clark captured (with considerable difficulty) one of these curious "barking squirrels" and sent it to Thomas Jefferson, and much to the president's delight, the hardy rodent survived the four-month trip to Washington.

Prairie dogs, the most social of all rodents, live in towns subdivided into wards and coteries. The latter are family groups consisting of an adult male and up to four adult females and their offspring. Neighboring families defend their coteries, but individuals within these units greet each other with "kisses" as a form of recognition, not courtship. Various postures and vocalizations transmit information within each social unit. A sharp two-syllable bark proclaims danger, which is followed by a prominent "jump-yip" display indicating "all clear" when the threat—perhaps a passing Coyote—has gone. Some calls in their repertoire identify specific predators. These and other behaviors and the messages they convey spread rapidly through the town.

Prairie dog towns form a maze of tunnels and dens constructed so that air circulates through the system. Dens serve specific purposes, including sanitation and nurseries. The mounds surrounding the entrances to tunnels provide lookout stations and keep out floodwaters. Some ecologists designate prairie dogs as ecosystem engineers—a type of keystone species that physically alters its environment in ways that significantly impact other species. In this case, their burrows provide shelter for other animals, improve soil aeration, lessen soil compaction resulting from herds of

hoofed grazers, improve water penetration, and reduce the cover that conceals predators. As one example, Mountain Plovers prefer to nest in active prairie dog towns and forage in the well-cropped vegetation. Horned Larks and Killdeer are often found in higher numbers in prairie dog towns than in the surrounding shortgrass prairie. Nonetheless, the association of prairie dogs and Burrowing Owls remains one of the closest on the shortgrass prairie.

Because Burrowing Owls remain active during the day, they are easily observed year round at most active prairie dog towns (fig. 8.7). These small but long-legged owls find refuge and nesting sites in unused tunnels and dens; no other species of owl nests underground. The birds line the tunnels and chambers they use with dried dung, and while the purpose of this strange behavior remains unclear, insects living in the dung may provide a local food supply. Bison were undoubtedly the primary source of dung when this behavior evolved, but livestock now fulfill this need. Whatever the reason may be, the owls remain serious about maintaining the dung lining, which they quickly replace if it is removed. The defensive behavior of young Burrowing Owls represents another intriguing adaptation to the species' lifestyle. When

disturbed in their nests, owlets mimic the buzz of rattlesnakes—one of several species often visiting the burrows—with such similarity that electronic comparisons with the buzz of real rattlesnakes show the two to be largely indistinguishable. The benefits of such protection are obvious for defenseless birds unable to escape predators in the confines of a small space.

Prairie dogs also form important links in the food chains of shortgrass prairies. Their predators include Golden Eagles, Coyotes, and, notably, Black-footed Ferrets. Regrettably, the last record of a Black-footed Ferret in Texas (in Bailey County) dates to 1963. When extensive control reduced or eliminated prairie dog populations, Black-footed Ferrets became indirect victims of the poisoning programs. Today conservation agencies raise and release the rare ferrets at well-protected sites, but Texas is not currently among the states involved with these efforts.

Epizootics of Sylvatic Plague—akin to the dreaded Black Death of European history—can destroy prairie dogs as well as humans. The first case of the flea-borne disease in Texas was discovered in 1946. In 2003, an epizootic occurred in a six-county area in the northwestern corner of the Texas Panhandle, including colonies at Rita Blanca National Grassland. When infected, individual prairie dogs seldom survive, and the mortality rate within colonies can exceed 95 percent. Because fleas infected with the lethal bacterium persist at a site for as long as a year after the last victim dies, colonies are slow to recover and may never regain their former densities. Humans can avoid the fleas by simply staying out of prairie dog colonies, but for those who contract the disease, early treatment with antibiotics provides relief.

AMERICAN BISON

Bison also represent a familiar and traditional icon of the North American prairies, earning their place as the emblem of the US Department of the Interior, sports teams, locations, and a host of other identities, including a moniker for hunter-showman William F. Cody (1846–1917). In 2016, they were formally recognized as the national mammal (see appendix A).

American Bison herds roamed on the Llano, where they foraged on the abundant shortgrasses, but these were relatively brief trips. The herds generally remained near the tributaries of the rivers arising just below the caprock in the rough terrain of the Rolling Plains (chapter 7). When heavy summer thunderstorms filled playa basins, the herds ventured onto the Llano for longer periods (fig. 8.8). These local migrations no doubt led to the Native American belief that bison originated underground, emerging from cave-like openings somewhere on the Llano Estacado.

Estimates of the presettlement size of the Great Plains bison herd vary widely—and wildly—from 20 to 60 million or more. Nonetheless, based on the carrying capacity of the range and other factors, 28–30 million likely represents a reasonable estimate for the overall bison herd, with 8.2 million of these on the Southern Plains, including those grazing on the Llano Estacado. Native Americans established a succession of bison economies, as reflected in the cultures revealed at the Lubbock Lake Site (infobox 8.4). *Ciboleros*—meat and hide hunters—also removed some 18,000–25,000 bison annually from the Llano Estacado between 1825 and 1850, but the worst was yet to come. With the

advent of the transcontinental railroad in 1869 and its accompanying settlement, herds everywhere quickly succumbed to overhunting. By the late 1870s, the bison herd and Comanche culture had faded into history—the victims of relentless exploitation—and the Llano today seems strangely empty for want of both.

Because of their dust-bathing activities, bison create wallows that can significantly alter the local ecology of prairie landscapes (fig. 8.8). Pawing, especially by bulls during the rutting season, further disturbs these sites. Once the wallows are initiated, continued use extends the disturbed areas into depressions up to 16.5 feet in diameter and 1 foot deep; some relict wallows remain in evidence, even in areas where bison were extirpated some 125 years ago. Unfortunately, the plow long ago erased most of the wallows etched into the flat Llano profile, but those that survived elsewhere reveal clear differences in their vegetation, soil texture and chemistry, and, notably, soil moisture when compared with the surrounding prairie. Depending on rainfall, the water accumulating in the wallows allows development of wetland plants, among them various species of rushes, water ferns, spikerushes, and smartweeds. The repeated disturbance in wallows still in use creates soil conditions favoring primarily fast-growing and short-lived "weedy" plants.

Even in death, bison continued their ecological influence on prairie environments. Given historical populations numbering in the millions, hundreds if not thousands of carcasses surely littered the landscape at any moment. Initially, toxic fluids associated with decomposition denuded the area surrounding a carcass, but these microhabitats eventually proved nutrient rich—especially in nitrogen. These "nutrient pulses" may have extended up to 8 feet beyond the carcass and favored the initial development of new vegetation. Together with wallowing, the effects of carcass decomposition once enriched the diversity of prairie vegetation with an ongoing cycle of patchy disturbance and recovery.

In 1864, at the western edge of the Texas Panhandle, a buffalo wallow served as a meager fortress for four soldiers and two scouts under attack by more than 100 Comanche and Kiowa

Figure 8.8. Once numerous on the High Plains, American Bison (top) exist there now only in small managed herds. An active bison wallow (bottom) at the Konza Prairie Preserve in Kansas illustrates those that once dotted the Llano Estacado. Photographs by Sandra S. Chapman and Jacquelyn Gill.

INFOBOX 8.4. LUBBOCK LAKE SITE

A site on Yellow House Draw once fed by springs developed into one of the more interesting archaeological sites on the High Plains. Its significance was realized in the 1930s when dredging efforts attempted to revive the water supply after the springs stopped flowing. Instead of reviving the flow, however, the excavations revealed strata with an essentially complete sequence of early Paleo-Indian cultures that continued into historical times. The meandering draw formed as an intermittent tributary of the Brazos River and, in time, cut deeply into the flat terrain, intercepting the water table and initiating seepage in the depression. The springs supplied enough water to maintain a wetland that attracted some of the earliest Paleo-Indian cultures known in North America. Starting with Clovis occupation around 11,500 BP, the cultural record also includes strata subsequently associated with Folsom, Plainview, and Firstview occupations as well as those of Early, Middle, and Late Archaic cultures; the latter ended about 2000 BP. A bison-based economy supported each of these cultures, but the site has also yielded the butchered remains of other animals, including those now extinct. Formally known as the Lubbock Lake National Historic and State Archeological Landmark, the 336-acre location features a visitor center, museum, and active "digs" worked by professional scientists assisted by volunteers. Guided and self-guided tours are available year round.

Clovis points, beautifully crafted from jasper, chert, and other brittle stones, represent the only evidence of their namesake culture. The lance-like, finger-length points were fashioned with grooves—"flutes"—running almost halfway up each side starting from the base. These distinctive features likely anchored the points as they were bound into the split end of a wooden shaft. To date, artifacts collected from paleocultures in Siberia have not revealed similar construction; hence the fluted points seemingly represent a uniquely American design. Discoveries of Clovis points include those embedded between the ribs of *Bison antiquus*, the now-extinct forerunner of the modern species. Clovis people were once regarded as the earliest of the Paleo-Indians to enter North America, but recent evidence collected from other sites in Texas and elsewhere now challenges that contention.

Nonetheless, their influence remains profound: based on genetic evidence, Clovis peoples were the ancestors of about 80 percent of the modern groups of Native Americans. Moreover, their role as hunters may have contributed to the extinction of the assemblage known as the North American megafauna, today known only as fossils. This assemblage of about 100 species of Pleistocene mammals—those with adult body weights exceeding 100 pounds—included herbivores, among them sloths, large rodents, and the ancestors of modern elephants, horses, and camels, but also huge bears and other predators such as the well-known Saber-toothed Cat. As mentioned earlier, work at the Lubbock Lake Site uncovered the butchered remains of several of these species, among them mammoth bones broken into pieces suitable for use as tools. The diversity and abundance of this fauna likely mirrored the fauna still extant on the plains of Africa, yet extinction somewhat suddenly claimed about two-thirds of the Pleistocene megafauna—an event that largely coincided with the arrival and expansion of Paleo-Indians in North America. Although the idea is much debated, overhunting likely contributed at least in part to the extinction process, presumably because the evolutionary history of the North American megafauna did not include exposure to humans. More specifically, these animals lacked the behavioral defenses necessary to deal with a new and potent predator armed with stone-tipped weapons. Other explanations focus on a glacial cycle that initiated climatic extremes and environmental instability. Whatever the cause, the Pleistocene extinctions cleared the way for the survivors—notably the modern species of bison—to expand into a grazing niche no longer occupied by competitors.

The Clovis culture lasted about 400 years and then faded as its peoples spread into new ecological realms. These groups soon developed new cultures, each adapted to its own niche in the unpopulated emptiness of the New World. But almost 12,000 years later, George Singer opened a store beside a small lake on Yellow House Draw where he supplied goods for settlers, ranchers, hunters, and military patrols—and in so doing started what became the modern city of Lubbock.

Figure 8.9. An aerial view reveals the abundance of playa lakes that dot the High Plains landscape after a summer thunderstorm. Photograph courtesy of the High Plains Underground Water Conservation District No. 1, Lubbock, Texas.

warriors. Luckily, rainwater from a sudden storm collected in the depression and helped the men endure the siege. Five survived, and along with the fallen trooper who was buried at the site, all received the Medal of Honor for their heroic action in what became known as the buffalo wallow fight in the Red River War (1874–1875). Years later the medals awarded to the two scouts were revoked since both were civilians. However, one of the scouts—Billy Dixon—refused to return his medal, which is now displayed at the Panhandle-Plains Historical Museum at West Texas A&M University. A granite monument marks the location in Hemphill County. In 1903, famed western artist Frederic Remington immortalized the event in the painting *Fight for the Water Hole.*

PLAYA LAKES: "ROUND LIKE PLATES"
GEOLOGY, GEOGRAPHY, AND VEGETATION

Whereas the Llano Estacado lacks streams and rivers, the vast tableland is not without surface water. Features known as playa lakes were first documented in Coronado's journals in the late spring of 1541: "Occasionally there were found some ponds, round like plates, a stone's throw wide or larger." *Playa,* in Spanish, translates as "beach," but the relationship between that meaning and the ecological reality of these landforms remains tenuous.

Some 25,000 playa basins occur on the High Plains of North America, and at least 19,000 of these are in Texas. When full, the lakes average about 15.6 acres in size, and about 87 percent cover less than 30 acres; each drains a closed watershed with an average area of 138 acres. They occur at a density of about one per square mile, and an observer might see dozens at one time from a low-flying airplane (fig. 8.9). Playas fill from rainwater dropped by strong but irregular thunderstorms. Because these storms—which typically peak during May and June, and again in September and October—are local, a swath of water-filled playa basins may cross a large area in which those on either side are completely dry. Without replenishment, evaporation and transpiration gradually dry the shallow lakes.

The geological origins of the playa basins have triggered both imagination and science, beginning with the rather implausible notion that the depressions formed as buffalo wallows (see fig. 8.8). That fancy aside, the relentless prairie winds were long regarded as the force etching the shallow basins in the flat landscape, thus giving rise to the playas as wind-deflated depressions. More recently, however, geomorphic and hydrologic processes seem to better explain the formation and expansion of the playa basins. In this model, water collecting in slight depressions forms carbonic acid from the oxidation of organic matter, and the acid slowly dissolves the underlying layer of caliche. As the caliche dissolves, the surface subsides and the basin slowly deepens and expands in circular fashion from a central point. In this theory, the size and rate of playa formation depend on organic matter provided by decaying vegetation and, ultimately, on the size of the watershed surrounding each basin and the runoff it supplies.

Two additional physical features define playa basins: their flat, platelike profile (Coronado got it right) instead of a bowl-shaped, ever-deepening inward slope; and a bottom covered with heavy clay that is markedly unlike the surrounding sandy loams. The clay sediments accumulate as the basins increase in size, forming nearly impermeable floors that retard percolation, thereby allowing the basins to retain water. Anyone glancing at a soil map quickly notices the irregular polka-dot appearance of the dark circular clay basins against a comparatively uniform backdrop of lighter soils. In dry periods, the clay floors in playas crack and allow a modest trickle of water to reach the Ogallala Aquifer when rains refill the basins.

Playas, whose total surface area represents only 2 percent of the landscape, nonetheless serve as keystone ecosystems and a primary source of biodiversity on the Llano. The regional flora of playas on the High Plains consists of nearly 350 species, with the more common species occurring in 12 associations. In general, pondweeds and Arrowhead occur in areas of open water, and spikerushes and smartweeds represent the emergent vegetation. Familiar wetland vegetation such as cattails and bulrushes is less common but often dominates sites where it does occur. Native and nonnative grasses such as Barnyard Grass and Red Sprangletop typically form the wet-meadow

vegetation along playa edges. These communities originate from seed banks persisting in the soil from the last time conditions fostered development of the same flora. All told, the rich playa flora provides the basic habitat for wildlife on the Llano, most of which is otherwise intensively cultivated.

WATERFOWL

Up to two million waterfowl overwinter on playas, whereas others stop over during migration. Most of these are ducks, primarily Mallards, Northern Pintails, American Wigeon, and Green-winged Teal; several hundred thousand Canada Geese and Snow Geese may also use the playas in winter. These numbers reflect high densities of waterfowl per unit area of surface water, which at times fosters epizootics of diseases such as Botulism and Fowl Cholera. Waterfowl become even more concentrated in dry years, which further increases the risk of disease-related mortality.

Playas also host nesting ducks—Mallards dominate the breeding population, which also includes Blue-winged and Cinnamon Teal, Northern Pintails, and Ruddy Ducks, among others. Waterfowl nesting on playas may add 50,000–100,000 birds to the population in the Central Flyway in years of average rainfall, but estimates reach 250,000 in wet years. For example, one survey estimated that more than 2,000 broods hatched in a 12-county area on the Llano Estacado, whereas the same area produced just 800 a year later. Such swings in brood production are typical of nesting efforts elsewhere in grassland communities. In fact, the wet-dry cycle provides a flush of nutrients when rainfall later fills a dry playa. When dry, the withered vegetation oxidizes, producing nutrients that immediately enrich the local ecosystem when a playa refills. The result is abundant regrowth of vegetation and, especially, a burst of invertebrate populations on which ducklings depend for protein during their initial growth.

SHOREBIRDS

The playas are by no means devoid of other birds. Among these are 30 species of shorebirds that rely on playas as resting and feeding stops during migration, some continuing in spring to breed as far north as the Arctic Tundra. Insects and other

invertebrates, supplemented by seeds, dominate the diets of birds such as Long-billed Dowitchers, American Avocets, and various sandpipers (fig. 8.10). In spring, the larvae of chironomid flies—midges—become a dietary staple, whereas in fall a greater variety of foods becomes available. Some species of shorebirds, among them small sandpipers, share similar foraging styles (probing) but apparently avoid interspecific competition by following different migration schedules. Dowitchers and others avoid competition by foraging at separate locations, usually segregated from each other based on water depth at the respective sites. For the most part, shorebirds prefer playas where the vegetation is sparse, with shallow water and exposed mudflats.

GRASSLAND BIRDS

Nor are waterbirds the only group associated with playas. Ring-necked Pheasants, introduced onto the High Plains by both public and private interests between 1933 and 1945, rapidly gained a foothold and doubled their range in the region between 1950 and 1977. After irrigation expanded on the High Plains after the Dust Bowl, grain crops provided the diet necessary to sustain the population. Today, cattails and semiaquatic vegetation in playa basins provide cover for pheasants, especially during the fall and winter months when upland vegetation lies close to the ground. With a mix of corn or sorghum fields nearby, the winter density of pheasants in

playa vegetation may reach nearly 4.4 birds per acre. Vegetated playas also provide pheasants with nesting cover in which nest densities average nearly one per acre and produce more chicks per unit area than other habitat on the High Plains.

Northern Harriers find favorable winter habitat on playas, where they hunt almost exclusively above vegetation that provides small rodents with shelter. Playas surrounded by croplands attract more harriers than those on rangelands, likely because the nutrients from irrigation runoff ("tailwater") promote both denser vegetation and increased seed production—both benefiting the resident rodent population. In winter, Northern Harriers often roost communally on dry parts of vegetated playas; one sighting recorded 66 birds at a roost used nightly for several months. Short-eared Owls, also daytime hunters, similarly overwinter on playas; their floppy flight pattern offers a simple means of distinguishing them from the buoyant gliding and hovering of Northern Harriers.

MAMMALS

The richness of the playas as oases of diversity includes a few larger mammals as well as rodents. Desert Cottontails, while also occurring in other habitats on the Llano (e.g., Muleshoe Sandhills), visit playa basins where they find a wide selection of plant foods in comparison with the fare available elsewhere on the semiarid Llano. A close look-alike, the Eastern Cottontail, likewise seeks food and cover in playa basins. Both the genetics and demography of Eastern Cottontails are influenced where intensive agricultural activities surround the basins. During the growing season when crops cover the intervening fields, the cottontails increase their movements between lakes, but during the months when the fields lie fallow, these movements diminish, which isolates the local populations and effects seasonal changes in their genetic features.

Coyotes visit playas with enough cover to shelter cottontails and other prey, and fall and winter censuses in the central region of the Llano discovered coyotes in 30 percent of the basins. Given the density of playas, these data indicate a substantial Coyote population, especially since the censuses did not include other cover types. The matter holds some ecological significance: a thriving Coyote population triggers a competitive hierarchy that checks Red Foxes, just as wolves elsewhere depress Coyote numbers. Compared with Coyotes, Red Foxes are particularly successful nest predators; hence fewer foxes on the Llano mean better nesting success for ducks, pheasants, and other ground-nesting birds. Indeed, Coyotes seem more likely to eat disease-killed waterfowl in winter than to destroy duck nests in summer.

TOADS AND SALAMANDERS

Three species of toads represent the most common anurans (toads and frogs) associated with playas: the Great Plains Toad, Plains Spadefoot, and New Mexican Spadefoot. Aptly named, spadefoots bear a single, sharp-edged black "spade" on each hind foot with which they dig vertical burrows in loose soils. In contrast, Great Plains Toads are more typical of the popular image of a toad (i.e., warty skin). Five additional species of anurans occur in playas, but their presence at a given playa varies from year to year, either because of movements or because some species simply skip breeding.

The three species breed during a short period in late spring or early summer when rainwater at least partially fills many basins. Because the water in shallow playas may soon evaporate, aquatic species such as these toads are molded by an evolutionary strategy that favors rapid reproduction and development. The toads thus adapt to a small

Figure 8.11. The small morphs of the Barred Tiger Salamander occasionally wander far from water.

"window" of favorable conditions in which to complete successful breeding (chapter 9).

Whereas the fauna of North America abounds with salamanders, only a single species—the Barred Tiger Salamander—occurs in playa lakes. Tiger salamanders extend from coast to coast, but in the West, these are represented by several closely related taxa. Our interest here concerns the strange life history of those adapted to semipermanent wetlands in a semiarid environment. These adaptations work well—a single playa may support thousands of Barred Tiger Salamanders, a case where abundance trumps diversity.

Barred Tiger Salamanders in playa lakes develop into one of three distinct types or morphs. First is a large morph that, when sexually mature, retains the immature features of its larval stage, a condition known as neoteny; its life cycle does not often include metamorphosis. This morph occurs where water remains permanently, typically in playas modified to receive treated wastewater. The second morph is smaller and undergoes rapid metamorphosis into an adult; it inhabits playas subject to periodic drying (fig. 8.11). The first and second morphs differ in their respective color patterns, diets, and reproductive cycles; they rarely interbreed and, when experimentally crossed, produce infertile offspring. The third is a rarer, cannibalistic morph whose larvae develop larger heads with much wider mouths and larger teeth than either of the other two morphs; their slit-like eyes are another distinguishing feature. After metamorphosis, the adults of this morph develop a protruding lower jaw, depressed snout, and color patterns resembling those of the large morph. So far as can be determined, the cannibal morphs arise from populations of small morphs, which they resemble in size.

The large morphs breed seasonally (late winter and early spring), an adaptation associated with their relatively stable environment. With stable water levels, they have time to grow larger and produce more eggs, all while remaining in a larval stage, in which they are likely to encounter conditions requiring an adult form. They feed selectively, preying primarily on small crustaceans and, when available, eggs of their own species. Because of neoteny, the large morphs retain their juvenile diet, which permits their populations to reach high densities.

The small morphs evolved to cope with the ephemeral nature of most playas, where they grow rapidly and quickly metamorphose into breeding adults, thus completing the aquatic phase of their life cycle before the lakes dry. They also breed year round, thereby taking opportunistic advantage of whatever playas are filled at any given time. The larvae likewise feed opportunistically, taking a variety of prey including aquatic insects and crustaceans. Small morphs occasionally wander away from their ponds, traveling by night and burrowing in loose soils or hiding beneath rocks or in the burrows of other animals before daylight. At times, gardeners and farmers unearth these morphs in burrows up to 2 feet deep.

The ecological role of the cannibal morphs remains unclear. They apparently develop when limited nutrient sources disrupt normal food chains. Hence, the cannibal morphs survive on others of their own species and, at least for a short time, prevail as the dominant form. Though the idea is speculative, the cannibal morphs seemingly represent a means of survival that establishes a nucleus for renewing populations of the small morphs when favorable conditions return.

OTHER WILDLIFE ON THE LLANO ESTACADO

MISSISSIPPI KITE

The Mississippi Kite, a sleek and graceful raptor grossly resembling a falcon, represents a relatively new member of the regional avifauna (fig. 8.12). Early explorers did not encounter the

Figure 8.12. First observed on the High Plains in the 1940s, Mississippi Kites now regularly occur in the regional avifauna. Photograph by Marc Stich.

bird until they entered the riparian forests along the Canadian River on the eastern edge of the Panhandle. Today, however, Mississippi Kites are regularly encountered on the Llano Estacado and in other parts of the Texas Panhandle, including locations with considerable human traffic. Reports of kites in the Panhandle became commonplace in the 1940s, apparently in response to the growth of shelterbelts planted in the preceding decade; the rows of trees offered suitable nesting habitat that was previously unavailable. Groves on golf courses likewise provide attractive nesting areas.

For the most part, Mississippi Kites feed on large-bodied insects. Like other raptors, these kites regurgitate "casts"—pellets of indigestible food matter, which offer biologists means of determining the birds' food habits. For Mississippi Kites, the exoskeletons of insects dominate the pellets. Grasshoppers and beetles are staples, but when available, cicadas rank high in the birds' diet. After seizing their prey, Mississippi Kites often tear apart the victim and ingest the choice "cuts" piecemeal while in flight. One biologist, momentarily puzzled by the discolored and somewhat sticky belly plumage he observed on some kites, suddenly realized that this was the dinner table for birds feeding in flight. The remains of prey accumulating in nests—small vertebrates such as lizards, mice, bats, and frogs—illustrate that Mississippi Kites are adaptable and opportunistic predators. The bones of box turtles and rabbits occasionally found among the remains are almost certainly obtained from road kills and underscore an important concept: diets reflect what is eaten and not necessarily what is killed.

Smoke rising from grass fires attracts Mississippi Kites, which stoop on insects escaping the flames and similarly feed on those disturbed by grazing livestock and even deer. With their insect-dominated diet, Mississippi Kites migrate in the fall to areas deep in the interior of South America where their prey remains active and available in winter.

Curiously, Mississippi Kites line their somewhat flimsy-looking stick nests with a thin blanket of green leaves, green moss, or Spanish Moss; the parents regularly add fresh leaves, sometimes every day. The nest-lining behavior of Mississippi Kites remains unexplained but likely reflects some aspect of sanitation.

LESSER PRAIRIE-CHICKEN

Shinnery Oak communities in Texas and adjacent New Mexico provide habitat for Lesser Prairie-Chickens at the southwestern edge of their distribution. The birds mate on sites formally known as leks, or more popularly as booming or dancing grounds (or arenas) because of the antics males employ during courtship. These courtship displays usually take place in natural openings in the brushy landscape and may be used annually for many years, even decades. The males dance and "boom"—displaying inflated air sacs on their necks—in the center of the lek in a courting ritual serving two purposes: first, to fend off competing males that, when subdued, are relegated to the sidelines; and second, to attract the females coyly watching the males' frantic behavior (fig. 8.13). Eventually, the females rush in and mate with the few victorious males remaining in the arena. Indeed, at least 90 percent of the mating is accomplished by 10 percent of the males, which

Figure 8.13. Male Lesser Prairie-Chickens face off while performing elaborate courtship displays to attract females.

serves as a fine example of natural selection based on the fittest individuals passing on their genes.

Shinnery Oak provides acorns, catkins, buds, and leaves, all seasonally important as food for Lesser Prairie-Chickens. In summer, however, the diet of adults shifts to insects, especially grasshoppers, and both the chicks and juveniles feed almost entirely on insects. Sites with abundant forbs produce more insects than other locations, which underscores the importance of floral diversity in the oak communities. Optimal nesting habitat includes both dense grasses and overhead cover of at least 50 percent. Still, many nests fail because of heavy losses to predators, among them Coyotes, skunks, and snakes, as well as several species of raptors that also prey on the rich community of small mammals present in shinnery communities.

AN ENDEMIC LIZARD

Ten species of lizards occur in Shinnery Oak in southeastern New Mexico, and the same species almost certainly occur on the Texas side of the border. None of these is restricted to shinnery communities, whereas in both states, blowouts in shinnery provide the exclusive habitat for Dunes Sagebrush Lizards. Blowouts develop where roots no longer anchor the sandy soil, which allows wind erosion to form depressions the lizards favor. Nonetheless, for food and cover, Dunes Sagebrush Lizards seldom venture more than 6 feet from the nearest clump of Shinnery Oak. Unfortunately, the species potentially faces hard times where anthropogenic activities threaten Shinnery Oak.

Figure 8.14. In the evening, Sandhill Cranes flock to Muleshoe National Wildlife Refuge, where they roost in Paul's Lake away from the threat of predators and sometimes linger in the morning before leaving to feed in nearby fields. Photograph by Jude Smith.

SANDHILL CRANE

About 90 percent of the midwinter population of Sandhill Cranes—at times more than 400,000 strong—visits the High Plains, mostly on the Llano Estacado. Muleshoe National Wildlife Refuge offers birders fine opportunities to see large numbers of sandhills, along with other species of prairie wildlife. At times, a Whooping Crane or two travel with Sandhill Cranes, so a scan of the gray flocks might reveal one of the rarest birds in North America.

Sandhill Cranes wintering in the region are closely associated with some 30 basins carved into the Llano in bygone times; about 20 others occur in New Mexico. Geologically, these lakes originated as the beds of streams that once flowed across the plains, and they differ from playa lakes. In places, the streams cut through the upper strata, leaving behind clusters of linear basins when subsequent geological events ended stream flow. Thereafter, springs at the edges of the basins flowed into many of them, supplemented by surface runoff from rainfall. These conditions present the cranes with ideal roosting sites; after feeding in the surrounding fields during the day, the birds return to the lakes at

night, first landing at the springs to drink and then walking into the lakes, where the alkaline water usually remains unfrozen all winter. Surrounded by water, the birds roost for the night in a zone buffered against predators. One census recorded as many as 200,000 cranes crowded into a single pluvial lake (fig. 8.14).

The lakes themselves are alkaline, the result of sodium sulfate or gypsum salts concentrated by evaporation. The saline lakes provide migrating shorebirds with feeding areas rich in invertebrates, although the abundance of these foods diminishes where springs no longer flow because of falling water tables. Snowy Plovers nest on the edges of these lakes, but the reduced flow of spring water again limits the suitability of this habitat. Because of the salinity, the flora at these lakes is significantly impoverished—just 49 species—compared with the far more diverse vegetation associated with playa lakes; however, it includes taxa unique to the region and absent in playas.

Sandhill Cranes mate for life; they court with a seemingly disorganized dance of leaps, flops, bows, and hops, along with an occasional flip of a stick—all worth having a camera ready to go. One may see this behavior throughout the year, but the tempo of their comical ballet increases somewhat in early March as the birds prepare to leave for their northern nesting grounds.

Unfortunately, the highly concentrated cranes at times experience epizootics, including lethal attacks of Fowl Cholera. Additionally, about 9,500 Sandhill Cranes wintering on the Llano succumbed during a five-year period to the toxins produced by moldy peanuts left in the fields after harvest; cold, wet weather stimulates production of the toxins. The story has a happy ending, however, as farmers thereafter willingly plowed their fields after the harvest, thereby burying the leftover peanuts and eliminating the risk of further epizootics.

HIGHLIGHTS

THE DUST BOWL

Prolonged drought visited North America's interior during a decade infamously known as the "Dirty Thirties." The iconic term "Dust Bowl" first appeared in a column written by Associated Press reporter Robert E. Geiger, who experienced "Black Sunday" on April 14, 1935—likely the worst of the era's dust storms (fig. 8.15). Environmental historians generally regard the Dust Bowl as one the three greatest ecological disasters in recorded history (the other two also concern degraded and abused soils, in China and around the eastern and southern edges of the Mediterranean Sea). Coincidentally, the Great Depression intensified the ravages of the drought on thousands of Americans, not the least of whom were farmers of the Great Plains. The enormity of the Dust Bowl spilled into American culture, including folk singer Woody Guthrie's memorable lyrics "A dust storm hit, an' it hit like thunder; it dusted us over, an' it covered us under."

The origins of the Dust Bowl began with the Homestead Act (1863) and the free land offered by the government to encourage settlement and the development of agriculture on the Great Plains—much of it already known for its aridity. Settlement began in earnest with the end of the Civil War and subsequent completion of the transcontinental railroad; it heightened further when Congress increased the size of allotments early in the twentieth century. A series of unusually wet years

Figure 8.15. A menacing wall of dust (top) bears down on Spearman, a small farming community in Hansford County at the tip of the Texas Panhandle on April 14, 1935, a date forever known as "Black Sunday." Windblown soil ironically subdued the very tractors that once plowed the land (bottom). Photographs courtesy of the US Department of Agriculture (top) and the National Oceanic and Atmospheric Administration (bottom).

provided a further catalyst, thereby enhancing the meteorological folly that "rain follows the plow." Meanwhile, other areas were overgrazed to the point that ranchers became farmers and tilled even more land. Two other events added to the mix: the demands of World War I increased wheat prices, and the advent of mechanized farm equipment increased production. On the Llano Estacado, the area devoted to farming doubled between 1900 and 1920 and then tripled between 1925 and 1930, nearly all of it without the benefit of irrigation.

Conditions for massive erosion were never better when the drought began. The wind-borne soils deposited on the plains thousands of years ago were still eolian, and when they dried without the holding power of sod-forming grasses, unyielding prairie winds moved the soil far and wide, even to the marble halls of Washington, DC. The same winds that powered windmills stripped that land of its topsoil. The environmental ravages of the Dust Bowl affected some one million acres of the High Plains. In Texas, the worst of the Dust Bowl struck the northwest corner of the Panhandle—Dallam County represents the epicenter of the affected region. Here the clouds of soil billowed over the High Plains like a dark shroud of agricultural hell.

By 1935, the Soil Conservation Service—renamed the Natural Resources Conservation Service in 1994—fought back by introducing contour plowing, shelterbelts, and other measures designed to halt rampant soil erosion. These activities came too late for many farmers—most famously the "Okies" portrayed in Steinbeck's hardscrabble novel *The Grapes of Wrath*—who abandoned the land they worked as either owners or tenants. As part of the New Deal, the government purchased land so badly eroded that it was essentially useless for farming and thus unwanted. These otherwise useless properties established the national grasslands, which are administered along with the national forests by the USDA Forest Service. One such site—Rita Blanca National Grassland—covers more than 77,000 acres in Dallam County. Unfortunately, much of the land reclaimed by the federal government was planted in nonnative grasses, which significantly reduced opportunities for recovery of the region's biodiversity.

Today, the restored grasslands at Rita Blanca steadily rebuild the fertility of their soils and, indeed, regain their qualities as functional ecosystems where Coyotes, prairie dogs, Pronghorns, hawks and songbirds, and a host of other animals and plants thrive in high-quality habitat. In addition to shortgrass prairie, the landscape at Rita Blanca—once part of the historic XIT Ranch—includes both marshes and some woodlands. The prairie winds still blow across the land and droughts still dry the soil, but wise stewardship manifested by the sod of deeply rooted

Figure 8.16. Swift Foxes inhabit the shortgrass prairies of the western High Plains. Photograph courtesy of the National Park Service.

grasses now diminishes the threat of another Dust Bowl—at least in one corner of Texas.

A LITTLE FOX

Swift Foxes have large ears but are otherwise little larger than well-fed Domestic Cats, and they are the smallest of the foxes in North America (fig. 8.16). Taxonomists recognize two subspecies, one of which—the Swift Fox—resides in the shortgrass prairie of the Llano Estacado where it subsequently adapted, in part, to pastures, fields, and fencerows. Overall, conversion of prairie to cropland, coupled with fire suppression, reduced the area of its former range by about 60 percent. In general, Swift Foxes avoid rough terrain and seldom venture near the escarpments bordering the Llano. A desert-dwelling subspecies, the Kit Fox, occurs in the Trans-Pecos (chapter 9).

Rabbits and rodents form the bulk of their diet, which also includes a variety of other foods such as insects, lizards, songbirds, and even fishes on occasion. Based on a study of their scats in Texas, Swift Foxes apparently pose little threat to game birds or poultry, even when they den near farmyards where chickens roam. Swift Foxes do not require drinking water; their foods fulfill these needs, although surface water is necessary to sustain their prey.

Coyotes are the major predator of Swift Foxes. Because Coyotes seldom eat the foxes they kill, predation in this case apparently represents an example of interspecific competition (i.e., intolerance) instead of a typical predator-prey relationship. Red Foxes, to a lesser extent, also seem intolerant of the smaller Swift Foxes. Perhaps in response to these social hierarchies, Swift Foxes construct a cluster of several dens of which only one is used for rearing pups, while the others provide an extra measure of protection from predators. Regrettably, these foxes sometimes den near roadways, and road kills are another source of mortality at these locations.

Swift Foxes, while certainly shy, lack the cunning nature attributed to other foxes and thus are easily trapped. Worse, they became unintended victims of control programs once designed to poison Coyotes, with the result that Swift Foxes were nearly extirpated from much of their range. In Canada, the damage was so extensive that it was necessary to reintroduce the species. In Texas, however, Swift Foxes gradually rebounded on their own when the poisoning programs ended, although their status remains watched by agencies concerned with endangered species.

CONTACT IN THE PANHANDLE

The Panhandle, including the Llano Estacado, represents another zone in Texas where the ranges of eastern and western species meet face to face and often overlap (a partially sympatric distribution). The results of these contacts may differ, however, as noted here with a few examples. Baltimore Orioles, a predominantly eastern bird, extend their distribution into Texas, including the eastern edges of the Panhandle, whereas Bullock's Orioles, a closely related species, represent a western counterpart that reaches into the western edges of the Panhandle (fig. 8.17). Both species occur in somewhat similar habitat (groves of trees along streams; shaded streets in towns, etc.). Taxonomists later decided these were races of the single species thereafter designated the "Northern Oriole," even though the bird's eastern and western populations are each distinguished by clear differences in head and neck coloration, tail pattern, and the size of the white wing patch. Based on more recent evidence, however, the single species designation was rejected, and Baltimore and Bullock's Orioles again gained recognition as separate species—one of several taxonomic seesaws in ornithological history. But hybrids

Figure 8.17. The distributions of Bullock's Orioles (top), a western species, and Baltimore Orioles (bottom), an eastern species, overlap on the High Plains, where their hybrids often occur. Photographs by George F. Moore.

muddy the appearance of these birds in the contact zone, where the offspring of mixed parentage show various degrees of differences in their plumage.

A somewhat similar situation may be developing in the Canadian River Breaks and other canyonland terrain on the eastern edges of the Llano Estacado where Red-bellied and Golden-fronted Woodpeckers come into contact. Golden-fronted Woodpeckers prefer somewhat more arid habitat than the closely related Red-bellied. Nonetheless, the ranges of the two species overlap in Texas, and in this zone hybrids represent nearly 16 percent of the population. Coincidentally, the two color phases of what were once regarded as two species are now recognized as races of the Northern Flicker. The Canadian River Breaks provide an avenue by which the two races meet on the plains and hybridize. In their "pure" state, the eastern race features yellow-shafted wing and tail feathers and, on the males, a black facial "mustache," whereas red-shafted wing and tail feathers and red mustaches distinguish the western form. As might be expected, the plumages of the hybrids exhibit a mix of these (and other) characteristics.

In contrast to the foregoing, Eastern and Western Meadowlarks, despite their remarkable similarity in appearance, each maintain a reproductively isolated gene pool that precludes hybridization. Significantly, the two meadowlarks differ in one important feature: their songs. The call of Western Meadowlarks consists of seven to ten notes often described as flutelike and gurgling, whereas Eastern Meadowlarks produce two clear, slurred whistles. The ranges of both species overlap, in part, on the Llano Estacado, but the dissimilarity of their calls is enough to keep one species from attracting the other, thereby preventing courtship, breeding, and hybridization. This isolating mechanism likely developed in the Pleistocene when glaciers divided the single ancestral population of meadowlarks into eastern and western components. Meadowlarks isolated on either side of the barrier retained similar plumages but developed species-specific vocalizations, not unlike the regional dialects of humans. Later, after the glaciers retreated, the eastern and western populations rejoined along a broad front, but their respective vocalizations effectively isolated what were now separate species and prevented them from interbreeding. Still, because the primary components of each species' song are learned instead of inherited, the potential remains for the males of either species to acquire the primary song of the other.

TUMBLING TUMBLEWEEDS

Tumbleweeds are as much a symbol of the plains as cattails are of freshwater marshes. They enrich the ballads and lore of the Old West and silently depict desolate, windswept settings in movie scenes. Yet these plants are invasive exotics, having first appeared in South Dakota in the early 1870s after their seeds traveled undetected in a shipment of flax seeds from the Ural Mountains of eastern Europe. Correctly known as Russian Thistle, Tumbleweeds—thanks to their effective dispersal of seeds—quickly expanded widely across the North American plains.

Ecologists regard the Tumbleweeds as "pioneers," a group of plants uniquely capable of gaining a foothold on disturbed lands, not the least of which are overgrazed ranges. Mature plants bear small, prickly leaves on woody stems and branches and can be as small as a basketball to as large as a small car. They also thrive on saline soils, thereby adding to the breadth of environments where they can invade with little competition. They also remove a lot of water from soils, a competitive edge that curries little favor with dryland farmers.

After maturing, Tumbleweeds wither and dry into brittle spheres, soon breaking away from their root systems to begin haphazard travels governed by the whims of the wind. Each plant, although dead, remains a warehouse of some 250,000 loosely attached seeds, some of which drop off as the ball bounces across the landscape. Seeds more firmly embedded germinate when a Tumbleweed lodges against an obstacle and drifting soil buries the seed-bearing stems. Tumbleweed seeds lack either protective covers or internal stores of energy; instead, they each contain a miniature, tightly coiled embryo already supplied with small doses of chlorophyll. Little wings on each seed further enhance wind dispersal and may even help absorb moisture. Thus outfitted, the next generation gains a rapid start.

Tumbleweeds offer little in return. They clog fence lines (fig. 8.18), even roadways, and while credited with saving cattle during the Dust Bowl, Tumbleweeds provide poor forage under most circumstances. Prairie dogs, however, readily include Tumbleweeds in their diet, and Pronghorns will nibble on the younger plants in years when abundant rainfall increases their succulence. Northern Bobwhites and other birds forage on the seeds, especially where Tumbleweeds pile up and offer cover as well as food. Somewhat uniquely, those Tumbleweeds that roll into playa lakes offer Barred Tiger Salamanders sites on which to lay their eggs, a fortunate circumstance because benthic vegetation is generally unavailable for this

purpose. Nonetheless, Tumbleweeds pose serious threats because their woody structure degrades slowly, becoming mobile tinderboxes that quickly spread range fires. All told, they represent a symbol of the American West that might better have stayed in the European East.

OGALLALA AQUIFER

The largest aquifer in North America lies under the High Plains in Texas and parts of seven other states extending north to South Dakota (fig. 8.19). In area, it covers 174,000 square miles with a lens of water-bearing sediments varying in thickness from 100 to 200 feet in the south to almost 400 feet in the north. The aquifer developed during the Pliocene when rivers flowing east from the Rocky Mountains crossed the plains, including the Llano Estacado, and deposited water-bearing strata of sand and gravel. Heavy rainfall during the epoch added more water to these deposits. Later, when geological events changed the drainage systems and isolated the Llano from the rest of the High Plains, the network of flowing streams disappeared. Deprived of its primary sources of recharge, that part of the aquifer underlying the Llano became a huge pocket of "fossil water," supplemented only by meager seepage from playa lakes.

With settlement, much of the Llano developed as dryland farming, but in the 1940s irrigation systems driven by powerful pumps began withdrawing large volumes of water from the Ogallala. The supply seemed unlimited, and under Texas law, the "right of capture" allowed landowners to take as much groundwater as they wished. Functionally, the Ogallala was now a mine, and eventually the "ore" started to peter out. During droughts, the demand increased even more, and during the dry years between 1993 and 1997, the aquifer dropped in places by more than 6.5 feet annually. Farmers eventually switched to center-pivot and drip irrigation systems, both water-saving methods compared with furrow irrigation, but even these changes did not stem the drawdown of the Ogallala. Today, dryland farming has returned to the Llano, and wheat and sorghum have commonly replaced corn and other water-demanding crops.

Playa basins, the only surface water on the Llano Estacado and key habitat for wildlife, interacted with the depletion of the Ogallala. Farmers dug pits in the basins to collect and store irrigation runoff,

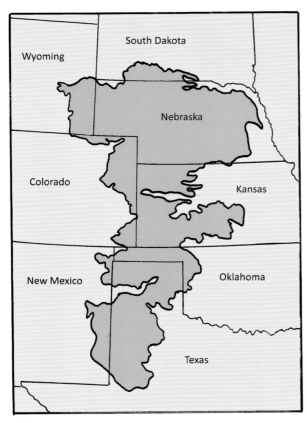

Figure 8.19. The Ogallala Aquifer lies underneath parts of eight states where recharge is minimal and usage is high.

thereby reducing the lakes' surface area subject to evaporation—the primary means by which the playas dried. Thus concentrated, the runoff water could be pumped for another round of irrigation, this time without the cost of pumping from a deep well. Indeed, more than 70 percent of the larger playas were modified with pits. Regrettably, even as the dependence on irrigation diminishes, the pits remain in place and continue to alter the inherent integrity of the playas.

CONSERVATION AND MANAGEMENT

The unique Shinnery Oak community faces unique problems. In spring, tannins in the oak leaves contain enough toxin to poison—and kill—cattle. Moreover, the accumulations of leaf litter provide winter shelter for Boll Weevils, the scourge of cotton farmers. Hence, in order to remove the source of toxin and replace the shrubs with grasses, Shinnery Oak becomes a target for herbicides. Thus treated, Shinnery Oaks have little chance to regain

their value as wildlife habitat for several species, particularly Lesser Prairie-Chickens. Nesting activities drop significantly in treated areas, as does the birds' store of body fat. Estimates suggest population declines of more than 90 percent as a result of habitat loss. Overall, Lesser Prairie-Chickens require at least 20 percent shrub cover; hence continued treatments with herbicides clash head on with the welfare of this and other species of wildlife, including the Texas Horned Lizard. Until recently, Texas allowed limited hunting of Lesser Prairie-Chickens, but as their populations diminished, hunting ended in 2014 when the birds were listed as threatened under the aegis of the Endangered Species Act. (The listing decisions for the Lesser Prairie-Chicken as well as the Dunes Sagebrush Lizard were subsequently vacated by court order.)

Because of unrestricted eradication programs once—but no longer—subsidized by state or federal agencies, prairie dog numbers are also far lower today than in the past. By 1922, government agents had poisoned about 90 percent of the prairie dogs in the Texas Panhandle. Indeed, the huge "megacolony" reported by Vernon Bailey in 1905 today remains as 183 fragmented sites collectively covering 4,546 acres. Once found in 114 counties across the state, colonies now persist in 73 counties, where 3,180 colonies occupy 146,536 acres, 84 percent of which occur on the High Plains. Of these, no more than six sites, including Rita Blanca National Grassland, are large enough (at least 5,000 acres) to support reintroductions of Black-footed Ferrets, but such efforts are not actively underway in Texas. The prairie dog management plan developed in 2004 by the Texas Parks and Wildlife Department strives to maintain 75 percent of the colonies once located within the species' historical range in the state. Recent efforts to add Black-tailed Prairie Dogs to the federal list of threatened species met with considerable opposition, and the US Fish and Wildlife Service dropped further consideration to list the species.

Agricultural activities damaged the hydrology of many playas when large pits were excavated in the basins to collect irrigation runoff for reuse. This somewhat lessened demands placed on the

Ogallala Aquifer but concurrently damaged wetland communities. New farming techniques have diminished the importance of these modifications, but the abandoned pits continue claiming water from the shallow edges of the playa basins. In time—measured in decades if not centuries—the pits will gradually fill with sediments, but meanwhile the ecological productivity of many playas remains impaired, thus diminishing the value of the lakes for wildlife. It would be beneficial if these were proactively refilled (berms of soil from the original excavations remain piled adjacent to the pits).

Ideally, an easement program that rewards landowners of still-pristine playas will preserve the ecological integrity of these unique landforms. Long-term easements, coupled with provisions of the Conservation Reserve Program (CRP), might offer an adequate incentive for many farmers. The easements should include a wide, untilled buffer zone around the perimeter of each playa to maintain a functional watershed. Buffer zones of natural vegetation slow the sediments that originate in croplands, flow into playas, and steadily lessen their depth and water-holding capacity. This form of deposition, known as anthropogenic sedimentation, currently occurs 8.5 times faster in playas without a buffer zone than in playas surrounded by grasslands. Efforts to vegetate disturbed sites should establish only native vegetation and avoid planting exotic species, which unfortunately is a common practice in some restoration projects, including the CRP. As a further inducement for participation in a conservation program, fee-based hunting and moderate grazing should remain options for landowners who place their playas and buffer zones under easement.

Many of the shelterbelts planted in the wake of the Dust Bowl are now aging and often decadent. Unfortunately, these have not always been replaced with younger trees, hence diminishing the protection they once provided against soil erosion. Additionally, the decline of shelterbelts precludes the continued presence of Mississippi Kites and other tree-nesting species in areas where these birds are now established. The solution is self-evident and should be encouraged by agencies concerned with soil, water, and wildlife conservation. In part to save on fuel costs, a practice known as minimum tillage also conserves soil moisture and protects against wind erosion. Instead of immediately plowing under the stubble remaining after the current year's harvest, with minimum tillage the residue remains in place until the planting season the following spring.

Protection of pluvial basins becomes a clear priority for the conservation and management of a large proportion of the North American population of Sandhill Cranes. Yet, of the 20–30 pluvial basins on the Llano Estacado, only one site—a unit of Paul's Lake at Muleshoe National Wildlife Refuge—currently falls under the protection of a natural resource agency. Other roosts remain vulnerable to unfavorable alterations of spring flows, water depths, or salinity regimes, each vital to the birds' winter survival. Once again, long-term easements with landowners seem the best remedy against the risk of habitat degradation at many of the important roosts in the region.

9
TRANS-PECOS
DESERT LEGENDS AND
RUGGED GRANDEUR

Wild, vast, and isolated, this land of desert,
river canyon, and rugged mountains is as close
to primeval as anything on this continent.

— RON C. TYLER

Trans-Pecos

For many, the vast, isolated, and largely untamed desert basins, rugged mountains, and deep canyons known as the "Trans-Pecos" represent the last frontier in Texas. US Army Lieutenant William Echols, when negotiating the region on camelback in 1860, described the Trans-Pecos as a "picture of barrenness and desolation," a characterization echoing Spanish explorers more than three centuries earlier who named the region *el despoblado,* the uninhabited land. According to Native American legend, after the Great Creator had completed the Earth and placed the stars in the sky, the remaining scraps formed a massive stack of stony debris thereafter hurled down on the Trans-Pecos, where it landed in scattered heaps. The geological origins of the Trans-Pecos are, of course, somewhat more complex and involve repetitive cycles of ocean submergence, drainage, mountain building, basin expansion, erosion, faulting, and volcanism. Consequently, one is tempted to accept the Native American explanation and meekly admire the serrated beauty that stretches to the horizon and beyond (fig. 9.1).

The Trans-Pecos extends to the Rio Grande westward from the Pecos River, the western border of the Edwards Plateau. The 32nd parallel—the borderline between Texas and New Mexico—forms the northern boundary of the region in Texas, although a sizable area with comparable physiognomy extends beyond this geopolitical limit. Encompassing about 32,000 square miles—an area about the size of South Carolina—this expanse of desert basins separating remote mountains has no rival within the state for geological complexity or spectacular scenery.

More than 600 million years of Earth's history are recorded in the mountains and rocks scattered across the Trans-Pecos. When the region was covered by seawater about 250 mya, reef-forming organisms flourished in warm shallow bays. In form and function, these reefs resembled modern-day coral reefs, but instead of coral, gigantic calcareous sponges, bryozoans, and encrusting algae deposited structural limestone. Eventually, thick deposits of sediments buried the reefs, including the largest, Capitan Reef. A period of upheaval (orogeny) approximately 60 to 35 mya created numerous mountain ranges and thrust the reefs about 1,000 feet skyward. Eventually, softer sediments eroded from the southern edge of the reefs, exposing El Capitan (elevation 8,085 feet), the "signature peak" of the Trans-Pecos (fig. 9.2). The period of orogeny preceded an era of extensive volcanic activity in the Trans-Pecos Volcanic Field, the easternmost Tertiary volcanic field in the United States. Lava, cinders, and ash spewed from enormous calderas during ten million years of volcanic activity. Tremendous blasts—the explosive power about 10,000 times that of the 1988 eruption of Mount St. Helens—created the Davis Mountains. Fiery eruptions tossed ash, cinders, pumice, and rocks widely, but more often, lava flowed quietly from volcanic fissures to blanket vast expanses or ooze quietly beneath the surface. In many places, ominous sentinel-like igneous columns—"hoodoos"—guard the rugged terrain as reminders of the violent past (fig. 9.3).

The region's complex geological history provided soils ranging from alkaline to highly acidic—that is, those derived from seabed limestones to some of volcanic ancestry. Today, the shallow soils are scantily clothed in low, scrubby vegetation adorned with thorns and spines, but forests develop on the tops of the highest peaks or deep inside the confines of shadowy canyons.

A HARSH ENVIRONMENT

An unforgiving climate enhances the chaotic ruggedness of a landscape that includes the northern reaches of the Chihuahuan Desert, the southernmost major desert in North America. The Chihuahuan Desert stretches from southwestern Texas across southern New Mexico and southeastern Arizona, and deep into central Mexico. It developed in a rain shadow between two large mountain ranges—the Sierra Madre Occidental and the Sierra Madre Oriental serve as barriers to moisture from the Pacific Ocean and the Gulf of Mexico, respectively.

(*overleaf*) Figure 9.1. This west-facing view from the Christmas Mountains near Big Bend National Park depicts the rugged nature of the Trans-Pecos region.

Figure 9.2. El Capitan, the exposed southern edge of an ancient limestone reef, rises abruptly from the floor of the Chihuahuan Desert. Now regarded as the "signature peak" of the Trans-Pecos, its prominence guided travelers crossing the region for centuries. Photograph courtesy of Guadalupe Mountains National Park, National Park Service.

Figure 9.3. Hoodoos—columns of volcanic rock exposed by erosion—stand as druid-like sentinels overlooking many slopes in the Davis Mountains.

The arid climate of the Trans-Pecos, in general terms, consists of cool, dry winters and hot, dry summers. The hottest and driest area of Texas lies in the western part of the Trans-Pecos, where annual precipitation rarely exceeds 8 inches and summer daytime temperatures often exceed 100°F. Low-lying basins usually experience the highest temperatures. For example, the temperature in Presidio, located in a basin near the Rio Grande, once reached 125°F, the highest recorded for the region. Mountaintops remain slightly cooler and receive greater amounts of precipitation. Although heavy winter snowfalls may blanket the highest peaks, lower intermountain areas seldom receive more than light dustings of snow (fig. 9.4).

Annual precipitation across the Trans-Pecos varies from 8 to 20 inches and generally increases from west to east. Most thunderstorms occur during the "North American Monsoon" season— the summer months from June to September— when warm, moist air often accompanies remnants of tropical storms. For the remainder of the year, scattered light showers and occasional light snowfalls, combined with high evapotranspiration rates, provide little surface or soil moisture. Evaporation is enhanced by strong summer winds whose intensity increases as they blow through narrow canyons. Periodic droughts, some lasting months but others nearly a decade, further complicate survival for the desert's organisms.

Rainfall is so rare in the desert that old-timers claim they "look forward to the single rainy day each year." An element of truth resides in all exaggerated sentiments, for downpours are indeed uncommon events. Rare deluges from intensive summer cloudbursts—especially those associated with remnants of tropical storms—cause flash floods, known locally as "gullywashers." Where wind erosion and heat create desert pavement, an amalgam of coalesced gravel and stones resistant to water absorption, sheets of rainwater flow across the hardened surfaces. Where the torrents flow into arroyos, walls of water race downstream, tumbling a phalanx of rocks and woody debris

Figure 9.5.
Spring rains
stimulate
germination,
rapid growth,
and flowering
in ephemerals,
producing a showy
spectacle across
desert floors and
mountain slopes.

and scouring the vegetation. Such events usually dissipate unnoticed in the emptiness of the desert, but some prove devastating. During the night of June 10–11, 1965, for example, remnants of Tropical Storm Charley dropped more than 12 inches of rain in the hills north of Sanderson, a small community in Terrell County. A 15-foot wall of water swept down a normally dry creek bed and into town, destroying houses, overturning cars, and drowning 26 residents.

Despite the severity of the environment, about 1 out of every 12 species of plants in Texas, including at least 1,000 found nowhere else in the state, occurs in the Trans-Pecos. The regional fauna is equally diverse. For example, with 482 recorded species, the avifauna in the Trans-Pecos alone is richer in comparison with that of all but three other states in the Union. Because of its diverse habitats, many regard the Trans-Pecos as the most species-rich desert area in the world.

ADAPTATIONS TO DESERT LIFE

The flora and fauna of the Trans-Pecos have adapted to two extreme conditions—heat and aridity—that threaten life. All desert organisms succumb when deprived of water for long periods, but animals, more so than plants, remain particularly vulnerable to the extremes of temperature and aridity. The thousands of desert organisms in the region have evolved adaptations almost as numerous as the species. With little more than a brief taste, the following paragraphs describe only a few of these.

PLANT ADAPTATIONS

Plants adapted to the severe restrictions of aridity are known as xerophytes, which survive long dry periods using one or more of three strategies: drought evasion, succulence, and drought tolerance.

Travelers crossing the Trans-Pecos experience long, monotonous stretches sparsely peppered with scrubby bushes. Yet within days after a significant warm-season rain, desert surfaces magically transform into colorful tapestries of flowering plants (fig. 9.5). Appearing as if from nowhere, these plants survive long droughts as seeds or as modified underground stems. Such plants are best described as ephemerals—plants that skip growing seasons whenever rainfall is insufficient to induce germination. The rare spectacle of showy wildflowers may occur only once every decade and does not last long—usually a matter of days—before the plants wither and disappear. After flowering, desert ephemerals such as Fendler's Bladderpod produce vast quantities of seeds, which can remain dormant for years in a seed bank until conditions favorable for germination return once again.

Seed coats of many ephemerals contain chemical compounds that impede water absorption and thereby constrain seedling development. In some, these inhibitors can be leached away by light rains, whereas others require soaking rains before water can penetrate the seed coat. The seeds of a small, mounding perennial, Golden Smoke, are coated with an elaiosome—a fleshy structure rich in proteins and fatty acids that emits odors mimicking those of dead insects. The odors repel granivorous rodents but attract harvester ants. Foraging ants carry the seeds to their colony, where their larvae devour the lipids and proteins of the elaiosome without harming the seed's viability. After being discarded in the colony's refuse pile, the seeds germinate in a bed of rich organic waste. Both species benefit in this superb example of mutualism—larval ants gain an enriched diet, and the plants' seeds germinate well dispersed in a fertile, protected environment.

The tissues of some plants, especially cacti, collect and store water when it is available and then use these reserves to survive droughts. These plants, called succulents, typically extend their shallow, dense root systems laterally near the soil surface to efficiently absorb water from light showers or heavy dew. Leaves of cacti, modified as prickly spines, reduce water loss by shading the plant surface, retarding air flow and transpiration, and protecting water-rich fleshy tissues from herbivores (fig. 9.6).

Drought-evading plants use various means to restrict water loss between periods of moisture availability. Most have extensive root systems, waxy cuticles, modified leaf structures, or other features adapted to cope with aridity. A desert tree, Palo Verde, reduces evaporative water loss by shedding its small leaves when moisture is limited. These trees survive because the photosynthetic tissues

Figure 9.6. Strawberry Cactus, like many of its kind, protects its succulent tissues with impenetrable arrays of long, sharp spines (top). Dense clumps of fleshy Candelilla stems that retard air flow, together with their waxy-coated surfaces, minimize water loss (bottom).

in their stems (*palo verde* means "green wood") continue manufacturing sugars.

Some desert species possess more than one adaptation for water absorption and retention. Limestone gravel slopes and igneous rocks in the Trans-Pecos support a low, shrubby plant distinguished by a thick waxy coating that retards water loss. Additionally, the fleshy stems of Candelilla, or Wax Plant, form dense clusters bearing only a few tiny leaves (fig. 9.6). The tight bundles of stems reduce the dehydrating effect of airflow, but even so, after rains the small leaves drop off within a few days, minimizing further water loss from their surfaces. In the past, Candelilla was harvested and rendered, roots and all, for wax to manufacture candles, lip balms, phonograph

INFOBOX 9.1. *CURANDEROS*
The Healers

Illness and injury were serious matters for the early inhabitants of the Trans-Pecos. Those living at far-off homesteads and campsites faced long journeys, often of weeks or even longer, over rough trails to reach the nearest town with a physician or medical facilities. Of necessity, inhabitants of this desolate region instead depended on local practitioners whose knowledge of the curative properties of certain plants of the Chihuahuan Desert flora had passed from generation to generation. Such knowledge required not only precise identification of the plants but also knowledge of how and what parts should be compounded into medicines. Those who gained a reputation for treating the sick were known as *curanderos* (men) or *curanderas* (women), but unlike modern medical professionals, these healers charged no fees for their services. Most were goat herders, trappers, or ranch hands who felt compelled to help all in need, whether neighbors or strangers.

When a patient seeking aid arrived at a curandero's doorstep, the healer normally sent one of his children to seek the seeds, leaves, or roots of the plants needed to effect a remedy. Most often, a powder or pulp made from these materials was mixed with some kind of animal fat to form a poultice or was boiled in water to brew a tealike elixir. Fortunately, the desert abounds with hundreds of plants with medicinal properties, a flora collectively identified today by ethnobotanists as a pharmacopoeia; a few of these plants are described here along with the afflictions they treat.

Creosotebush

A liquid derived from leaves of Creosotebush steeped in a small amount of water provided an antiseptic lotion for treating cuts and bruises. The tonic was especially renowned for its ability to cure saddle sores on horses. To treat arthritis, a patient soaked for 30 minutes each day for several days in a tub filled with the strong-scented water produced by boiling large quantities of Creosotebush leaves.

Tarbush

Boiling the leaves and small outer stems of the American Tarbush in water yielded a powerful laxative. The potent mixture produced desired results within three hours.

Mormon Tea

A tea steeped from the stems of any of the three species of *Ephedra* found in the Trans-Pecos treated several maladies, including yellow jaundice and most liver or kidney disorders. After drinking the tea, the patient adhered to a vegetable diet for about a week. Mormon Tea is still used today in many parts of Mexico to treat gonorrhea.

Leatherplant

When the leathery stems for which this plant is named are broken off, drops of reddish fluid ooze from the exposed ends of the severed stems. The sap is massaged directly from the stem onto canker sores, gum infections, and other oral ailments. These treatments are repeated several times daily until the infections are cured.

Sunflowers and Desert Tobacco

To treat sunstroke, a common malady in the desert, curanderos made a thick paste from ground sunflower seeds combined with animal fat or cottonseed oil. The concoction was then spread on Desert Tobacco leaves, applied paste side down on the patient's forehead and temples, and replaced about every 30 minutes for several hours. Meanwhile, the curandero prepared a tonic from boiled sunflower blossoms. After cooling overnight, the brew was poured over the patient's head, massaged into the scalp, and allowed to soak in for half an hour.

Treatments for pneumonia and colds were similar to those prescribed by modern-day physicians—bed rest and time. Notably, however, instead of the hot lemonade often prescribed today, curanderos recommended a drink of hot tequila—an age-old treatment still of obvious value.

---

records, sealing wax, mascara, chewing gum, shoe polish, and food additives. During World War I, Candelilla wax also waterproofed munitions and insulated electric appliances; additionally, the plant's white sap was touted as a treatment for sexually transmitted diseases (infobox 9.1).

Creosotebush, a distinctive shrub of the Trans-Pecos, usually appears in a regularly spaced pattern resembling that of orchards (fig. 9.7). The spacing was once attributed to allelopathy—a condition whereby plants produce chemicals that inhibit the growth of competitors, including their own seedlings. However, scientists later discovered that the extensive root systems of Creosotebushes are so efficient that seedlings cannot obtain enough water to germinate or survive in the vacant spaces. On the other hand, seedlings of prickly pears and several other cacti increase their chances of survival

Figure 9.7. Creosotebushes effectively outcompete most other plants for water, which creates regular spacing between the shrubs. Note the water-retarding surface of small interlocking stones known as desert pavement and the Chisos Mountains rising above the harsh environment of the Chihuahuan Desert.

in the shelter of a "nurse plant." When birds deposit cactus seeds in the soil beneath Creosotebushes, the shady environment allows cactus seedlings to take root. Thereafter, shallow cactus roots intercept rainwater before it penetrates to the shrub's deeper roots, which allows the cactus to thrive while the Creosotebush slowly dies for lack of water.

ANIMAL ADAPTATIONS
All desert organisms experience survival problems when water is scarce, but animals are more susceptible than plants to temperature extremes. Most internal biological processes function only within relatively narrow boundaries, and when this range is exceeded, the animal dies. Summer temperatures in the Trans-Pecos frequently exceed the limits endured by most animals, and periods of punishing heat often coincide with periods of water scarcity. Most chemical reactions in an animal's

body depend on water, which also functions as the primary coolant. As might be expected, most desert animals have evolved both behavioral and physiological adaptations that enhance their survival during lengthy hot and dry periods.

Many desert animals alter their behavior to avoid excessive heat. Phainopeplas leave the region to spend the summer months in higher, cooler elevations in Mexico, while others spend hot days in the shade, tucked in cool burrows, buried deep in soft soils, or concealed in shadowy rocky crevices. Great Plains Narrow-mouthed Toads frequently share shelters with tarantulas, where, in an unexpected relationship (another example of mutualism), the toads apparently devour ants that could damage the spiders' legs while the tarantulas protect the toads from predators. Many animals are crepuscular, limiting their activities to the relatively cooler hours of dawn or dusk. Kit Foxes, Banner-tailed Kangaroo Rats, and other animals become active only at night when the desert cools and the relative humidity increases.

Desert animals that remain active during hot summer days have evolved ways of dissipating heat. The elongated ears of Black-tailed Jackrabbits and the long, slender appendages of many desert lizards and mammals (fig. 9.8) dissipate heat into the environment when they rest in shady locations. Evaporative cooling, which results when water evaporates from moist surfaces, is the most common method for reducing heat in birds and mammals. When Coyotes and foxes pant or birds gape with open mouths while fluttering their throat region, air moving across the wet membranes carries off heat. Because of their black plumage, Turkey Vultures in the desert tend to absorb substantial heat, which they dissipate in two ways. First, and as unsavory as it may seem, perched vultures defecate on their legs. Cooling results when the moist excreta evaporate and the cooled blood in their legs circulates throughout the body—a cooling method called urohydrosis. Turkey Vultures also escape searing desert temperatures by soaring upward on thermals to reach cooler air.

Surface water is isolated and inaccessible to most desert inhabitants. Most birds simply fly to dependable water sources, but larger mammals such as Coyotes, Mule Deer, and Desert Bighorn Sheep typically include at least one dependable water hole in their home range. However, those desert animals lacking access to water fulfill their physiological requirements in other ways. Some animals, such as kangaroo rats and most pocket mice, rarely drink and instead survive by obtaining water from the insects or seeds they ingest. The Black-throated Sparrow, which may represent the most desert-adapted songbird in North America, requires almost no drinking water; it excretes highly concentrated—and nearly dry—urine and feces.

The Desert Cicada has evolved a unique means of obtaining water from plants and dissipating heat. Its needlelike mouthparts puncture the water-conducting tissues—xylem—of desert plants. Water ingested from this source is swiftly routed to abdominal and thoracic pores, where it exits and evaporates from body surfaces in a process not

Figure 9.8. Long ears and legs of Black-tailed Jackrabbits aid in detecting and escaping predators as well as in dissipating heat (top). Similarly, long tails and narrow appendages in the Big Bend Tree Lizard and other desert lizards serve several functions (bottom). Note also the lizard's protective coloration.

unlike sweating. Evaporative cooling allows Desert Cicadas to maintain body temperatures up to 9°F below ambient temperatures.

Few amphibians occur in the Trans-Pecos—most of these are found in riparian zones or near other water sources, including seasonally available ones. Only a few, mostly toads, venture far from water, where they tend to be more abundant on soils that retain moisture. When moisture is available, Red-spotted Toads absorb water through a specialized patch of skin on their rear.

The rear feet of the Texas Toad and Couch's Spadefoot possess hardened projections that serve as shovels. They bury themselves up to 35 inches belowground, where they create a protective chamber from mucus or skin cells before entering a resting state called estivation. The low-frequency sound of raindrops pounding the desert floor stimulates desert toads to emerge from entombment. After exiting, the toads immediately seek pools of water where they breed. Couch's Spadefoot eggs hatch within 24 hours and the tadpoles develop legs within 10 days. Rapid development allows the tadpoles to gain a competitive advantage for food resources, lessen their exposure to predators, and shorten their dependence on temporary pools. In contrast, the tadpoles of most toads in mesic (moist) habitats require 30 days or more to metamorphose. At high densities in drying pools, Couch's Spadefoot tadpoles sometimes resort to cannibalistic oophagy—they consume the eggs of other spadefoots. This behavior provides energy-rich nutrients that allow tadpoles to complete metamorphosis more rapidly, further reducing the danger of desiccation as pools dry.

Many temporary ponds explode with life within hours or days after a desert thunderstorm fills a long-dry depression. Desiccation-resistant eggs of some invertebrates remain in diapause, viable in the dry soil, sometimes for years. When rains again fill the depression, the eggs of tiny crustaceans (phyllopods) develop through several larval stages into adults and reproduce, thereby completing their life cycle in a short-lived desert pond in about 16 days. Like the sudden burst of ephemeral flowers, phyllopods offer another example of "boom and bust" in a water-starved environment.

INTERDEPENDENCE OF PLANTS AND POLLINATORS

"Timing is everything"—so the saying goes—and it is especially true in severe habitats where the essential elements for survival and reproduction appear infrequently and unpredictably. Some of the most fascinating ecological phenomena in deserts occur among interacting species, especially in the associations between flowers and their pollinators.

A classic example of plant-animal interdependence involves the relationship between yuccas and yucca moths (fig. 9.9). Each of the seven yucca species dotting the Trans-Pecos

Figure 9.9. A single species of yucca moth acts as both pollinator and seed predator of yuccas in the Trans-Pecos. Shown here are a panicle of Soaptree Yucca blossoms (top) and a pair of yucca moths (bottom) gathering pollen inside one of the flowers. Photographs by Brian R. Chapman (top) and M. J. Hatfield, US Fish and Wildlife Service (bottom).

landscape is pollinated by a single species of yucca moth. When ready to lay eggs, a female yucca moth gathers heavy, sticky pollen from several flowers of one plant, rolls the pollen into a large ball, and carries it to the stigma of another flower of the same species. Whereas the simple transfer of pollen completes the process in other flowering plants, successful pollination in yuccas requires one more step—pounding the pollen into the flower's stigma with a force that only a yucca moth can muster.

After the flower is fertilized, the moth lays eggs in the flower's ovary, which later develops into a large capsule containing hundreds of yucca seeds mixed with moth eggs. After hatching, moth larvae feed on yucca seeds until they mature enough to chew holes and escape through the capsule wall. Thus, yuccas are precisely fertilized by pollen from plants of the same species while the moths concurrently gain safe nurseries with plentiful food for their developing larvae. However, the interdependent relationship works only if the moth lays only a few eggs in each flower. Yuccas somehow monitor the weight of the pollinated ovary—if a moth lays too many eggs, the overburdened flower drops off, eliminating both seeds and moth larvae.

Many flowering plants, such as the Globe Mallow, depend on several species of tiny, often metallic-colored bees for pollination. Because the bees likewise depend on abundant flowers for nectar and pollen, they have evolved drought-evading strategies that synchronize their life cycle with flower production. Just as seeds of ephemeral plants remain in a seed bank during droughts, the larvae of desert bees arrest their development and await adequate summer rainfall buried in the soil as protection from excessive heat, low humidity, predators, and pathogens. In time, rainfall triggers the concurrent events of emerging bees and germinating seeds, producing a flush of colorful flowers ready for busy pollinators.

In spring, the migratory schedules of several nectar- and pollen-feeding animals closely match the flowering chronology of some desert plants. For example, the northward route of migrating Mexican Long-nosed Bats from wintering grounds in Mexico is coordinated with the availability of nectar and pollen in the flowers of Parry Agave and approximately 21 other species. Similarly, the annual spring migration of hummingbirds coincides with the flowering of Ocotillo in the

Trans-Pecos. Unlike most desert plants, Ocotillo does not always flower in response to rainfall. Instead, the increasing hours of sunshine each spring trigger production of colorful Ocotillo blossoms, even in the absence of rain, at the tips of the plants' tall, leafless stems (fig. 9.10). The timing is critical for hummingbirds—without a dependable source of nourishment just when the birds cross the desert, they might lack enough energy to reach their breeding grounds farther north. Ocotillos, however, are not equally dependent on hummingbirds for pollination—they have a "backup." Female carpenter bees, although too large to enter an Ocotillo flower, nonetheless serve as effective pollinators by opening a slit in the base of an Ocotillo blossom. Then, after sipping nectar from inside the cut flower, the bee moves on to Ocotillo blossoms elsewhere, carrying pollen attached to its abdominal hair.

Figure 9.10. Even without rainfall, tubular Ocotillo flowers bloom in spring in time to nourish migrating hummingbirds.

BIOPHYSIOGRAPHIC ASSOCIATIONS

With elevations ranging from 1,000 to 8,750 feet and extreme variations in topography, soils, precipitation, and vegetation, ecologists enjoy great latitude in describing ecological subunits of the Trans-Pecos. A recent classification of plant associations in the Trans-Pecos, for example, recognized no fewer than 117 vegetation cover types. Our treatment characterizes the habitats ranging from the forests crowning the tallest peaks to the scrubby plateaus, intermountain grasslands, salt basins, and riparian zones at lower elevations

Figure 9.11.
Yucca-studded
grasslands carpet
the wide basins
separating
mountain ranges
in the Trans-
Pecos. Viewed
from Marfa Flat, a
rare rain shower
moistens the
Davis Mountains.

(fig. 9.11). Still, the confines of a single chapter preclude a fuller treatment of such a remarkable biological paradise as the Trans-Pecos.

STOCKTON PLATEAU

The Stockton Plateau extends west of the Pecos River, the physiographic boundary separating the arid Trans-Pecos from the hilly Edwards Plateau. Some authorities, however, regard the Stockton Plateau as an extension of the "Hill Country" because both areas formed during the Cretaceous Period and share some similarities in vegetation. Despite these similarities, the much-drier environment west of the Pecos River supports vegetation that reflects the character of the Chihuahuan Desert forming the flanks of the Stockton Plateau. Here we adopt the traditional view and include the Stockton Plateau as part of the Trans-Pecos.

When the first Euro-Americans traversed the Stockton Plateau, grasses dominated the vegetation and freshwater springs flowed in many areas. But after years of overstocking and intensive grazing in the last decades of the nineteenth century, erosion thinned the soils and the grasslands gave way to a shrub-savanna association dominated by Creosotebush, Broomweed, Tasajillo, prickly pears, and American Tarbush, a stunted, woody daisy. Honey Mesquite, Redberry Juniper, and shrubby oaks dot the flat highlands, whereas sparsely distributed clumps of grama grasses, Sotol, and Lechuguilla occupy lower areas. Runoff from overgrazed rocky substrates diminished the amount of water percolating into underlying aquifers and inevitably eliminated many of the once-numerous springs. Comanche Springs, for example, once spawned a stream that gushed through the desert for more than 30 miles and supported an atypical desert fauna of Common Muskrats, Texas Spiny Soft-shells, and Comanche Springs Pupfish. When the springs dried in the 1960s, the pupfish, now listed as endangered, was relocated to San Solomon Spring at Balmorhea State Park (Reeves County).

A few springs still flow from aquifers beneath the Stockton Plateau. One of the largest perennial springs, Caroline Springs (Terrell County), issues about 3,000 to 5,000 gallons of clear freshwater per minute into Independence Creek, a tributary of the lower Pecos River. With additional inflow from other springs downstream, Independence Creek contributes at its confluence with the Pecos about 40 percent of the Pecos River's volume and provides refugia (places where animals can survive unfavorable conditions) for several rare desert fishes.

Numerous fissures, sinkholes, and caverns perforate the limestone of the Stockton Plateau. Small, shallow caverns like Trash Barrel Cave (Val Verde County), named for its proximity to a garbage receptacle alongside a highway, descend in a twist of tight passageways and provide cool day roosts for bats and other desert creatures. Sorcerer's Cave (Terrell County), one of the deepest caves in Texas, winds downward to the subterranean Sirion River flowing 558 feet beneath the plateau, and Fern Cave (Val Verde County) harbors a maternity colony of several million Brazilian Free-tailed Bats. Fossils and other remains discovered in Cueva Quebrada (Val Verde County) provide a glimpse of the climatic conditions, flora, and fauna present about 14,000 years ago.

CHIHUAHUAN DESERT
Ecologists often define physiographic boundaries using the presence or absence of certain plants known as indicator species (the presence of such species reflects a specific combination of environmental conditions). It is not possible, however, to precisely delineate the extent of the Chihuahuan Desert in western Texas; because of overgrazing and erosion, indicator plants for the Chihuahuan Desert have encroached into other types of vegetation in the Trans-Pecos, thus blurring boundaries between the desert and adjacent communities. Creosotebush is by far the most abundant plant in the Trans-Pecos, but this species is also a major component in the flora of both the Sonoran and Great Basin Deserts. Other species that characterize the Chihuahuan Desert include Lechuguilla, American Tarbush, Ocotillo, Sotol, yuccas, Catclaw Acacia, and Catclaw Mimosa. Of these, some authorities select Lechuguilla as the best indicator species.

The most unsullied example of Chihuahuan Desert vegetation occurs in the severe environment surrounding the base of the Chisos Mountains in southern Brewster County (see fig. 9.7). From Big Bend, this distinctive association extends northwestward on the Rio Grande lowlands to the southern slopes of the Davis and Sierra Diablo Mountains—virtually all vegetative associations

in the lower elevations of the Trans-Pecos contain substantial elements of the Chihuahuan Desert. If the climate continues its present warming trend, the Chihuahuan Desert will undoubtedly continue to expand.

DESERT BASINS AND SALT FLATS
A common physiographic feature of the Trans-Pecos is basin-and-range topography, characterized by elongated mountain ranges isolated by wide, flat valleys. As wind, water, and ice gradually grind down mountains in the Trans-Pecos, thunderstorms wash the eroded sediments downslope, spreading large outwash fans—bajadas—that gradually fill the basins with deposits of clay, silt, sand, and gravel. Sediments in the Hueco Basin east of El Paso, for example, are more than 9,000 feet thick.

Often lying in the rain shadow of surrounding ranges, many basins receive less than 14 inches of rainfall per year and represent some of the hottest and driest habitats in Texas. Because some basins lack streambeds or arroyos to carry off the limited rainwater, runoff collects in local depressions, forming bolsons. These broad, shallow lakes soon evaporate, leaving extensive deposits of gypsum and salt—harsh conditions devoid of vegetation. Salt-tolerant species such as Fourwing Saltbush, Quinine Bush, and Alkali Sacaton typically fringe the margins of dry lakebeds where the water table lies just beneath the surface. Several lizards— among them the Side-blotched Lizard and Little-striped Whiptail—and the Black-throated Sparrow are so attuned to survival in severe basin habitats that they serve as indicator species.

The most extensive bolson in the Trans-Pecos, the Salt Basin, consists of five extensive flats, which drain surface water northward to the extensive Salt Flat, a bolson west of El Capitan. During the Pleistocene Epoch, a broad, shallow lake formed in the lowest portions of the Salt Basin, where water fluctuated between 18 and 37 feet in depth and percolated into the underlying strata, forming a series of aquifers. Later, when the climate dried, the lake evaporated and left a series of beaches as reminders of the former shorelines. Periodically, the usually dry lakebed receives enough runoff to re-create a shallow ephemeral lake (fig. 9.12). Evaporation rapidly saps the lake, leaving deposits

Figure 9.12. After heavy rains, runoff flowing to the normally dry Salt Basin near the southwestern base of the Guadalupe Mountains forms an ephemeral shallow lake. Guadalupe Peak and El Capitan loom in the distance. Photograph courtesy of Guadalupe Mountains National Park, National Park Service.

of sandy gypsum that the prevailing westerly winds transport to the edges of the Guadalupe Mountain bajada. Here, white gypsum collects in a blindingly bright dune field covering about 2,000 acres, the third-largest deposit of gypsum sands in the world, and supports an endemic plant, Gypsum Broomscale. Hairy Crinklewort and Gypsum Grama blanket some gypsum mounds.

Nearby, reddish dunes formed from quartz sands color the scene. These reach heights exceeding 60 feet in the southern part of the dune field, but dunes elsewhere in the field are much smaller. Vegetation such as Broom Dalea, Onion Blanketflower, Warnock's Ragwort, and thin cryptogamic crusts stabilize the large dunes; slacks between dunes support low clumps of mesquites and Creosotebushes. The dune fields

offer habitat for several animals, including Desert Cottontails, Black-tailed Jackrabbits, Yellow-faced Pocket Gophers, Ord's Kangaroo Rats, Texas Horned Lizards, and Western Diamond-backed Rattlesnakes (fig. 9.13).

The extensive salt deposits in the Salt Basin's dry lakebeds represented a valuable commodity for early Native Americans and the Euro-American settlers who followed. In the late 1860s, a dispute over ownership of these deposits escalated into a bloody conflict that attracted national interest. A series of skirmishes—known as the "San Elizario Salt War"—between the feuding groups lasted nearly 12 years. During the prolonged struggles, a detachment of 20 Texas Rangers gave in to a mob (the only time in history that Texas Rangers ever surrendered), approximately 30 men died in battle,

Figure 9.13. The Texas Horned Lizard (top) and Ord's Kangaroo Rat (bottom) inhabit the sandy dunelands near the Guadalupe Mountains. Photographs by Andrew Brinker (top) and Troy L. Best (bottom).

and looters ransacked the town of San Elizario. The war ended abruptly in 1877 when US Army troops quelled the fighting.

TRANS-PECOS GRASSLANDS

Although it is difficult to imagine today, early accounts of the Trans-Pecos described a region of extensive grasslands interspersed with few shrubs. Lieutenant Echols encountered "good grass" during the summer of 1860 as he traveled from the Pecos River to Fort Stockton. Botanist H. J. Cottle once characterized the Trans-Pecos as "the largest remaining area of native grassland in the United States." Unfortunately, the lush "waist-high grass" that Echols encountered along Terlingua Creek—at the time a running stream lined with cottonwood trees and full of beavers—has vanished. Much of the original grasslands has given way to desert

shrubland, likely resulting from overgrazing during the past 150 years.

Elevation plays a role in defining three grassland associations in the Trans-Pecos. Semidesert grasslands occur on mesas, plains, plateaus, bajadas, and the lower slopes of mountains at elevations between 3,500 and 4,000 feet. Plains grasslands, composed of shortgrasses (grasses that rarely exceed 2 feet in height), dominate sites on plateaus and midelevation slopes between 4,000 and 5,200 feet. At higher elevations, plains grasslands intergrade with juniper-oak-pinyon savannas and woodlands and contact the semidesert grasslands at the lower elevations. Where precipitation is greater at sites between 4,000 and 7,000 feet, mountain grasslands carpet the uplands virtually to the exclusion of desert shrubs. A high valley at the foot of Sawtooth Mountain on the west side of the Davis Mountains offers a good example of midelevation montane grassland (fig. 9.14).

Hispid Cotton Rats, perhaps the most abundant wild mammal in Texas, occur in both the semidesert and plains grasslands of the Trans-Pecos. Extremely prolific, they produce as many as nine litters per year, each with two to ten young. Following wet cycles, Hispid Cotton Rat populations often increase exponentially—in a year's time, a single female and her descendants can potentially add more than 15,000 individuals to the population. A related species, the Yellow-nosed Cotton Rat, is confined to grasslands at higher elevations, on rocky slopes, and in bunchgrass habitats along perennial mountain streams within the region. However, the range of this species, easily identified by the color of its namesake nose, may be expanding into lower elevations in the Trans-Pecos.

Although the American Bison was common on the High Plains, the Pronghorn was the most abundant large herbivore in the Trans-Pecos when Lieutenant Echols visited the region. The Pronghorn—popularly known as "antelope"— represent a monotypic family endemic to North America and unrelated to the true antelope of Africa. "Pronghorn" describes the unique horns adorning both sexes. The horns are composed of an outer sheath of fused hairs covering a bony

Figure 9.14. With fire suppression, junipers, Tree Cholla, and other vegetation often invade the extensive mountain grasslands of the Trans-Pecos. Sawtooth Mountain, on the western edge of the Davis Mountains, provides evidence of the area's volcanic origins.

and permanent inner core. The running ability of Pronghorns, the fastest land mammal in North America, likely developed to escape the now-extinct American Cheetah, a swift grassland predator that evolved from a cougar-like ancestor. Although they can easily outrun existing land predators, Pronghorns were sadly unprepared for the arrival of ranchers who jealously coveted grass for their livestock.

The availability of extensive grasslands and the construction of a railroad did not go unnoticed by ranchers who brought large herds of cattle to the Trans-Pecos by the 1880s. Shortly thereafter, most of the land in the region was bought, leased, or claimed for grazing the ever-increasing herds of cattle, sheep, and goats or for businesses supporting railroads and ranches (infobox 9.2). Believing that the Pronghorn competed with cattle for forage, early ranchers soon waged a range war against Pronghorns, and by the turn of the twentieth century, few remained. Only later did biologists determine that Pronghorns prefer other foods—less than 4 percent of their year-round diet includes grasses. To protect the remaining herd, the Texas legislature closed the hunting season for Pronghorns in 1903, but the law was largely ignored in the far-off Trans-Pecos. Restocking programs begun in the 1940s augmented a population increase, but today the Pronghorn population is again declining. Curiously, while Pronghorns can run swiftly, they rarely jump. Hence, fences—especially the mesh fences used on sheep ranches—become major barriers for Pronghorns seeking forage or escaping predators, whereas they can pass deftly *under* the bottom of strand-wire fences. Given that, some ranchers have replaced the bottom strand of their barbed-wire fences with a smooth wire. All told, however, degraded rangelands, recurring droughts, and injurious fences together have contributed to the current decline of Pronghorn populations in the Trans-Pecos. Currently, the Trans-Pecos Pronghorn population is being augmented by "surplus" Pronghorns relocated from the Panhandle to ranches owned by cooperating landowners.

LOW MOUNTAINS AND BAJADAS

Rocky ridges share rugged features such as extensive areas of exposed bedrock, precipitous canyons, and towering cliffs, but despite similar topography, the geological history of adjacent ranges often differs substantially. In some ranges,

INFOBOX 9.2. LAW WEST OF THE PECOS

Deep in the thorny Trans-Pecos stands the Jersey Lilly, the one-time bar and courtroom of Judge Roy Bean and now a state historical site. Phantly Roy Bean Jr. (1825–1903) stands alone among the adventurers, eccentrics, charlatans, scoundrels, lawmen, rustlers, murderers, barkeeps, and entrepreneurs coloring the history of the Trans-Pecos.

Roy Bean came to Texas as a teenager after leaving his birthplace in Kentucky and getting into trouble in New Orleans. With his brother, Sam (1819–1903), Roy held a variety of jobs in Texas, Mexico, California, and New Mexico, often relocating after involvement in nefarious incidents. For example, executioners left Roy on a horse with a noose around his neck, intending that he hang for killing a Mexican army officer. Luckily for Bean, the horse did not bolt and the young lady for whom he had committed the murder cut him down.

For the next 20 years, Roy engaged in several disreputable businesses in San Antonio. He cut and sold firewood from land he did not own, butchered unbranded cattle he rustled from area ranchers, and used water from a nearby stream to dilute milk in his short-lived dairy business. After customers found minnows in the thinned liquid, his response aptly demonstrated the personality traits for which he later became famous: "By Gobs, I'll have to stop them cows from drinking out of the creek."

Bean, looking for new opportunities—and perhaps hoping to avoid legal worries—purchased a tent and ten barrels of whiskey and moved to Vinegaroon, a tent city of railroad workers located where the Pecos River joins the Rio Grande. The town is the namesake of a species of whip scorpion that, while somewhat frightening, is a rather harmless arthropod whose reputation stems from its likeness to true scorpions. Instead of a scorpion's venom-loaded stinger, the tail of a Vinegaroon sports a threadlike structure that functions only as a sensory organ of touch. Vinegaroons are not defenseless, however. When irritated, they spray a mist of nearly pure acetic acid that discourages the source of unwanted attention. For amusement, barroom patrons and schoolboys alike matched Vinegaroons with tarantulas in makeshift arenas, always with the same outcome—a victorious Vinegaroon. The spray, unless it hits the eyes, does not harm humans.

Because of its remote location—some 200 miles from the nearest court—the town of Vinegaroon needed a judge, so Texas Rangers recommended Bean as Justice of the Peace for the new Precinct 6, in Pecos County. With his tent saloon doubling as the courthouse, Judge Bean dispensed liquor and justice with equal enthusiasm and referred to himself as the "Law West of the Pecos."

As the railhead continued westward, Judge Bean moved his court and saloon to Eagle's Nest, where the tracks emerged from the canyon of the Rio Grande. A local rancher had sold his land there to the railroad with the restriction that no part of it could be sold to, or used by, Roy Bean. Undaunted, the judge simply built his saloon and billiard parlor on railroad property and remained an illegal squatter for 20 years. He named his saloon the Jersey Lilly in honor of Lillie Langtry (1853–1929), but he misspelled her name. Lillie was an English singer and actress Bean admired but never met. He also constructed an opera house and changed the town's name to Langtry.

Judge Bean became a legend for his unusual court proceedings and rulings. Hung juries were not allowed, and jurors were required to buy drinks at the saloon during breaks. Sentences included work assignments or fines. Fines, retained by the judge, conveniently matched the amount in a defendant's pockets. Once, when a dead man was found with a pistol and $40 in his pockets, the judge famously fined the corpse $40 for carrying a concealed weapon, which he likewise confiscated. The rest of the story is often overlooked: Bean used the money for a grave, casket, and headstone for the deceased, whose pistol became the judge's gavel. Despite his shady past, Judge Roy Bean was locally appreciated in his later years for his generosity and kindness toward the poor.

The life and legend of Judge Roy Bean lives on in fact and fiction in books, comics, movies, and television. The Texas Department of Transportation maintains the Judge Roy Bean Visitor Center in Langtry, which includes the original Jersey Lilly saloon and a replica of the opera house (actually Bean's home). Additionally, visitors can stroll through a garden of cactus, yucca, agave, and other desert plants, each identified and characterized with local lore. One of these, the Langtry Rainbow, a whitish cactus with large, white-centered magenta flowers, occurs only on limestone outcrops in the Trans-Pecos. Many of the species in the garden provided Native Americans and settlers with food, fiber, or medicines, including a few still in use for these purposes. At nearby Dead Man's Gulch, the tracks of the Southern Pacific joined New Orleans with southern California in 1883, an event marking completion of the second transcontinental railroad and commemorated by a silver spike.

rocks and soils of differing origins overlap, but in general, only thin veneers of soil cover the Trans-Pecos mountains. Most of the rain falling on limestone slopes runs off the hard surfaces or flows through fissures into aquifers far below. Consequently, the shallow, stony soils covering these ranges remain dry for most of the year, creating conditions where few plants can survive. Soils on volcanic slopes are just as thin and sparse, but the water often percolates through cracks to emerge downslope as mountain springs. More than 80 perennial springs and numerous intermittent seeps emerge in the Bofecillos Mountains, a volcanic range located west of Big Bend National Park. Vegetation developing in canyons provides the best indication of available water—canyons in ranges formed from limestone or other sedimentary deposits tend to be sparsely vegetated, whereas those in basaltic mountains produce plant communities of greater diversity.

Strong thunderstorms generate runoff torrents that carry rocks, gravel, and finer sediments to lower elevations. Water cascading downhill after a storm loses velocity when the flow disperses over alluvial fans at the mouths of canyons. Larger rocks and gravel fall out at upper levels and finer materials move farther and collect at the lower, front edges of the bajada, creating unique conditions for community development. Because the surface tension of soil water varies in proportion to the diameter of the particles, the availability of water steadily diminishes downslope, reaching a point where few plants can absorb water from the fine, so-called tight soils deposited at the lower end of an alluvial fan. Similarly, soil salinities also increase in the lower reaches of an alluvial fan. As a result, vegetation on alluvial fans develops along a gradient, with both cover and species diversity steadily diminishing downslope from high to low elevations.

The grasslands on mountain slopes between 4,000 and 5,000 feet include many species common in the shortgrass communities of the High Plains (chapter 8). The principal species in both locations include Blue, Sideoats, Hairy, and Black Grama, Cane Bluestem, and Silver Bluestem. A meager collection of succulent plants and rough grasses dominates the steeper slopes and rocky bajadas.

Desert shrubs, such as Lechuguilla, Ocotillo, Sotol, American Tarbush, and prickly pears, share rocky slopes and upper bajadas with scattered clumps of Black Grama and Bush Muhly. Mock Orange, Silktassel, and Mountain Mahogany, which occupy protected sites on slopes and in canyons, provide forage for Mule Deer and reintroduced populations of Desert Bighorn Sheep and introduced Rocky Mountain Elk. Clumps of junipers and Pinyon Pine occur only on the few peaks that exceed 6,000 feet in elevation.

Crevices and caves in rock outcrops provide roosting habitat for the Pallid Bat, which gets its name from its pale, light yellow coloration. After emerging in the evening, the bats prey on large insects, most of which exceed 0.7 inch in length. Selection of such large prey is unusual among insectivorous bats, but their method of hunting is even more unique. Instead of capturing and ingesting their prey in flight as most insectivorous bats do, Pallid Bats devour large insects after forcing them to the ground. Furthermore, Pallid Bats often forage on the ground for beetles, crickets, scorpions, lizards, and rodents. Recent observations indicate that Pallid Bats also consume nectar from flowering Parry Agaves, thereby potentially competing with nectar-feeding Mexican Long-nosed Bats, which migrate annually to the Trans-Pecos from Mexico.

Collared Peccaries, Bobcats, and Gray Foxes roam widely through the region, but Desert Bighorn Sheep represent the icon of the mammalian fauna in the low mountains of the Trans-Pecos (fig. 9.15). Because of their concave hooves, these heavy-bodied sheep can rapidly—and gracefully—traverse steep rocky slopes with ease. Their preference for rugged habitat, their agility, and their ability to withstand dehydration—all adaptive strategies—allow their survival in areas too dry for many of their predators. More than a century ago, Vernon Bailey recorded bighorns in many of the more isolated mountain ranges: "Here the sheep find ideal homes on the open slopes of terraced lime rock or jagged crests of old lava dikes." Just 60 years after that passage was penned in 1890, bighorns vanished from the Trans-Pecos. Although hunting pressure no doubt contributed to their extirpation, the diseases of domestic sheep and goats and

Figure 9.15. Reintroduced Desert Bighorn Sheep now inhabit many mountain ranges in the Trans-Pecos. Photograph by Froylan Hernandez.

competition with introduced Aoudad Sheep were likely the major decimating factors. In 1954, the Texas Parks and Wildlife Department began restocking the Trans-Pecos with bighorns captured in other states. The agency also established several propagation centers, and the offspring produced at these facilities were released into seven mountain ranges in the region. By 2008, the population of Desert Bighorns exceeded 1,200 individuals.

SKY ISLANDS AND MADREAN PINE-OAK WOODLANDS

The Chisos, Chinati, Davis, and Guadalupe Mountains, the Sierra del Carmen, and the Maderas del Carmen form links in a chain of mountains connecting the Sierra Madre Oriental of Mexico to the southern Rocky Mountains in New Mexico. These six ranges rise well above 7,000 feet—high enough to encourage cloud formation when air blowing over the mountains cools and loses its capacity to retain moisture. Consequently, the highest elevations receive 18 to 20 inches of rainfall annually, or about twice the precipitation falling on the lower slopes. In response, the summits and highlands bear forest communities of hardwoods and conifers resembling those of the Rocky Mountains, with some representatives of the Sierra Madre Oriental in Mexico. The vegetation on these mountaintop "sky islands" scarcely resembles that of the surrounding desert "sea."

The montane woodlands above 6,000 feet consist of oaks, junipers, and Pinyon Pines. Of the three species of Pinyon Pine in the region, Colorado Pinyons are restricted to the Guadalupe Mountains and Sierra Diablo, while Mexican Pinyons occur in the Chisos, Davis, and other mountain ranges. A third species, Papershell or Remote Pinyon, occurs on limestone soils in Big Bend's Dead Horse Mountains and in the Glass Mountains (Brewster County). The highest elevations of the Guadalupe, Davis, and Chisos Mountains support parklike forests of Ponderosa Pine, Southwestern White Pine, and Douglas-Fir, sheltering a diverse array of seasonal wildflowers and lush grasses. Small patches of closed-canopy forests develop on some peaks and in the highest, wettest canyons, but an understory of grasses and brushy trees occurs more often. Associated hardwoods include Gray Oak, Bigtooth Maple, Alligator Juniper, Gambel Oak, Texas Madrone, Mountain Mahogany, and others. Scattered stands of Quaking Aspen grow on north-facing rocky slopes in all three of the highest ranges of Texas.

Rugged Trans-Pecos mountaintops provide refugia for several animals. Mountain Lions, Bobcats, Collared Peccaries, and Mule Deer wander

Figure 9.16. A Mountain Lion relaxes within a "sky island" safely inside Big Bend National Park (top). Under the watchful eye of their mother, Black Bear cubs frolic on a Pinyon Pine high in the Chisos Mountains (bottom). Photographs courtesy of Reine Wonite (top) and Big Bend National Park, National Park Service (bottom).

Figure 9.17. Madrean pine-oak woodlands develop on igneous soils in canyons high in the Davis Mountains.

freely through the peaks and canyons clothed by montane forests (fig. 9.16). Decades after their extirpation from all sky islands in the Trans-Pecos, Black Bears moved northward from Coahuila to recolonize the Chisos Mountains (infobox 9.3; fig. 9.16). Several birds more commonly associated with Mexico, such as the Mexican Jay, Colima Warbler, and Painted Redstart, reach the northern limits of their distributions in the Chisos and Davis Mountains. The Davis Mountains Cottontail occurs at elevations of about 1,400 to 2,500 feet in pinyon-oak-juniper woodlands of the Chisos, Chinati, Davis, and Guadalupe Mountains and on Elephant Mountain. A true endemic, and the only chipmunk in Texas, the Gray-footed Chipmunk dashes from the shadows in search of seeds, berries, and acorns in the coniferous forest of the Sierra Diablo and Guadalupe Mountains. The Trans-Pecos Black-headed Snake hides itself in cracks and crevices

and emerges to feed on centipedes and small insects on summer nights following rains.

Island-like communities described as Madrean pine-oak woodlands also develop on the igneous soils of the Davis and Guadalupe Mountains at sites shielded from extreme winds and heat (fig. 9.17). These microhabitats—cliff bases, sheltered canyons, and north-facing slopes—occur at elevations between 4,500 and over 8,000 feet. The unique flora is dominated by an association of Silverleaf and Gambel Oaks, Ponderosa, Southwestern White, and Pinyon Pines, and Alligator Juniper, which may form a closed-canopy forest at higher elevations receiving greater moisture. Like the canopy species, most of the

INFOBOX 9.3. RETURN OF THE BLACK BEARS

Black Bears, which Vernon Bailey recorded as "common" in the mountains of the Trans-Pecos at the dawn of the twentieth century, soon fell on hard times. Recreational hunters unrestrained by game laws shot bears at will, but predator control on behalf of livestock owners likely proved at least as devastating. One method—deployment of strychnine-laced horse or cow carcasses—indiscriminately poisoned ravens, badgers, and foxes along with Coyotes, Bobcats, Mountain Lions, and, of course, the once common Black Bears. Indeed, the practice could claim any scavenger or predator, whether large or small, living where livestock grazed. By the 1950s, bears were all but completely extirpated from the region. Conversely, on the other side of the Rio Grande, Mexican ranchers were far more tolerant of predators and, especially in the case of bears, often honored their presence. Thus in Mexico, the Sierra del Carmen/Maderas del Carmen—the largest range of sky islands in the Chihuahuan Desert—became an undeclared refuge for Black Bears. Back in Texas, the state initially listed Black Bears as endangered in 1987, but modest gains in their numbers, as noted below, allowed an upgrade to threatened in 1996.

Early in the 1980s, visitors and staff at Big Bend National Park occasionally saw bears in the Chisos Mountains. Sightings occurred more frequently thereafter, and a small population of bears seemed established by the end of the decade. By 2000, 29 bears, including cubs, had repopulated the vacant habitat quite without human assistance, a rather uncommon feat for extirpated wildlife; the census likely represented about 90 percent of the actual population. To reach the sky islands in the Chisos range, the founders of this population had to travel considerable distances across rather unfriendly bear habitat—the Chihuahuan Desert.

All seemed well, but drought, aided by infestations of Variable Oakleaf Caterpillars, devastated the acorn crop between 1999 and 2000. Lacking this staple in their diet, the bears headed south, several reaching Mexico, some dying en route, and a few returned to the Chisos, as determined from animals previously outfitted with radio collars. Overall, the population had once again been devastated, this time from natural causes. Still, thanks to the reservoir population in Mexico, immigration continues, and, helped along with on-site reproduction, the bear population in the Chisos may be increasing once more. Importantly, bears from the Sierra del Carmen/Maderas del Carmen serve as more than colonists—they introduce new genes, thereby reducing the threat of inbreeding depression that so often harms isolated populations. In the future, the bears in the Chisos may themselves provide immigrants to repopulate sky islands in the Davis, Glass, and El Norte Mountains elsewhere in the Trans-Pecos. In fact, some recent sightings in the Davis Mountains indicate that bears have already used the Chisos Mountains as a stepping-stone for recolonizing the region.

understory plants are also evergreen, and many are restricted to the alluvial soils at the bottom of high-elevation ravines. Although orchids might seem unlikely in desert mountain habitats, moist canyons in high-elevation oak-juniper-pinyon habitats are inhabited by five colorful saprophytic species. For example, a brilliant rose-pink stalk signals the presence of Giant Coral Root in fern grottoes, pine needle beds, and stream banks along protected canyons high in the Davis Mountains. Sheltered canyons also provide nesting sites for two of the rarer hawks in Texas, the Zone-tailed Hawk and Common Black-Hawk, while in the canyons and rocky bluffs, Black-tailed and Rock Rattlesnakes repose in the jumble of boulders and leafy debris. Because of their rich composition, Madrean pine-oak woodlands represent "hot spots" of biodiversity and are targets for protection by several international conservation organizations.

SPRINGS, CIÉNAGAS, AND TINAJAS

Surface water is seldom permanent in the Trans-Pecos, and even the heavy runoff from thunderstorms or snowmelt quickly evaporates. Nonetheless, somewhat more permanent sources of water exist as springs, seeps, ciénagas, and tinajas—all surprisingly more abundant in some areas of the Trans-Pecos than might be imagined. These sources not only provide drinking water during the hottest and driest seasons of the year but also form secluded microhabitats that sustain moisture-loving communities.

Springs and seeps, which differ only in the volume of outflow, result when surface water

Figure 9.18.
A seep in Hell's
Canyon drips
water into a small
pool (top), which
provides water
for wildlife and
habitat for the
Canyon Treefrog
(bottom) in the
Davis Mountains.

Many springs in the Trans-Pecos issue cool water from shallow underground aquifers. However, just below these cool water-bearing strata lies a geothermal gradient where temperatures increase with depth. Hot springs, of course, occur more commonly in volcanic regions, but even in nonvolcanic areas, warm to mildly hot outflows may emerge from springs formed where water penetrates deeply below the surface. Although "hot spring" lacks a universally accepted definition, water temperatures exceeding 97.7°F occur at many natural outflows in the Trans-Pecos. Chinati Hot Springs in the Chinati Mountains and Boquillas Hot Springs in Big Bend served as "baths" for Native Americans; centuries later, Euro-American entrepreneurs commercially promoted both sites as sources of "healing waters."

Spring-fed streams are rare in the Trans-Pecos. Instead, the discharge of many springs collects to form shallow saline wetlands, or "ciénagas." Marsh vegetation develops at these sites, and because of their isolation from rivers and streams, ciénagas often contain endemic species. For example, the endemic fauna at ciénagas in Pecos and Reeves Counties includes Leon Springs Pupfish, Comanche Springs Pupfish, Pecos Gambusia, and the Pecos Assiminea Snail, all of which are listed as endangered because their small populations exist in just one or two locations. The ancestral populations of these species likely reached these locations in streams flowing during the wetter Pleistocene. When the climate warmed and dried, the streams withered, concurrently isolating both ciénagas and some fish. Without further exchange of genetic material and now in a suite of conditions unlike their former environment, fish isolated in ciénagas evolved features unlike those of their parent populations.

Springs feed cool water into McKittrick Creek, a small, discontinuous stream flowing through limestone canyons on the east side of Guadalupe Mountains National Park (fig. 9.19). The creek, however, provides habitat for only three species of fish, none of which are native to the area. Previous landowners stocked the creek with Yellowbelly Sunfish and Rainbow Trout. Green Sunfish, the third species, likely arrived unintentionally when the other species were released. Rainbow Trout

infiltrates the surface and travels through networks of fissures and cracks that function as underground drainage pipes. Often intermittent and with minimal rates of flow, seeps generally emerge as small pools or marshy sumps on the faces of slopes or rocky surfaces (fig. 9 18). Despite their diminutive flow, seeps provide crucial moisture for many species of small mammals, birds, amphibians, a suite of endemic snails, and insects (fig. 9.18). Several endemic butterflies, for example, obtain essential amino acids and salts from muds at seeps.

Figure 9.19. Spring-fed McKittrick Creek sometimes flows underground, but surface pools near the springs provide habitat for three species of introduced fishes. Photograph courtesy of Guadalupe Mountains National Park, National Park Service.

spawn in the pools of cool water near the outlets of springs, but during periods of high water their offspring sometimes wash downstream to warmer parts of the creek.

In desert canyons and arroyos where water flows occasionally, tinajas become vital sources for stored surface water. Sand and gravel carried by floodwaters cascading over waterfalls scour out deep pockets in the bedrock below, leaving pools of standing water in the well-like depressions. Sheltered from the heat and wind at the base of cliffs or ledges, tinajas epitomize miniature oases of lush vegetation and centers of wildlife activity. However, because of heavy flows of floodwater in steep-walled canyons, some depressions at these sites become so deep that animals seeking a drink fall in and cannot escape. For example, the sheer walls of Ernst Tinaja (fig. 9.20) in Big Bend National Park have claimed the lives of many thirsty deer, Collared Peccaries, and Mountain Lions—and at least one man.

Unique geological conditions and rainwater create shallow aquatic reservoirs called *huecos* (Spanish for "cracks" or "hollows") in clusters of low igneous outcrops about 32 miles east of El Paso. Most huecos develop naturally, but early inhabitants, including Native Americans who adorned the rocks with pictographs and petroglyphs, built small dams to impound additional water in the huecos. Later, the Butterfield Overland Mail used the site as a relay station where horses could be exchanged, watered, and rested. The shelter, shade, and moist, fertile soils in huecos provide microhabitats for species not normally associated with desert biota. As examples, the only known population of Erect Colubrine in the United States thrives here, as do fairy shrimp when rainwater fills the depressions. Because of their historical and biological significance, the huecos and rock paintings are preserved at Hueco Tanks State Park and Historic Site.

RIPARIAN ZONES

Riparian vegetation occurs on the banks of the only two rivers—the Pecos River and the Rio Grande— that drain the region as well as along the few permanent and many ephemeral watercourses in the region. Dense groves of cottonwoods, willows, Desert Willows, Screwbean, and two cane grasses,

Figure 9.20. Torrents of falling water carve depressions known as tinajas, such as Ernst Tinaja in Big Bend National Park, shown here, and become important reservoirs of water for wildlife during dry periods. Photograph courtesy of Matthew Yarbrough and Big Bend National Park, National Park Service.

Common Reed and Giant Reed, crowd the banks of the Rio Grande wherever a floodplain exists (fig. 9.21). Cottonwoods were once more abundant—most fell to woodcutters in the late 1800s. Today, the flows in both rivers are greatly reduced from their historical rates, and floods, which once occurred annually, no longer displace rocks and sand bars or scour stream-bank vegetation. As a consequence, introduced species, especially Saltcedar, Russian Olive, and Giant Reed, invaded the bottomlands bordering long stretches of these rivers. Saltcedar, considered an undesirable phreatophyte, has a deep root system that penetrates permanent sources of water. A single tree can effectively translocate up to 30 gallons of water per day into the atmosphere.

Narrow bands of riparian trees signal the beds of arroyos and ephemeral streams in mountain foothills and alluvial fans. Riparian vegetation in mountain and foothill areas forms an inconsistent mix of oaks, cottonwoods, willows, Little Walnut, and Seepwillow Baccharis. Although some arroyos seem dry, water often flows surreptitiously through rocks, gravels, and sands just below the parched surface. The narrow lapels of most arroyos meandering across desert basins are sparsely vegetated with mesquites, Catclaw Acacia, and other thorny shrubs, but springs, seeps, and depressions that hold water are flagged by cottonwoods that appear to glow amid the stark surroundings.

Although less than 5 percent of the Trans-Pecos land area consists of riparian zones, these green ribbons of life support an amazing diversity of animals and serve as nesting habitats for birds and movement corridors for many species. Shady,

cool understories provide a moist habitat enabling the growth of sensitive plants found nowhere else in the desert. Wild Turkeys rely on the protective cover in these habitats for their winter roosts. Riparian environments also provide ecosystem benefits, among them filtering out excess nitrogen and other toxic compounds, reducing soil erosion, and stabilizing stream banks. Though riparian zones may be small in area, they may be the most valuable natural communities in the Trans-Pecos.

HIGHLIGHTS

ROCK OUTCROPS

Talus slopes, cliffs, clusters of rocks exposed by erosion, and other rocky environments are widespread in both limestone and igneous mountains in the Trans-Pecos. Most exposed rocks, especially those of volcanic origin, wear colorful crusts of lichens. Although usually small and isolated, rock outcrops create and shelter microhabitats occupied by unique communities of lithophilic ("rock-loving") ferns and shrubs. Several endemic species, such as Tufted Rockmat and Yellow Rock Nettle, exist only in the rock fissures and crannies of western Texas. The soft greens of delicate Maidenhair Ferns and the rich colors of columbines temper the harsh surroundings at springs and seeps nestled among the stones. Hidden within the crevices are day roosts of bats, such as the Western Perimyotis and the Big Brown Bat, nests of Rock Wrens, and dens of Ringtails.

The surfaces of boulders, faces of cliffs, and other rocky exteriors sheltered from exposure often develop thin films of desert varnish. This coating consists primarily of clay minerals mixed with iron and manganese oxides, other trace elements, and small concentrations of organic matter. Desert varnish rarely forms on limestone surfaces, but rocks that are less water soluble and more resistant to weathering, such as basalts, fine quartzes, shales, and some sandstones, often develop rich black or brownish patinas. Some

Figure 9.21. In places, dense stands of Giant Reed (foreground) choke the banks of the Rio Grande. The Chisos Mountains loom in the distance. Photograph by Cookie Ballou and Big Bend National Park, National Park Service.

Figure 9.22. Native Americans carved petroglyphs of wildlife as messages into the desert varnish coating boulders and canyon walls (top). They also dabbed pigments on surfaces of sheltered rocks to create pictographs (bottom).

Figure 9.23. Resurrection Plants appear withered and lifeless during dry periods, when they curl into tight balls to protect tender tissues (top), but within hours after a rain, they uncurl, turn green, and resume photosynthetic activity (bottom).

petroglyphs in the Trans-Pecos were created when Native Americans chipped away the dark varnish to expose the lighter-colored underlying surface (fig. 9.22).

RESURRECTION PLANTS

Desiccation kills most desert plants, but several species known as "Resurrection Plants" can survive prolonged periods without water. Growing in rock outcrops or in the absence of soil moisture, these plants gradually dry and enter a state of dormancy.

While drying, the stems curl inward, forming a tight ball that limits the amount of exposed surface area and conserves the remaining internal moisture. Appearing dry, brown, and dead, a Resurrection Plant may remain dormant for years (fig. 9.23). Within hours after a rain, however, the plant absorbs water, uncurls, and grows rapidly, producing a flat rosette of scaly green stems and giving the appearance of a miraculous return from the dead (fig. 9.23).

The most common Resurrection Plant in the Trans-Pecos, known locally as Siempre Viva, is not a flowering plant but a member of a primitive group of lycopods, or club mosses. Species in this group occur worldwide, typically in moist locations where ferns and mosses are common. Demonstrating their ancient ancestry, lycopods lack flowers, fruits, and leaves but produce roots, stems, and leaflike extensions of the stem. Low growing, they reproduce by generating single-celled spores in club-like structures called strobili.

Dry stems of Resurrection Plants can be brewed as an herbal tea for use as an antimicrobial remedy for colds and sore throats, and the plant is often sold as a novelty to tourists. When Spanish friars entered the region, they used the plant to demonstrate to the natives the concept of rebirth. The plant's ability to seemingly return from the dead certainly justifies its common name and provides another exceptional example of adaptation to harsh conditions.

Figure 9.24. No larger than 0.5 inch, this *Angelito*—a Giant Red Velvet Mite—searches for Desert Termites, its main food.

DESERT TERMITES

The cycle of life—reproduction, growth, and death—functions in desert ecosystems just as it does in mesic or aquatic habitats. The processes associated with decomposition, however, take longer in the desert. The primary decomposers of plant tissues in moist environments, bacteria and fungi, experience somewhat greater difficulties in carrying out decomposition in arid climates. Consequently, the breakdown of dead plant tissues such as leaves or stems falls largely to the desert arthropods, particularly mites and termites.

Rarely seen and often mistaken for ants, termites are among the most common insects in desert environments. Like most species of ants, Desert Termites are social insects with a caste system. Several thousand individuals in a miniature kingdom share shelter and food and divide the responsibilities among individuals—workers, soldiers, and reproductives—anatomically adapted for specific purposes. The most abundant of these, worker termites, forage for food, care for the young, and construct and repair the nest. Termite soldiers defend the colony from intruders. Reproduction is the primary responsibility of a king and a queen, which mate and produce eggs. The nymphs hatching from the eggs can potentially develop into any of the castes, but their fate is determined by members of the reproductive and soldier castes, which secrete chemicals that shape further development. Thousands of winged reproductives— kings and queens—leave the colony in mass flights, or "swarms," just after sunset following a summer rainstorm to mate and establish new settlements. Most, however, swiftly fall prey to opportunistic predators (fig. 9.24; infobox 9.4).

Heat and dehydration present formidable challenges to the survival of fragile and soft-bodied insects. To persist, Desert Termites construct large subterranean galleries or gather under the shelter of rocks, wood, or pads of cow manure. Termites mix fine soil particles with their feces and saliva to construct hardened tubes, called "cartons," around living or dead grasses, forb stems, leaf litter, or dry manure. Protected within the walls of cartons, Desert Termites feed on cellulose, which is digested by bacteria in their gut.

Despite their small size and relatively low numbers, Desert Termites function as a keystone species in the Chihuahuan Desert—a fine example of nature endowing little guys with big punches. As major decomposers of plant material, these termites help regulate the flow of carbon and nitrogen in the desert ecosystem. Most organisms cannot digest cellulose, but termites convert this material into their protein-rich bodies, which in turn become a food source for many desert organisms. Additionally, their tunneling activities increase soil porosity and improve water retention and plant growth. In one area, the cover of perennial shortgrasses disappeared after a period of termite control. Quiet and easily overlooked, Desert Termites nonetheless play a significant role in the desert's cycle of life.

LAKE CABEZA DE VACA

The concept of "permanence" erodes both literally and figuratively when one considers the immense span of geological time—in all, few things are permanent. To illustrate, the Rio Grande and its magnificent canyons that now define the southern edge of the Trans-Pecos were once, quite simply, not part of the landscape. Instead, a series of

INFOBOX 9.4. *ANGELITOS*
The Little Red Angels

Flying low in a small airplane, a biologist was astonished to see a swash of crimson on the desert floor below. The color was not a bloom of an unknown desert lichen, as first suspected, but rather the massive emergence of Giant Red Velvet Mites. Indeed, closer inspection revealed that some 40–50 mites occupied each 4-inch square of desert surface; about three to four million mites were scuttling over every 2 acres of the area examined. Mysteriously, the red horde disappeared within three hours.

Giant Red Velvet Mites are the largest mites in the world. Most mites are about the size of the period ending this sentence (some are even smaller), but *angelitos*—regional lingo for "little angels" in parts of the Desert Southwest—reach a colossal 0.5 inch in length. A dense, velvetlike coat of setae covers their bodies, whose bright scarlet coloration likely serves an aposematic (warning) function. In experiments with insect-seeking predators, Giant Red Velvet Mites are either avoided or immediately spit out—a clear indication that their unpleasant taste acts as an effective defense mechanism.

For most of the year, Giant Red Velvet Mites remain content in their underground burrows, nowhere to be seen. However, on a clear, warm morning following a spring thunderstorm, the huge red mites climb cautiously to the entrances of their burrows and extend their forelegs to test the air temperature, and if it is suitable, they emerge to search for prey. With remarkable timing, the mites emerge just when their primary prey—Desert Termites—appear. Winged termites also disperse after warm rains and, after a short mating flight, alight on the ground, where they shed their wings and tunnel into the soil. Consequently, the predatory mites often have less than an hour during their surface excursions to locate, subdue, and devour their victims. Most of the mites emerge only once each year, but a few also attend termite landings associated with monsoonal rainstorms in late summer.

While roving in search of prey, those males that encounter a prospective mate construct a loosely spun, somewhat circular web on the ground nearby. Anchored to sand grains, each web is about 2 inches in diameter and sometimes encircles termites. Males approach females cautiously, beginning courtship by gently touching the posterior parts of the female's body, but the rest of the species' mating behavior remains unobserved.

After an hour or two of surface activity, Giant Red Velvet Mites excavate new burrows at sites with sandy soils. Conical mounds of quarried sand steadily grow next to the entrances as tunnels are extended to maximum depths of 5–10 inches during three to four days of digging. About 75 days after retreating underground, females lay a clutch of tiny, salmon-colored eggs at the bottom of their burrow. The eggs incubate uncovered for another 35 days. After hatching, the minute larvae scramble from the burrow in search of a suitable host, usually grasshoppers. Once attached to their host, the larvae develop as ectoparasites until they metamorphose into the first of several nymphal stages and then drop back onto the ground. The complete cycle of nymphal development is poorly known but likely occurs in the soil. After the first warm rain the following spring, the young mites join the adults on the desert surface, again blanketing the desert in a swath of red.

Unfortunately, some visitors to the Chihuahuan Desert and other arid regions of the southwestern United States may never witness an eruption of Giant Red Velvet Mites. Still, for those who happen to visit the right place at the right time, the abrupt appearance of Little Red Angels offers yet another memorable experience in a desert remarkably full of equally remarkable events.

---

unconnected small lakes occupied the basins between the mountain ranges in what is today the westernmost corner of Texas, southeastern New Mexico, and adjacent areas in Mexico. Then, during the wetter climatic regimes of the Pliocene and early Pleistocene, runoff from surrounding mountains flooded the basins, thereby connecting the isolated lakes to create Lake Cabeza de Vaca. The huge pluvial lake—named for one of four survivors of the ill-fated Narváez Expedition of 1527–1528—eventually covered an area of 8,880 to 10,040 square miles.

Geological evidence for the lake's existence—ancient beach lines and wave-notched escarpments on slopes of the surrounding mountains—enabled the outline of the ancient lake to be mapped. It was then possible to superimpose range maps showing the current distribution of species and subspecies of local vertebrates over the geological map. The overlay suggested that Lake Cabeza de Vaca separated the populations of ancestral taxa into subpopulations that, because of their long genetic isolation, subsequently differentiated into the taxa present today. Hence, "new" species or subspecies developed in about 46 percent of the reptiles and 20 percent of the mammals in the fauna currently occupying the area once covered by water. For example, the distributions of two subspecies of Long-Nosed Snakes converge in the ancient lakebed. Moreover, two species of banded geckos remain spatially separated as if the lake were still there. In time, sediments gradually filled the ancient lake, and volcanic-induced faulting in the mid-Pleistocene pushed the lake water southward, where it found an outflow channel. The resulting stream eventually etched the modern channel of the Rio Grande through El Paso Canyon before joining with the Rio Conchos farther downstream. Some geologists believe the Rio Conchos, which drains the Mexican Plateau to the south, carved the iconic canyons of the Big Bend region long before it merged with the Rio Grande.

TEXAS BONES AND CALIFORNIA CONDORS

California Condors once soared widely across much of North America, a determination sadly made from their fossil remains instead of living birds. With wingspans extending 10 feet tip to tip, California Condors are the largest bird in North America, and although they were never abundant, their numbers dropped precipitously—and mysteriously—in the twilight of the Pleistocene. The small, residual population further declined as Euro-Americans settled the continent, and the species faced the finality of extinction.

In 1932, fossil condor bones were discovered in a cave on a cliff face in the Mule Ear Peaks (Brewster County). The fossil bones of yet another condor were retrieved decades later from a packrat midden discovered in Maravillas Canyon, also in Brewster County. The discoveries provided the first evidence that condors once lived in Texas, and the abundance of the bones indicated that their presence in the Trans-Pecos was not an isolated event. Accurate fossil dating techniques

indicated that the birds died at least 10,000 years ago. Moreover, most of the dates coincided with the wholesale extinction of the Pleistocene megafauna in North America—mammoths, Ground Sloths, Short-Faced Bears, and about 100 other mammals weighing 100 pounds or more—of which the American Bison and Musk Ox represent two survivors.

The relatively sudden extinction of the megafauna remains much debated but most likely resulted from one of two events acting alone or together. The first of these was a change at the end of the cool, moist Pleistocene to a warmer and drier climate and, in turn, environments to which the animals could not adapt quickly enough in response and thus perished. Second, Paleolithic hunters may have simply killed too many animals in a blitzkrieg-like event known as the Pleistocene Overkill. This explanation suggests that because megafauna in North America had evolved independently of humans, the animals lacked a means of coping with the hunting techniques of the potent predators that entered North America some 10,000 to 12,000 years ago. Quite possibly, the changing climate had reduced the megafauna to the point that hunting became the "final straw." Whatever the cause, it seems clear that as the Pleistocene megafauna melted away, so did the staple food of California Condors—the carcasses of large animals.

By 1987, just nine birds remained at large on the Pacific coast; these were captured and added to those already serving as breeding stock in zoos and other rearing facilities. Eventually, their offspring were released at three locations, one in California, another in Arizona, and the third on the Baja Peninsula in Mexico. Alas, Texas is not scheduled for releases, so a few musty bones must suffice for bragging rights—at least for now. Meanwhile, some 400 California Condors, about half now soaring in the wild, currently represent the entire population.

THE MARFA LIGHTS

The lonely, dark hours of desert nights are often punctuated by mysterious sounds and bizarre sightings. When cowboy Robert Reed Ellison saw flickering lights on an uninhabited desert flat east of Marfa in 1883, he was not the first to observe the illuminations, but he did provide the first account of what eventually became known as the "Marfa Lights." Native Americans and early settlers had marveled at the lights for many years, attributing them to campfires (even though ashes were never found), falling stars, or the wandering ghosts of Spanish conquistadors.

The Marfa Lights, which are always seen at a distance, appear as glowing or twinkling orbs of various colors that seemingly float, bob, dance sideways, disappear and reappear, or split into two. Although their appearance is unpredictable—verified sightings occur on only 10 to 20 nights annually—the lights have been observed at all hours of the night, including dawn and dusk, in all types of weather, and in all seasons of the year. On one occasion, the lights persisted for more than three hours and were documented with video surveillance cameras, which enabled physicists to measure the visible light flux of the event.

Numerous hypotheses have been proposed to explain the glowing orbs that appear near Marfa and at other places in the Trans-Pecos, but the phenomenon remains a mystery. Periodic sightings of the Marfa Lights on Mitchell Flat continue from a public observation platform erected for just this purpose about 15 miles east of Marfa on US Highway 90.

CONSERVATION AND MANAGEMENT

The Trans-Pecos is home to numerous and diverse habitats and many uniquely adapted plants and animals, all the products of eons untrammeled by human interference. All changed, however, with the arrival of American settlers whose dreams and plans for irrigated farms and well-stocked ranches necessarily required altering the region's native vegetation and water resources. These disturbances varied considerably; in comparison with terrestrial sites, the limited aquatic habitats in the Trans-Pecos undoubtedly experienced the greatest modifications during the past 100–150 years.

Historically, more than 50 springs flowed throughout the Pecos Basin, which in Texas includes much of Reeves and Loving Counties, but no more than 8 of these remain active. In the past, freshwater from the once-numerous springs downstream diluted the saltwater added at Malaga Bend in New Mexico except in times of extreme drought. When most of the springs dried, however, salinity increased in the lower Pecos, as did the frequency of mass mortalities of fish. Oddly, the fish kills resulted from "blooms" of a golden alga, *Prymnesium parvum,* which released a toxin producing effects similar to those of the red tides caused by dinoflagellates in coastal waters. Golden algae normally occur in estuarine habitats but apparently can thrive equally well in the saline water of the Pecos.

To stabilize stream banks in the Trans-Pecos, nonnative Saltcedar trees were introduced in the 1800s. These invasive trees now flourish along the banks of the Pecos River and the Rio Grande, but they have also found their way upstream on the smaller tributaries in the watersheds of these two larger rivers. Saltcedars not only consume sizable amounts of water and choke out native vegetation in riparian areas; they also draw salts from belowground that are eventually deposited at the surface in leaf litter. When floods flush riverbanks, the salts in the litter dissolve and further increase the salinity of the stream water. Collaborative efforts organized by the World Wildlife Fund and involving citizens and agencies from both Mexico and the United States are underway to eliminate Saltcedars using applications of herbicides, biocontrol, fire, and manual labor. In reaches of the Rio Grande within Big Bend National Park where predictable stream flow has been restored, the US Fish and Wildlife Service has established experimental populations of the Rio Grande Silvery Minnow, one of the most endangered fish in North America.

Ciénagas perhaps represent the most heavily impacted wetlands in the Trans-Pecos. The ciénaga fed by San Solomon Springs, along with its wildlife community, was destroyed years ago when the spring water was diverted, or "channelized," into concrete-lined irrigation canals. After Comanche Springs at nearby Fort Stockton went dry in 1961, the residents of Balmorhea, located near San Solomon Springs, took note of the economic impacts on agriculture and tourism that followed the loss of the spring in their neighboring community. With help from local citizens, federal agencies, and faculty and students from Sul

Figure 9.25.
Restoration
efforts at San
Solomon Springs
near Balmorhea
resulted in a
ciénaga that
functions like the
original wetland.
Note the Texas
Spiny Softshell
in the water near
the bank on the
left. Photograph
by William I.
Lutterschmidt.

Ross State University and the University of Texas (infobox 7.4), conservation biologists from the Texas Parks and Wildlife Department led an effort to create a new wetland at San Solomon Springs that looks and functions like a natural ciénaga (fig. 9.25). It is now a permanent wetland that provides habitat for several endangered fishes, yet the outflow also provides local farmers with irrigation water. Balmorhea State Park, where some of the spring water maintains a large public swimming area, protects the site from a fate all too common at ciénagas elsewhere.

Large expanses of the Trans-Pecos are managed and protected by state and federal agencies as well as by nongovernmental organizations. Of these, Big Bend National Park, the namesake of the region in the Trans-Pecos where the Rio Grande turns abruptly northeast, protects the largest area of Chihuahuan Desert in the United States. Within its boundaries lie more than 810,000 acres of every vegetative association and topographic feature described in this chapter, all available to intrepid travelers willing to explore backcountry landscapes far from paved roads. The park includes habitat

for more than 1,200 species of plants, including at least 60 species of cacti, more than 600 species of vertebrates, and in excess of 3,600 species of insects.

Guadalupe Mountains National Park, which adjoins the Texas–New Mexico border east of El Paso, contains Guadalupe Peak—the highest point in Texas at 8,751 feet—and the Bowl, which offers an excellent example of high-elevation montane forest. The park remains open year round for hiking on more than 80 miles of improved trails, but vehicular access to the park's interior is limited. One trail leads to the summit of Guadalupe Peak, while another winds upward through McKittrick Canyon following the creek of the same name. Montane areas in the park contain floral elements more closely associated with those of the Rocky Mountains in Arizona and New Mexico, but the lowlands surrounding the Guadalupe Mountains reflect grasslands, scrublands, and salt flats more representative of desert communities.

The largest park of its kind in Texas, Big Bend Ranch State Park protects 311,000 acres surrounding the Bofecillos Mountains west of Big Bend National

Park. The rough terrain in the park, accessed primarily by hiking, includes more than 100 streams, springs, and seeps as well as Madrid Falls, the second-highest waterfall in Texas. Uniquely, the park is managed as a working cattle ranch that boasts a trademark herd of Texas Longhorns. The Texas Parks and Wildlife Department recently translocated a small number of Desert Bighorn Sheep to Big Bend Ranch in hopes of reestablishing a self-sustaining population in the park.

An independent nonprofit organization, the Chihuahuan Desert Research Institute (CDRI), promotes public awareness and appreciation of the natural diversity of the Trans-Pecos. Its research and educational programs focus on a 500-acre tract in the foothills of the Davis Mountains south of Fort Davis. The CDRI property features an extensive and well-labeled botanical garden of plants native to the region and a network of nature trails.

Several other state parks, wildlife management areas, university-owned lands, and various protected areas (most notably those acquired or protected by conservation agreements developed by The Nature Conservancy, which owns a 33,000-acre preserve including the highest elevations of the Davis Mountains) also ensure the wilderness heritage of the Trans-Pecos. Protected lands of all stripes stand at the core of what writer Wallace Stegner famously identified as the "geography of hope." Still, protected areas alone cannot fully conserve biological diversity. The latter requires the participation of private landowners as stewards of the region's natural resources, not the least of which concerns managing surface and subsurface waters in ways that achieve ecological as well as economic benefits.

10
SOUTH TEXAS BRUSHLAND
THE LAST GREAT HABITAT

This entire region is covered by chaparral, somewhat scattered and stunted on many of the stony hills, but rank and dense on most of the lower areas.

— HARRY C. OBERHOLSER

South Texas Brushland

Despite recurrent droughts and its long history of land use, the South Texas Brushland hosts a surprising number of rare plants and animals, many of which occur nowhere else in the state or nation. Often called the "Tamaulipan Thornscrub" or simply "Brush Country," the region offers a mix of tropical species that extend northward from Mexico, desert species typical of the Trans-Pecos, grassland species representing the Edwards Plateau and the Coastal Prairies, and infestations of introduced Old World grasses. The region is sometimes referred to as the "Last Great Habitat" because of its biodiversity and extensive tracts of contiguous and relatively undisturbed wildlife habitat (fig. 10.1).

Essentially a vast undulating plain, the region covers approximately 20.5 million acres in the southwestern corner of the state. Unlike the Edwards Plateau (chapter 7), which is defined by geological boundaries, the South Texas Brushland is delineated largely by the extent of thorny scrub. This association stretches northward to the Balcones Escarpment (chapter 7) and the southernmost extensions of the Cross Timbers and Prairies (chapter 5) and Blackland Prairies (chapter 4). The Rio Grande (known in Mexico as Rio Bravo del Norte) forms the southern and western geopolitical boundaries of the natural region in Texas, although similar habitat extends deeply into northeastern Mexico. To the immediate east, an indistinct boundary forms where the brushland habitat intergrades with the grasslands of the Gulf Prairies, but some biogeographers extend the region much farther up the coast (fig. 10.2).

Elevations range from 1,000 feet to sea level. High temperatures, infrequent killing frosts, and erratic, unpredictable rainfall characterize the climate. With average annual temperatures exceeding 73°F, the growing season sometimes lasts for 365 days in the southernmost portion of the area. Sea breezes from the Gulf of Mexico carry moisture-rich air inland, but a prevailing zone of upper-level subtropical high pressure

usually prevents the formation of thunderstorms and showers. Annual rainfall varies from 35 inches in the east to 16 inches in the west, but greater amounts (59 inches in 1896) occasionally fall from tropical storms; droughts sometimes last two to five years.

Rainfall in most years does not overcome the mean annual precipitation deficiency—the amount of water plants lose through transpiration. Near the Edwards Plateau in Medina and Wilson Counties, plants transpire up to 12 inches more moisture than the total annual rainfall, whereas in the drier southwestern brushlands of Starr and Zapata Counties, the deficit exceeds 35 inches. Precipitation deficiencies in the western brushlands, in fact, exceed those of the Chihuahuan Desert in Brewster County and thereby engender similar thorn-studded, desertlike environments (infobox 10.1).

Permanent streams are rare, but the terrain is dissected by numerous arroyos—dry, deeply etched gullies that flood rapidly with heavy rainfall. Thorny thickets capable of withstanding the forces of fast-moving water and long dry spells line the watercourses (fig. 10.3). Dense overstories of woody plants, such as Granjeno, Retama, Huisache, Brasil, and mesquites, provide travel corridors along the arroyos and offer shady microenvironments that protect wildlife from extreme temperatures.

South Texas Brushland forms a sizable part of some of the largest privately owned ranches in the United States—notably, the 825,000-acre King Ranch and the 500,000-acre Kenedy Ranch—and thereby excludes human intrusions into much of the region. This protection, along with the relatively low density of human populations elsewhere, clearly helps maintain the Brushland's extraordinary diversity of animals, plants, and habitats.

VEGETATIVE ASSOCIATIONS

Early accounts of the regional vegetation appearing in the journals of Spanish explorers vary somewhat, likely because of annual variations in rainfall (wet vs. drought years), season, and travel routes. Nonetheless, most records indicate that much less brush existed four centuries ago than occurs today. Indeed, just a century ago, cattlemen claimed they

(*overleaf*) **Figure 10.1. After a recent rain, wildflowers enliven open spaces in the South Texas Brushland.**

Figure 10.2.
A vast expanse
of undisturbed
terrain defines
the "Last Great
Habitat."
Photograph
by Timothy E.
Fulbright.

INFOBOX 10.1. HELL IN TEXAS

The harsh, thorn-studded environment of southern Texas offers difficult challenges for plants and animals, including those relatively fragile beings known as humans. Extreme summer temperatures, frequent and long-lasting droughts, and the winds and floods of tropical storms, topped off by an arsenal of organisms armed with stingers, fangs, and venom, each offer a special brand of misery. In short, the region provides little succor and takes no prisoners, which an anonymous and apparently battle-scarred veteran aptly captured in verse. "Hell in Texas" thus imagines the South Texas Brushland as a land created by an ill-disposed demon.

He began to put thorns on all the trees,
And he mixed the sand with a million fleas,
He scattered tarantulas along all the roads,
Put thorns on the cacti and horns on the toads;
He lengthened the horns of Texas steers,
And put an addition on jackrabbit ears.
He put devils in the bronco steed,
And poisoned the feet of the centipede.
The rattlesnake bites you, the scorpion stings,
The mosquito delights you by buzzing its wings.
The sand burs prevail, so do the ants,
And those who sit down need soles on their pants.
And all would be mavericks unless they bore,
The marks of scratches and bites by the score.
The heat in the summer is a hundred and ten,
Too hot for the devil and too hot for men.

Figure 10.3. Flash floods carve arroyos, also known as washes or gulches, in the South Texas Brushland and other arid landscapes in western North America. Between sporadic rains, the dense brushy vegetation bordering arroyos provides wildlife with sheltered corridors.

could only tether the reins of their horses to cow chips—and then bury the chips. Tallgrass prairie, dominated by Little Bluestem and Switchgrass, blanketed the eastern half of the region, which included scattered clumps of mesquites, prickly pears, and other woody species. Shortgrass prairie covered the drier western half, and small mottes of dense, brushy thickets formed along riverbanks, creeks, and arroyos. The periodic natural fires raging across the prairie controlled brush invasions, enriched the soil, and improved grazing conditions for limited populations of American Bison, Pronghorn, and White-tailed Deer.

Early settlers soon increased the grazing pressure on these grasslands, but their herds of cattle and goats roamed free, mixing with native wildlife and the wild descendants of horses left by the Spaniards. By the mid-1800s, however, ranchers began fencing their lands and thereby concentrated their growing herds into overcrowded pastures. Sheep far outnumbered cattle by the late 1880s—the two million sheep grazing in the region represented more than 45 percent of all the sheep in Texas. When concentrated, sheep clip grasses down to the roots and thereby significantly change

the composition of the vegetation—some claim that sheep literally grazed themselves out of South Texas.

Overgrazing also reduced the biomass and density of the grasses, which limited the fuel necessary to carry wildfires across the landscape. As competition for the remaining grasses increased, sheep, cattle, and White-tailed Deer foraged more intensively on the sugar-rich seedpods of mesquites and other shrubby legumes. In this way, seeds distributed in animal excreta germinated, took root, and gradually converted the grassland into mesquite-acacia savanna—a semiopen thornscrub of woody vegetation less than 10 feet tall. Droughts and rare prairie fires periodically reduced the thorny brush in local areas, but the once-large expanses of grassland eventually gave way to other types of habitat.

NORTHERN SEMIARID THORNSCRUB

Gently sloping terrain fashioned by broad alluvial fans—bajadas—washed from the brim of the Balcones Escarpment characterize the northern boundary of the South Texas Brushland. The soils in this zone tend to be deep, with textures ranging

Figure 10.4. Overgrazing encourages invasion of prickly pears and Blackbrush in the South Texas grasslands. Photograph by Timothy E. Fulbright.

from fine to medium. Much of the land is too dry for crops, but an irrigated floodplain in Maverick County supports agriculture. The corn, cotton, winter vegetables, and small grains produced in this area now form part of the "Winter Garden" of Texas. Although livestock graze the brushy grasslands in the region, local ranchers often smile when referring to cattle production as their "hobby"—the majority of their income originates from hunting leases for White-tailed Deer, Mourning Dove, and Northern Bobwhite.

Vegetation on the bajadas varies from north to south. In the northwestern corner, where conditions are the driest, mesquite-acacia savanna covers large tracts not subjected to overgrazing. Native grasses such as Slim Tridens, Plains Bristlegrass, Sideoats Grama, and Silver Bluestem predominate in large, open patches between clumps of thorny brush. Overgrazing increases the density of mesquites and encourages invasions of rough vegetation such as Whitebrush, Cenizo, Blackbrush, other acacias, Agarita, and prickly pears (fig. 10.4). Grasslands with scattered mottes dominated by Honey Mesquite and Granjeno occur in the southern part of this region, whereas woody thickets in the middle zone more likely consist of mesquites, Plateau Live Oak, and Brasil. Deeper soils support swards of Little Bluestem, Sideoats Grama, Lovegrass Tridens, and introduced grasses, but on shallow soils, scattered, low-growing brush, including Guajillo, Creosotebush (in the western part of the region), and Blackbrush intersperse the grasslands.

Cool-water springs erupting from aquifers beneath the Edwards Plateau (chapter 7), together with streams descending from the plateau, nourish distinctive habitats—and provide welcome relief—in an otherwise water-parched region. The Nueces River and its tributaries, the Frio and Leona Rivers, are dependable, spring-influenced streams. The Nueces, the southernmost major river in Texas northeast of the Rio Grande, follows a tortuous 315-mile route through the South Texas Brushland from its headwaters on the Edwards Plateau to the Gulf of Mexico. Named for the native Pecan trees along its banks—*nueces* is Spanish for "nuts"—the river is bordered by riparian gallery forests dominated by American Sycamore, willows, hackberries, Huisache, and, of course, Pecan. Like many rivers in dry areas, parts of the Nueces flow underground.

American Alligators range far upstream to reaches where Largemouth and Smallmouth Bass, catfishes, Redbreast and Green Sunfish, and Rio Grande Perch attract anglers. Overhead, the forest canopy provides nesting habitat for several rare birds, including White-tipped Doves, Great Kiskadees, Brown-crested Flycatchers, and Common Black-Hawks.

TAMAULIPAN MEZQUITAL

The largest part of the South Texas Brushland forms a gently rolling plain cut by dry arroyos and blanketed by mesquite-dominated thornscrub known as "Mezquital." Because this habitat extends into Mexico, many refer to the entire region as the Tamaulipan Mezquital. Drought-tolerant vegetation—mostly deciduous small-leaved trees and shrubs armored with thorns or spines—characterizes the regional flora (fig. 10.5). Depending on rainfall and soil type, the dense shrub canopy may be occupied by codominant species that include any of several acacias, Cenizo, Chapotillo, Mountain Torchwood, Texas Paloverde, prickly pears, and other cacti. With suitable conditions, Colima, Brasil, Granjeno, Coyotillo, or Texas Persimmon may dominate dense, motte-like thickets. On drier sites in the western Mezquital in Texas, stands of Red Grama, Texas Grama, Buffalograss, and Curly Mesquite fill openings between clumps of thornscrub. Elsewhere, Cane Bluestem, Silver Bluestem, Multi-flowered False Rhodesgrass, Sideoats Grama, and Pink Pappusgrass become common understory grasses. Native grasslands, however, are increasingly being displaced by aggressive exotic grasses such as Guinea Grass, Kleberg and King Ranch Bluestems, and Buffelgrass.

Heavily vegetated drainages—ramaderos— separate low hills in Starr and Zapata Counties. Although the deep soils in the beds of these defiles seem dry most of the year, water in fact slowly seeps downstream beneath the surface; hence ramaderos appear as sinuous green ribbons snaking through the dry brushland. These verdant corridors vary in width, but in length some extend eastward into the Mezquital for many miles. Tall mesquites, Huisache, Texas Ebony, and Guayacan line the ramaderos, which may provide pathways for wildlife. Monarch

Figure 10.5. Sharp spines lie hidden beneath the sweet-smelling blossoms of Blackbrush (top) and Huisache (bottom).

Butterflies also follow these corridors en route to and from their wintering grounds in north-central Mexico. Hungry White-tailed Deer and cattle enter ramaderos in search of Guayacan, the root system of which may persist for 1,000 years. Other common wildlife includes White-winged Doves, Plain Chachalacas, and Texas Indigo Snakes.

Dense thickets provide cool microenvironments beneficial to both plants and animals. At the peak of summer, surface temperatures in the open may exceed 140°F, whereas temperatures drop to 104°F or lower under the canopies of nearby mesquites. White-tailed Deer, Collared Peccaries, Black-tailed Jackrabbits, and Eastern Cottontails bed during the day in the relatively cool shade of the thickets to avoid heat stress. Similarly, Northern Bobwhites and Wild Turkeys survive extreme summer temperatures concealed beneath dense thornscrub canopies.

Mesquite-Granjeno communities commonly develop on loamy or sandy upland soils. The environment under mesquite canopies provides an important element in the life cycle of

Figure 10.6. Cenizo is an indicator of caliche soils on the Bordas Escarpment.

Granjeno. Extreme soil surface temperatures in exposed sunny areas inhibit the germination of Granjeno seeds, but those on the soil beneath mesquite canopies readily germinate and survive. Consequently, Granjeno clusters often develop beneath mesquite nurse trees rather than in open areas. The fleshy red fruits of Granjeno provide food for several species of birds and mammals, and White-tailed Deer often forage on the leaves.

Somewhat surprisingly, numerous small wetlands or potholes dot the dry brushland terrain of the Tamaulipan Mezquital. Some of these are resacas—small oxbow lakes isolated when segments of the Rio Grande streambed changed course. Others develop in shallow depressions that likely originated when strong winds carved "blowouts" in drought-parched soils. Rains from tropical storms or seasonal monsoons periodically fill these basins, which become temporary green-tree reservoirs attractive to wintering waterfowl and other wildlife. The region also contains hypersaline lakes, such as La Sal Vieja, nourished by underground salt springs.

THE BORDAS ESCARPMENT

The Bordas Escarpment forms the west face of a low limestone cuesta extending in a wide arc from Live Oak County southward to Starr County. This notable topographic feature bisects the Tamaulipan Mezquital and rises to heights exceeding 150 feet above the surrounding plain. Approached from the east, however, the gentle slope leading to the crest offers little indication of the change in elevation. In some areas, stream erosion breaks through the face of the cuesta, whereas at higher locations, two or more lower escarpments develop as a "staircase" along the face of the primary escarpment.

Shallow calcareous soils top the escarpment and overlie caliche or limestone bedrock where alkaline conditions support a short, open shrub canopy dominated by Cenizo, Acacia, Allthorn, and yuccas (fig. 10.6). At many locations along the outcrop, however, the shrubs provide 70 percent or more ground cover, comparable to densities of the thornscrub vegetation growing on deeper soils to either side of the formation. During the spring and summer months, a low, spreading shrub bejeweled with bright clusters of small yellow flowers adorns exposed caliche substrates on the escarpment. Although the showy flowers of Stinging Cevallia might be tempting to collect, it is best to avoid touching the plant. Stems and leaves of the nettle are armored with stinging hairs, each of which resembles a tiny glass pagoda when viewed under magnification. When touched, the tip of each hair penetrates the skin, breaks off, and releases enough formic acid through a hypodermic-like pore to cause stinging sensations and a skin rash. Evidently immune to the stinging hairs, butterflies, bees, and flies swarm around the blooms when the plants flower.

In the past, the thin gravelly soils and rocky ledges supported Peyote—a small spineless cactus containing hallucinogenic chemicals, including mescaline—which did not go unnoticed by Native Americans, who valued the plant as part of their spiritual heritage. After railroads penetrated the area in the late 1800s, Peyote was collected, dried, packed in barrels, and shipped by rail to markets in the north and west. Nowadays, Peyote is a controlled substance (it is illegal to reap or possess the cactus or its derivatives), but lawful exemptions allow limited harvest for the traditional ceremonies of Native Americans.

A community known as the "barretal" occurs where the Bordas Escarpment meets a narrow band of gravel, white clay, and caliche near the southeastern corner of Starr County. This vegetation, which forms small copses, represents an ecotone between the thorny thickets of the escarpment and the riparian forest along the Rio Grande. The site is the only place in the United States with a thicket of native citrus trees, which grows amid dense stands of Blackbrush, Allthorn, Junco, and other shrubs. The citrus trees, Baretta, develop as slender evergreens that reach heights of 20 feet. When blooming in the spring, Baretta is recognized by branch-tip clusters of tiny greenish-white flowers that produce late-summer seeds in small, dry, winged fruits. When crushed, Baretta leaves emit a pungent odor frequently described only as "not unpleasant." A second species of citrus, Limoncillo, persists as fewer than 20 scattered trees at just one site in Texas (Cameron County) but occurs more extensively in Mexico. The small population in Texas may have originated from seeds dispersed in floodwaters from tributaries of the Rio Grande. The dense vegetation of the barretal provides habitat for the Mexican Burrowing Toad, Reticulate Collared Lizard, and Elf Owl. White clay outcrops containing fossil oysters, *Ostrea georgiana,* and other invertebrate deposits are now included in the Lower Rio Grande Valley National Wildlife Refuge (NWR).

RIO GRANDE RIPARIAN FORESTS

Inhabitants of the South Texas Brushland often describe the stretch of the Rio Grande from Del Rio to the Bordas Escarpment in Starr County as the Rio Grande Valley, or simply as the Upper Valley. For much of the way, low limestone walls or gentle slopes border the river. Then, downstream from the Bordas formation, the river enters a flat plain known affectionately as the Lower Rio Grande Valley (LRGV). The original bottomland forest for most of the river's course from Del Rio southward included a unique mix of species representing subtropical Central America and southeastern North America. Since the 1920s, however, clearing has claimed more than 90 percent of the riparian forests on the US side of the Rio Grande, leaving only isolated remnants scattered in narrow strips along the riverbank or on a few slightly wider floodplains.

Construction of eight major dams on the Rio Grande and many others on its tributaries, as well as increased water consumption on both sides of the river, significantly reduced the river's historical flows in the twentieth century. Several large diversions below Rio Grande City move water inland for irrigation, thus reducing the flow released from Falcon Dam. To illustrate, above Rio Grande City, the river's average flow measures 3,504 cubic feet per second, whereas below the diversions, the rate drops to just 899 cubic feet per second. Currently, only 20 percent of the Rio Grande's natural flow typically reaches the Gulf, and the flow essentially fell to zero during the droughts of 2001 and 2002. For the first time in recorded history, a wide, dry sandbar formed across the mouth of the river.

Riparian ecosystems evolve in response to periodic ups and downs in water levels—an aptly named "flood-pulse cycle." If this normal cycle ends or becomes significantly reduced, the successional patterns under which the native biota developed no longer prevail and the streamside vegetation loses its diversity, soon followed by similar decreases in the richness of the fauna. Comparisons of the riparian forest vegetation at Santa Ana NWR between 1973–1978 and 1994–1996 clearly illustrate these effects on the composition and structure of riparian forests bordering the lower

Rio Grande. In the 1970s, the closed-canopy riparian woodland represented a subtropical evergreen forest dominated by Texas Ebony, Cedar Elm, Sugar Hackberry, and other trees. Twenty years later, however, a broken canopy on average about 33 feet lower in height, many dead trees, and a denser understory of shrubs had replaced the previous forest. Consequently, the breeding population of Red-billed Pigeons, which depend on dense, mature stands of riparian forest for nesting and roosting habitat, significantly declined along the river. During the intervening years, construction of Falcon Dam eliminated the flood-pulse cycle on the lower Rio Grande—circumstances that almost certainly account for the diminished forests downstream.

Elsewhere along the Rio Grande, bottomland hardwood forests composed of Cedar Elm, Berlandier Ash, Granjeno, and Honey Mesquite line the riverbanks in small, intermittent patches. Early Spanish inhabitants called these riparian woodlands "bosques"—a term still used in some communities along the river. Where dense, tall canopies occur, many birds—Green Jays, Altamira Orioles, Hooded Orioles, Common Pauraques, and Plain Chachalacas—reach their greatest densities (fig. 10.7). Farther downstream, the riparian forests may be variously dominated by Huisache, Sugar Hackberry, Cedar Elm, Berlandier Ash, Texas Ebony, or Anacua. The composition of the understory, which often includes extremely dense shrubby or herbaceous vegetation, also varies considerably. Where clearing removed the riparian forest, Giant Reed invaded and formed thick, virtually impenetrable barriers to the river.

The largest undisturbed remnant of tropical thorn woodland in the United States occupies about 24,000 acres of the riparian zone below Falcon Dam. Known as the Rio Grande–Falcon Thorn Woodland, this unique streamside community of dense desert scrubland extends south for about 19 river miles, ending where the Bordas Escarpment intercepts the river valley. The only known grove of Montezuma Baldcypress in the United States shares the riverbank with Black Willow, Berlandier Ash, Texas Ebony, and Honey Mesquite. Cenizo, Texas Wild Olive, and a rare understory plant, Gregg's Wild Buckwheat, occur on the exposed caliche bluffs bordering the valley.

Figure 10.7. Green Jays (top) add color to dense bosque thickets, but secretive Plain Chachalacas (bottom) are usually heard rather than seen. Photographs by Timothy E. Fulbright and Brian R. Chapman.

The Falcon Thorn Woodland provides habitat for more than 300 species of birds, 50 species of mammals, 50 species of reptiles, and 20 species of amphibians (fig. 10.8). The only nesting populations in the United States of the Brown Jay, Audubon's Oriole, and Hook-billed Kite occupy this woodland and nearby riparian forests where many other

Figure 10.8. The Thorn Woodland features a rich diversity of birds, represented here by a Curve-billed Thrasher (top) and a pair of Pyrrhuloxias (bottom), a brief taste of the region's attraction to birders. Photographs by Timothy E. Fulbright.

peripheral species reach the northern limit of their range. The Plain Chachalaca, Gray Hawk, Ferruginous Pygmy-Owl, Elf Owl, and Green Kingfisher also nest in the thorn forest, and one of the largest toads in the world, the Cane Toad, resides here. Reticulate Collared Lizards and Rose-bellied Lizards scamper noisily through the litter beneath the trees. White-nosed Coatis and Ocelots occasionally cross the border to forage in the forest, and several endangered birds, including the Aplomado Falcon and Peregrine Falcon, frequent the woodland.

LOWER RIO GRANDE VALLEY

Although known locally as the LRGV, the six-county area at the southern tip of Texas actually formed as a river *delta*. The waters of the Rio Grande, after completing the 1,896-mile journey from the mountains of south-central Colorado, once regularly flooded and deposited rich alluvial soils on the low plain at the river's mouth. Each flood altered the river's course, generating new twists and bends while isolating channels shaped by previous floods. The abandoned channels—oxbows, or more eloquently in Spanish, "resacas"—retain water and become important wetlands in the hot climate of the delta region. Then, beginning almost 100 years ago, construction of numerous dams throughout its length altered the natural flow of the Rio Grande. The rich soils remain, but in the absence of the historical cycle of flooding, resacas have declined in both quality and quantity.

The climate of the delta region resembles that of a tropical to subtropical desert. Rainfall patterns are irregular, and average annual precipitation amounts vary from 15 to 30 inches. The greatest amounts of rain usually fall from tropical storms and may cause local flooding. Periodic droughts, however, are almost as common as floods. One natural watercourse, the Arroyo Colorado, an ancient distributary of the Rio Grande, parallels the river and carries floodwaters from the upper delta to the Laguna Madre. In the late 1940s, the US Army Corps of Engineers dredged and channelized the lower 25 miles of the Arroyo Colorado to accommodate commercial barge traffic, and polluted runoff unfortunately lowers the levels of dissolved oxygen enough to kill fish and other aquatic organisms.

The subtropical environment and deep delta soils of the LRGV once produced dense Honey Mesquite–Granjeno thickets mixed with Anacua, Brasil, Texas Wild Olive, Texas Ebony, and many other species. Drapes of Spanish Moss and tufts of another epiphytic bromeliad, Giant Ball Moss, festoon Cedar Elms and other trees near wetlands. Prior to settlement, this region and its vegetation would have been considered an extension of the Tamaulipan Mezquital. Now, only about 2 percent of this lush vegetation remains in Texas, and clearing has likewise claimed a large percentage of similar habitat on the Mexican side of the border. Vast areas now produce melons, cotton, sugarcane, sweet potatoes, and various vegetables. Citrus orchards alone cover about 27,000 acres, devoted to several varieties of oranges and sweet red

Figure 10.9. Dense vegetation bordering many resacas provides habitat for a variety of wildlife including White-winged Doves (inset). Photographs by Robert Anzak and Greg Lasley (inset).

grapefruit; production of the latter represents about 7 percent of the nation's annual supply. Scattered remnants of the original flora—usually in tracts of less than 100 acres—often surround arroyos and resacas, where they persist as living monuments to a unique blend of biogeographical elements (fig. 10.9).

Brush-bordered resacas throughout the delta attract many of the waterfowl, shorebirds, and Neotropical migrants funneling through South Texas on their journeys to and from Central and South America. Resident tropical birds such as Plain Chachalacas, Altamira Orioles, Green Jays, and Ferruginous Pygmy-Owls represent the more than 500 species of birds that occupy the wetlands, thick brush, and riparian forests during the year (infobox 10.2). Mexican Burrowing Toads, Sheep Frogs, Blue Spiny Lizards, and Texas Indigo Snakes reside in shady sites near resacas. Several other subtropical species, including Tamaulipan Hook-nosed Snakes, Mexican Spiny-tailed Iguanas, and Mexican Spiny Pocket Mice, reach their northernmost range limits in the Rio Grande delta.

An estimated population of more than 12 million White-winged Doves once nested in the thornscrub of the LRGV. Whitewings, unlike the more familiar and slightly smaller Mourning Doves, nest in dense colonies, but extensive agricultural development in Texas and Mexico cleared significant areas of their brushy nesting habitat. By 1939, the whitewing population in Texas had fallen to about 500,000 birds, and significant reductions were also noted in the agricultural areas of Mexico. Fortunately, colonies of whitewings began nesting in citrus groves in the Valley and gradually expanded their breeding range to include urban environments as far north as Dallas–Fort Worth and Tulsa, Oklahoma.

Unfortunately, other birds proved less adaptable. As the brush and oak mottes vanished, breeding populations of several species declined and eventually disappeared. Casualties include the Blue-gray Gnatcatcher, Yellow-breasted Chat, Summer Tanager, Orchard Oriole, Rose-throated Becard, Gray-crowned Yellowthroat, and White-collared Seedeater. These were replaced by birds better adapted to environments modified by agricultural and urban development, including Altamira Orioles, Western Kingbirds, Loggerhead Shrikes, and Lesser Goldfinches. Squadrons of

INFOBOX 10.2. A PARADISE FOR BIRDERS AND BUTTERFLIERS

Small wonder that the LRGV stands as one of the top birding destinations in the United States—it lies at the convergence of temperate and tropical climatic zones. Vastly different habitat types, each with temperate, subtropical, desert, or coastal affinities, occur within less than 100 miles in any direction, thereby mingling the major floral and faunal elements of each region. The result is an incredibly rich trove of species—more than 1,200 plants, 900 beetles, and 300 butterflies as well as approximately 700 species of vertebrates dwell at least seasonally in just four counties—Cameron, Willacy, Hidalgo, and Starr—at the southernmost tip of Texas. Bird-watchers, nowadays more commonly identified as "birders," flock to the area in search of more than 500 species of birds and a sure way to expand their life lists—prized sightings include Hook-billed Kites and White-collared Seedeaters.

For avid enthusiasts, birding is an outdoor hobby requiring travel, and those visiting the LRGV each year add more than $100 million to the regional economy. This form of ecotourism, like most others, also places little or no burden on the local infrastructure: no additional children to school, no significant demands on utilities or threats to the quality of life—just a welcome influx of cash. To ensure the viability of this income, communities in the Valley formed a partnership—the World Birding Center (WBC)—with the Texas Parks and Wildlife Department and the US Fish and Wildlife Service. Rather than just one location, however, the WBC consists of nine distinctly different birding sites along a 120-mile historical road bordering the Rio Grande. This unique arrangement attracts dedicated birders from all over the country and simultaneously protects and restores native habitat, as well as increasing public awareness and appreciation for biodiversity.

Bentsen–Rio Grande Valley State Park, headquarters for the WBC, enjoys a well-deserved reputation as a gold mine for tropical birds that occur nowhere else in the United States. Surrounded by 1,700 acres of federal land, the 760-acre state park may be the richest birding area north of the border. Where else can a birder see 50 to 60 species a day, including three species of kingfishers hunting from perches at a single resaca? The other eight WBC sites offer equally rewarding experiences.

Although not included in the WBC system, Santa Ana NWR offers another birding hot spot. Bordering the Rio Grande just southeast of McAllen in Hidalgo County, the 2,088-acre refuge includes a complex of resacas, thornscrub, and riparian habitats where birders may observe more than 400 species of birds. Known as "the jewel of the refuge system," the location forms a funnel through which thousands of migratory birds in both the Central and Mississippi Flyways annually travel to and from their wintering grounds in Central and South America. In spring, as many as 35 species of warblers visit the refuge while kettles of migrating hawks swirl overhead in great spirals. Nocturnal visits to the refuge are just as interesting—the red glow reflected from the eyes of hundreds of pauraques becomes a sight never forgotten.

Another outdoor nature hobby has recently emerged as a partner with birding. Enthusiasts known as "butterfliers" trek to the WBC sites and Santa Ana NWR from all over the world to see and photograph the amazing variety of butterflies that visit the region. About half of all the species in North America have been seen at Santa Ana, where observers once recorded 65 on a single day in October. Mexican Bluewings, Malachites, and Guava Skippers are among the many tropical butterflies to watch for. The National Butterfly Center at Mission includes a visitor center and 100 acres of grounds expressly managed to attract butterflies; the organization also sponsors a well-attended Butterfly Festival every fall.

Green Parakeets and Red-crowned Parrots, perhaps the most conspicuous of the recently established residents, chatter and screech noisily as they zoom through riparian forests and above city streets in the LRGV.

Where the thornscrub meets the narrow strip of Coastal Prairie bordering the Laguna Madre in Cameron County, the Laguna Atascosa NWR preserves a large patch of native brushland. There,

thorny entanglements of trees and shrubs extend over "lomas," natural silty-clay dunes that often reach heights of 36 feet. With soil salinities lower than those of the nearby salt prairies, lomas support "islands" of vegetation dominated by an overstory of Granjeno, Honey Mesquite, and Texas Ebony. Beneath this canopy lies a nearly impenetrable understory of Snake Eyes, Desert Olive, Crucita, Cenizo, Whitebrush, and Texas Lantana. Ocelots,

INFOBOX 10.3. THE LEOPARD CAT

As one of the most beautiful wildcats in the world, the Ocelot has long been appreciated more for its fur than for its ecological role. Ocelots, although much smaller than leopards, have similar grayish to golden brown fur marked with many brown spots bordered in black.

A black line above each eye extends to the back of the head, and its striking pelage features parallel black stripes on the neck that sometimes continue as rows of elongated spots down the back. Some spots on the back and sides resemble the rosettes marking the pelage of Jaguars and Clouded Leopards—a similarity that prompted the Nahuatl to call the animal *ōcēlōtl*.

About the same size as a Bobcat, the more elegantly appointed Ocelot tends to occupy thorny habitats often described as "impenetrable." In the dappled tangle of mesquites, Lotebush, Granjeno, and Blackbrush, an Ocelot's spotted coat affords camouflage during the day when it sleeps on mesquite branches or in shallow depressions. At night, it hunts rabbits, rodents, and birds at sites where shade from the dense canopy has reduced the ground cover of herbaceous vegetation (see fig. 10.10). Most Ocelots regularly patrol a home range of 1–4 square miles, but some wander farther to find prey.

Although once occurring as far west as the desert grasslands of Arizona and New Mexico and in the southeastern forests of Texas, Louisiana, and Arkansas, the species was steadily extirpated from most of its former range in the United States. Initially, the decline resulted when hunters and trappers filled demands for luxurious coats made from spotted cat skins. Ocelot populations continued to decline as agriculture and urban development steadily claimed about 98 percent of their habitat in South Texas—habitat losses coincident with highway expansion in the region. Today, just two remnant populations—with 50 or fewer Ocelots in each—persist in Texas, one in the brushlands on private ranches (in Willacy County), the other

in the thornscrub habitat at Laguna Atascosa NWR (located largely in adjacent Cameron County). A few Ocelots from larger populations in Mexico occasionally cross the Rio Grande to hunt in the narrow, dense strips of riparian brush bordering the river (now protected as the Boca Chica Unit of the Lower Rio Grande Valley NWR), but these locations are removed from the two isolated populations mentioned previously, with many treacherous roads to cross.

The two populations in the South Texas Brushland remain separated by expanses of agricultural lands long cleared of the dense cover that once harbored a much larger population of these "Leopard Cats." Consequently, the smaller, isolated populations—in effect, islands—face a new threat: inbreeding depression, which occurs when the odds favor matings between related individuals. With repeated inbreeding, genetic diversity erodes, commonly reducing birth rates, birth weights, disease resistance, and ability to adapt to changing environments; inbreeding depression may also cause physical abnormalities. Some evidence indicates that less diversity now exists in the Willacy County population in comparison with genetic samples obtained from museum specimens more than 100 years old.

Still, hope endures. Some ranchers in South Texas are now restoring brushlands on their properties as additional habitat for the cats—one thoughtful rancher recently set aside 3,500 acres for this purpose. Additionally, the Texas Department of Transportation is constructing wildlife-friendly underpasses on highways where vehicles frequently hit Ocelots. Plans also call for capturing a small number of Ocelots in Mexico for release in Texas, thereby infusing new genetic material to mitigate the adverse effects of inbreeding. Biologists working with Ocelots—each cat has its own distinctive pattern of spots—maintain a photo catalog that enhances field observations of individuals with known histories. Most of us may never glimpse an Ocelot, yet when we contemplate an expanse of dense thornscrub, we may find comfort in knowing that these beautiful cats still prowl in Texas.

an endangered species, rest during the day in these thickets and emerge at night to prowl in search of prey (infobox 10.3; fig. 10.10).

BOSCAJE DE LA PALMA

The "thicket of the palm"—perhaps the most unique habitat in Texas—survives at the

southernmost bend of the Rio Grande, a point some 10 miles southeast of Brownsville. This remnant community of Mexican Palmetto—locally known as Sabal Palm—includes a remarkably diverse collection of tropical and subtropical flora and fauna. Similar arborescent palm forests occur in only three other areas of the continental United

Figure 10.10. An Ocelot searches for prey at night where the dense overstory reduces ground cover. Photograph by Michael Tewes.

States outside Florida—in the California and Arizona desert, on the Mississippi Delta, and on the southeastern Atlantic Coast.

When Spanish explorers initially visited the mouth of the Rio Grande, they encountered a dense palm forest of at least 40,000 acres that extended approximately 80 miles upriver. Impressed by the forest, Spanish mapmakers originally designated the waterway *Rio de las Palmas*—River of the Palms. Although *palmetto* translates as "little palm," Mexican Palmettos are scarcely diminutive. They reach heights of 26 feet, and dense clusters of large, fan-shaped fronds extending from long spineless petioles form crowns that span 15 feet. As the trees grow, mature fronds gradually droop downward before dying, thereby encasing the upper 9 feet of most trunks with a thick mantle of overlapping dead leaves. This cloak provides roosting sites for the Southern Yellow Bat, a year-round resident whose pelage blends with the yellowish brown of the decadent fronds.

The interlaced canopies of adjacent palms restrict the amount of sunlight reaching the forest floor. Nonetheless, the shadowy environment beneath the trees supports an understory community of more than 80 native species. The major woody plants include Tepeguaje, Anacua, and Texas Ebony, while tangled growths of herbs and vines such as Runyon's Water-Willow, Tropical Threefold, and Slender Passionflower

restrict passage across the forest floor. Because of its location, the palm forest provides habitat for many tropical species of both plants and animals at the northern limit of their distribution. The species richness resulting from the integration of tropical and temperate organisms exceeds essentially all but that of a tropical rain forest. For example, the understory of the palm forest alone provides habitat for more than 900 species of beetles, and the list of birds exceeds 380 species. Plain Chachalacas, more often heard than seen, rustle through the dense undergrowth while Green Jays, Altamira Orioles, and Great Kiskadees make brief but colorful appearances in forest openings. The thick leaf litter conceals Tamaulipan species not found elsewhere in Texas: Speckled Racers, Regal Black-striped Snakes, and Mexican White-lipped Frogs. Vegetation surrounding water-filled resacas in the palm forest provides perches for Green Kingfishers, Buff-bellied Hummingbirds, and Groove-billed Anis during the day, and calling perches for Ferruginous Pygmy-Owls at night.

Clearing the land for agriculture reduced the palm forest to its current size by the late 1930s. In 1971, the National Audubon Society established the Sabal Palm Sanctuary when it purchased the most extensive tract of palm forest still remaining. Today, the 32-acre grove of Sabal Palmetto is protected within a 527-acre buffer zone. Open to the public, the sanctuary includes hiking trails and an information center housed in a historic mansion, which provides visitors with historical and ecological details. A private foundation provided the funding necessary to protect the area for the foreseeable future.

DISTINCTIVE FAUNA
COOPERATIVELY HUNTING HAWKS

If Black-tailed Jackrabbits, Eastern Cottontails, and Hispid Cotton Rats have occasional nightmares, their dreams likely envision terror-filled pursuits by a team of large, dark hawks. Such visions mirror reality in the South Texas Brushland, where groups of Harris's Hawks hunt together in tandem, one after another taking turns chasing their prey from thicket to thicket until the pursued animal darts into a burrow or succumbs to the relentless attack.

With plumage unlike that of any other North American raptor, Harris's Hawks may have served as the living model for the powerful Thunderbird, the sacred spirit of several Native American cultures. Always keenly aware of their surroundings, Native Americans likely noticed the birds' unusual hunting methods. Cooperative hunting is well known in social mammals such as Gray Wolves, Coyotes, and African Lions, but the tactic is rare among hawks. In North America, Aplomado Falcons and mated pairs of Golden Eagles practice group pursuit, but only occasionally and often unsuccessfully.

Using the most sophisticated cooperative hunting strategies known in birds, Harris's Hawks usually initiate their attacks when two to six birds assemble on the branches of a large mesquite or thornbush where they scan for prey. Some periodically fly to new perches, alighting briefly as they leapfrog across the landscape. Hunting begins when a lead bird swoops toward the quarry as the others fly overhead, ready to attack should the first bird fail. The birds are well equipped with both rapid acceleration and agile flight, which often includes tight turns. They fly low when hunting, often landing and hopping toward their prey, and relentlessly flush their target from one patch of cover to another. Members of the hunting party frequently switch positions—alternately attacking and following, but always in pursuit. When the hunt ends successfully, the dominant bird eats first, but all participants eventually share the meal. Their diet commonly includes Black-tailed Jackrabbits, small mammals, Northern Bobwhite, and small birds, but lizards and small snakes sometimes add to the fare.

Cooperative hunting confers a number of adaptive advantages. Tactics that distract and harass soon exhaust prey, and once cornered, a fatigued victim cannot mount an effective defense—all of which improves capture success. This may be particularly advantageous when taking jackrabbits, which are several times larger than an individual Harris's Hawk. Finally, the group guards the carcass until all edible portions are consumed. Team hunting represents a key element in the social life of Harris's Hawks, but a living nightmare for their prey.

COLLARED PECCARY

Of the three species of peccaries occurring in the Americas, only one occurs in the United States. Scientists refer to these animals as the "Collared Peccary," but to most residents of the South Texas Brushland—and elsewhere in the state—they are "Javelinas." Almost any discussion about Javelinas elicits exaggerated tales of ferocious attacks by herds of vicious, piglike animals. Their inflated reputation for ferocity likely prompted a regional university, Texas A&M University–Kingsville, to adopt the Javelina as its mascot when the institution was established in 1925. In actuality, Javelinas are social animals that usually travel quietly in herds of 5 to 20 members. Aggressive encounters are extremely rare and usually occur only when one is cornered, wounded, or attacked by dogs.

Frequently mistaken for a pig, which is a distant relative, the Collared Peccary belongs to a unique family (Tayassuidae) of hoofed animals and, despite a similar snout, differs from pigs in many respects. The straight, javelin shape of their long, daggerlike canines—tusks—provides the basis for the name "Javelina," whereas the tusks of Feral Pigs are curved rather than straight. Both Feral Pigs and Javelinas, however, maintain razor-sharp edges on their tusks by the occlusion of the upper and lower canines—a honing action that occurs each time the mouth is closed. A thick coat of bristly gray-black hair gives Javelinas a grizzled appearance, and a collar-like band of white hair enhances the size of their heads. When excited, the animals erect a mane of long bristles on their backs, which aggrandizes their size and ferociousness. Weighing up to 60 pounds and with shoulder heights reaching 24 inches, adult male Javelinas do not approach the weight (up to 400 pounds) or size (shoulder heights of 36 inches) of mature male Feral Pigs.

Herds—family groups—often travel single file within a home range that may vary from 180 to 556 acres, depending on the availability of food, cover, and water. Although Javelinas mark home-range boundaries with secretions from scent glands on their rumps, herds often intrude on each other, and each herd defends a small territory near the core of its own home range. The territories include bedding areas in depressions typically protected by thickets of dense, thorny brush. Activity patterns shift during the year in keeping with changing temperatures. During the heat of summer, Javelinas become more active in late evening and at night, but they gradually shift to daytime activities when the days shorten and temperatures drop. On cold nights, the herds often cluster tightly for warmth and security.

Unlike other wild ungulates in the Western Hemisphere, Javelinas breed year round. The dominant male in each herd performs most of the mating; other adult males in the herd maintain subdominant roles that preclude breeding. Pregnant females temporarily separate from the herd to give birth but soon return when their young can walk. Older sisters often protect their younger siblings. In the wild, the average life span is about 7.5 years, but in captivity some animals live for up to 24 years.

Collared Peccaries have poor eyesight and forage opportunistically by smell for a varied diet, including animals and carrion, but plants, especially prickly pears, make up about 95 percent of their diet. Their highly adapted digestive system features three stomach compartments designed to break down cellulose-rich plant foods, with semidegraded food later returned and rechewed as cud. Although Javelinas drink when water is available, they secure enough moisture from succulent cactus pads to survive extended periods between rains.

Bobcats and Coyotes prey on juvenile Javelinas. Before gaining status as a game animal in 1939, Javelinas were hunted commercially for their hides, which were shipped to the eastern United States and Europe for gloves and hairbrushes. Today, landowners collect sizable fees from Javelina hunters, but the real value of the animals may center on their ability to help control the density and spread of prickly pears on brushy rangelands.

IF IT'S AN INDIGO, LET IT GO

Fear of snakes is learned at an early age, especially in South Texas, where venomous species abound. Such fears often result in dead snakes, dangerous or not, yet one species gains the approval—and

Figure 10.11. This juvenile Texas Indigo Snake may eventually attain a length of 6 feet. Photograph by Michael S. Price, *Wild about Texas.*

protection—of residents of the Brushland who believe it preys exclusively on rattlesnakes. Texas Indigo Snakes are also appreciated because they lack aggressiveness and rarely bite when handled. Thus, instead of fear, area schoolchildren learn, "If it's an Indigo, let it go."

Texas Indigo Snakes are easily recognized by coloration and size (fig. 10.11). Shiny black with an iridescent bluish-purple sheen on the rear half of their body, adult indigos usually attain lengths between 5 and 6.5 feet. Only two other species in Texas, the Western Diamond-backed Rattlesnake and the Bullsnake, reach comparable lengths, although rattlesnakes of similar size often weigh considerably more than indigo snakes.

Within the semiarid thornbrush country in southwestern Texas—the northern limit for the species—Texas Indigo Snakes occur in mesquite savannas, open grasslands, and lightly vegetated sites near water. These include streams, ponds, resacas, irrigation canals, and windmill seeps, along with riparian habitat as far north as the Balcones Escarpment in Uvalde and Medina Counties and as far northwestward as Edwards and Val Verde Counties.

Although large and heavy, these snakes frequently bask atop dense bushes, on low tree branches, or on the stick nests of Southern Plains Woodrats. They do not excavate burrows but instead seek refuge at night in rodent dens, deep crevices, or other protected hollows, often returning to the same sanctuary each evening. During the day they forage opportunistically for frogs, toads, lizards, juvenile turtles, birds, bird eggs, and small rodents. Contrary to popular belief, Texas Indigo Snakes do not feed exclusively on rattlesnakes (a fact that need not be broadcasted) but do so occasionally. To kill a rattler, an indigo snake slithers beside its victim and swiftly bites its head, which gives the rattlesnake no chance to deploy its lethal fangs. After chewing vigorously, the indigo snake swallows the struggling prey headfirst. When careless, indigo snakes sometimes endure repeated strikes but show no effects from the venom. Unlike constrictors that wrap and suffocate their victims, indigo snakes use their massive bodies to subdue larger prey by pinning it against the ground.

Fortunately, many South Texas farmers and ranchers welcome Texas Indigo Snakes as partners in rattlesnake control—some even stock indigos imported from other areas. Although their reputation as a killer of rattlesnakes may be overstated, indigos do remove large numbers of rodents, which alone more than justifies their status as a protected species under state law. All told, "If it's an Indigo, let it go" remains an effective motto for the conservation of this gentle giant of Texas snakes.

TEXAS TORTOISE, THE "POSTER CHILD"

Arid thornscrub-grassland habitat extends southward from the Balcones Fault Zone (chapter 7) southward into the Mexican states of Tamaulipas, Nuevo León, and Coahuila. Although many species occur within this region, the range of one—the Texas Tortoise—corresponds almost exactly with these boundaries. Thus, biologists Francis Rose and Frank Judd recently characterized the Texas Tortoise as the "poster child" of the region.

Of the 30 species of turtles in the state, only 3—the Texas Tortoise, Eastern Box Turtle, and Ornate Box Turtle—are "dryland" turtles; that is, they inhabit terrestrial habitats, avoid wetlands, and lack webbed feet. Box turtles, true to their name, fold up a hinged area of their plastron to enclose their head, tail, and legs within their shell when attacked. In contrast, neither the Texas Tortoise nor any of the other four species of tortoises in North America

Figure 10.12. Texas Tortoises often excavate a shallow pallet beneath a prickly pear clump. Photograph by Francis L. Rose.

possess this unique feature. Instead, tortoises fully retract their head and appendages under the edge of their carapace, where they hide their head behind forelegs armored with thick scales.

Within the region, these tortoises occupy a variety of vegetative zones and soil types, but they tend to be more abundant where softer soils support thornscrub grasslands embellished with clumps of prickly pears. They remain active during mornings and evenings when temperatures are moderate, but they retreat during the heat of the day and before nightfall to "pallets"—shallow scrapes they clear beneath mesquite trees or clumps of cacti (fig. 10.12). Some scoot backward to shelter under the edges of woodrat middens, often at the center of prickly pear clumps.

The pallets and refuges in woodrat middens offer easy access to the tortoise's main foodstuff—prickly pears. Texas Tortoises also graze on a variety of herbs and grasses and occasionally feed on insects and the fecal pellets of rabbits, Raccoons, and Javelinas, but the pads, tunas (fruits), and flowers of prickly pears provide the bulk of their diet and their main source of water. Ripe, reddish-purple tunas are especially preferred—they do not contain the organic acids found in cactus pads and they have a higher sugar and water content. The relationship between Texas Tortoises and prickly pears may represent an example of mutualism—the cactus provides food, water, shade, and protection from predators, and the tortoises in turn disperse cactus seeds in their droppings. With adequate moisture, seeds scarified as they pass through a tortoise's digestive tract have a much greater chance of germinating than those that have not been ingested.

As adults, Texas Tortoises rarely drink, whereas younger animals occasionally require intakes of freestanding water. To drink, they extend their neck, immerse their head, and then suck water through their nose (external nares). Obviously, water retention poses a serious problem in hot, arid environments, but in light of their impermeable skin and shell, tortoises limit most of their water losses to breathing and defecation and lose little through urination—instead, the bladder absorbs water and efficiently precipitates uric acid to rid the body of toxic wastes.

Courtship occurs from May through October but becomes more intense during July and August. Competing males sometimes joust in a manner resembling two ponderous sumo wrestlers attempting to flip their opponent. Encounters between individuals, regardless of sex, often involve head bobbing, which may be a form of species identification. Mating occurs after a male bites and rams a smaller female until she succumbs to his advances. Females dig pits in soft soil in which they lay and cover clutches of two to five eggs before departing. After emerging from late August to early November, the soft-bodied hatchlings immediately seek the protection of dark environments. They reach sexual maturity in 4 to 8 years; few Texas Tortoises live for more than 20 years.

Populations of the Brushland's "poster child" have declined in many areas and have disappeared in others. Consequently, the Texas Parks and Wildlife Department lists the species as "threatened," and a solid argument can be made for federal protection under the aegis of the Endangered Species Act. Regrettably, the enigmatic Texas Tortoise faces an uncertain future from both habitat destruction and mortality associated with the region's increasing vehicular traffic.

HIGHLIGHTS

FOR KING AND CURLEWS

The thornscrub and grasslands in Hidalgo County surround three large salt lakes—La Sal del Rey,

La Sal Viejo, and East Lake—each sitting atop a huge underground salt dome. Salt domes form where thick salt deposits, the ancient remnants of evaporated seas, lie under 500 to 6,000 feet of rock or dense sediments. The salt in these deposits has a lower specific gravity than the overlying sediments. Hence, when tectonic stresses produce cracks in the overlying strata, the weight of the overburden pushes the salt upward through the fissures. Over time, salt domes near the surface form depressions, which collect runoff that absorbs salt from the strata below and thereby become lakes filled with briny water. The water in La Sal del Rey, for example, is seven times saltier than seawater—at such concentrations, salt crystals form on the shoreline as the water evaporates in the searing heat of a South Texas summer.

The crystal-studded shores of La Sal del Rey have attracted both humans and animals since prehistoric times. Valued for its nutritional importance, salt also served to preserve meat and animal skins. Archaeological evidence suggests that early hunter-gatherers obtained salt at the site not only for their own use, but also for trade with tribes as far away as central Mexico. When Spain ruled the land, salt miners—*salineros*—harvested the rich deposits, but only under special agreement—under Spanish law, all mineral resources belonged to the king. La Sal del Rey—"the King's salt"—earned its name as the most productive of the three lakes. Mining at the lake continued into the 1930s, but an estimated four million tons of salt still remain in the massive dome at the site.

The salt lakes provide significant wintering habitat for migrating shorebirds and waterfowl. For example, at least 2,200 Long-billed Curlews—more than 10 percent of the global population—overwinter annually at the salt lakes in Hidalgo and Willacy Counties. The Long-billed Curlew, the largest shorebird in North America, shares winter shorelines and shallow waters with large flocks of Sandhill Cranes, Snow Geese, Redheads, Lesser Scaup, and Northern Pintails that stop to rest or feed on Brine Shrimp and salt-tolerant aquatic insects. Thousands of Wilson's Phalaropes, American Avocets, and Eared Grebes also visit during the winter (fig. 10.13). During the spring and summer months, the sparsely vegetated shoreline

provides nesting habitat for Killdeer, Black-necked Stilts, Black Skimmers, and Least and Gull-billed Terns. The salt deposits also attract many species of colorful butterflies and tiger beetles.

In 1992, a 5,400-acre tract surrounding La Sal del Rey became part of the Lower Rio Grande Valley NWR. The Western Hemisphere Shorebird Reserve Network designated this area, which includes the 530-acre La Sal del Rey and the slightly larger East Lake, as a "Site of International Significance" for migrating shorebirds. Additionally, the historical significance of the site is recognized in the National Register of Historic Places. A fence surrounding East Lake reduces nest predation, and human access is prohibited year round. To minimize disturbance to nesting birds, La Sal del Rey remains off limits during the spring and summer, but a trail to the lake is open during the winter months. Guided tours are occasionally scheduled for visitors wishing to see the large flocks of wintering shorebirds and waterfowl. The third salt lake, La Sal Viejo, remains privately owned.

CHIGGERS

Waving softly in the wind, luxuriant stands of grass often tempt hikers to stroll in the lush vegetation. Lurking in the grass, however, are hundreds of tiny hitchhikers waiting unseen for a ride—and a meal. Even the shortest hike in early summer can produce within hours an eruption of itchy, red welts that commonly persist for a week or more. The evidence is unmistakable—chiggers, the minute larval stage of a mite in family Trombiculidae, have proffered their own version of a South Texas "Howdy."

After spending the winter just below the soil surface, female mites emerge to lay about 15 eggs per day on tall grasses and vegetation in shady sites beneath bushes. Prelarvae hatch from the eggs within six days and then transform six days later into larvae—minuscule red chiggers approximately 0.008 inch in diameter and covered with hairs. Chiggers aggregate in small clusters on vegetation with access to potential hosts and, despite their size, swiftly attach themselves to passersby when an opportunity arises. Various animals, including small mammals, toads, lizards, Ornate Box Turtles, Texas Tortoises, Northern Bobwhites, and some insects serve as hosts for the ectoparasitic larvae; humans are accidental hosts.

After boarding a host, chiggers do not burrow into the skin or suck blood, as is often assumed. Instead, the mites chew into their host's skin while injecting cell-dissolving enzymes into the wound. Once an opening is created, they sink a stylosome, a tubular extension of the mouth, into the host's skin to extract the liquid contents of the digested cells. In humans, intense itching signals the appearance of a red welt at the site within 2 to 48 hours. Chiggers remain attached for three to five days before dropping off their host to become sexually mature adults after developing through three nymphal stages. In South Texas, where it is warm and humid for much of the year, trombiculid mites may repeat this cycle every two months. The adults that initiated the cycle in the spring die before winter begins, replaced by offspring that overwinter in the soil.

Folk remedies for chigger bites abound—among these are polka dots of fingernail polish that cover the welts on a hapless victim. Other treatments involve salt baths, latherings of alcohol, oils, shampoos, or solid deodorants, and coatings of clay or baking soda. Some remedies offer short-term comfort and a few may cause a chigger to drop off, but the intense itching can be lessened only with over-the-counter corticosteroids or antihistamines. Fortunately, chiggers in South Texas do not carry diseases, but impulsive scratching may cause skin damage and result in secondary infections. Chigger bites can be prevented only by avoiding habitats where the mites likely occur or by dousing clothing and exposed skin with sulfur dust or a formulated insect repellant before going afield.

MORNING JEWELS

At dawn on most summer mornings, grass stalks and spiderwebs on the South Texas Brushland glisten with millions of tiny diamonds. These sparkling orbs—dewdrops—form during the night when warm, moist air seeping inland from the Gulf Coast envelops cool surfaces. Many objects, including grass blades, cactus pads, and spiderwebs, radiate heat efficiently at night, thus becoming much cooler than the surrounding air and causing condensation on their surfaces (fig. 10.14). The fleshy tissues of prickly pears, always in want of moisture, absorb and store the drops formed in this manner. Although ephemeral—dew evaporates almost as quickly as it forms—this daily moisture may be a principal source of water for some plants and animals during dry periods.

When lacking access to freestanding water, thirsty Mexican Ground Squirrels, Northern Grasshopper Mice, Southern Plains Woodrats, and probably other rodents instead harvest dewdrops. Texas Horned Lizards and Pinacate Beetles represent other well-documented examples of animals using dew as a water source. Texas Horned Lizards lick dew from plants but also drink the beads of moisture that form on their backs; they arch their bodies upward to direct the droplets to their mouths in a manner similar to rain harvesting

Figure 10.14. Glistening dew droplets on this prickly pear provide moisture for both the cactus and animals that can negotiate the spines.

(appendix A). Pinacate Beetles, a species of darkling beetle, collect dew in grooves on their wing covers (elytra), elevate their posteriors, and let the droplets trickle forward to their mouths.

Animals walking through dew-drenched vegetation periodically stop and shake off water accumulating on their fur. Sometimes the feathers of roosting birds collect dew that must be removed before or just after taking flight. Even tiny insects such as mosquitoes shake themselves dry before flying on dewy mornings. Cicada wings, however, have a self-cleaning mechanism that removes dew droplets—rows of microscopic waxy cones form a superhydrophobic surface on cicada wings. When small water droplets merge on the water-repellent coating, the single large bead that develops has less surface area than the combined total of the droplets that formed it. Consequently, this releases the energy once used to flatten the water across the surfaces occupied by the smaller droplets and pops the large bead upward. Tiny particles of soot and other grime are also ejected with the droplet.

The quantity of dew formed on plants is not well known, and accounts of animals, especially small mammals, using dew as a water source originate more from observation than measurement. In fact, most physiological studies of water requirements take place in laboratories and hence overlook the contributions made by dew in natural environments. In the realm of natural history, relationships between "morning jewels" and survival in arid habitats thus beg additional study.

CONSERVATION AND MANAGEMENT

Most of the South Texas Brushland is privately owned and managed for livestock production and hunting leases. In particular, the significant income resulting from hunting and outdoor recreation provides landowners with strong incentives to operate their properties in ways that sustain unspoiled ecological conditions. In the LRGV, however, population growth, land clearing, irrigated agriculture, and reservoir construction have already eliminated most of the original habitat. These impacts, in turn, have produced other degrading impacts, among them eutrophication, salinization, soil modification, and invasions of nonnative species. In the face of a growing human population in the LRGV, the most significant management and conservation efforts involve purchasing, restoring, and protecting those few remnants of relatively pristine habitat.

Fortunately, a tract of subtropical riparian forest and thornscrub brushland in the Tamaulipan Mezquital already enjoys shelter within Santa Ana NWR in Hidalgo County. As with many other remnants of native habitat, expanses of farmland surround the 2,088-acre refuge. Nonetheless, Santa Ana NWR harbors many threatened and endangered species while protecting more than 450 species of plants, 260 species of butterflies, 50 species of reptiles and amphibians, 400 species of birds, and 30 species of mammals (fig. 10.15). By day, birders can reliably see Plain Chachalacas, Great Kiskadees, and Green Jays, and at night, the beam of a flashlight catches the red-eye shine of

Figure 10.15. A dazzling assortment of butterflies, represented here by a Malachite (top) and a Zebra Heliconian (bottom), await visitors to the Lower Rio Grande Valley. Photographs by Luciano R. Guerra, National Butterfly Center.

Common Pauraques that linger along the trails and roadways cutting through the thick brush.

One of the most ambitious conservation efforts began in 1979 with the establishment of the Lower Rio Grande Valley NWR. Unlike most others in the federal system, this refuge consists of isolated tracts that are owned by the US Fish and Wildlife Service; other, complementary tracts are managed by private landowners, nonprofit organizations, and the state of Texas. The refuge tracts, totaling more than 80,000 acres of protected habitat, form a loosely connected corridor for wildlife along the last 275 river miles of the Rio Grande. Santa Ana and Laguna Atascosa NWRs anchor this unique system.

The Nature Conservancy, a nonprofit conservation organization, conducted biological surveys on more than 40,000 acres of private land in the LRGV and successfully established three nature preserves in the area. The 415-acre Las Estrellas Preserve (Starr County) protects the federally endangered Star Cactus and 15 other rare plants and animals. In Hidalgo County, Chihuahua Woods Preserve guards a small area packed with eight species of cacti. Southmost Preserve, a 1,023-acre sanctuary on the Rio Grande in southern Cameron County, embraces a stand of native Mexican Palmetto.

Rapid growth of human populations and economic developments poses enormous challenges for the Last Great Habitat. Fortunately, ecotourism has emerged as a significant component of the regional economy. Birders and butterfliers visiting the World Birding Center complex and area refuges each year leave behind millions of dollars—an incentive perhaps strong enough to conserve and manage additional areas of South Texas Brushland.

11
COASTAL PRAIRIES
TALL GRASSES AND SEA BREEZES

As I looked about me, I felt that the grass was the country, as water is the sea.

— WILLA CATHER, *MY ANTONIA* (1918)

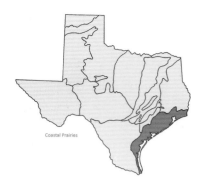

Coastal Prairies

Paralleling the coastline between the eastern border of Texas and Brownsville lies a strip of prairie separating marshes and lagoons from the brushland and savanna communities of the interior. The swath extends inland from the coast for 30 to 80 miles, although some sources claim greater widths. Like grasslands elsewhere, these prairies are now much altered from their original state. Indeed, a subunit popularly identified as the "rice prairies" reflects the influence of one form of agriculture that laid claim to the region's fertile soils and water regime and erased the original vegetation from all but a few areas. South of Kingsville, a long history of grazing on large ranches has also changed the composition of the original prairie community (fig. 11.1).

THE VEGETATION, PAST AND PRESENT
THE GRASSLANDS

Tallgrasses dominated the original community, which included many of the same species that characterized the fabled tallgrass prairie once stretching from Illinois to eastern Nebraska and now forming the "corn belt." Among these were Little Bluestem, Brownseed Paspalum, Big Bluestem, Indiangrass, Switchgrass, and Eastern Gamagrass, supplemented by Gulf Muhly, Pan American Balsamscale, and others. Two varieties of Little Bluestem were also codominants in the climax community, but both have decreased under grazing pressure. One of these, Seacoast Bluestem, is associated with deep sandy soils, where the plants develop a prominent network of rhizomes. Overall, the Coastal Prairies represent a Little Bluestem–Brownseed Paspalum association. In places, monotypic stands of Gulf Cordgrass cover large areas of moist saline flats subject to inundation; these stands of rank growth require prescribed fires to yield good forage for cattle and wildlife.

Most of the soils in the region are clays or clay loams, and except where deep sands occur, nearly all are subject to poor surface and internal drainage. Prior to settlement, Mima mounds occupied as much as 25 percent of the surface area of grasslands in Galveston County and elsewhere along the upper coastline of Texas. These landforms perhaps originated when islets of marsh grasses trapped Quaternary sediments that were then filling in coastal lagoons and thereafter persisted as mounds when the shoreline emerged (chapter 4). The vegetation and landscape in a large area of sandy soil merit separate treatment.

Mesquites are among the several invasive woody species, which also include Huisache, nonnative Chinese Tallow, Macartney Rose, and others. However, mesquites are not newcomers to these grasslands but instead occurred widely in a shrubby, stunted form that, likely because of fire suppression, later increased in density and assumed its present treelike stature (fig. 11.2). Post Oak savannas cover upland areas, and Pecans, hackberries, and ashes characterize the species-rich riparian forests along the several rivers flowing across the prairies to the Gulf of Mexico. Oaks, hickories, and other eastern trees become more common in the riparian communities at the northeastern end of the coastal strip of grasslands.

The endangered Black Lace Cactus holds on in Refugio and Kleberg Counties where the prairie forms ecotones with shrubland communities. Another population once occurring in Jim Wells County has not been documented since the 1980s, and some plants may persist in McMullen County, but their precise taxonomic status awaits further study. The distribution of Black Lace Cactus likely extended at one time into nearby counties, especially those that represent gaps in the present range (Nueces and San Patricio), but the matter remains speculative. The small, columnar plants bear pink flowers that peak in mid-April and early May (fig. 11.3). They prefer sunlit openings in mesquite brush near streams and tolerate saline soils in these areas. Much of the life history of Black Lace Cactus remains unknown. Ants may play a role in seed dispersal, and other agents may fulfill the same function.

The Coastal Sand Plain, or South Texas Eolian Sand Plain (often called the "Sand Sheet" for short),

(*overleaf*) **Figure 11.1. This colorful mixture of grasses and forbs resembles the flora of the original Coastal Prairies. Photograph by David J. Rosen.**

presents something of a dilemma in a description of the ecological regions in South Texas. Some ecologists, in light of the extensive development of thornbrush (and the likelihood of its further expansion), consider the sandy plain as part of the larger South Texas Brushland (chapter 10). Others, however, emphasize the area's open grasslands, which indeed share many of the same species that characterize the vegetation of the Coastal Prairie tallgrass communities on coastal barrier islands and sand ridges farther north. In this view, the sandy plain is considered unique but nonetheless floristically related to a prairie environment. Both views have merit, as befits a dilemma, and while we have necessarily made a choice, we acknowledge that others may not agree with our decision. Mother Nature is clearly not always willing to have her works easily sorted by mere mortals.

COASTAL SAND PLAIN, THE LLANO MESTEÑO

Quaternary deposits of eolian sands define a large plain extending inland across Kenedy and Willacy Counties and as far west as parts of Jim Hogg County—estimates vary, but the area covers at least 2.47 million acres. It is bordered by the Laguna Madre on the coast, Baffin Bay to the

Figure 11.3. The endangered Black Lace Cactus today occurs in just two counties in the Coastal Prairies. Photograph by Pat Clements, US Fish and Wildlife Service.

COASTAL PRAIRIES

225

Figure 11.4.
Wind-driven
dunes march
across the Llano
Mesteño in Brooks
County. Note
the ponds at the
upper margin
of the dunes.
Photograph
by Timothy E.
Fulbright.

north, and soils associated with the Rio Grande
delta to the south; about 60 miles to the west, the
sands give way to Tertiary marine sediments and
a shrubland of thorny vegetation (chapter 10). The
terrain of dunes and swales alternates with broad
flats and a scattering of small seasonal lakes (fig.
11.4). Locals refer to the location as the "Wild Horse
Desert" because of an observation recorded in
1846 of 2,000–3,000 mustangs gathered en masse
on the dunes and sandy plains in Kenedy County.
General Philip Henry Sheridan (1831–1888), after
crossing a desertlike part of the region, remarked,
"If I possessed both Texas and Hell, I'd rent out
Texas and live in Hell." The Spanish name for the
area likely arose from *mesta,* which refers to a
powerful collective of sheepherders that influenced
commerce in the early history of the Castile region
in Spain.

Formally known geologically as the South Texas
Eolian Sand Plain—and ecologically as the Coastal
Sand Plain—the sands were originally deposited
as either a barrier island or strand in ways similar
to the formation of Padre Island or the Matagorda
Peninsula (chapter 12). Remnants of the extensive
sheet of sand, known as the Ingleside Barrier,
are visible elsewhere—for example, Flour Bluff, a
community south of Corpus Christi, stands atop
a stranded barrier island. As the climate dried at
the end of the Quaternary, persistent southeasterly
winds during the following Holocene Epoch,
which began about 11,000 years ago, propelled
sands inland from an ancient barrier island south
of Baffin Bay and from Padre Island. These events
resulted in a grassland landscape starring a cast of
soil types, dunes, swales, flats, and wetlands unlike
any other in Texas.

Historically, Seacoast Bluestem, supplemented
by Switchgrass and other tallgrasses, dominated
the community, but more than a century of
continuous grazing has lessened the frequency
with which these species now occur. Nonetheless,
the area is still predominantly grassland or savanna
and remains one of the last expanses of intact
prairie habitat in Texas. Additionally, Gulfdune
Paspalum covers the swales between the dunes,
and a forb, Camphor Weed, characterizes the
dune ridges (fig.11.5). In all, a greater occurrence
of forbs distinguishes the ridge community from
the vegetation in the swales. A large number of
plants endemic to Texas occur in the Coastal Sand

Figure 11.5. This sand sheet dune is in the early stages of developing a stand of Camphor Weed. Later in the season, the dune will be ablaze with colorful flowers. Photograph by Frank W. Judd.

Plain, and of these, Rio Grande Greenthread and 13 other species occur exclusively in this area. Across most of the plain, the surface resembles a large washboard, but dunes west of the Laguna Madre in Kenedy County may reach heights of 50 feet. As the dunes migrate, they smother grassland vegetation, but the blanket of sand fulfills the habitat requirements for other species, including Amelia's Sand Verbena—a rare endemic with a showy flower—as well as the Keeled Earless Lizard, Gulf Coast Kangaroo Rat, and Merriam's Pocket Mouse.

Wetland vegetation develops in small lakes (*copitas*) and swales lying between the dunes; these intermittent wetlands receive rainwater from seasonal or tropical storms, but groundwater also nourishes a few of the deeper sites (fig. 11.6). Most contain freshwater, but salt spray influences those near the coast, which thus develop saline characteristics. Clay layers beneath the sands slow percolation and extend the hydroperiod for the surface water longer than might be expected. Jointed Flat Sedge and a variety of spikerushes, bulrushes, and water-tolerant grasses thrive in the freshwater pools and, at times, develop enough cover to obscure the surface water. Arrowhead, Elegant Waterlily, Water Stargrass, and Sago Pondweed occur in deeper, more permanent ponds. Where saline conditions prevail, halophytes such as Shoregrass, Carolina Wolfberry, Sea Oxeye, Sea Purslane, White Spikerush, and Saltgrass rim the ponds; this vegetation typically forms concentric bands surrounding the depression in keeping with the changing levels of salinity as the water evaporates, with the most salt-tolerant species forming the inner rings. Clay dunes (*lomas*) often develop on the downwind edges of the larger ephemeral saline ponds. Gulf Cordgrass occurs in association with saline conditions, often becoming especially prominent on broad, low sites known as "flats."

When the ponds begin to dry between rains, the resident amphibians—Western Narrow-mouthed Toads, Black-spotted Newts, and Rio Grande Lesser Sirens—burrow deeply into the clay bottoms, where they enter estivation. During these times, most remain protected from desiccation within an egg-like chamber of dried mud where they wait out extended droughts. With the return of enough rainfall to refill the ponds, the amphibians end their slumber, emerge, and quickly engage in intensive breeding bouts—a process designed to produce a new generation before the ponds again dry.

Patches of trees known as mottes dot the grassland landscape. Live Oaks dominate these sites, which also include a number of other trees of lesser stature. Spanish Moss often drapes the Live Oaks, which are also decorated with feathery Giant

Figure 11.6. Roseate Spoonbills, Snowy Egrets, and Reddish Egrets feed at a salty copita recently filled by rainwater from a tropical storm.

Ball Mosses that can grow to the size of soccer balls. Neither of these are true mosses; both are flowering plants in the pineapple family. Nor are they parasites; they are epiphytes ("air plants") that draw their nourishment from air and rainwater, and they seldom harm the trees on which they grow. Mottes with dense canopies provide nesting habitat for several tropical birds that reach their northern limits in this region, among them Hooded Orioles, Red-billed Pigeons, Northern Beardless Tyrannulets, Couch's Kingbirds, and Ferruginous Pygmy-Owls. A tiny warbler, the Tropical Parula, constructs its nest in the beards of Spanish Moss dangling from oak limbs. Mottes also serve as critically important resting sites for Neotropical migrants.

Elements of the South Texas Brushland creep into the Coastal Sand Plain, likely encouraged by years of overgrazing in a semiarid zone. The woody plants, collectively known—for good reason—as thorn shrubs, consist of drought-tolerant, small-leaved species, many of which are legumes, notably Texas Paloverde and several acacias, including Blackbrush and Guajillo. Others species include Colima, Lotebush, Granjeno, and Texas Persimmon, as well as Honey Mesquite and Brasil. Red Grama and Curly Mesquite are among the grasses invading overgrazed sites.

BOTTOMLAND FORESTS AND OAK MOTTES

The Coastal Prairies are not without woodlands, distinguished in places by types known as bottomland forests and oak mottes. Bottomland forests develop along rivers where floods deposit rich soils that remain moist in all but the severest droughts. Thus endowed, these riparian environments support stands of hardwoods, several of which produce mast and a rich understory while maintaining a diverse community of wildlife. Beginning with the Nueces River to the south and extending eastward to the Sabine River, these forests cross the Coastal Prairies as woody paths between the Gulf and interior landscapes (fig. 11.7). The riparian forests bordering the San Antonio River historically marked the border between two adjacent biotic provinces.

The bottomland forests associated with the Brazos, Colorado, and San Bernard Rivers form a complex known regionally as the Columbia Bottomlands, which extend inland about 93 miles and include parts of seven counties. These forests today cover about 172,600 acres, or just 25 percent of their former extent, with the balance lost or degraded by development, logging, overgrazing, and infestations of exotic plants. Moreover, the remaining stands are highly fragmented, which interrupts their effectiveness as corridors for terrestrial wildlife. San Bernard National Wildlife

Figure 11.7. Riparian forests lining rivers traversing the Coastal Plain add habitat diversity and serve a diverse wildlife community.

Refuge, at the mouth of its namesake river in Brazoria County, protects segments of these communities, but a satellite unit—Dance Bayou—represents a rare location where the climax vegetation has remained essentially unaltered (fig. 11.8).

Although small (750 acres), the bottomland forest in the Dance Bayou Unit displays a diverse community of uneven-aged trees, including extremely large individuals, but also snags and logs, vines, and gaps where trees have fallen—all indicating the old-growth structure of an undisturbed forest. About 300 species of native plants occur at Dance Bayou beneath a canopy dominated by Cedar Elm, Green Ash, Plateau Live Oak, Sugar Hackberry, and Water Oak, among a few others; the remaining species form a complex of subcanopy, shrub, and herbaceous communities. Dwarf Palmetto is particularly common in the understory in many of these forests. A nearby tract protects an isolated stand of Corkwood, a rare endemic shrub, with disjunct populations of a closely related taxon found in Missouri and Arkansas. The species reproduces by seeds and rhizomes, which often produce thickets of clones; because of the limited dispersal of its pollen and seeds, the shrubs remain reproductively isolated.

The richness of the flora in undisturbed bottomlands, coupled with its structural complexity, produces important habitats for a wealth of forest-dwelling birds. In addition to Sugar Hackberry and Cedar Elm, Southern Dewberry, Rough-leaved Dogwood, Yaupon, Carolina Laurel Cherry, and Trumpet Creeper are among the native plants that serve as major food sources for resident and migratory birds. In spring, the latter seek food and shelter in bottomland forests and other coastal woodlands after a long and arduous nonstop flight across the Gulf of Mexico. Woody plants in these forests withstand periodic floods, assisted by deep root systems and, in some cases, by stems that flex with strong currents and recover when the high water recedes.

Two exotic plants threaten the integrity of bottomland hardwood forests, including the essentially pristine community at Dance Bayou. Chinese Tallow trees colonize many types of habitat along the Texas coast, including tree-fall gaps and seasonally flooded sites within the larger confines of bottomland forests. In the understory, Deeprooted Sedge—an invader from South America—spreads from roadsides and other disturbed locations into undisturbed sites beneath the forest canopy, where it may form monotypic stands. Species richness and native herbaceous ground cover both diminish in proportion to the increasing presence of this invasive species. Large volumes of tiny, easily dispersed seeds with high viability, combined with the ability to reproduce asexually and to tolerate shade, give Deeprooted

Figure 11.8.
Dance Bayou
(Brazoria County)
typifies the dense
vegetation that
once existed in
the bottomland
forests in the
Coastal Prairies.
Photograph by
David J. Rosen.

Sedge a competitive advantage that enables exclusion of native plants.

Oak mottes—isolated groves of Live Oak—appear throughout the Coastal Prairies; sculpted canopies of those near saltwater identify mottes known as maritime forests (chapter 12). Whereas Live Oaks dominate the canopy, the mottes also include Texas Persimmon, Water Oak, Post Oak, Brasil, and Honey Mesquite, with Yaupon, Ground Cherry, Texas Mallow, American Beautyberry, Mustang Grape, and Greenbrier among the common plants in the understory. Oak mottes serve as valuable wildlife habitat, not only for migrating birds but also for resident game species such as White-tailed Deer, Wild Turkeys, and Collared Peccaries. In addition to cover, the oaks provide an abundant food source in most years. In bygone days, the strong, curved limbs of Live Oak were prized in Yankee shipyards as ribs for whalers and other wooden vessels.

UNWANTED WOODLAND

Whereas the plow long ago turned under much of the Coastal Prairie, a newer threat looms across the region—and we have no less a figure than Benjamin Franklin (1706-1790) to thank. Historians credit Franklin—an inveterate entrepreneur—with introducing Chinese Tallow to North America, presumably because of the tree's usefulness for making soap and candles. Today, the invasion of Chinese Tallow is so extensive on the Coastal Prairie that it has become a new type of woodland, as well as a potential invader of floodplain forests, cheniers, and wetlands. Between 1970 and 2001, for example, tallow woodlands expanded from 5 acres to 30,000 acres in Galveston County alone (fig. 11.9).

The trees, which may reach heights of about 50 feet, are easily identified by their ovate, often heart-shaped leaves—bright green above and somewhat paler beneath—whose stalks (petioles) each bear two prominent glands at their base. In fall, the leaves turn to a blaze of red, orange, yellow, or

purple. The greenish-white inflorescence develops as an 8-inch drooping, tassel-like spike, with female flowers at its base and males at the top. The female flowers mature into three-lobed capsules, with each lobe containing a round seed that remains on the tree for several weeks after the capsule walls break away. A white tallow-rich covering gives the seeds a popcorn-like appearance.

Chinese Tallow is remarkably adapted to invade and establish itself in new locations. It produces seeds in large numbers when just three years old—once established, a single tree may yield 100,000 seeds per year and do so for a century. Moreover, the seeds remain viable for several years and are widely dispersed by birds and floodwaters. Root fragments also produce new plants, which facilitates dispersal when hurricanes strike the Texas coast. Neither drainage nor soil type, including moderately saline soil, limits the species, and few herbivores of any kind forage on the plants. The leaves contain toxins that, when decaying on the ground, may inhibit the seedlings of other species from gaining a foothold, although the shade from the quick-growing tallow trees may be an even more important deterrent.

Figure 11.9. These young Chinese Tallow trees invading this coastal prairie will grow rapidly into dense woodlands that shade out the native grasses (top), a problem accelerated because birds widely disperse the fat-rich seeds (bottom). Photographs by Guy N. Cameron and Brian R. Chapman.

Indeed, a closed canopy may develop in just ten years to the exclusion of native species requiring direct sunlight. Whereas migrant songbirds visit patches of Chinese Tallow after crossing the Gulf of Mexico, the small populations of arthropods in

these trees, including the absence of caterpillars, offer scant food for the hungry birds; as such, the trees perhaps become ecological traps that provide the birds with cover but offer little chance to refuel before they continue northward. All told, Chinese Tallow rapidly develops as monocultures that not only lack ecological virtues but transform native grasslands into alien woodlands.

HIGHLIGHTS

RICE CULTURE AND GEESE

Rice culture developed in four major areas between the Sabine River and Matagorda Bay; these straddle Trinity Bay, with the Beaumont Prairie lying to the east and the Katy, Lissie, and Garwood Prairies to the west. Riparian woodlands bordering the major rivers flowing to the Gulf separate these units, but several smaller rice prairies in the region lack distinct boundaries. Rice culture in Texas began near Beaumont in 1850 and then expanded westward to Eagle Lake, with the areas of native prairie that were eventually transformed becoming known as "rice prairies." The nearly flat topography and slowly permeable subsoils that reduce seepage, coupled with a dependable water supply and long growing season, together provide a near-perfect setting for large-scale rice production—and habitat

for wintering waterfowl. In particular, many geese visit these fields each winter, preferring these to the coastal wetlands where they once foraged (fig. 11.10). The rice prairies lie at the core of the wintering grounds for waterfowl in the Central Flyway.

Two factors account for the expansion of rice production that was soon exploited by clouds of hungry geese. First, mechanized farm equipment replaced hand and animal labor at the close of World War II, which was followed by widespread use of effective fertilizers, chemicals for controlling plant and insect pests, and the development of plants with improved yields. These advancements led to a new cultural practice known as second cropping, which relies on early-maturing varieties to produce two harvests per growing season. In places, up to 905 pounds of waste rice per acre remain in the fields after the second harvest in November, but this declines rapidly by mid-December. White-fronted Geese are the first to arrive, followed by Canada Geese and, most numerous of all, Lesser Snow Geese, including those known as "Blue Geese." The latter, once considered a separate species and later a subspecies, are actually a natural color variant— a "morph" or "phase"—of the polymorphic Lesser

Figure 11.11. Clouds of wintering waterfowl, here represented mostly by Northern Pintails, feed in rice fields during the fall and winter. Photograph by Steve Balas.

Snow Goose, just as humans vary in hair and eye color. Other polymorphic species include Arctic Foxes (white and blue) and Screech Owls (brown, red, or gray). None of these variants warrant separate taxonomic distinction, and an individual of one phase may successfully interbreed with a mate of another color. In the case of Lesser Snow Geese, however, mate selection reflects imprinting of the goslings on the parental color type and, in part, facilitates assortative matings.

With a surfeit of winter food from both rice and other grains, the midcontinental population of Lesser Snow Geese mushroomed in the last decades of the 1900s. Mortality, normally at a high point during winter, decreased in response to the additional source of nutritionally rich food. Moreover, the birds returning to their tundra nesting areas arrived in peak condition, thereby improving their reproductive success. Because of these advantages, Lesser Snow Geese soon exceeded the carrying capacity of their habitat in northern Canada, turning their feeding areas into barren mudflats ("eat outs"). The bills of Snow Geese feature a so-called grinning patch, a horny area on the edge of each mandible adapted for uprooting ("grubbing") the belowground parts of plants, a process that kills the vegetation (as

opposed to clipping surface foliage). Broods, unable to fly elsewhere for food, often starve or fail to acquire the body reserves necessary to sustain their fall migration. Still, the population continued growing, fueled by the abundance of winter food available in agricultural lands far to the south. Today, large areas of highly degraded habitat scar the tundra, and the damage may be irreversible— or at best, it may take decades or longer to recover.

In recent years, recurring drought has significantly lessened the acreage devoted to rice production in Texas. Meanwhile, Snow Geese have altered their winter distribution, and many now overwinter elsewhere in the Central Flyway and in parts of the adjacent Mississippi Flyway (Missouri and Arkansas). Together, these factors have reduced the size of the goose population that once flourished on the rice prairies. Where a million birds were once commonplace, the Snow Goose population now about numbers about 300,000 to 400,000 each winter.

Ducks, too, visit the rice prairies, during the fall and winter. Northern Pintails, Green-winged Teal, and other species often reach numbers approaching one million (fig. 11.11). The seeds of weeds growing in either fallow or harvested rice fields provide ducks with additional food sources—

Barnyard Grass, nutgrasses, and smartweeds are among the more important of these. In summer, Fulvous Whistling-Ducks breed in rice fields, building their well-concealed nests as thick, floating platforms over water or on drier sites at the edge of levees and dikes; when a nest is situated over water, ramps sometimes lead to the edge of the nest. The birds prefer fields heavily infested with weeds for both nesting and feeding. They eat water-sown rice seeds, but drilling, now a common method of planting, prevents these depredations; additionally, tillering replaces much of the loss and fills in gaps where the birds remove seeds. Later in the summer, Fulvous Whistling-Ducks feed almost entirely on weed seeds even when rice grains remain available after harvest. In the past, rice seeds treated with chlorinated hydrocarbons such as Aldrin poisoned the birds, but this peril ended in 1974 when use of these pesticides became illegal in the United States.

GOPHERS AND GRASSLAND ECOLOGY

The taxonomy of pocket gophers remains a hotbed for ongoing revisions—seemingly a pastime for those armed with calipers and DNA sequencers—but most mammalogists currently recognize three species as residents of the Coastal Prairies. Of these, Baird's Pocket Gophers occupy the easternmost segment of the region, Attwater's Pocket Gophers the central area south to Corpus Christi Bay, and Texas Pocket Gophers the segment from Nueces Bay south to the Lower Rio Grande Valley, including Padre and Mustang Islands (fig. 11.12). Altered taxonomic designations of course require corresponding changes in the distribution of each taxon. For example, Attwater's and Baird's Pocket Gophers are taxa carved from the gene pool of the Plains Pocket Gopher. As a result, Attwater's Pocket Gophers occupy a range extending from the Brazos River southward to the San Antonio River in Karnes and Goliad Counties and then into San Patricio County to the coast. The northern boundary on the Brazos River includes a hybrid zone where Attwater's and Baird's Pocket Gophers interbreed, an occurrence likely facilitated by historical changes in the river's course.

Gophers influence their environment in two basic ways, first by their dietary preferences

Figure 11.12. Mounds marking the burrows of pocket gophers reduce moisture and provide bedding for seed germination. Ruffled soil and an incompletely plugged opening indicate freshly deposited mounds.

(selecting some species and not others), and second by disturbing soils with their burrowing activities. Because of the high energetic demands of burrowing, gophers consume considerably more vegetation than other small, nonburrowing mammals; hence they sometimes reduce the abundance of certain plants, usually the result of eating the roots of palatable species. Indeed, burrowing sometimes requires up to 3,400 times more energy than traveling a comparable distance aboveground. Herbivory also lessens the biomass of the vegetation, including plants growing aboveground that gophers pull into their tunnels.

Attwater's Pocket Gophers prefer the roots of perennial grasses such as Little Bluestem but overall follow the pattern of a generalist when selecting foods. Nonetheless, some seasonal preferences occur, among them increased consumption of forbs during the summer dry season, likely for their water content as much as for their nutritional benefits. Likewise, reproductively active females avoid annual grasses (e.g., Rescuegrass) and instead seek perennial forbs (e.g., Widow's Tear), a choice presumably reflecting a still-unknown nutritional benefit associated with breeding. Nonbreeding females show neither dietary preferences nor avoidances.

The mounds of these gophers also serve as microsites, especially on burned-over grasslands, where forbs readily germinate. The availability of disturbed soil for the growth of pioneering vegetation is significant; at least 9 percent of the area occupied by gophers consists of their mounds, which represent soil deposits of more than 75,119 pounds per acre per year. Water loss from these sites may be reduced and thus favor new growth; the mounds rise about 4.7–6.0 inches above the surrounding surface and create cooler, moister microenvironments in the underlying root zone. Moreover, vegetation buried by the mounds often forms a mat that may also retard evaporation. All told, the activities of pocket gophers—herbivory and mound building—likely maintain the diversity and biomass of forbs in the composition of the Coastal Prairies. Periodic fires influence these attributes as well, adding to the complexity of a highly coevolved ecosystem in which pocket gophers play a key role.

THE COASTAL BEND: A REALM OF DIVERSITY

The coastlines of six counties (Calhoun, Aransas, Refugio, San Patricio, Nueces, and Kleberg) form an area known as the Coastal Bend—the deep recess in the arc of the Texas coast where it turns south toward Mexico. The area warrants separate mention because it stands as a crossroads where biota with eastern or western distributions, or northern or southern, occur for at least part of their annual cycle, and some, such as Black Lace Cactus, are found nowhere else. Here, for example, lie the easternmost reaches of populations of Collared Peccaries that otherwise occur in the desert and scrublands of the American Southwest and Mexico. The Coastal Bend also lies in the zone where Masked Ducks reach their northern limits; this species is a tropical relative of the widely distributed Ruddy Duck in temperate North America. Nesting, although rarely recorded in Texas, may be more common than supposed because Masked Ducks generally prefer habitat where thick beds of aquatic vegetation make it difficult to determine their activities. As many as 25 Masked Ducks have been recorded nesting in the Coastal Bend at sites where visibility was unimpeded by vegetation. Both Ruddy and Masked Ducks belong to a group of waterfowl known as "stifftails" in recognition of their upright fan of tail feathers; they also lay unusually large eggs in proportion to their body size.

The Coastal Bend also marks the northernmost location for Brown-crested Flycatchers, which are replaced to the east and north by a related species, the Great Crested Flycatcher. Similarly, a western relative, the Ash-throated Flycatcher, extends its range to the Coastal Bend and no farther. For these species, the Coastal Bend becomes a biogeographical meeting ground where the genetic integrity of each remains intact (there are no hybrids). Likewise meeting in the region are two pairs of closely related species, with one of each pair representing affinities with different climatic regimes. These are the subtropical Black-crested Titmouse and Golden-fronted Woodpecker and the corresponding temperate pair, the Eastern Tufted Titmouse and Red-bellied Woodpecker. The northern boundary for both of the subtropical species ends at the San Antonio River on the northern border of Refugio County, although the river itself presents no barrier for either bird and instead coincides with a change in biotic provinces (chapter 1). However, a warming environment is steadily affecting the dynamics of this and other boundaries traditionally separating biotic provinces.

Curiously, for an area otherwise so rich in avifauna, the Coastal Bend generally lacks even the most common corvids; Blue Jays rarely extend south of the Nueces River, and the southern range of American Crows ends at the Aransas River. Fish Crows, regularly seen on the upper coast, rarely venture south of Houston, whereas Chihuahuan Ravens remain in arid regions to the west and south of the Coastal Prairies. To human eyes, the vacant habitat for these species seems entirely suitable, and their limited occurrence in the Coastal Bend remains an enigma of biogeography. However, while historically associated with the semitropical area of the Lower Rio Grande Valley, the brilliantly colored Green Jay has recently expanded its breeding range north and east and may soon become "regular" in the avifauna of the Coastal Bend and beyond.

The range of the endemic Maritime Pocket Gopher, one of seven subspecies of the more widely distributed Texas Pocket Gopher, lies exclusively along the mainland coasts of Nueces and Kleberg Counties between Baffin Bay and Flour Bluff, where it is closely associated with native Coastal Prairie and deep sandy soils. Its range today is patchy because of urbanization, agricultural activities, and the transition of native prairie to shrublands. Maritime Pocket Gophers avoid stands of exotic grasses (e.g., Buffelgrass) and clay soils, and the latter act as a barrier to expansion of their distribution. Collectively, these conditions potentially threaten the persistence of the taxon and reduce its genetic variability and gene flow, which may soon result in its listing for federal protection (it is currently regarded as a species of concern).

The blend of habitats—Coastal Prairie and its fringe of marshlands, brush, and oak savannas—likewise favors a wealth of grasses, with a total of 72 genera and 218 species (see infobox). Another 1,085 flowering plants and 7 ferns and fernlike species occur in the same region. This flora develops, sometimes exclusively, on an equally diverse series of soils, among them deep sands, loams of several types, and sticky clays, some of which are saline. Knife-edged ecotones mark sites where one soil type sharply gives way to another, as where deposits of deep sands abut heavy clays. During wet periods, ponds and swales commonly develop, and the region is crossed by several rivers and their tributaries. In places, the meandering paths of these watercourses have left behind oxbow lakes as reminders of their former channels. When regularly replenished with surface runoff, these sites sustain marsh and aquatic vegetation, among them Tule, Lotus, and hornworts, as well as a rich variety of waterbirds, including Purple and Common Gallinules and Least Grebes.

Reptiles, too, meet in the Coastal Bend. In two instances, bays projecting into the coast separate the subspecies of two venomous snakes. The range of the Western Massasauga extends south from Central Texas to the eastern shores of Nueces Bay (in San Patricio County), whereas the distribution of the Desert Massasauga extends into West Texas from the western and southern edges of the same bay (in Nueces County). Similarly, the Southern Copperhead ranges eastward from the Trinity-Galveston Bay complex in Chambers County, and

its close relative, the Broad-banded Copperhead, extends south from Copano Bay in Aransas County. Intermediate forms of these copperheads occupy the intervening area from Galveston to Refugio Counties.

Changes are underway for some segments of the avifauna in the Coastal Bend, particularly for species with tropical, subtropical, or warm-desert affinities. "Before and after" comparisons four decades apart indicate northward extensions of 25–137 miles in the breeding ranges of at least 68 species. These include nine subtropical species, with birds as diverse as the Great Kiskadee, Buff-bellied Hummingbird, Green Kingfisher, and White-tipped Dove. Of several possible causes for the range extensions, changes in significant components of the habitat (but not in vegetational structure), such as seasonal food availability, offer the best explanation. Altered patterns in the amounts and temporal distribution of precipitation, together with moderation in nighttime temperatures during the breeding season, are among the driving forces in this scenario—climate change is affecting a region now warmer and drier than before. To date, the immigration of subtropical species has not been matched with a similar shift in the temperate species—instead, these avian communities now overlap, which may produce genetic consequences, for example, matings between closely related species such as Black-crested and Tufted Titmice.

AN IMMIGRANT EGRET

Few exotic species have successfully invaded the North American fauna without some human involvement, whether voluntary or not, but among those that have managed to do so, the Cattle Egret stands alone for its ability to establish itself, rapidly extend its range, and seamlessly integrate with native species. From their Old World homeland in Africa, the colonizers reached northeastern South America in the late 1800s, quite likely assisted by the winds of an Atlantic storm. From there, some moved north through Central America, but others island-hopped across the Caribbean, eventually reaching Florida in 1942 and thereafter continuing along the Atlantic and Gulf Coasts. They reached the Texas coast in 1954, and nesting was confirmed

INFOBOX. ROB AND BESSIE WELDER WILDLIFE FOUNDATION

"I here create a foundation . . . to further the education of the people of Texas and elsewhere in wildlife conservation [with] a place for research." With these words at the core of his bequest, the will of rancher, oilman, and conservationist Robert H. Welder (1891–1953) established both a physical setting and the financial means for maintaining wildlife resources for future generations. The foundation came into existence in 1954 as a private, nonprofit, tax-exempt operating institution funded by oil royalties, income from cattle sales, and an endowment, later supplemented by generous donations.

Some 7,800 acres carved from the Welder ranch near Sinton provide the physical setting, and it would be hard to find a more suitable environment in which to achieve his goals. The refuge lies in the heart of the Coastal Bend astride a transitional zone between the Coastal Prairies and South Texas Brushland. Specific habitats include oxbow lakes formed from meanders in the Aransas River—the property's northern border—bunchgrass prairies, live oak and chaparral communities, mesquite grasslands, riparian forests of Sugar Hackberry and Pecan, and several other communities adapted to a variety of sandy, loamy, and clay soils. This mix of vegetation in turn provides diversity for abundant populations of both resident and transient wildlife—all contributing to a marvelous outdoor laboratory. On-site facilities include a headquarters building with offices, library, museum, and laboratory space. An expansion later added a wing housing the foundation's vertebrate and plant collections, as well as an extensive egg collection donated by the family of Roy Quillin, an accomplished amateur oologist. The facilities also include staff residences, student housing, and a rotunda. All of the buildings were designed and constructed in a style reflecting the heritage of the original Mexican land grant (1834), which later became the core of the extensive Welder holdings.

The foundation's program took shape in the hands of director Clarence Cottam and assistant director W. Caleb Glazener. With consent from the trustees, the team of Cottam and Glazener initiated a program that focused on research conducted by students seeking graduate degrees at leading colleges and universities in Texas and elsewhere in the United States. The first phase focused on detailed surveys of the vegetation, the key to understanding the form and function of wildlife populations. The focus then turned to White-tailed Deer and Wild Turkeys, two of the more prominent species on the refuge grounds. Thus began a series of studies dealing with the population dynamics and behavior of these two popular game species. A major component of this research included a large area fenced to exclude predators, mainly Coyotes, yet not confine the movements of deer. The studies revealed the role of Coyotes in regulating deer numbers and, in a larger sense, the role of predation as a part of nature. A wealth of other projects targeted, among other subjects, Collared Peccaries, Northern Bobwhites, American Alligators, Bobcats, waterfowl and migratory songbirds, and the impacts of fire and grazing systems on both vegetation and wildlife. Reproductive success, disease and parasites, the food habits of wildlife, and the impacts of Red Imported Fire Ants represent some of the topics under investigation. Techniques and strategies for dealing with invasive vegetation rank high in the current research agenda. The foundation also supports a number of studies in other parts of the state and nation.

An active program of conservation education and outreach complements the research efforts. Thousands of tourists visit the refuge each year as well as classes of K-12 schoolchildren and university students on field trips, which feature hands-on experiences with plants and wildlife and instruction about bird banding and other techniques. The staff also conducts field days for teachers, youth leaders, and landowners; topics include prescribed fire, plant identification, and range ecology. The foundation is one of seven national training centers for Conservation Leaders of Tomorrow that hold workshops for college students and resource managers, including training about the role of hunting in conservation. An annual youth hunt is also part of the outreach program.

The graves of Rob and Bessie Welder lie side by side in an oak-shaded grove near the headquarters building. Standing there, one realizes that Rob Welder's foresight and generosity have been bountifully fulfilled, much to the benefit of future generations—just as he wished.

Figure 11.13. A Cattle Egret in breeding plumage looks alert while its mate incubates eggs (right). A young Crested Caracara feeds on carcass remains left by Turkey Vultures (above). Photographs by Brian R. Chapman (right) and Richard L. Glinski (above).

in 1959 (fig. 11.13). Once established on the coastal prairies, the birds moved inland and now occur across virtually all of Texas and beyond (stock on the Atlantic Coast likewise moved inland, and the species today occurs throughout the United States and into southern Canada).

Cattle Egrets nest in heronries, but not alone; along the Texas coast they move into heronries already established by Snowy Egrets and Tricolored Herons. Inland, however, Cattle Egrets seek heronries occupied by Snowy Egrets and Little Blue Herons, species that apparently select their nesting locations based on the availability of crayfish, a major food for their nestlings. In contrast, Cattle Egrets feed primarily on grasshoppers and crickets and usually do so in association with grazing cattle, but the distribution and abundance of neither these insects nor cattle seem to influence which heronries they select. Some of the larger heronries exceed 15,000 pairs (of all species), but others consist of fewer than 100 pairs.

Continued use of the larger heronries produces accumulations of guano that kill or thin vegetation at the site and, depending on the trees, sometimes cause the birds to move elsewhere. Sugar Hackberry, for example, tolerates the ammonia-rich guano, whereas oaks and Pecans do not.

To date, no evidence has emerged to suggest that Cattle Egrets have adversely affected other species.

They share common heronries with other egrets and herons, but on average they nest about three weeks later than the native birds, hence precluding interspecific competition. Likewise, the insect-based diet of Cattle Egrets differs considerably from the diets of other herons, particularly in the case of nestlings. Recent surveys indicate population declines of nearly 3 percent in some areas of Texas, but this may indicate that Cattle Egrets have fully saturated their nesting sites, and it seems certain that these immigrants are here to stay.

A SCAVENGING FALCON

For a member of the falcon family, Crested Caracaras hardly fit the mold typical of their kin. They forgo the sleek features of falcons, notably lacking long, pointed wings and rapid flight, and instead present a bulkier appearance accompanied by massive bills, long legs, and more deliberate movements. Still, caracaras are handsome birds and sport a black crest, a red unfeathered face, a white throat, and a barred breast; the remainder of the body is black except for a white, dark-tipped tail. Their range extends from Mexico into Texas along the Gulf Coast, where they reside in both brushlands and the Coastal Prairies; sightings also occur occasionally as far inland as Waco and North Texas. A small isolated population occupies the tip of Florida, where it is listed as threatened. Originally, the Mexican flag featured a Crested Caracara clutching a snake atop a cactus, but this so-called Mexican Eagle was later replaced with a Golden Eagle, a raptor with a relatively limited distribution in Mexico that preys largely on small mammals.

Their foods and feeding behavior, as much as anything, set caracaras apart from falcons, whose diets typically focus on capturing other birds. Overall, caracaras are opportunistic generalists and will feed on carrion; as such, they become scavengers. They patrol highway rights-of-way early in the day in search of roadkills, from which they may aggressively chase away vultures. At other times, however, caracaras mix with groups of Black and Turkey Vultures feeding on a carcass, so the relationship between these species varies somewhat, but they remain the dominant species in these associations. Caracaras cannot open up large carcasses and thus rely on vultures to do so. Indeed, a caracara sometimes swoops down to harass vultures feeding on a carcass to the point of driving them into flight and then pursues one of the birds until it regurgitates. Remarkably, the caracara then flips over into a dive and deftly snatches the secondhand meal in midair, often landing to pick up scraps missed in the aerial feeding maneuver.

In addition to scavenging (fig. 11.13), caracaras also hunt for live prey, either searching from perches or flying low over the ground. However, they do not attack prey from the air but instead land and walk to their targets. Insects are common foods, and the availability of beetles at carcasses may be part of the reason for their interest in carrion. Other prey includes reptiles, small rodents, young rabbits, and birds and their eggs, and they at times wade in shallow water for fish and crustaceans. They also turn over dried cow patties in search of insects, dig up turtle eggs, and walk in fields behind tractors or near grazing cattle to forage on insects disturbed by these activities. From perches, caracaras intently watch songbirds

tending their nests, which they raid for either eggs or nestlings. As further evidence of their versatile feeding behavior, caracaras respond to prairie fires, where they may remain for several days foraging behind the flames for dead animals. Larger prey is torn apart while held falcon-like in one foot.

On the Coastal Prairie, caracaras often nest in tall growths of Macartney Rose, an exotic that was introduced from China in the 1800s for use as living fences but is now an invasive pest on rangelands. In prairie settings, the birds nest in the tallest vegetation available; hence clumps of Macartney Rose that reach heights of 10 feet or more offer prime nesting locations. Caracaras build their nests below the canopy, which obscures detection; some nests are reused in subsequent years, whereas some pairs refurbish the nests of other raptors, notably White-tailed Hawks. Overall, caracaras and other raptors of similar size partition their nesting habitat where they coexist on the Coastal Prairies, including the Coastal Sand Plain.

A SQUEALING DUCK
The shrill voice of a Black-bellied Whistling-Duck resembles a squealing "pee-che-che-ne," which hardly conveys a "quack." Indeed, whistling-ducks align taxonomically not with true ducks but with geese and swans, with which they share common features in addition to some arcane physical characteristics such as their syrinx structure; the sexes look alike and remain paired for life, with both jointly rearing their broods. But unlike all but one species of their larger brethren, both male and female Black-bellied Whistling-Ducks share incubation duties at nests in tree cavities or, at times, on the ground in well-concealed locations. In either case, the nests may be some distance from the nearest water. The plumage pattern of the downy whistling-ducks at hatching is also unlike that of other waterfowl (fig. 11.14).

The perching abilities of Black-bellied Whistling-Ducks are admirably displayed when the adults search for nesting cavities. Clutches average about 13 eggs, but far larger "dump nests"—101 eggs in one case—occur when several females add to the original clutch. Surprisingly, many of these "superclutches" are incubated, typically by the founding pair, and many hatch. Hence, large

Figure 11.14. The closest relatives of Black-bellied Whistling-Ducks include geese and swans, but the pattern of their downy plumage (inset) differs from that of all other waterfowl. Photographs by James F. Parnell and Frederick F. Knowlton (inset).

broods are not uncommon. At hatching, the day-old ducklings drop without harm from heights of 20 feet or more. Nests in tree cavities often fall victim to Raccoons and ratsnakes, so to be effective, boxes erected to supplement nesting habitat should be protected against the same predators; without such guards, the nesting boxes offer no advantage and may in fact become liabilities. A relative, the Fulvous Whistling-Duck, also occurs in Texas but nests at ground level in wetland vegetation, including rice fields; six other whistling-duck species occur in warmer regions worldwide. Because of their varied nesting habitats, the name "tree ducks" was replaced with an auditory descriptor befitting the entire group.

At the beginning of the twentieth century, Black-bellied Whistling-Ducks seldom ventured beyond the Lower Rio Grande Valley. By midcentury, however, sightings were occurring farther north, and the birds soon became regular summer visitors in the Coastal Bend. Black-bellied Whistling-Ducks banded at a nesting area near Corpus Christi turned up the following winter in northern Mexico, but it appears that only birds at the northern edge of their range display these migratory movements; farther south, they remain in place as year-round residents. In the following years, Black-bellied Whistling-Ducks steadily moved up the Gulf Coast and now nest across the southern third of Texas and into Louisiana. Some also remain in Texas throughout the winter in numbers large enough to be included in the legal waterfowl harvest.

ATTWATER'S PRAIRIE-CHICKENS: NEXT TO GO?

Largely because of habitat loss, hard times have befallen many species, but the Greater Prairie-Chicken might well stand alone as a poster child for the plight of grassland birds. Ornithologists recognize three subspecies, one of which—the Heath Hen of the Eastern Seaboard from New England to Virginia—long ago succumbed to extinction. Another, the Greater Prairie-Chicken of the interior tallgrass prairies, persists greatly reduced in both range and numbers. Unfortunately, the third subspecies—the Attwater's Prairie-Chicken—like its eastern kin seems almost certain to travel a dead-end path toward extinction.

Attwater's Prairie-Chickens once thrived on the Coastal Prairies when the tallgrass communities still prevailed—indeed, the birds occurred nowhere else. Estimates peg the original population at about one million birds. Unregulated hunting—daily bags of 200–300 birds were not uncommon—in the late nineteenth and early twentieth centuries took a heavy toll, and all the while settlement and farming steadily claimed the grasslands. Urbanization and development of the petrochemical industry, fire suppression, and invasions of woody species, especially Chinese Tallow, made further inroads on the population. A small population at the eastern edge of Louisiana totally disappeared in 1919. By 1937, about 8,700 birds remained in Texas, a number low enough to trigger the end of the hunting season for Attwater's Prairie-Chickens. Still, the decline continued, with 456 birds counted in 1992, 158 in 1994, and just 42 in 1996 (ironically, with so few birds remaining to count, the censuses gained greater precision). Meanwhile, Attwater's Prairie-Chickens were formally listed as endangered in 1967 when the population numbered about 1,000 birds, and in 1972 lands purchased by The Nature Conservancy and the World Wildlife Fund became the core of the Attwater Prairie Chicken National Wildlife Refuge (Colorado County) near Eagle Lake, which at 10,541 acres is one of the largest protected examples of upland prairie left in the region.

Like other species of prairie grouse, Attwater's Prairie-Chickens court on sites known as leks (or booming grounds). Such sites are widely recognized as the social center of prairie chicken ecology and may remain in use for generations. Vegetation at leks is sparse, as befits their function as display arenas where visibility is paramount. In spring, a dozen or more males gather at dawn and again before sunset to engage in mating displays that mimic those described for Lesser Prairie-Chickens (chapter 8).

Species beset with greatly diminished numbers often fall into a downward spiral known as an extinction vortex—a fatal whirlwind of events that winks out small populations. Attwater's Prairie-Chickens faced just such a situation when overhunting and habitat loss significantly reduced their numbers and set the stage for the disproportional impacts of events both environmental (fires or storms) and biological (predation, disease, or genetic maladies). To illustrate, today the birds necessarily nest in remnant, poorly drained grasslands unsuitable as croplands. Downpours from storms easily flood the terrain, severely reducing reproduction and devastating a population already clinging to survival. Even the loss of a few individuals to predation, normally of no consequence, now becomes a threat in an extinction vortex. Last-ditch efforts for their recovery continue, but lacking a miracle, Attwater's Prairie-Chickens may soon become yet another epitaph etched on the tombstone of extinction.

CONSERVATION AND MANAGEMENT

Less than 1 percent of the original 65 million acres of Coastal Prairies in Texas remains, with the balance lost to agriculture, commerce, and residential development. Still, a few precious remnants are intact, and conservation agencies control some key areas. Among these is the 296.5-acre Nash Prairie in Brazoria County, where the presence of Mima mounds provides a reliable indicator that the land has remained unplowed. This area, one of the largest and best examples of native prairie still remaining, includes a native flora of 301 species of vascular plants, of which 59 are grasses. Rare species include Coastal Gayfeather, Buttonbush Flatsedge, and Houston Meadow-rue. Because its flora closely represents the composition of the original Coastal Prairies, the site provides a source of seeds for restoring other coastal locations. The smaller 101-acre Mowotony Prairie, also in Brazoria County, presents a native flora of 195 species, which includes 41 species not found at Nash Prairie. Together, the flora of the two areas well reflects the richness of the vast Coastal Prairies that once bordered the coast of Texas.

The Texas City Prairie Preserve, a 2,300-acre tract of prairie, marsh, and Post Oak mottes, until recently protected one of the few remaining wild populations of Attwater's Prairie-Chickens, but even that population winked out in 2012. Prairie restoration, including prescribed burning, light grazing, and control of invasive vegetation, especially Deeprooted Sedge, remains the primary management activity at the preserve; perhaps the site, when restored, may become a location where the endangered birds can be reintroduced. Together with Nash and Mowotony Prairies, the Texas City Prairie Preserve benefits from the stewardship of The Nature Conservancy.

The Texas Parks and Wildlife Department established the Candy Abshier Wildlife Management Area in 1990. The site includes prairie and other communities—diversity that has established it as a popular stop on the Great Texas Coastal Birding Trail. In particular, the area offers opportunities to observe large numbers of migrating hawks of several species. The same agency recently acquired the Powderhorn Ranch, a large area of marsh, oak forest, and grasslands bordering Matagorda Bay, as the future site of a state park and wildlife management area. With management, the grasslands can be restored as prairie representative of the original vegetation.

The newest native prairie preserve lies tucked in the midst of the Houston Metroplex, where it was saved from development, in large part, by contributions from the general public. In 2014, the 51-acre Deer Park Prairie Preserve in Harris County became a reality when the Bayou Land Conservancy turned over the property to the Native Prairies Association of Texas after raising almost $4 million to purchase it from a private developer; together, these organizations signed a conservation easement that would forever protect the prairie from development. Amateur naturalists discovered the species-rich site, already surrounded by developments and a cemetery, shortly before it was to disappear under yet another complex of multifamily housing. Similarly, the Katy Prairie Conservancy strives to acquire, restore, and protect native prairies on the western edges of greater Houston.

The precarious situation befalling Attwater's Prairie-Chickens stimulated several courses of action to head off their impending extinction— fewer than 100 birds have survived in the wild since 1995. In addition to establishing a national wildlife refuge dedicated expressly to their protection, management-based research started to determine ways to restore the population both on and off the refuge. The release of hand-reared captives failed—few of the broods produced by the released birds survived for more than two weeks and fell well short of the requirements for establishing and sustaining a wild population. Further research revealed that Red Imported Fire Ants, which invaded the Coastal Prairies about 1970, reduce the abundance of native insects—a critically important food for chicks—to levels where the young birds starve. As shown by field experiments, reduction of the fire ant infestations more than doubled brood survival, an improvement that can sustain a wild population of prairie-chickens in otherwise suitable habitat. In addition to controlling fire ants with lethal baits, other actions such as patch burning, disking, and reduced grazing also increase insect populations; these treatments increase the percentage of early-successional forbs in the vegetation on which insects flourish.

In addition to the refuge, the Coastal Prairie Conservation Initiative includes an 80,000-acre project area of privately owned lands that seems well suited for controlling fire ants in a large-scale effort to restore Attwater's Prairie-Chickens. Initiated in 1998 as the Refugio-Goliad Prairie Project, a consortium of landowners and both public and private agencies, including, in part, the US Fish and Wildlife Service, Texas Parks and Wildlife Department, and The Nature Conservancy, formed a partnership designed to maintain and restore Coastal Prairie using tools that will reduce brush encroachment and exotic vegetation, including prescribed burning, and a fact-based program of grazing and wildlife management. A "Safe Harbor" agreement encourages landowners in the project area to allow releases of endangered species on their property. Thus, if managed to reduce fire ants, a sizable area of Coastal Prairie may again include viable populations of Attwater's Prairie-Chickens and thwart their march toward oblivion.

Bottomland hardwood forests, also fragmented and degraded by human disturbances, require protection and management, especially those associated with the San Antonio, Brazos, and other major rivers extending inland 60 miles from the Gulf of Mexico. Few, if any, of the once-forested bottomlands will be restored; hence those that remain relatively intact gain precedence as targets for a viable conservation strategy. In addition to their value as wetland systems controlling floods and improving water quality, bottomland hardwood forests along the coast provide migrating songbirds with food and shelter following an energy-demanding nonstop flight across the Gulf. These stopovers both renew lost energy reserves and prepare the birds with the nutrition they need to continue their migration and enhance their reproductive success; moreover, the same sites provide food resources for fall migrants. Because individual birds may not make landfall at the same location each year, management geared toward species-specific requirements seems less important than maintaining the inherent complexity of forest composition and structure that fulfills a spectrum of needs: vegetation that not only produces fruits and flowers but also harbors abundant insect populations. In this light, invasions of exotic plants that exclude native vegetation pose a serious threat. All told, maintenance of a permanent mosaic of bottomland forests lies at the heart of a conservation strategy designed to ensure their role in the annual cycle of migrant songbirds.

As with most other invasive species from other lands, control—not eradication—remains the only feasible goal for challenging the invasion of Chinese Tallow. Unfortunately, the species' adaptations, among them prolific seed production and sprouting, together form a near-perfect arsenal designed to ensure survival. With adequate fuel, fires kill trees less than 3 feet tall, but it becomes less effective on others because some top-killed trees resprout. Some evidence, however, suggests that annual burns conducted during the growing season may prove more effective than previously supposed and may curtail additional invasions. Biological control, while desirable, remains elusive; no insects or other organisms effectively damage the trees, nor has a fatal, species-specific disease been discovered. Hand clearing, while labor intensive, works on small trees or small areas, but mechanical treatment on larger trees and areas is costly and the stumps will sprout unless treated chemically. Of the various herbicide treatments, basal bark spraying proves most effective—the chemicals are applied completely around the lower trunk of each tree. Still, viable seeds persist in the soil after any of these treatments and make further management necessary when seedlings emerge. The battle against Chinese Tallow remains at best no more than a stalemate, and these unwanted woodlands seem likely to persist as one more injury to the integrity of the Coastal Prairie.

Largely because waste grains provide a food subsidy leading to increased winter survival, the midcontinent population of Lesser Snow Geese increased 5–7 percent annually in the last decades of the twentieth century—a rate grossly incompatible with the carrying capacity of their northern nesting areas. To preclude catastrophic starvation, wildlife agencies initiated liberal hunting regulations—increased or no bag limits and, in some areas, spring hunting seasons—to reduce the growth of the population to replacement levels. In addition to directly increasing the impact of hunting mortality, spring hunting also produced an indirect effect: enough harassment to diminish the accumulation of fat and protein necessary for successful reproduction. Whereas the goose harvest has more than doubled since enactment of the liberal regulations, the desired result has yet to be achieved and the challenge thus remains a work in progress.

The factors that endanger the Black Lace Cactus include the clearing and replacement of native vegetation with pastures of exotic grasses that compete with the smaller cactus for light and other resources. Oil, gas, and mineral extraction pose additional threats, especially in McMullen County. Moreover, a catastrophic event befalling any one of the remnants would present the species with an even greater risk of extinction. Fortunately, owners of the lands where the species persists have taken a proprietary interest in the plants and have unofficially set aside areas for their protection. In addition, seeds kept in long-term storage at the Desert Botanical Garden in Arizona ensure protection of the species' genetic identity. Meanwhile, the South Texas Botanical Gardens and Nature Center is creating a refugium for the species at its facilities in Corpus Christi.

12

TEXAS GULF COAST

MARSHES, BAYS, AND
BARRIER ISLANDS

Long before we saw the sea, its spray was

on our lips, and showered salt rain upon us.

— CHARLES DICKENS

The Texas coast descends southwestward from the Sabine River—the geopolitical boundary with Louisiana—and then curves southward near Baffin Bay to the Mexican border at the mouth of the Rio Grande (fig. 12.1). Along this 367-mile coastline, variations in rainfall, evaporation rates, inflow from freshwater rivers, and seawater exchanges with the Gulf of Mexico fashion at least three ecologically distinct regions.

Rainfall and runoff exceed evaporation in the "Upper Coast" region, which extends from the Sabine River to the Colorado River and Matagorda Bay. Near the Louisiana border, for example, annual rainfall frequently exceeds 50 inches, and large quantities of freshwater flow throughout the year into the bays in this region. In this area of temperate climate, the inflow of freshwater dilutes the saltwater, hence creating "positive" estuaries. In the "Middle Coast," which runs south from Matagorda Bay to Corpus Christi Bay, the climate changes from temperate to subtropical, and runoff and precipitation largely balance with evaporation.

Along the "Lower Coast," evaporation exceeds both precipitation and runoff. Near the Rio Grande, an annual average rainfall of 16 inches and subtropical temperatures create desertlike conditions with few permanent streams; freshwater inflows seldom occur except after tropical storms. Consequently, hypersaline, or "negative," conditions prevail in the Laguna Madre, a 120-mile long estuary paralleling the coastline from Corpus Christi Bay south to the Rio Grande.

CLIMATES AND GULF CURRENTS

Generally, the warm waters of the Gulf of Mexico moderate the climate along the Texas coast and blur seasonal changes. The average frost-free period lasts for 245 days on the Upper Coast, where temperatures sometimes drop to freezing or below during the winter months. Although the temperature once dipped to 10°F at Corpus Christi, the Lower Coast rarely experiences freezing weather. Summerlike temperatures along the Texas coast frequently extend well into October before

(*opposite*) **Figure 12.1. Sandy beaches line the Gulf of Mexico along the entire Texas coastline.**

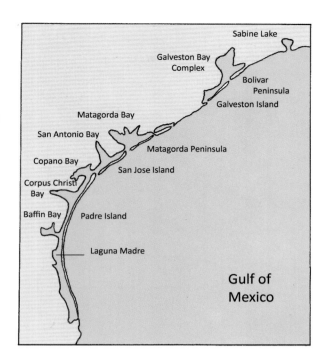

the first cold front arrives, and winters are short and mild.

The timing and height of high (flood) and low (ebb) tides vary daily in keeping with the relative positions of the Earth, moon, and sun. Although tides elsewhere in the world may vary daily up to 30 feet, the average magnitude in Texas is only 1.5 feet—slightly greater on the Upper Coast, but much less on the Lower Coast. Wind also influences tidal range, especially if it pushes water in the same direction as the tidal flow. Winter winds from the northwest usually create the lowest tides, whereas the highest tides often occur during southeasterly storms.

The flow of the Gulf Stream northward along the Atlantic Coast of North America influences the predominant current in the Gulf of Mexico. The Loop Current, which forms from eddies circulating clockwise in the Gulf Stream, transports warm water from the Caribbean Sea through the Yucatán Channel into the Gulf of Mexico. The current— about 125–190 miles wide and 2,600 feet deep— carries the warm water northward. When it nears the southeastern Gulf Coast, the current loops back southward along the coast of eastern Florida and then exits the Gulf below the Florida Keys to rejoin the Gulf Stream in the Atlantic Ocean. Once or

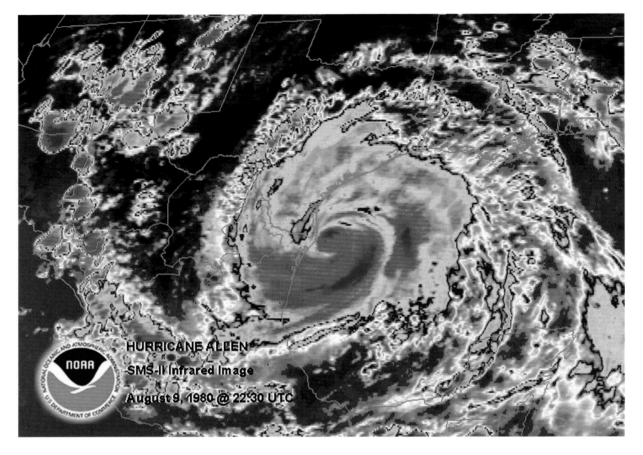

HURRICANE ALLEN
SMS-II Infrared Image
August 9, 1980 @ 22:30 UTC

twice annually, the Loop Current spins off a "Loop Current Eddy" (or "Warm Core Ring") that forms an enormous clockwise gyre circulating warm water westward in the Gulf.

The Warm Core Ring generates high relative humidity in coastal regions bordering the Gulf. If the circulation of a ring coincides with the path of an approaching tropical storm or hurricane, the results can be disastrous. Hurricanes feed on the energy stored in warm water and absorb additional water vapor from the moisture-rich air above the current. In 2005, infamous Hurricane Katrina swelled abruptly to a Category 5 storm as it crossed over a Warm Core Ring just before slamming into New Orleans. For this reason, weather forecasters pay particular attention to the formation and progress of Warm Core Rings during the hurricane season.

Most hurricanes develop between early June and late fall when summertime heat warms the Atlantic Ocean lying between 10° and 20° north latitude. Convections of warm air laden with water vapor rapidly ascend, generating a low-pressure area—

a tropical depression—that may sustain winds circulating counterclockwise at surface speeds of less than 39 miles per hour (mph). Most tropical depressions soon disintegrate, but those that continue to intensify with wind speeds between 39 and 74 mph become tropical storms and are given names.

Hurricanes develop from tropical storms when surface winds exceed 74 mph and typically form a cloud-free "eye" at their center (fig. 12.2). The Saffir-Simpson scale divides hurricanes into five categories based on sustained wind speed. The winds in Category 1 hurricanes, the weakest, reach speeds of 74–95 mph, whereas those in the strongest, Category 5, exceed 157 mph. Some of the strongest hurricanes battering the Texas coast in the past bore winds of more than 170 mph, but tropical storms or hurricanes of any intensity may cause significant disruptions, flooding, and structural damage along the coast, and often far inland as well.

Storm-generated waves often create ecological havoc in shallow lagoons and on beachfronts.

Stirred sediments increase turbidity and sometimes smother seagrass beds and oyster reefs. Wind-driven tides during nesting seasons may inundate shoreline nests of terns, Black Skimmers, and sea turtles, and robust waves frequently uproot seagrasses and mangrove clumps. Substantial inflows of freshwater from storm-generated rains may temporarily alter salinities in bay systems. Fish and invertebrate populations tend to fare well in hurricanes, but substantial changes in salinity may initiate changes in species composition and affect food chains.

COASTAL ECOSYSTEMS

Geological events extending as far back as the Tertiary Period sculpted the present coastline of Texas. For at least 66 million years, rivers supplied successive layers of sediments that extended the shoreline more than 250 miles seaward into the Gulf of Mexico. Evidence for these events abounds in a series of sediment wedges whose slopes mark the depositional history of the Coastal Plain. During the Pleistocene Epoch, the sea level rose and fell in keeping with the expansion and melting of the great ice sheets that once covered much of the Northern Hemisphere. The sea level dropped when prolonged cold tied up precipitation as ice and thereby lessened the flow of river water into the oceans. Conversely, during warm interglacial periods, the ice thawed and the heavy influx of meltwater caused the sea level to rise.

During an unusually warm period about 135,000 years ago, the sea level rose about 25 feet, forming a coastline some 20 miles inland from its present location. The rivers, in turn, cut new channels and deposited their sediments on floodplains that matched the existing coastline. Then, during an extremely frigid glacial period that followed, the coastline retreated seaward, causing rivers flowing to the Gulf to cut even deeper into the former floodplain. Still later, about 18,000 years ago, continental glaciers again started melting. Rising seawater drowned river mouths, creating coastal bays and lagoons before reaching its present level about 5,000 years ago.

Geologically, these events represent the blink of an eye, but they also establish the Texas coast as a relatively young and perhaps unusually

Figure 12.3. The waves rolling shoreward in this image are breaking at a slight angle to the shoreline, indicating that the longshore current flows to the right. Photograph by Sandra S. Chapman.

dynamic ecological frontier. The plants and animals that invaded and successfully colonized the new coastal habitats evolved elsewhere; hence they have had only 5,000 years to solidify the complex interrelationships characteristic of older ecosystems. Thus, for future generations, some of the portraits that follow may indeed be little more than yellowed snapshots in a dusty album of memories.

GULF BEACHES

Windswept beaches rim the Gulf along the entire Texas shoreline, but their sands originated far inland, transported to the shore as river-borne sediments. Currents in the Gulf eventually deposit the suspended sediments near the coast, although waves make the final delivery to the shoreline. After traversing the Gulf, waves approaching the gently sloping seafloor along the coast increase in height and steepness as the water is forced upward. While the bottom of the wave drags and slows, the crest continues to move at its original speed and stumbles forward as a "breaker" (fig. 12.3). This action lifts and redeposits the sediments lying beneath the foamy breakers, building an offshore sandbar. Once past the bar, the subdued wave continues transporting loose sediments shoreward before exhausting itself as a ripple on the shoreline.

Because the Texas coast scribes an irregular east-to-south arc, waves usually break at an angle to the shoreline and generate a longshore current.

Figure 12.4.
Coppice dunes
originate behind
obstacles that
block blowing
sand, after which
Seaoats, Goatfoot
Morning-glory,
and other plants
gradually stabilize
the loose sand.

The longshore current along the Upper and Middle Coasts generally drifts southward, whereas it flows northward along the Lower Coast. Longshore currents redistribute sediments and mollusk shells, and where the opposing currents meet—at a segment of Padre Island known as "Big Shell"— sand and shells accumulate. Over time, the finer sand grains here blow inland, leaving behind the larger shells and their fragments that distinguish this beach.

Sand accumulates on the beachfronts where waves die, forming a slight berm at the high tide line. During low tides, onshore winds drive dry sand from the berm to the backshore. Any obstacle on the beach—driftwood, debris, plants—that impedes the wind allows the sand to settle in a mound behind the barrier. Some mounds enlarge to become coppice dunes, which may be colonized and stabilized by hardy pioneer plants such as Goatfoot Morning-glory, Beach Tea, Seashore Dropseed, and Sea Purslane (fig. 12.4). Strong winds periodically shift loose sand farther landward on the backshore, where larger dunes merge into a foredune ridge that often extends along the entire length of a beach or barrier island. Grasses that trap blowing sands—notably Seaoats and Bitter Panicum—often crown foredunes on the Texas coast. As the entrapped sand buries their bases, the plants respond with upward growth but also expand their rhizome and root systems, which anchor the dune. Unfortunately, human disturbances such as foot or vehicular traffic can damage this vegetation, often producing wind-generated gaps, or "blowouts," in the barrier, thereby lessening its protective function (fig. 12.5).

Waves dissipate much of their energy in the surf zone but continue landward into the swash zone—the intertidal realm on the beach where their travels end. The water then retreats in a veiled rip current (or "undertow") that flows swiftly seaward beneath incoming rollers. Lunar tides and wind direction frequently alter the position of offshore bars and regularly expose parts of the swash zone to sun and heat.

At first glance, the relentless pounding and scouring of waves would appear to preclude life in the surf and swash zones, but hundreds of species in fact thrive in these environments. Beachgoers commonly see small, short-legged Sanderlings and Snowy Plovers rushing seaward following the wash of a retreating wave, and then as quickly returning shoreward in advance of an incoming surge—travels that essentially mark the limits of the swash zone where these and other shorebirds busily search for food exposed by the endless cycle. The interstitial spaces between sand grains harbor meiofauna—bacteria, foraminifera, diatoms, and other minute forms of life. Larger

Figure 12.5. Disturbances that eliminate plant cover render dunes susceptible to wind-generated "blowouts," thereby lessening their protective value.

organisms—macrofauna—such as filter-feeding Variable Coquina, polychaete worms, and Mole Crabs burrowing in the sandy swash zone provide other foods. Black Skimmers glide silently just above the swash zone, their lower mandibles slicing through the shallow backwash to capture tiny fish feeding on food particles in the disturbed water, and Least Terns, seeking the same prey, fearlessly dive headlong into dying waves. Willets, Black-bellied Plovers, and other longer-legged shorebirds wade chest-deep probing for Smooth Duck Clams, worms, and other foods in the sands beneath the churning water (fig. 12.6).

At the upper limits of the intertidal zone, flotsam forms a wrack line composed of seashells, stranded seaweeds, driftwood, discarded trash, and the bodies of dead marine animals, all of which accumulate to bake and decay in the sun. During the day, flies, isopods, gulls, and other scavengers feast on this wealth of organic matter, and Ruddy Turnstones uncover prey as they flip shells and poke through clumps of seaweed. At night, Coyotes rummage in the wrack line, joined by Ghost Crabs that emerge from backshore burrows, first wetting their gills in the surf.

Beach profiles constantly change. Waves generated by hurricanes or winter storms generally erode shorelines, which reduces the width of the backshore, and sometimes storm tides cut deeply into the foredune ridge. After heavy rains inland, swollen rivers transport more sediments to the sea and ultimately replenish the beaches. In spring, the beaches on the Gulf Coast attract five species of sea turtles that lay their eggs in nests constructed in the sandy backshore. Of these, Kemp's Ridley Sea Turtle is the most common, but all species of sea turtles remain endangered.

BARRIER ISLANDS AND PENINSULAS

Narrow fingers of sand—barrier islands and peninsulas—parallel much of the coastline from East Bay (Galveston Bay) to the Rio Grande and form the outer edges of numerous bays and lagoons. Foredune ridges on these landforms shield the coast from storm-generated waves and tidal surges, but the barriers themselves are fragile, and strong storms often erode the dunes (fig. 12.7).

Although geologists do not agree on how barrier islands and sandy peninsulas form, those along the Texas coast apparently develop by spit accretion. This hypothesis suggests that sand spits form across the mouths of drowned river valleys (bays) from sediments deposited by longshore currents. Over time, the spits become peninsulas

Figure 12.6. A Willet searches in the surf zone for exposed marine invertebrates (top left). Seaweeds and the bodies of marine organisms such as By-the-Wind Sailors accumulate at the wrack line (top right), providing food for Ruddy Turnstones, Sanderlings, and other shorebirds (bottom left). Near dusk, a Coyote slinks across coppice dunes to forage at the wrack line on the beach (bottom right).

that continue elongating until they block the rivers. Then floodwaters or storm surges breach the dam, splitting the former spit into one or more islands. Another hypothesis proposes that offshore bars in the surf zone eventually accumulate enough sediment to emerge as islands.

Because of their constant exposure to offshore winds, foredune ridges steadily grow until they reach heights where sand can no longer be lifted to the crest. In time, sand blown from the foredunes accumulates leeward, creating rows of lower dunes separated by shallow swales known as slacks; dense mats of grasses and forbs develop in these protected areas. Storm surges and high waves associated with tropical storms or hurricanes sometimes cut wide channels ("washover passes") through the foredune ridge. When this occurs, the surge carries large quantities of sand to the back side of the beach, forming "washover fans" that smother vegetation and ponds far inland. Sediments deposited by longshore currents

usually fill in these passes a few days later, whereas dredged channels must be maintained by jetties and regular dredging.

Leeward of the dune systems, broad "flats" appear as a mosaic of wetlands and barren sands interspersed with lush carpets of Little Bluestem and Gulfdune Paspalum accented by sedges, prickly pears, wildflowers, and shrubs. Seaside Heliotrope, White-tipped Sedges, spikerushes, and Southern Cattails surround the edges of freshwater or brackish ponds and swales where Ruddy Ducks and American Coots often appear in winter. In summer, Black-necked Stilts conceal their skimpy nests in the sedges and Green Treefrogs perch on cattails, always alert for Desert Massasaugas. Unvegetated "marching dunes" migrate across the flats and bury meadows and ponds in their path. Pale-colored Gulf Coast Kangaroo Rats forage along dune margins near dense vegetation where they can escape Barn Owls and other predators (fig. 12.8). The grassy flats provide homes for a

Figure 12.7. Storm waves eroded, but did not breach, these foredunes on Padre Island (top). Hurricanes and other strong storms generate high tides and large waves that break down the foredune ridge and create washover passes that later close across the beachfront (bottom). Sediments deposited by longshore currents refill most passes within days after a storm.

Figure 12.8. Grassy flats blanket most barrier islands, but windswept openings in the dense vegetation provide habitat for a variety of plants and animals (top). This group of stunted Live Oaks (bottom), affectionately known as Padre Island National Forest, existed for many years as the last survivors of a maritime forest that once covered the northern part of the island. Like their predecessors, these remnants eventually yielded to the harsh environment.

variety of animals, including Keeled Earless Lizards, Western Diamond-backed Rattlesnakes, Eastern Moles, Texas Pocket Gophers, and Black-tailed Jackrabbits. Coyotes and American Badgers dig cave-like dens into the leeward sides of vegetated dunes.

Padre Island is the largest of the Texas barrier islands. It is often connected to its counterpart, Mustang Island, but the two are now separated by a jettied artificial channel, Packery Channel. Historically, these two islands have been separated and joined many times by shifting natural channels; together they form the world's longest barrier island and, by area, the second-largest island in the contiguous United States. The island stretches 115 miles from Aransas Pass, the jetty-protected channel providing access to Corpus Christi Bay, southward to Brazos Santiago Pass, which separates it from Boca Chica and the Rio Grande. An elongated lagoon, the Laguna Madre, lies between Padre Island and the mainland. Near its midpoint, a dredged pass—the Mansfield Channel—provides for the exchange of Gulf seawater and navigable access to the upper portion of the Lower Laguna Madre and Baffin Bay. Long ago, a maritime forest of Live Oaks likely covered the middle of Padre Island, but today only a few dense mottes of stunted oaks still crown some low, isolated dunes at its northern end. Stands of larger oaks once extended down the center of the island but have gradually

disappeared, probably victims of drought or saltwater overwash during hurricanes.

During World War II, scientists briefly considered Padre Island as a potential location for testing the first atomic bomb. Fortunately, they eventually selected White Sands Proving Grounds in New Mexico as the test site, but for many years the US Navy practiced aerial bombing on the middle of the island. (Most bombs contained flour or other nonexplosive materials to mark locations of "hits" relative to a target.) Both ends of the island have been developed with numerous hotels, boat channels, and residential sites, but a central 70-mile segment is protected as Padre Island National Seashore. Mustang Island State Park, located south of Port Aransas, and an adjacent Nature Conservancy Preserve limit development on 4,630 acres, including 5 miles of coastline.

COASTAL MARSHES

Marshes along the Texas coast arise in protected areas bordering bay shorelines or on the inner margins of peninsulas and barrier islands. The emergent vegetation in these wetlands varies from narrow strips less than 20 feet wide to extensive stands covering large areas (fig. 12.9). Whereas periodic flooding influences all of these marshes to some extent, the composition of the plant communities at each location develops in relation to local conditions—the duration and frequency of flooding by saltwater, freshwater, or some combination thereof.

Freshwater marshes, or "fresh marshes," along coastal rivers and creeks occur more widely on the Upper Coast than elsewhere, especially west of Port Arthur in Jefferson County. Dense stands of Saltmeadow Cordgrass, spikerushes, and Maidencane dominate these habitats, while mats of American White Water Lily cover open areas with calm water. Brackish water may invade some areas when high tides back up the inflow of freshwater, and the dominant species then include Smooth Cordgrass, Marshmillet, Maidencane, Camphor Daisy, and bulrushes.

An extensive system of freshwater, brackish, and saltwater marshes exists in association with the Chenier Plain, which stretches from the southwestern Louisiana coast to Galveston Bay. The plain began forming more than 3,000 years ago when the mouth of the Mississippi River shifted westward and periodically carried huge loads of sediment to the Gulf. Marsh vegetation developed on shoreline deposits of these sediments, but at times when the sediment loads diminished, sandy beaches formed at these sites. The cycle of deposition and redistribution produced a series of high sandy or shelly ridges—cheniers—typically separated by 1–2 miles of salt or brackish marshes. *Chenier*, which in French means "oak," recognizes the stands of Live Oaks that dominate the ridges.

The Chenier Plain intercepts the flow of the Sabine and Neches Rivers, creating a 90,000-acre estuary known as Sabine Lake on the Texas-Louisiana border (fig. 12.10). At the northern end, the two rivers contribute enough flow to sustain freshwater marshes. However, brackish or salt marshes develop in the low areas between the sandy ridges nearer the coast where they connect with tidal channels at Sabine Pass, a 5-mile-long outlet to the Gulf.

Smooth Cordgrass and other salt-tolerant species (halophytes) generally dominate intertidal salt marshes. Plant diversity is relatively low in these "low marshes" because the flora must tolerate saline conditions as well as complete or partial inundation and anoxic root zones. On the mud bottom, Marsh Periwinkles graze on periphyton, a velvety carpet of epiphytic algae, diatoms, bacteria,

Figure 12.10. Sabine Lake originated when a series of cheniers—the sandy ridges visible between the lake and the Gulf of Mexico—dammed the mouth of a shallow bay and blocked the flow of the Sabine and Neches Rivers, which can be seen at the northern end of the lake (top). Marshes develop along tidal channels between tree-covered dunes (bottom). Photographs by the National Oceanic and Atmospheric Administration (top) and Brian R. Chapman (bottom).

and other microorganisms. To escape predation by Blue Crabs and Red Drum that forage in the marsh during high tides, these small snails migrate up cordgrass stalks and remain attached to leaves above water until the tide ebbs. Plicate Hornshells, also small snails that feed on periphyton, remain dispersed on the marsh floor at high tide but cram by the hundreds into shallow pools at low tide.

As elevations gradually increase landward, low wet marsh usually grades into drier "high marsh" on sites that often remain free of tidal influences for extended periods. Halophytes at the margins of these marshes develop in zones corresponding to the salt tolerance of each species. Beds of glassworts usually dominate the most saline soils, followed by ascending zones of Saltwort, Sea Blite, Sea Lavender, Carolina Wolfberry, and Sea Oxeye (fig. 12.11). Around pools, these zones often form a target-like pattern of concentric circles. Elsewhere, the purple-flowered spikes of Seaside Heliotrope embellish tangles of Saltgrass and Salt-flat Grass, and Gulf Cordgrass abounds in low, poorly drained swales near the coast.

The Whooping Crane, a critically endangered species, stands as the "poster child" for the value of coastal marshes in Texas. The species breeds in the wetlands of Wood Buffalo National Park in northern Canada and migrates south to overwinter in the salt marshes of Aransas National Wildlife Refuge and elsewhere on the central Texas coast— a journey of more than 2,400 miles. Present in Texas from November through March, groups of adults with juveniles disperse across 22,500 acres of protected salt flats and marshes to feed on Blue Crabs, minnows, other small animals, and berries. From a low of just 18 birds in the 1930s, the population now numbers more than 300 birds— testimony to the effectiveness of coordinated international management of the species (fig. 12.12).

OPEN BAY BOTTOMS

The shallow waters in most Texas bays remain too turbid to establish submerged, rooted vegetation; hence detritus from rivers and neighboring salt marshes, mudflats, and seagrass beds forms the base of food chains in these habitats. Some bays develop dense benthic communities composed mainly of bivalves, snails, and their predators. The complexity of these communities depends on the amount of freshwater inflow, and the constant fluctuation may not favor diversity. With suitable conditions, some mollusks, such as the Dwarf Surf Clam, pack tightly into the sediment and leave no space for other organisms. These dense

Figure 12.11. Glassworts emerge as the first zone of halophytes ascending the shoreline behind the Snowy Egret.

Figure 12.12. A family of Whooping Cranes wades in a shallow estuary at the Aransas National Wildlife Refuge. Note the juvenile's distinctive head plumage. Photograph by David Rein.

concentrations serve as significant food sources for Black Drum and benthic-feeding waterfowl such as Lesser Scaup.

SEAGRASS MEADOWS

Extensive underwater beds of seagrasses can logically be called "meadows," but the term "seagrass" represents a slight misnomer. These plants in fact do not belong in the family of true grasses (Poaceae), nor are they "seaweeds" (marine algae), another common misconception. Instead, seagrasses are highly adapted flowering monocots that complete their entire life cycle underwater, and like true grasses, they produce true roots, stems, and leaves with photosynthetic pigments and maintain high primary productivity. Seagrasses occur in most Texas bays and lagoons, but the most extensive seagrass meadows develop in the shallow marine systems south of Matagorda Bay.

Of the 50 or so species of seagrasses identified worldwide, only 5 occur in Texas bays and lagoons. By far, Shoal Grass has the most widespread distribution in Texas, becoming especially abundant in the Laguna Madre (fig. 12.13). Turtle Grass occurs in large, dense patches in Redfish Bay and the southernmost part of the Laguna Madre, but the other three species, Widgeon Grass, Clover Grass, and Manatee Grass, develop in relatively small patches, often in association with Shoal Grass.

Seagrass meadows buffer water movements, reduce turbidity, stabilize lagoon bottoms, and promote sedimentation. Their periphyton-coated leaves and the floating suspensions of drift algae restrained among the dense foliage add to the overall productivity of the meadows (see fig. 12.13). Research elsewhere has revealed that the primary productivity from these epiphytic sources contributes more carbon to the food chain than do the seagrasses themselves, though this has not been studied in Texas bays. Few herbivores feed directly on seagrasses (exceptions include Striped Mullet, Sheepshead, Redhead ducks, and Green Sea Turtles), but many species consume drift algae and periphyton. The Muddy Cerith, for example, grazes on algae growing on benthic sediments, while the tiny Grass Cerith and Beautiful Caecum climb grass stalks to forage on epiphytic algae. These snails, among others, enter the food chain as prey of Blue Crabs and Black Drum. Bay Scallops abound on the floor of seagrass meadows and, when necessary, rapidly clap their valves to scoot away from predators.

Several commercially important nektonic organisms, notably Brown, Pink, and White Shrimp, complete all or part of their life cycle in seagrass meadows. Postlarval and juvenile Spotted Seatrout and Red Drum also remain in or near the protection of seagrass meadows for up to five years before moving into the Gulf. The structural complexity of these habitats provides optimal concealment for many small, cryptic fishes, and

Figure 12.13. Periphyton coating the leaves of Shoal Grass (shown here) enhances the productivity of seagrass beds (top). Manatee Grass (bottom) and other seagrasses often occur in small patches. Photographs by Roy L. Lehman.

some, such as the Lined Seahorse and Dusky Pipefish, complete their entire life cycle sheltered in seagrass. Large schools of Tidewater Silversides, Gulf Menhaden, Bay Anchovy, and Rainwater Killifish flash through the waving foliage of seagrass meadows as they forage on plankton and periphyton, thus providing a link in the food chain between primary producers and higher consumers.

Dead foliage adds nutrients to the detrital food chain in seagrass meadows and adjacent sites. Decomposition enriches the meadows, and storms and high tides transport rafts of decaying leaves elsewhere, where they often accumulate in smelly mounds on shorelines. The overall productivity of seagrass meadows in Texas ranks second only to salt marshes and compares with coral reefs and oceanic upwellings. At times, however, the

INFOBOX 12.1. COLORED WATERS
Red and Brown Tides

In the Gulf of Mexico, sediments churned up in the surf zone along the Texas coast often add a yellowish tint to the nearshore waters. Under certain conditions, however, "blooms" of algae color these waters red or reddish brown, producing a condition known as "Red Tide." The algae responsible, a dinoflagellate known as *Karenia brevis*, not only tint the water but also release toxic compounds that generate substantial fish kills and also cause the demise of sea turtles, birds, and Atlantic Bottle-nosed Dolphins.

The Spanish explorer Cabeza de Vaca, who was marooned in Texas after a shipwreck in 1528, was the first to report Red Tide and fish kills. The cause of the phenomenon then was likely sand from dust storms in the Sahara Desert of Africa. Carried aloft across the Atlantic, the iron-rich dust nourished nitrogen-fixing cyanobacteria when it fell into the Gulf of Mexico. The influx of iron allowed these organisms to flourish and convert massive amounts of dissolved nitrogen into nitrates, creating optimal conditions for dinoflagellates. The frequency and distribution of Red Tide blooms has increased steadily during the past 100 years, concomitant with increases in agricultural and urban runoff, especially phosphorus-bearing fertilizers, and other types of pollution carried downstream by rivers. The metabolic processes of *Karenia brevis* release potent neurotoxins known as brevetoxins, which contaminate the waters and kill marine organisms. As fish decompose, they become new sources of phosphorus and nitrogen and prolong the Red Tide outbreak.

Harmful phytoplankton in the Laguna Madre occasionally generate blooms known as "Texas Brown Tides." As the name suggests, the blooms create a brownish discoloration in the water, but the toxins that are released kill only juvenile fish—adult fish and certain invertebrates are not affected. Brown Tide blooms sometimes reach densities great enough to restrict light transmission, thus blocking photosynthesis and killing seagrasses.

Red and Brown Tides are not actually associated with tidal conditions and do not always color the water; therefore, most scientists refer to outbreaks of these organisms as "harmful algal blooms" (HABs). State agencies monitor the occurrence of HABs along the Texas coast because some humans experience burning eyes, irritation of the nose and throat, and breathing difficulties when exposed to airborne toxins released by the algae. The agencies also restrict shellfish harvests in areas near HAB outbreaks because oysters and clams concentrate brevetoxins in their tissues and, when eaten, can cause human illness or death. On a more positive note, research designed to develop better ways of identifying the toxins led to the discovery that derivatives of the poison may eventually treat cystic fibrosis, a severe lung disorder.

Dead fish and other marine organisms litter the Padre Island shoreline following a red tide "bloom" in the Gulf of Mexico.

concentrated nutrients nourish "blooms" of harmful algae that cause fish kills similar to those resulting from "Red Tides," though this aspect has not been studied (infobox 12.1).

OYSTER REEFS

Many species of mollusks dwell singly and widely scattered on the bottoms of estuaries. Conversely, Eastern Oysters build large, multifaceted reefs where more than 300 other marine species find refuge, anchorage, and food. Often the only hard substrate in soft-bottomed bays, an oyster reef creates a surface area about 50 times greater than a mud bottom, with thousands of microhabitats offering havens for marine worms, amphipods, shrimp, crabs, and small fishes. Winds and tides occasionally expose parts of oyster reefs, which then become temporary feeding locations for shorebirds and resting sites for pelicans, gulls, and terns. Oyster reefs once flourished in every shallow bay in Texas where salinities normally fluctuated between 10 and 30 parts per thousand and water temperatures rarely exceeded 68°F–86°F. Eastern Oysters can withstand higher temperatures and

TEXAS GULF COAST

salinities, but not for long, and thus do not survive in the hypersaline Laguna Madre.

As a bivalve, the Eastern Oyster features two opposing shells (valves) connected by a hinge. A single, large adductor muscle clamps the valves together when adverse conditions pose threats, including attacks from predators such as Blue Crabs, Gulf Stone Crabs, and American Oystercatchers. Eastern Oysters open their valves slightly to filter food and circulate about 50 gallons of water across their gills each day. Most oyster reefs lie perpendicular to currents associated with tidal movements, which facilitates their exposure to nutrients and suspended particles. Inside the mantle, cilia direct plankton and other tiny food particles captured on the mucus-coated gills to the mouth, whereas larger or inedible materials are sorted out and expelled. Oysters represent a keystone species because of their dual roles of maintaining water quality and clarity and creating habitat for other marine creatures. However, oysters cannot distinguish between the types of particles they filter. Consequently, their tissues may concentrate bacteria or heavy metals that pollute marine systems, and oysters harvested from contaminated waters may infect humans with diseases such as typhoid fever and hepatitis.

After reaching sexual maturity as males—usually at one year—Eastern Oysters commonly transform into females after two to three years. Warm water in late June to late October stimulates spawning, which begins when a few oysters at a reef release their sperm or eggs. Thus prompted, the remaining oysters follow suit, resulting in a cloud of gamete-filled water near the reef (a single male may emit up to 2.5 billion sperm and a female often releases more than 5 million eggs). After fertilization, zygotes undergo three planktonic larval stages, the last of which settles onto the bottom in search of a hard surface, or "cultch." Only about one larva out of a million fertilized eggs survives to this point, but those that do "set" by gluing themselves to a cultch surface, where they remain permanently. Now called "spats," the juveniles grow in three months from the size of a pepper grain to about the size of a dime. The shells of some oysters attain lengths of 6 inches, and many develop irregularly, shaped by

topography as they compete for space on a crowded reef.

Humans have long enjoyed eating oysters, as shown by the shell middens left by Native Americans along the Texas coast, but other predators and parasites also target oysters. Black Drum locate living oysters with their sensitive chin barbels and then crush the shells with pharyngeal teeth to obtain the meat inside. American Oystercatchers pry open the valves with their heavy, chisel-shaped bills. Predatory marine snails such as Hays' Rocksnail bore through oyster shells with their rasp-like radulas (mouthparts) and chemical secretions and then insert and extract food through a tubular proboscis. Similarly, tiny parasitic snails called Oyster Mosquitoes, positioned in seams between the valves, suck blood from oysters by inserting needlelike tubes into the soft tissues.

Oyster reefs build slowly as successive generations continually attach to the shells of their ancestors. Some reef complexes in Texas bays were quite large—in 1907, for example, one in Matagorda Bay covered 494 acres. Reefs of this size developed for more than a century, but many later disappeared or were so reduced that they no longer functioned as ecosystems. Dams on rivers reduced freshwater inflow into bays, thereby upsetting the critical salinity levels required for oysters to flourish. Siltation resulting from the construction of the Gulf Intracoastal Waterway (GIWW) smothered many reefs in the 1940s, but even greater losses resulted between 1922 and 1983 from commercial dredging on live and dead oyster reefs—the shells served many purposes, including pavement for roads, parking lots, and neighborhood driveways (fig. 12.14).

Efforts are underway to restore oyster reefs in several bay systems along the Texas coast. The Texas Parks and Wildlife Department has deposited more than 79,000 cubic yards of cultch (river rock and crushed limestone) at eight sites in Galveston Bay and Sabine Lake. An ingenious research project funded by the Texas General Land Office involves both private and public cooperation—three well-known restaurants in Corpus Christi save discarded oyster shells, which are collected by students from the Harte Research Institute at Texas A&M

Figure 12.14. Oyster shells dredged from Nueces Bay are being transferred from a barge to a truck for transport to a road-building site in 1966.

University–Corpus Christi (infobox 12.2). After aging for six months at a site owned by the Port of Corpus Christi, a project participant, the shells are packed in mesh bags and submerged at sites in the Mission-Aransas Bay estuary. The project, entitled "Sink Your Shucks," will create reefs of more than 5 acres as new habitat for fish, crabs, and a flourishing community of marine life.

LAGUNA MADRE
A shallow sliver of water known as the Laguna Madre, or "Mother Lagoon," separates Padre Island from the mainland. Extending southward approximately 115 miles from Corpus Christi Bay to the Rio Grande Delta, the lagoon is divided near its midpoint by an 11-mile-long sandy barrier. Formed by storm washovers and windblown sands, the low, flat land bridge is locally known by several names, including Saltillo Flats, Salt Flats, and the Laguna Madre Flats; it connects with the Sand Sheet in the Coastal Prairie (chapter 11). The northern half of the lagoon, Upper Laguna Madre, stretches for 47 miles with an average width of about 3.7 miles. Slightly longer and wider, the Lower Laguna Madre, isolated in the past, now connects with the

Upper Laguna Madre via the GIWW, a navigational channel that crosses the Saltillo Flats.

The Laguna Madre lies in a semiarid region with rare and unpredictable rainfall, long and hot summers, and mild winters. Moreover, evaporation exceeds precipitation, and long droughts limit the inflow of freshwater. Consequently, the Laguna Madre generally remains hypersaline, one of only six such lagoons in the world. Whereas the water salinities in the Gulf average 35 parts per thousand (ppt), those in the lagoon normally exceed 42 ppt. After long droughts, salinities sometimes reach 150 ppt, and once they attained a record high of 295 ppt. However, when torrential rains associated with tropical storms or hurricanes flood the region, the salinity in the lagoon approaches freshwater levels (less than 5 ppt). Extremely high salinities are now less common than before because of several dredged channels that connect with the Gulf and increase water circulation.

Except for dredged channels, water depth in the Laguna Madre averages about 3.3 feet. The shallow waters remain relatively clear, which allows enough sunlight penetration to develop extensive seagrass meadows. These highly productive environments provide nursery areas for more species of finfish and shellfish than anywhere else along the Texas coast. Despite the high salinities, large numbers of resident and migratory waterbirds also depend on the resources of the Laguna Madre. For example, nearly 80 percent of the Redhead ducks in North America regularly overwinter in the lagoon along with large populations of Lesser Scaup, Northern Pintails, Red-breasted Mergansers, and Buffleheads. Foraging shorebirds crisscross exposed salt flats in search of invertebrate foods.

Only a few natural islands existed in the Laguna Madre prior to 1940, and these provided nesting sites for large numbers of herons, egrets, gulls, and terns. White and Brown Pelicans nest on

North and South Bird Island in the Upper Laguna Madre (fig.12.15). The nesting colony of White Pelicans represents an example of a discontinuous distribution—the main population nests in far-off Great Salt Lake, Utah, and in Canada's prairie provinces. During the 1940s, many new islands were formed when dredges cut channels for the GIWW, access for oil and gas exploration, and aquatic runways for US Navy seaplanes. The process piled the dredged material, or "spoil," along the edges of the channels, where it dried and gradually developed vegetative cover. Some of these islets attracted squatters who built fishing cabins and piers (now mostly removed), but many became rookeries for nesting waterbirds.

Figure 12.15. White Pelicans nest in open areas on South Bird Island, a natural island in the Laguna Madre (top). Islands of dredged material also serve as rookeries for nesting waterbirds (bottom). The Gulf Intracoastal Waterway in front of these islands was dredged through extensive seagrass meadows (dark areas).

Much of the lagoon's shoreline remains sparsely populated and largely undeveloped, including large ranches, and a significant area falls under the protection of Padre Island National Seashore and Laguna Atascosa National Wildlife Refuge. As a result, the Laguna Madre remains reasonably unspoiled, and the dredged channels provide access to a remarkable marine ecosystem that exemplifies nature at its best.

WIND-TIDAL FLATS

A series of featureless flats borders the backshores of barrier islands and lagoon shorelines from Aransas Pass to Port Isabel. These wind-tidal flats, sporadically inundated by windblown tides or storm surges but rarely by astronomical tides, generally stay dry and exposed. Because of high soil salinity, essentially no rooted vegetation survives on the sandy and muddy soils in these areas, but the shoreward margins of some flats support sparse stands of salt-tolerant vegetation such as glassworts and Saltwort. Although the surface appears barren, a 0.4- to 1.2-inch-thick crust of filamentous cyanobacteria, or blue-green algae, carpets the flats and binds the sediments.

Primary productivity of the leathery crust almost equals that of the seagrass meadows. Where winds frequently drive water onto the flats, tenacious communities of benthic invertebrates convert the algal biomass into animal protein that passes through the food chain to higher consumers, including Blue Crabs, bottom-feeding fish, and shorebirds. Although not well documented, the invertebrate community of wind-tidal flats includes such obvious species as Fiddler Crabs. Wind-tidal flats provide significant foraging areas for migrating and wintering shorebirds and wading birds. At times, flocks with hundreds of Snowy Plovers, Piping Plovers, Least Sandpipers, or Willets feed or rest at these sites, joined by lesser numbers of Hudsonian Godwits, Long-billed Curlews, and Sanderlings. Facing the wind in tight formations, squadrons of Royal Terns and Black Skimmers often loiter on the flats. When submerged, the flats attract Great Blue Herons, Tricolored Herons, and Reddish Egrets that stalk fish in the shallow water.

MANGROVE THICKETS

Winter freezes normally restrict mangroves to tropical estuaries, and, of the four species of mangroves occurring on the Texas coast, freezing weather indeed periodically eliminates three— Red Mangrove, White Mangrove, and Buttonwood. Black Mangroves, however, tolerate a slightly wider range of temperatures and represent the most abundant of the group in Texas (note here that "mangroves" is a collective term for many unrelated taxa). Small mangals (mangrove forests) have survived for centuries in the subtropical climate at South Bay (Cameron County), and since the 1930s, a population of Black Mangroves has persisted near Harbor Island in Aransas Bay (Aransas County; fig. 12.16). The density and extent of this population increase during intervals with warm winters but contract when subfreezing temperatures kill many

Figure 12.16. A thicket of Black Mangrove lines the Aransas Channel, shown here in front of the Matagorda Island lighthouse (top). Dense stands of leafless pneumatophores jutting upward from the substrate stabilize the plants and provide oxygen to their submerged roots (bottom).

inundated by high tides, where they face little competition from other woody species. Most mangroves are stabilized by numerous prop roots, which also transport oxygen to roots buried in anaerobic mud, but Black Mangroves have few of these. Instead, an extensive fan of buried roots stabilizes these plants. Gaseous exchange is enabled by dense beds of pneumatophores—soda straw–like "air roots" surrounding each plant and extending upward from the substrate 4–12 inches above the water level at high tide. Black Mangroves eliminate salt through specialized glands on their leaves. The upper surfaces of their leaves are green and shiny, but the undersides have a dense fleece of grayish hairs on which salty excretions collect and crystallize before dropping off. Black Mangroves produce terminal clusters of yellow flowers, which are pollinated by bees. Their fruits carry a single seed, which can float in seawater for more than a year without rooting.

Like saltmarsh communities, mangrove thickets provide vital sources of energy. Structurally, the networks of pneumatophores form protected nursery areas for numerous marine organisms, while the crowns of the shrubs serve as convenient perches for Brown Pelicans, Neotropic Cormorants, and Laughing Gulls. Mangroves also stabilize shorelines and serve as buffers against storm-generated waves, and some observers suggest that with continued climate change, mangroves may displace salt marsh communities along the entire coast. Mazes of pneumatophore spires trap sediments more efficiently than salt grasses, and mangrove root systems elevate and change shoreline profiles. Furthermore, mangroves may absorb more carbon dioxide than the marsh grasses they might replace—a prospective benefit in a world threatened by increasing concentrations of greenhouse gases, and the subject of current research.

MARITIME FORESTS

Imagine the plight facing a tiny Indigo Bunting—one of thousands of Neotropical migrants leaving their wintering grounds in South America each spring. With luck and a gentle tailwind, the little bird will reach land again on the Texas coast after a nonstop 18-hour flight across the Gulf of Mexico.

of the shrubs. During the past 85 years, mangroves gradually expanded northward into salt marshes along Nueces Bay (Nueces County), Matagorda Bay (Calhoun County), and Pelican Island in Galveston Bay. Many attribute this range extension to climate change and expect it to continue along the Texas coast in the years ahead.

Few woody plants tolerate seawater, but mangroves thrive on muddy soils periodically

Figure 12.17. Typical of other Neotropical migrants, an Indigo Bunting (left) rests and refuels in the shelter of a salt-trained maritime forest (above) after a grueling, nonstop flight across the Gulf of Mexico. Photographs by Thomas G. Barnes, US Fish and Wildlife Service (left), and Brian R. Chapman (above).

Unfortunately, these migratory journeys often encounter energy-sapping headwinds or late cold fronts. Some overfatigued birds lose their battle with the elements and later wash ashore as brilliantly colored flotsam; others "fall out" on sandy beaches or land exhausted on boats or offshore drilling rigs. If luck holds, however, the bunting and its companions reach the coast in search of trees—the prominent feature in a landscape of marshes and bays (fig. 12.17). The weary birds, seeking food, water, and shelter, tuck into maritime forests for rest and fuel for the next leg of their long journey. When recovered, the Indigo Bunting, like many other species, leaves this five-star sanctuary to continue northward, but for some, the maritime forest serves as their final destination and nesting habitat (infobox 12.3).

Maritime forests hug Atlantic and Gulf coastlines from Virginia to Texas. Along the shoreline, constant exposure to salt-laden winds sculpts a characteristically sloping canopy. Carried inland by onshore winds, the salt aerosol kills generations of terminal buds on the windward side of the canopy, whereas those on the leeward side continue growing normally. The airborne salt also imports calcium, potassium, and other cations that enrich nutrient-poor sandy soils in maritime forests.

Maritime forests on the Texas coast develop on sites such as ancient beach ridges, the mouths of natural rivers and streams, barrier islands, and Native American shell middens—locations where the elevation allows trees to develop a solid root structure. Live Oaks, the dominant species of maritime forests, are particularly well adapted to their role as shoreline trees. The gnarled and knotty structure of the wood strengthens their limbs to withstand storm winds without breaking, and their thick, heavily cuticled leaves resist salt damage and the erosive effects of windblown sand. The dense, tightly closed canopy provides a protective, salt-free environment for understory vegetation, including Saw Palmetto, American Beautyberry, Red Bay, and Yaupon. Vines such as Virginia Creeper and Mustang Grape reach for the sun and sometimes cap parts of the canopy.

Contiguous stands of bottomland hardwood forests along major rivers and low-order streams near the coast likewise provide significant stopover sites for migrating land birds, and those within 12 miles of the coastline become especially important in this regard. Exhausted birds reaching a coastal refuge rarely move more than 1,650 feet from their initial landing site during a stopover.

Maritime forests along the Texas coast were never as extensive as those once occurring along the Atlantic shoreline—and they are even less so

INFOBOX 12.3. CONNIE HAGAR
The Bird Lady of Rockport (1886–1973)

Martha Conger Hagar (née Neblett), although born and raised in a well-educated family imbued with a genteel appreciation of literature, history, and the fine arts, nonetheless was something of a tomboy who relished contact with nature. Known as Connie, she and her father, the mayor of Corsicana, took long walks together on which he would point out the trees and herbaceous plants they encountered. A younger sister, known simply as Bert, later joined these outings, thereby beginning the sisters' long partnership as outdoor enthusiasts.

After a failed marriage to a naval officer, Connie eventually married—after years of courtship—Jack Hagar, a transplanted Bostonian who had moved to Corsicana to pursue his interests in oil and real estate. In the years that followed, the sisters started a nature club in Corsicana that met biweekly to study a range of interests from plants to stars, and of course birds. In 1934, Connie learned that the US Bureau of Biological Survey—the predecessor of today's US Fish and Wildlife Service—was seeking volunteers for a project that involved banding and counting birds—an opportunity not to be missed by Connie and Bert.

Meanwhile, the nature club entered into an affiliation with the National Audubon Society. Field trips were naturally very much a part of the club's activities, which often featured visits to the Big Thicket and far-off Big Bend National Park. Connie and Bert supplemented these adventures with their own activities, which included speaking to nature groups in nearby counties. Lacking the convenience of visual aids such as a slide projector to complement these lectures, Connie collected bird nests to use as props for acquainting the audience with the region's birds and their breeding ecology. The diminutive naturalist—she stood no more than five feet tall—maintained a busy schedule, but even in the field, she wore well-starched and ironed dresses, faithfully upholding her standards of what was "ladylike."

The sisters also went on their own field trips, perhaps the most memorable of which was a vacation in 1933 to the small coastal town of Rockport on Aransas Bay, where they marveled at the diversity of shorebirds. The experience proved unforgettable and they returned at every opportunity. However, these infrequent trips only whetted her appetite, and in 1935, Connie prevailed on Jack to pull up stakes in Corsicana and move to Rockport, where she immediately intensified her love and study of birds, aided immeasurably by the recent publication of the first of Roger Tory Peterson's famed field guides. Thus armed—and by now carrying her first pair of binoculars—Connie Hagar and her discoveries soon gained notice among the growing cadre of birders both in Texas and well beyond.

Connie added more than 20 species to the checklist of birds in Texas, including several migratory species, some of which were thought to be extinct. Among her discoveries was the migration of nine species of hummingbirds along the Texas coast. In all, she reported more than 500 species of birds in the Rockport area, which represents almost 75 percent of the entire avifauna between Canada and Mexico. Rockport is an important landfall for Neotropical migrants crossing the Gulf of Mexico each spring, an event that was a highlight in Connie's calendar of ornithological observations.

The roster of notables who visited her, many staying in the tourist cottages she and Jack owned and operated, included nature writer Edwin Way Teale, biologists Clarence Cottam and Harold Mayfield, many other ornithologists, presidents, and other officers of the National Audubon Society, and of course Roger Tory Peterson, who later published a field guide devoted exclusively to Texas birds that was duly enriched by Connie's observations. J. Frank Dobie and Roy Bedichek, both revered names in the literature of Texas, also visited her. Occasionally, a few of the well-known experts challenged her records but always left convinced of the validity of her sightings—she once pointed out a bird in question through a window just over the visitor's shoulder. Renowned photographer Albert Eisenstaedt also visited Connie, with the result that she appeared (in 1954) on the cover of *Life* magazine. For birders, Connie Hagar had become a star and Rockport a Mecca.

Connie Hagar received numerous honors, not the least of which was a special citation presented to her at the Audubon Society's national meeting in Corpus Christi in 1962. In 1943, the Texas legislature designated a wildlife sanctuary in her honor along the shoreline of Little Bay and adjacent areas at Rockport. Another refuge was established on the site where she and Jack owned rental cottages (now removed); in spring, the location is flush with newly arrived migrant birds. A local organization, Friends of Connie Hagar, is dedicated to her memory and contributions to ornithology as well as promoting public awareness, appreciation, and conservation of birds and their habitat in the Coastal Bend of Texas.

now, with only remnants remaining in isolated patches sandwiched between houses, gasoline stations, and golf courses. The "Big Tree," a huge Live Oak in Goose Island State Park near Rockport, stands as a sentinel for a remnant in Aransas County. Until the recent discovery of a larger specimen elsewhere, the Big Tree held the title of Texas State Champion Live Oak for nearly 35 years.

HIGHLIGHTS

ROCKS AND DOCKS

No natural rocky seashores exist on the Texas coast, except for several small relicts of serpulid worm reefs exposed on the shoreline of Baffin Bay. These consist of aggregated calcareous tubes of polychaete worms and formed when South Texas enjoyed a wetter climate and the lagoon was less saline. No longer growing, most of these reefs exist today in submerged patches that may be diminishing in size and distribution.

Despite a scarcity of naturally occurring rocky habitat, jetties constructed of large boulders or chunks of concrete, as well as bridge piers and other hard vertical surfaces, provide excellent substitutes (fig. 12.18). In particular, the rocky fingers shielding channels or shorelines experience ecological conditions mimicking those affecting rocky seashores along the Atlantic or Pacific Oceans and therefore develop comparable zones of biological assemblages.

Organisms inhabiting the uppermost zone—the supratidal zone—must endure periodic exposure to wind, sun, and extreme temperatures. Spray from breaking waves moderates these desiccating effects, but seawater washes the higher areas only when storms generate extremely high tides. An assortment of cyanobacteria, green algae, snails, limpets, barnacles, isopods, and crabs typically occupies the harsh supratidal zone.

Below lies the intertidal zone, where life is alternately inundated or exposed with each tidal cycle. Some of the organisms adapted to the supratidal zone just above intergrade with species in this area. To withstand the rigors of wave turbulence, organisms here either attach themselves firmly or move with the changing tide. Small, cryptically colored fishes known as blennies survive in small holes in rocks or narrow interstices.

Figure 12.18. Jetties constructed of granite rocks from the Llano Uplift on the Edwards Plateau protect many ship channels on the Texas Coast, as shown here at Port Aransas (top). The supratidal zone on these rocks, marked by snails, keyhole limpets, and green algae, mimics that occurring on rocky seashores elsewhere (bottom).

Lacking scales, these small fishes have a thick layer of mucus coating their bodies that enables them to move swiftly in and out of refugia. Tubelike sea anemones creep slowly to exposed locations where they firmly seat themselves and then wave a bouquet of tentacles equipped with stinging cells (cnidocytes). When touched, the cnidocytes discharge harpoon-like nematocysts that inject venom into prey or painfully discourage predators. The tentacles shift immobilized prey to the mouth and a gastrovascular tube for digestion; the same opening expels waste.

Figure 12.19. Red Lionfish, recent invaders of the Texas coast, although beautiful, quickly devastate fish and invertebrate communities; when mishandled, venom glands in their fins deliver a painful sting. Photograph by A. Sterne, provided by Michelle A. Johnston, Flower Garden Banks National Marine Sanctuary, National Oceanic and Atmospheric Administration.

Strata of different kinds of algae (known collectively as "seaweeds") develop on the rocks in keeping with the amount of light they require. Water turbidity and narrow tidal limits further affect the zonation of marine algae on hard surfaces. Green algae require more light and typically occur in the supratidal and upper intertidal zones, whereas brown algae, which require less light, extend below these to about 6 feet. Some brown algae exposed to relentless water turbulence reinforce their surfaces with calcareous material. Enough light extends deeper to allow red algae to prosper farther down, but some species occupy all three zones.

A recent invasion of Red Lionfish, native to the western Pacific Ocean, poses a threat to biodiversity on Texas jetties. Amazingly fecund, they spawn weekly throughout the year, and a single female may lay more than two million eggs annually. Lionfish are also aggressive and voracious—they can swallow prey up to two-thirds their size and often eat more than 30 juvenile fish and small crabs per hour. Waving in the current, their elongated multicolored fins are quite beautiful, but bright colors in nature often serve as warning signals (fig.12.19). Grooves on the fins of Red Lionfish contain venomous glandular

tissue capable of delivering a painful but nonlethal sting. Consequently, Red Lionfish have few natural predators, and their numbers continue to increase in the coastal waters of Texas.

WATERBIRD COLONIES

Someone walking through a forest in springtime might never spot the nest of a songbird, but it is not hard to locate the nests of most waterbirds. In suitable habitat, they concentrate their nests so that the distance between adjacent nests lies just beyond the "pecking distance" of each neighbor. Waterbirds can pack tens to hundreds of nests on shell-strewn shorelines or in copses of trees, where the noisy bustle and smell make these colonies hard to miss.

Several species often share a colony (also called a rookery), but slight differences nonetheless segregate the location of their nests (height in trees, shoreline substrate, or distance from water), and the immediate area surrounding each nest represents a vigorously defended territory (fig. 12.20). Colonial life accrues important benefits, of which safety in numbers may be foremost—hundreds of eyes stand ready to detect predators or invaders, including humans. Intruders experience vigorous attacks, including mobbing, the daunting "white rain" of aerial defecation, and the equally repugnant regurgitation of malodorous crop contents. Additionally, egg laying and other nesting activities within a colony typically occur at about the same time—a synchrony that limits the effects of predation. Predators can consume only a small percentage of the available prey during a constricted period; hence many eggs or nestlings survive. Furthermore, the short-term abundance of prey limits the size of predator populations.

Fish-eating waterbirds rely on a highly mobile food resource, so to gain information about the current location of their prey, nesting birds may monitor arrivals and departures of others in the colony. In other words, colonies may serve as "information centers." When food supplies become inadequate, some herons, egrets, and pelicans rely on a callous survival strategy. In these species, incubation begins when the first egg is laid; hence the nestlings hatch asynchronously, and the first to hatch grows for several days before the last egg

Figure 12.20. Great Blue Herons, Roseate Spoonbills, and three species of egrets share this colony in Saltcedars by placing their nests at different heights in the canopy (top). Royal Terns, Sandwich Terns, and Laughing Gulls segregate themselves in a mixed-species colony by selecting slightly different nest substrates and surroundings (bottom).

sites remain in use for many successive years—the result of strong "site fidelity"—the birds eventually abandon some locations when their nitrogen-rich droppings kill the trees in which they nest. The same unsanitary conditions may also encourage outbreaks of parasites and diseases. The raucous clamor and lure of abundant prey at a colony may actually attract predators, including humans, and led to dire consequences for some species late in the nineteenth century. Women's fashions of the period featured adornments of feathers, most notably the showy plumes of breeding egrets. To sate the demand, plume hunters raided waterbird colonies, especially in Texas and Florida, killing the adult birds but leaving the nestlings behind to starve. Once-thriving colonies soon became graveyards of empty nests. Fortunately, public outrage led to passage of the Lacey Act (1900), federal legislation that restricted interstate trade in wildlife (including feathers); development (in 1903) of the National Wildlife Refuge System, which initially protected areas with waterbird colonies; and a change in fashion styles. The waterbird populations eventually recovered, and today colonies can be observed—from afar during the nesting season—along many parts of the Texas coast.

AN IMPROBABLE FISH

Visitors to public aquaria often doubt that seahorses are really a type of fish. The confusion arises because their body shape and upright posture depart dramatically from the familiar torpedo-like profile common to most fish. Their name originates from the horse-shaped outline of their head, a morphological similarity also reflected in their generic name, *Hippocampus,* the Greek word for "horse." The fins on the sides of a seahorse's head beat up to 50 times per second and provide stability, while the small dorsal fin creates forward momentum. The body terminates in a long, prehensile tail used to grasp a seagrass stem or other upright anchorage (fig. 12.21).

Seahorses evolved their unique shape and posture when seagrass communities expanded during the Late Oligocene. Because of their upright posture, seahorses maneuver effectively among vertical foliage in seagrass meadows,

hatches. If fish are abundant, the entire brood likely survives, but when food is insufficient, the older, larger chicks snatch away most of the food brought to the nest, leaving the youngest chicks to starve. Sometimes the oldest nestlings bludgeon their smaller siblings, killing or tossing them from the nest. These adaptive strategies—asynchronous hatching and siblicide—match survival and population growth with the food supply available in a given year.

Colonial nesting, despite its benefits, also faces potential shortcomings. As examples, parasites and diseases readily transfer among the densely packed birds, and oil spills, boat traffic, and other human disturbances may disrupt the nesting cycle and cause colony abandonment. Although some colony

Figure 12.21. Tiny at birth, Dwarf Seahorses (top) begin to immediately feed on even smaller marine copepods (inset). As adults, the colors of Lined Seahorses (bottom) vary from black to yellow and sometimes change to match their environment or behavioral situation. Photographs by Bradford J. Gemmell.

where they take advantage of the cover and camouflage such habitats provide. In effect, seagrasses afford seahorses both protection from predators and opportunities to ambush their prey. Of approximately 50 species of seahorses found worldwide, most live in seagrass meadows. Three species commonly occur along the Texas coast—the Dwarf Seahorse prefers seagrass beds in shallow lagoons, whereas the other two species seek deeper waters.

The equine shape of their snouts creates little water disturbance and allows seahorses to stealthily approach their primary prey—copepods. These tiny crustaceans rely on disturbances to evade predators and respond almost instantly when approached. Within 2–3 milliseconds after detecting a movement—one of the shortest response times of all animals—a copepod can blast away from danger at speeds exceeding 500 body lengths per second. Moving cautiously, however, a seahorse can approach within 0.04 inch of a copepod without causing alarm. Once in position, Dwarf Seahorses require only 1 millisecond to suck in their prey— much faster than a copepod's getaway speed. Curiously, seahorses lack stomachs and hence feed almost continuously.

The courtship behavior and reproductive traits of seahorses are just as unique as their morphology. Males, although otherwise mobile, usually remain within an 11-square-foot zone, whereas females range over a much larger area. When females approach, nearby males tail-wrestle to attract the attention of a potential mate. When a victor emerges from the mock battle, courtship activities begin and involve color changes, holding tails while swimming side by side, and twirling around a seagrass stalk the pair holds jointly with their tails. When pair formation seems assured, the male pumps water to expand an egg pouch on his abdomen. The pair then spirals upward snout to snout to a point above the meadow, where the female squirts up to 200 eggs into the male's pouch before swimming away. The male fertilizes the eggs as he sinks back to an anchorage in the seagrass bed. The female returns daily to maintain the pair bond with the male during his "pregnancy," whereas the male remains on station, providing an incubator with a controlled environment for the developing embryos. After a 21-day gestation period, male dwarf seahorses "give birth" at night to about 200 young. When the female returns the next day, the pair replays the courtship ritual, beginning anew the unique reproductive sequence.

Male "pregnancy" poses an unusual phenomenon that has generated considerable speculation about its possible adaptive advantages. The energetic costs associated with producing eggs require about a third of a female's body weight. Therefore, females may select only those males that seem robust enough to successfully care for the developing young while the females recover from egg production and ready themselves for the next breeding effort. Whatever the strategy may be, seahorses assuredly represent one of the most fascinating animals on our planet and are well worth a visit to a public aquarium.

CONSERVATION AND MANAGEMENT

Estimates indicate that nearly 40 percent of all Americans live within 50 miles of a coastline—an area that is less than 10 percent of the nation's total landmass. A somewhat similar relationship holds for Texas, where more than 1.1 million of its citizens lived in the coastal floodplain in 2010, and the numbers continue to swell in coastal counties at the rate of about 2–3 percent per year. The development of housing, resorts, and marinas steadily degrades vulnerable marine communities, especially where dredging or filling destroys coastal marshes and seagrass meadows. Other damage occurs less obviously, as when boat propellers scour benthic communities—such damage has impacted up to 98 percent of the seagrass beds in Corpus Christi and Redfish Bays. Meanwhile, downstream flows from urban areas well inland also contribute to the diminished water quality in some bays and estuaries, further reducing productivity in the affected ecosystems.

The future of barrier islands remains especially precarious. These ribbons of sand face threats from rising sea levels, increasing frequencies and intensities of storms, and reductions in sand supply wherever dams trap sediments before they can reach the Gulf. Today, most barrier islands are gradually migrating landward and diminishing in size. Nonetheless, the human population living on barrier islands continues to expand—from 1990 to 2000 the average density increased by 14 percent. Both ends of Padre Island, once the secluded haunts of fishermen and a few eccentric beachcombers, have become dense warrens of human occupation ranging from condos to curio shops.

Some conservationists regard seagrass communities as one of the world's most threatened ecosystems. Globally, anthropogenic perturbations—from propellers to pollution—

contribute to an annual loss rate averaging about 1.5 percent. Seagrass beds in Texas fare no better. For example, all the seagrass beds in West Galveston Bay had disappeared by 1982, and between the mid-1970s and 1988, dredging activities in the GIWW destroyed approximately 35,000 acres of Shoal Grass. In response, the Texas Parks and Wildlife Department developed a monitoring plan to assess such losses and initiated numerous attempts, some successful, to restore seagrass meadows, which provide invaluable nursery habitat for fish and shellfish, including commercial and recreational species.

Land subsidence poses a major concern along the Upper Coast. Many areas surrounding Galveston Bay sank by more than 1.5 feet—some by as much as 10 feet—between the 1950s and the 1990s. With the removal of groundwater, oil, and gas, the clay layers under the bay and surrounding areas collapsed and compacted, causing shifts in the shorelines. For example, the hallowed site at San Jacinto Battleground State Historic Site where Texas won its independence now lies partially submerged. By 1990, regulations reduced the drilling activities, which lessened subsidence in the area, but accelerated wave action continued to erode the exposed shorelines and destroy the marshes.

A related concern—sea-level rise—threatens many low-lying areas along the coast. The combined effects of erosion and sea-level rise may increase water depth by nearly 4 inches by 2050. Among a host of other ecological impacts, such changes may flood the nesting and feeding areas of colonial waterbirds and eliminate many seagrass and marsh communities.

Oil and chemical spills remain an ongoing—and very real—hazard on the Texas coast. Natural seeps on the floor of the Gulf of Mexico regularly discharge small blobs of oil and tar that wash up rather harmlessly on beaches, but industrial mishaps on deepwater drilling rigs produce ecological calamities. In 1979, for example, oil from the IXTOC I spill in the Bay of Campeche drifted onto Padre Island, coating hundreds of wintering shorebirds and other marine organisms and fouling some beaches with thick layers of

Figure 12.22. In 1979, thick, asphalt-like deposits of oil collected on the shores of Padre Island from the IXTOC I spill in the Bay of Campeche (top left). Oil coating a Ghost Crab (top right), Willet (bottom left), and Pied-billed Grebe (bottom right) demonstrates the devastating impact of the spill on marine organisms.

tar (fig. 12.22). Another, the Deepwater Horizon oil spill (2010), largely spared the Texas coast but seriously damaged Louisiana's marine ecosystems, again serving as ample warning to prevent more such events. Barges carrying chlorine and other hazardous chemicals regularly pass through Aransas National Wildlife Refuge and other coastal refuges and preserves along the GIWW. An ill-timed accident might kill some or all of the Whooping Cranes overwintering at the refuge, and a summer spill would prove disastrous for thousands of nesting waterbirds. In all, the marine organisms and their habitats on the Texas coast remain highly vulnerable to both natural and human disturbances.

APPENDIX A
STATE SYMBOLS
NATURAL HISTORY OF TEXAS ICONS

The official designation of items and objects as much-loved objects of Texas should represent the entire state and not just one region or locality.

— GOVERNOR RICK PERRY

Inspiration for adopting official state symbols began in 1893 with a display prepared for the World's Columbian Exposition. The exposition, also widely known as the Chicago World's Fair, celebrated the 400th anniversary of Christopher Columbus's arrival in the New World. The fairgrounds covered more than 600 acres outfitted with more than 200 new buildings where visitors could learn about the cultures, achievements, and natural resources of 46 countries along with those of the United States and its territories. Among the thousands of exhibits, the "National Garland of Flowers" presented a spectacular display of America's floral diversity, with a showy flower representing each state.

History does not record which flower represented Texas in the National Garland or how it was selected, but the display undoubtedly ignited the concept for designating state icons. In fact, legislative action in several states had already designated a "State Flower" specifically for the occasion. Texas, however, had begun its deliberations seven years later when state legislator Phil Clements promoted the "White Rose of Commerce"—the Cotton Boll—as the State Flower. Another legislator, John Nance Garner, strongly advocated the flower of a prickly pear, thus earning him the nickname "Cactus Jack," a moniker that followed him to national prominence as a two-term vice president of the United States in the administration of Franklin Delano Roosevelt. Then, with support from the Texas chapter of the National Society of the Colonial Dames of America, state Representative John Green of Cuero nominated "Buffalo Clover"—later known as the "Bluebonnet"—as the State Flower. The bill designating the Bluebonnet as the State Flower of Texas—and the state's first officially recognized icon—was approved by the 27th legislature and signed into law by Governor Joseph D. Sayers on March 7, 1901.

Despite its slow start, Texas soon caught up with—or surpassed—most of the other states in designating species or things as official icons. The current list includes 67 representations, ranging from the State Cooking Implement (cast-iron Dutch oven) to the State Domino Game (42) and State 10K Race (Texas Round-up 10K). The list of state icons includes 22 species of plants or animals, including one "herd" (American Bison). However, only eight of those on the list were formally adopted by legislative action and written into the Texas statutes. The remainder gained official status from a concurrent resolution approved by both chambers in the legislature and signed by the governor. Icons designated in that manner are not incorporated into the state's statutes but will be if later approved by specific legislation.

The accounts below, listed in order by their date of adoption, briefly describe the natural history of a native species, or suite of species within a single genus, proclaimed by legislative action as an official state symbol. The descriptions do not include exotic species or domesticated and cultivated varieties, no matter how long ago they were introduced or how dear they may be to the hearts of certain Texans. Alas, these criteria omit Texas Longhorns, the Blue Lacy (a dog breed), and Nymphaea Texas Dawn (a waterlily cultivar), among others.

THE STATE FLOWER OF TEXAS: BLUEBONNETS (1901, 1971)

Legislative support for adoption of a bluebonnet as the State Flower in 1901 was overwhelming, but it was not long thereafter that a new controversy regarding the selection erupted in the halls of the state capitol. As a result of periodic and sometimes quite heated debate, the bluebonnet story did not reach a conclusion until 1971. The disagreements centered on the initial legislation that specified *Lupinus subcarnosus* (the Sandyland Bluebonnet) as the State Flower. The distribution of this species is restricted to sandy soils in the rolling hills of southern (near Cuero) and coastal Texas. Through the years, many legislators advocated changing the state-designated species to *L. texensis* (now known as the Texas Bluebonnet), a more widely distributed taxon in the state and one with larger, showier flowers. Compromise was eventually achieved in a concurrent resolution, approved on March 8, 1971, which included *L. texensis* and "any other variety of bluebonnet not heretofore recorded" as the state symbol. Thereafter, all bluebonnets in the state, regardless of their taxonomic identity and color, collectively represented the State Flower.

Thus, even a maroon-colored variety developed by horticulturists at Texas A&M University qualifies as the "State Flower," likely to the chagrin of other state universities.

The Texas Bluebonnet, one of the six species of bluebonnets in the state, is also known in some areas as *el conejo* ("the rabbit") because each of its 50 or so violet-blue flowers is tipped with white, like a bunny's tail. The blossoms of this self-pollinating annual species peak in late March to early April. The Texas Department of Transportation seeds Texas Bluebonnet and other native wildflowers along roadsides to produce colorful displays and adjusts its mowing schedules to ensure both long-lasting color and seed production. Occasionally, the species produces a rare pink or white variant, which provided a prominent university with the genetic materials to produce bluebonnets bearing maroon flowers. At times, maroon bluebonnets miraculously bloom on the campus of a long-standing university rival—a happenstance more likely linked to spirited students than to "Mother Nature."

In addition to being a state symbol, the State Flower is honored in five other state icons: Flower Song ("Bluebonnets"); Tartan (Bluebonnet Tartan); Bluebonnet City (Ennis); Bluebonnet Festival (Chappell Hill Bluebonnet Festival); and Bluebonnet Trail (Ennis). Concerns for the visible and well-known icon mounted when it appeared that an invasive annual known as Turnip Weed— or more aptly as Bastard Cabbage—might threaten bluebonnets throughout the state. The weed, introduced many years ago from Europe, grows faster and larger than most native wildflowers. Because it germinates in the fall and overwinters as a basal rosette, Bastard Cabbage gains a head start the following spring; the rosettes, already established, rapidly mature and produce flowers and seeds while transforming into dense, largely monospecific patches of vegetation—all at the expense of bluebonnets and other native wildflowers. Fortunately, research conducted at the Lady Bird Johnson Wildflower Center (infobox 7.1) determined that sowing areas dominated by Bastard Cabbage with the seeds of Indian Blanket, another native wildflower, might reduce the weed population by 80 percent. With the threat thwarted, bluebonnets will continue adding their color to the palette of spring wildflowers in Texas.

THE STATE TREE OF TEXAS: PECAN (1919)

The 20th governor of Texas, James S. Hogg (1851–1906), lives on in Texas lore primarily for naming his only daughter "Ima" (a false urban legend claims he had a second daughter named "Ura"). But he is also remembered for his admiration of Pecan trees and fondness for their tasty nuts. Indeed, before he died, Governor Hogg requested that a Pecan tree be planted at his grave. Although many of his contributions to the state remain dimmed by the veil of history, his affection for the Pecan motivated his admirers in the legislature to enact Senate Bill No. 317, naming the Pecan the State Tree of Texas.

The native range of Pecan includes most of the Mississippi Valley and extends westward to eastern Kansas and Central Texas. Long before Euro-Americans arrived, Native Americans ate and traded pecans and, mimicking squirrels, often stored the nuts in hollow trees.

Spanish explorers were the first Europeans to discover pecans, which they introduced to Europe. Pecan trees were also planted widely in North America. For example, Thomas Jefferson planted Pecan trees at Monticello and gave pecan nuts to George Washington to seed his orchards at Mount Vernon.

Pecan trees grow best in a humid climate on well-drained, loamy soils that are not subject to prolonged flooding. Mature trees usually reach heights between 70 and 100 feet, but stately old Pecans develop trunks more than 3 feet in diameter and grow as tall as 150 feet. Pecan trees often live for more than 300 years and can bear fruit for most of that period. Although called a nut, pecan fruit is technically a drupe—an outer husk surrounds a thin-shelled stone (or pit) protecting a seed. The green husks turn brown or black as the seeds ripen, and then dry and split open along four sutures to release the nut in late fall (September through December). For foragers of mast, the nuts usually remain edible for nearly a year.

When eaten fresh, the nutritious meat of a pecan nut itself presents a buttery snack, but more often it becomes the core ingredient of a sweet dessert (e.g., a pie or praline candy) or a finely chopped topping for ice cream. Horticulturists have cloned more than 100 varieties—each selected principally for characteristics associated with commercial nut production. Pecan wood, however, makes fine furniture and flooring as well as flavoring for smoked meats.

Texas remains the largest producer of native pecans, but commercial growers in Georgia harvest more nuts from hybrids. Pecan nuts provide a rich source of dietary fiber and a variety of nutrients, including protein, iron, and B vitamins. For that reason, in 2001 the pecan was honored with a second designation—the State Health Nut. No winter holiday in Texas is complete without at least one slice of pecan pie, and in 2013 the legislature acknowledged this traditional holiday treat when it named pecan pie as the official State Pie of Texas. (Some members of the legislative body wanted to make it illegal to bake pecan pies with nuts from any other state, but the proposal failed.) Altogether, three entries on the official list of state icons recognize pecans.

THE STATE BIRD OF TEXAS: NORTHERN MOCKINGBIRD (1927)

Few Latin names that identify organisms are more apt than *Mimus polyglottos,* the official scientific handle for the Northern Mockingbird. Literally meaning "many-tongued mimic," the name reflects the remarkable ability of mockingbirds to learn and repeat the calls of other birds and, occasionally, other sounds as well (e.g., the croaks of frogs, the meows of cats, and even sirens). These consist of whistles, rasps, trills, and scolds uttered in phrases, each repeated several times before beginning another call.

Both sexes learn the calls of other birds, but males typically become masters of the art. New sounds are added continually, and some individuals learn at least 150 distinct calls, which may be performed in separate spring and fall repertoires. Unmated males are exceptionally persistent callers, including at night, often when the moon is full. Behavioral studies suggest that females opt for males with a particularly rich repertoire. Nonetheless, other species of birds are not fooled and can distinguish "real" calls from the learned calls of mockingbirds. Mockingbirds also utilize four calls of their own to convey intraspecific information such as the presence of predators or a territorial defense.

Northern Mockingbirds have other quirks, particularly the stereotypic behavior known as wing flashing. This display typically occurs when a mockingbird runs across a lawn or other exposed site in search of food and then abruptly stops while spreading and lifting up both wings, thereby exposing—or flashing—the white patch on each wing that contrasts conspicuously against the bird's predominantly gray plumage. Of the several explanations for this behavior, the most common suggests that flashing helps flush insects. Another idea proposes that the behavior is innate and subsequently develops into a defense against predators. Interesting ideas, but none fully explains the function of wing flashing.

Northern Mockingbirds can also recognize individual humans, a rare feat for untamed animals. In a controlled experiment, the birds remembered humans who had previously threatened their nests, and they attacked when the same people approached their nests days later. In contrast, the birds ignored other humans who had not bothered their nests. Such recognition required only two 30-second exposures for the birds to identify humans in these situations and may explain why mockingbirds adapt so well to urban environments.

Northern Mockingbirds occur coast to coast across the width of North America, for the most part south of Canada. Those nesting at the northern edges of their range migrate, but the remainder, including those in Texas, are year-round residents throughout the state. They reach their greatest abundance on the Edwards Plateau (chapter 7) and in the South Texas Brushland (chapter 10). The birds construct loose, twiggy nests—sometimes supplemented with paper, shredded cigarette filters, and other human artifacts—typically 3 to 10 feet high in trees and shrubs and usually lay four or five spotted blue-green eggs. Only females incubate, but both parents share building the nest and caring for the young. Pair bonds may last for several years, occasionally for life, and both sexes vigorously defend their nestlings from cats, hawks,

and other predators. A single pair may produce up to four broods per year. Northern Mockingbirds are omnivores; they feed on fruits and berries as well as on insects, snails, earthworms, other small invertebrates, and occasionally small lizards, but they rarely visit seed-filled feeders. Their range expansion into northern states and the southern edges of Canada may be related to plantings of Multiflora Rose, whose fruit is a favored winter food.

Such widespread and interesting birds have not escaped cultural recognition. They appear in the title of the classic novel (and subsequent movie) *To Kill a Mockingbird,* and in an equally classic folk tune, "Listen to the Mockingbird Sing." One of Audubon's famous paintings features several mockingbirds confronting a rattlesnake coiled in a shrub; while rather fanciful in its composition, as snakes of any kind only rarely prey on incubating birds, the painting represents some of his better artwork.

At Monticello, Thomas Jefferson kept a mockingbird named Dick as a household pet. The singing abilities of Northern Mockingbirds developed a market for selling young birds as cage pets in Houston for 25 cents each, and the local newspaper in Marshall reported that an accomplished songster commanded $40. Conversely, many were killed because they damaged fruit, including one record of 500 birds poisoned or shot in a single year (1890) at a vineyard on the Medina River. Some, along with other species, ended up as the main ingredient of "bird pie." Various state laws were intended to protect songbirds from such fates, but only when Congress passed the Migratory Bird Treaty Act in 1918 did the weight of federal law fully protect songbirds from further exploitation.

In 1927, at the suggestion of the Texas Federation of Women's Clubs, the legislature formally adopted the species as the State Bird of Texas, a distinction also awarded by Florida, Mississippi, Tennessee, and Arkansas. The proclamation notes, in part, that a Northern Mockingbird "is a singer of distinctive type, a fighter for the protection of his home, falling if need be in its defense, like any true Texan." Well said.

THE STATE STONE OF TEXAS: PETRIFIED PALMWOOD (1969)

Petrified palmwood poses something of a misnomer. Although certainly rock hard, these and other petrified organisms in scientific terms represent fossils. The palms that formed these fossils once lived in what is now eastern Texas during the Oligocene Epoch (43–23 million years ago), when the Texas shoreline lay about 100 miles farther inland than its current position. The palms thrived in the warmth of the lush, tropical-like swampy forest then blanketing the region.

Most fossil palms in these deposits represent the genus *Palmoxylon,* a group that resembles modern palms. As with palms today, the trunks of these trees formed from fibrous tissues that surrounded and strengthened numerous hollow tubes that transported water and nutrients upward and the products of photosynthesis downward. When the trees died, some fell into decay-resistant graves of mineral-rich mud and stagnant water. Slowly, silica and other dissolved minerals gradually replaced the plant's organic tissues, including the distinctive structural features of the original wood.

Petrified palmwood may be colorful, and some opalized specimens are especially beautiful. Depending on how artisans cut the fossil trunks, the rodlike structures that once supported the nutrient-carrying tubes appear as prominent dark dots or tapering or continuous lines. These patterns, and the rainbow of colors and designs in the matrix, make petrified palmwood a favorite among rock collectors and jewelry designers. Composed of silica, the stony fossils can be highly polished and remain strong enough to withstand the rigors of daily use. These characteristics prompted the Texas Gem and Mineral Society to prepare the necessary resolutions for legislative approval of petrified palmwood as the official State Stone of Texas, an action completed when Governor Preston Smith signed the House Concurrent Resolution.

Texas is not the only state to misclassify a fossil—Washington also regards petrified palmwood as its official State Gem. Louisiana and North Dakota, however, got it right—both identify petrified palmwood as their official State Fossil.

THE STATE GEM OF TEXAS: TEXAS BLUE TOPAZ (1969)

While preparing the resolution that proposed petrified palmwood as the State Stone, industrious members of the Texas Gem and Mineral Society developed a second resolution, which nominated Texas blue topaz as the state gemstone. In a unique action, the Texas House of Representatives included both proposals in a single concurrent resolution that ultimately won the approval of both the House and Senate.

Topaz crystals form in intrusive igneous rocks that cooled and solidified as granite, but others develop inside vapor cavities in silica-rich lava that resembles granite. In its pure form, topaz is a colorless crystal composed primarily of aluminum silicate, although it is often tinted by various impurities. Found worldwide in deposits of volcanic origin, topaz varies in color, commonly wine red, violet, yellow, pink, or light blue. Orange topaz is the traditional November birthstone. Topaz colors can be enhanced or altered artificially by exposing a crystal to heat and radiation.

Light blue topaz, the rare form recognized as the State Gem of Texas, occurs in Precambrian granite deposits of the Llano Uplift (chapter 7) in Mason County. Topaz is not quite as hard as a diamond or ruby and can be distinguished from similar-appearing stones by its clarity, hardness, and crystalline structure. Unlike many other gemstones, topaz can be cut into a variety of shapes because of its perfect cleavage. Most cut topaz stones display pleochroism, the appearance of several colors in a crystal depending on the viewing angle. Two Mason County gemologists cut a topaz in a way that reflects a five-pointed star in the light blue stone. In 1977, this unique design—the "Lone Star Cut"—became the official State Gemstone Cut.

THE STATE GRASS OF TEXAS: SIDEOATS GRAMA (1971)

Long before Euro-Americans settled in Texas, grasslands carpeted about two-thirds of the state's landscape. Every natural region from the Piney Woods to the Trans-Pecos included extensive prairies or lesser areas dominated by grasses. Of approximately 723 species of grasses in Texas, Sideoats Grama represents one of the most widely distributed and abundant of the group. This deep-rooted and fire-adapted species—a midgrass 15–30 inches in height—tolerates a broad range of environmental conditions (soil types) and flourishes in drier areas along with Blue Grama and other perennial grasses.

The inflorescence of Sideoats Grama easily distinguishes it from other grasses. In summer, the plants send up slender, zigzagged stalks on which evenly spaced florets produce small flowers tipped with a subtle but noticeable pink to purplish tinge. The seed heads of Sideoats Grama, which at maturity resemble miniature clusters of oats, all hang from one side of the stalk, an arrangement clearly reflecting the origin of its common name. In winter, the grass can be recognized by a series of long, evenly spaced hairs extending from the edges of the leaves near the plant's base.

Sideoats Grama provides nutritious forage for deer and livestock and thus represents an important range grass. It usually begins growing earlier in the spring and remains green later in the fall than other grama grasses. Although slightly less palatable than Blue Grama, Sideoats Grama compensates by producing a much larger volume of foliage that maintains a relatively high value as forage throughout the winter. Sideoats Grama played a significant role in the recovery of drought-ravaged grasslands following the destructive Dust Bowl of the 1930s. It was one of the few grasses that survived to compete with invasive weeds. Today, more than nine vigorous cultivars are available to rapidly cover sites with specific soil and environmental conditions.

In recognition of its many uses and adaptability to more soil types than other grasses in the state, the Texas Council of Chapters of the Soil Conservation Society of America and the Texas Section of the American Society of Range Management jointly recommended naming Sideoats Grama the official State Grass of Texas—a designation formalized when Governor Preston Smith signed a Senate Concurrent Resolution.

THE STATE SHELL OF TEXAS: LIGHTNING WHELK (1987)

For many beachcombers, the ultimate prize is a perfectly formed Texas Lightning Whelk. The distinctive spiraling shell of this mollusk protects a marine snail that lives in the warm nearshore waters of bays and lagoons from Louisiana to Mexico. Using the edge of its shell as a wedge and exerting pressure with its muscular foot, a Lighting Whelk pries open the valves of clams, oysters, and scallops encountered on sandy or muddy bay bottoms. The whelk then inserts its tonguelike radula through the opening and scrapes out the meat inside. In the absence of living prey, Lightning Whelks subsist on carrion.

Lightning Whelk shells are appreciated for both their sculptured beauty and their size, which can sometimes exceed 12 inches in length. Narrow, brownish "lightning" stripes streak across their off-white shells from behind horn-like points on the shoulders, the widest end of the pear-shaped shell, to the tapered bottom. However, the opening of the shell represents the most unique feature of this Texas icon—it opens to the left, whereas the opening faces to the right in most single-shelled mollusks, including other subspecies of Lightning Whelks.

Mildred Tate (1917–1990) first championed the Texas Lightning Whelk as the State Shell when she became interested in conchology after moving to Brazosport from Oklahoma in 1951. She started the first shell club in Brazosport and then cofounded the Brazosport Museum of Natural Science, which now houses an excellent shell collection in the Mildred Tate Hall of Malacology. At the age of 70, Ms. Tate discussed her idea for a State Shell with several Texas legislators, backed by support from several other Texas shell clubs. When Representative John Willy presented the resolution for a State Shell to the House of Representatives, he noted that the topaz represented the Texas mountains and the bluebonnet symbolized the Texas plains—points made to emphasize that the coastal areas of the state lacked such an icon. In response, the Texas Senate and House designated the Texas Lightning Whelk, a regional subspecies of a more widely distributed organism, as the State Shell in a bill signed by Governor Bill Clements.

The Texas Lightning Whelk deserves its honor as the Texas State Shell for its cultural history as well as for its physical merits. The whelks provided Native Americans with food and material for cups and bowls as well as for scrapers and gouges. For some tribes, the unique, left-handed spiral suggested that the whelks represented sacred objects. Sailors may have bathed using the long strands of the whelk's horny, disk-shaped egg cases as sponges. The US Postal Service selected a Lightning Whelk to appear on one of the first stamps commemorating an American shell.

THE STATE FISH OF TEXAS: GUADALUPE BASS (1989)

In colorful tree-lined canyons etched deeply into the Edwards Plateau limestone, cool springs give rise to sparkling streams. Here, where the currents bounce over rocks and swirl into deep eddies around Baldcypress roots, lurks an endemic fish renowned for challenging the skills and patience of even the most experienced anglers. Once hooked, the relatively small Guadalupe Bass responds with long runs, sharp turns, energetic jumps, and attempts to entangle the line on cypress knees or other obstructions. Fishermen tested by these wily bass compare their battles with those waged by Rainbow Trout or Smallmouth Bass.

When discovered in Central Texas, Guadalupe Bass shared streams in common with Largemouth and Spotted Bass. Initially, the new taxon was described as a subspecies of the Largemouth Bass, but it was later reclassified as a subspecies of the similar-looking Spotted Bass. Finally, after biologists determined that Spotted and Guadalupe Bass sharing the same streams did not produce intermediate forms—an indication that the two species did not hybridize—the Guadalupe Bass gained recognition as a distinct species. Within a stream system, the three species of bass may select distinct habitats and thereby enforce the reproductive isolation necessary for maintaining their respective genetic integrity; Guadalupe Bass prefer small streams whereas Largemouth Bass favor sluggish waters and Spotted Bass seek fast-moving water in larger streams.

Guadalupe Bass and Spotted Bass, although similar, are distinguished by careful examination of the dark markings extending down their sides (10–12 transverse bars in Guadalupe Bass vs. more than 13 large spots in the aptly named Spotted Bass) and the number of scales along the lateral lines of

each species (65 on average in Guadalupe Bass vs. more than 70 for Spotted Bass). Guadalupe Bass do not grow as large as Largemouth or Spotted Bass— the largest ever recorded, recently caught in the Colorado River below Austin, weighed 3.71 pounds and measured 17 inches in length. For comparison, record books list more than 50 Texas Largemouth Bass with masses and lengths exceeding 15 pounds and 24 inches, respectively. Spotted Bass reach similar lengths but weigh slightly less than largemouths.

Historically, Guadalupe Bass occurred in rivers and streams on the northern and eastern Edwards Plateau, especially in parts of the Guadalupe, Brazos, Colorado, and San Antonio river basins. The fish were later introduced into the Nueces and Llano Rivers and several smaller streams before the natural range of the species could be fully determined. After a population was well established in the Nueces River, state authorities established a refuge for Guadalupe Bass in the Sabinal River within the Lost Maples State Natural Area. The refuge provides dependably pure stream water, but more important, it protects the native species from hybridizing with the Smallmouth Bass that were introduced into many Hill Country rivers to enhance recreational fishing. Because of the same threat, the Texas Parks and Wildlife Department also stocked purebred Guadalupe Bass in the tributaries of the Guadalupe River as a means of restoring and protecting this unique Texas endemic.

In 1989, the Texas legislature recognized Guadalupe Bass as a unique sport fish by passing House Concurrent Resolution No. 61. The resolution, signed by Governor William P. Clements on May 10, 1989, confirmed the Guadalupe Bass as the official State Fish of Texas. As a protected game fish, the bass can be caught legally only with a pole and line. Although many anglers practice catch-and-release—a way to both enjoy and pass on to others the thrill of action-packed fishing— limited harvests are allowed in many stream and river systems. All told, conservation now provides Guadalupe Bass with a future—and Texans with a future of good fishing.

THE STATE REPTILE OF TEXAS: TEXAS HORNED LIZARD (1993)

Three species of horned lizards occur in Texas, one of which—the Texas Horned Lizard—became the official State Reptile in 1993. The three species differ in several ways, but the number and length of the "horns" projecting from the rear of their heads offer the best clues for their identification. True to its name, the Greater Short-horned Lizard has horns that are little more than a crest of about eight stubby nubbins, whereas the Round-tailed Horned Lizard sports four horns, all of about the same length. The two central horns of the Texas Horned Lizard, however, are much longer than any of the surrounding spines. The range of each species extends beyond the borders of Texas, but the Texas Horned Lizard is the most widely distributed in the state, as its range overlaps somewhat in West Texas with those of the other two species.

At some unknown time in history, horned lizards were considered toads or frogs—a designation that not only stuck but remains widely used. They are, of course, not amphibians that fertilize their eggs externally, often in water where they typically hatch in a larval form (e.g., tadpoles). Instead, they are true reptiles that, among other differences, fertilize their eggs internally, often in arid or semiarid environments far from pond or stream. Such zoological distinctions aside, "Horned Frogs" remain the mascot—and pride—of Texas Christian University. Earlier in history, these lizards were symbols of health and happiness for Native Americans.

When breeding, Texas Horned Lizards lay 13–30 eggs in shallow pits excavated in loose soil (in contrast, Greater Short-horned Lizards bear their young alive). Parental care ends when the eggs are covered. Depending on soil temperature, eggs hatch five to nine weeks later, producing miniature replicas of the adults less than 1 inch long. For food, horned lizards seek ants, termites, spiders, and other invertebrates. Harvester ants are particularly important, and 12 species of this group occur in Texas. They denude vegetation from a circular area about 39 inches or more in diameter from which foraging trails radiate into the surrounding vegetation; an entrance to extensive underground chambers is located in the disk's center. A Texas Horned Lizard readily eats 70–100 of these large ants daily. Regrettably, harvester ants, as well as other native species of ants, often disappear in areas invaded by Red Imported Fire Ants. Moreover, pesticides used to control Red Imported Fire Ants also affect native species. Illegal pet trade and habitat loss pose other threats to horned lizards.

Despite their spiny armor and cryptic coloration, horned lizards become prey for snakes and mammals as well as birds, including hawks and roadrunners. When threatened, they inflate their bodies somewhat to appear larger. Well and good, but the defenses of Texas Horned Lizards rise to still another level—they squirt a stream of blood from the corners of their eyes when attacked by foxes, Coyotes, or dogs, but not by other predators. The blood originates not from their eyes but from ruptured vessels in sinuses around their eyes (i.e., the eyes serve as an outlet, not a source). In laboratory experiments, Kit Foxes reacted adversely when squirted by Texas Horned Lizards, which they then avoided, thereafter selecting other foods. As a further test, mice were smeared with the blood from horned lizards, and the foxes again selected other foods. Tests with Coyotes produced similar results. These experiments determined that the active ingredient serving as a distasteful chemical defense flows in the bloodstream of horned lizards and does not originate from special glands near their eyes. Likewise, another ingredient detoxifies the venom of harvester ants, their principal food, and suggests that the ants' venom may play a role in the chemistry of the canid-repelling properties of the lizards' blood. Blood from the bodies of those species of horned lizards that do not squirt blood produced no adverse effects on canids; nor do those lizards eat venomous ants.

Texas Horned Lizards use rain harvesting to collect drinking water. Stimulated by rain, they flatten their bodies to increase the surface area on which the drops fall; they concurrently assume an arched position that allows gravity to carry the rainwater to their mouths. The lizards drink by opening and closing their jaws (not by lapping), a behavior also triggered by the stimulus of rainfall.

Drinking water is also absorbed, even from light rainfalls and dew, by capillaries in a network of minute grooves running between the scales; some of these also channel water directly to the mouth. Hence, the scaly skin of horned lizards not only acts as armor but also collects and transports drinking water.

Because the species is declining in numbers and distribution—and thus appears on the state list of threatened species—the Texas Parks and Wildlife Department initiated a statewide program known as the Texas Horned Lizard Watch. Citizen volunteers survey local populations, whether in a backyard, park, or ranch—an easy and rewarding warm-weather project for which details appear online. Unfortunately, horned lizards are difficult to maintain in captivity, so breeding and stocking efforts are not a feasible recovery technique. But here is another way to help: when registering your car, opt for a Texas Conservation License Plate featuring the state's official reptile. Funds generated by these plates, in part, directly support the Texas Horned Lizard Watch and other programs involving native species.

THE STATE PLANT OF TEXAS: PRICKLY PEAR CACTUS (1995)

Ninety-four years after John Nance Garner earned the nickname "Cactus Jack" for his enthusiastic but unsuccessful attempt to name prickly pear the State Flower of Texas, the state legislature designated the cactus as the State Plant of Texas. To avoid controversy, the legislature lumped all of the species in Texas together instead of honoring just one. Thus, prickly pear cacti—all species of *Opuntia* with broad, pad-like stems—collectively became the official State Plant of Texas in 1995 when Governor George W. Bush signed House Concurrent Resolution No. 44. Shortly after prickly pear cacti became the official State Plant, another resolution proposed to name the cactus the official State Vegetable Fruit, but the idea died without further action.

At the time, the genus *Opuntia* also included Tree Cholla and Tasajillo—cacti with round, jointed stems—but the language in the resolution clearly excluded these forms from sharing the designation as the State Plant. Genetic research later assigned round-stemmed cacti to another genus, *Cylindropuntia,* thus scientifically validating the legislature's exclusion.

Prickly pear cacti represent a perceptive choice because at least one species—and usually more than one—occurs in each natural region of the state. They thrive in a variety of ecological conditions, ranging from the Piney Woods, where the Eastern Prickly Pear grows in shady sites receiving more than 50 inches of rainfall per year, to the sand-blasted, oven-like Chihuahuan Desert of the Trans-Pecos, where the Spiny-fruited Prickly Pear subsists on as little as 8 inches of annual rainfall. Regardless of their individual differences, however, all species of prickly pears are easily recognized—their spine-studded stems develop as succulent, flat pads (cladodes) either joined end to end or branched into upright or prostrate clusters. Some tangled clumps form impenetrable barriers over large areas, although individual plants rarely exceed 8 feet in height.

Perhaps because of their typically bluish-green color, the fleshy cladodes may be mistaken for leaves. Numerous wart-like projections called areoles or tubercles develop on the surfaces of each pad. Each tubercle gives rise to a cluster of tiny, barbed, hairlike spines (glochids) and long, stiff spines that are highly modified leaves. The spines vary by species in length and color, but regardless of size, they effectively deter most animals from feeding on the succulent pads. Almost invisible, the glochids readily detach when touched, often causing irritation and occasional infections before being destroyed by an animal's immune defenses.

Showy yellow, orange, red, purple, or subtly bicolored aromatic flowers, which sometimes vary in color on a single plant, originate from buds on the margins of the stems. Inside a large bowl-shaped flower, rings of stamens facilitate pollination by small bees. In response to the stimulation of a pollinator, the thigmotactic stamens bend inward, which forces the bee deeper into the flower toward the main store of pollen, which coats the bee in readiness for transfer to another flower. As they ripen, the cylindrical fruits (tunas) of prickly pears change from green to a deep red or purple, and tufts of glochids on the skin discourage some foragers. Sugary-sweet when ripe, the tunas each contain several hundred hard-coated seeds that are dispersed in the droppings of Collared Peccaries, Black-tailed Jackrabbits, and other animals that eat the tunas.

Depending on the outlook of a landowner, a prickly pear may represent a noxious "weed," an attractive ornamental, or a source of food and income (chapter 6). According to some ranchers, dense stands of prickly pear compete with herbaceous forage plants for space, water, and nutrients and thus interfere with livestock operations. Others value the cacti as beneficial for wildlife and livestock, especially in South Texas, where the pads (after the spines are burned off) provide emergency fodder for cattle during winter and droughts.

Early explorers and Native Americans alike depended on prickly pears as a source of moisture, food, and medicine. Over the years, various preparations made from the plant's tissues were believed to regulate blood sugar and weight, facilitate childbirth, heal burns, and treat asthma, fatigue, corns, diarrhea, gastritis, sexually transmitted diseases, and numerous other conditions. Many grocery stores in Texas sell ripe tunas as a treat to eat raw (after the skin and glochids are removed) or as a base for jellies, candies, or sweet drinks. The recipes in cookbooks featuring southwestern or Mexican foods commonly include young cladodes (sold as nopales), which, when finely chopped, provide an ingredient for soups, salads, vegetable dishes, tacos, and breads.

Although hardy, prickly pears are not immune to parasites and diseases. In South Texas, cochineal, a sessile scale insect, feeds on the nutrients and moisture in cactus tissues. After covering itself with waxy, cotton-like white plates, the insect secretes carminic acid to deter predators. The acid can be extracted from both the insect and its eggs to produce cochineal dye, whose red pigments were esteemed by native peoples and today color foods and cosmetics. A number of other insects feed on prickly pears, but the Cactus Moth poses the greatest threat. This South American moth was purposefully released in Australia in 1925 as a biological agent to control *Opuntia*. Following success there, they were released on several islands in the Caribbean and somehow later reached Florida. Currently, the moths are moving westward across the southern states at a rate of about 100 miles per year and no doubt will eventually reach Texas.

THE STATE INSECT OF TEXAS: MONARCH BUTTERFLY (1995)

Twice each year, hundreds of thousands of birds funnel through Texas traveling between their summer breeding areas and wintering grounds. These spectacular migrations attract legions of birders to the state, but in fact, a small insect—the Monarch Butterfly—accomplishes a far more complex and amazing feat of migration.

Butterflies abound in Texas—more than 400 species and subspecies add color and delicate beauty to all of the natural regions in the state. Most butterflies complete their annual life cycle—egg, larva, pupa, and adult—within a relatively small area, including winter shelter in hollow trees, crevices, or dense clumps of vegetation. However, a few species—notably including the Monarch Butterfly—migrate long distances to avoid the rigors of winter.

Two geographically distinct populations of Monarch Butterflies exist in North America, each distinguished by separate migratory pathways and overwintering sites. A western population overwinters clustered in trees at approximately 300 locations dotting the coast of southern California. A much larger population, which remains east of the Rocky Mountains, disperses widely across the midwestern and eastern United States and into southern Canada before returning each fall to one of two wintering grounds. A small group overwinters in Arizona, but the majority of the eastern population travels one or the other of two pathways across Texas en route to a montane forest in south-central Mexico. Early migrants follow a wide aerial corridor from Wichita Falls to Eagle Pass during the last days of September, whereas later travelers follow the Texas coastline. The wintering location in Mexico remained unknown until discovered by Catalina Aguado Trail and Kenneth Bruggers in 1975.

Monarchs headed for Mexico arrive by mid-November. One by one they congregate in dense clusters on the boughs of Oyamel Firs and other trees and then enter diapause—a condition resembling hibernation. Thick fogs permeate the forest, but the tree canopies serve as umbrellas and, together with the structure of the clusters, protect the butterflies from excess wetting. Additionally, the moisture-laden air in the valleys provides a thermal blanket that retains heat and protects the insects from freezing temperatures. Interestingly—and unlike migratory birds, which are cued to changes in photoperiod—monarchs must first experience cold temperatures before beginning their northward migration when springtime temperatures return. Presumably, exposure to cold switches their internal sun compass from a southern to a northward bearing. In controlled experiments, Monarch Butterflies that were not exposed to cold resumed their southbound journey instead of heading north the following spring.

After rousing from diapause by early March, the monarchs mate before heading north. They pause frequently to sip nectar from flowers and mineral-laden moisture from seeps and other sources. By April, they reach northern Texas or southern Oklahoma, where the females lay egg clusters on the underside of milkweed leaves, on which the small caterpillars (larvae) feed after hatching four days later. The tissues of milkweeds contain cardiac glycosides, alkaloids that are bitter, emetic, and toxic to most animals, but not to the caterpillars. The concentrated alkaloids in fact protect the caterpillars from hungry predators, and this chemical armor later transfers to the adults. The bright colors on monarch wings are likely aposematic—an easily recognized warning of their undesirable taste and toxicity. In a well-known example of mimicry, the defense of noxious taste is shared with the similar-appearing Viceroy Butterfly, whose tissues contain salicylic acid.

About two weeks after hatching, a fully grown yellowish-green caterpillar, marked with thin black stripes, attaches to a stem or leaf with silken threads and spins a cocoon (chrysalis). Inside these enclosures, the caterpillars undergo metamorphosis—transformation to butterflies—the first of four generations of monarchs produced each year. Soon thereafter, the Monarch Butterflies that left Mexico just weeks before die.

First-generation offspring continue the journey, dispersing to the north and northeast before stopping in May or June to lay their eggs on milkweeds. Second, third, and fourth generations of monarchs originate as each new group of butterflies progresses farther northward. Each generation, except the fourth, lives only two to six weeks before reproducing and dying. Adults of the fourth generation appear in September or early October and instinctively head south. In doing so, Monarch Butterflies rely on genetic memory to retrace the entire route traveled by three generations of their forebears to reach a destination they have never seen. Some of these flutter southward for more than 4,800 miles from a milkweed plant in Canada to a distant forest in the state of Michoacán, Mexico.

In recognition of their coloration (likened by some to a Texas sunset), their reliance on milkweed for protection, and their remarkable migration, the Monarch Butterfly became the State Insect of Texas when Governor George W. Bush signed the House Concurrent Resolution. Four other states also honor the Monarch Butterfly as their State Insect, and two others as their State Butterfly. Unfortunately, the US Congress did not approve legislation nominating the monarch as the National Insect of the United States.

Plagued by many issues, populations of Monarch Butterflies have declined significantly in recent years. Although designated a Monarch Butterfly Biosphere Reserve in 2000, the winter habitat in Mexico still suffers from logging and subsistence farming. In the United States, the erosion of monarch numbers correlates with the development of genetically modified crops designed to resist herbicides; milkweeds are among the weeds eliminated in or near treated fields. In response, Texas and several other states initiated programs to restore milkweeds on roadsides, transmission line rights-of-way, and untilled fields, and several organizations, including Monarch Watch at the University of Kansas, regularly monitor monarch populations. With increased public awareness of its remarkable abilities, the Monarch Butterfly will long remain celebrated as the State Insect of Texas.

THE STATE FLYING MAMMAL OF TEXAS: MEXICAN FREE-TAILED BAT (1995)

While the Texas House of Representatives wrangled over which mammal should serve as the official symbol for Texas, an Austin-based conservation organization realized that the two leading candidates—the Texas Longhorn and the Nine-banded Armadillo—were each four-legged terrestrial species. Having recently won public support for protecting the large bat colony roosting under the Congress Avenue Bridge in downtown Austin, Bat Conservation International proposed the Mexican Free-tailed Bat to state Senator E. Jeffery Wentworth as the "State Flying Mammal of Texas." The senator enthusiastically sponsored the project—his district in the Texas Hill Country included, among others, Bracken Cave, widely regarded as housing the world's largest bat colony. Following its legislative approval, Governor George W. Bush signed Senate Concurrent Resolution No. 95 designating the Mexican Free-tailed Bat the official flying mammal several weeks before the House finally resolved the dilemma about the terrestrial mammal.

The designation is appropriate; few other bats—or mammals, for that matter—are as abundant or as widely distributed in Texas. Sometimes known as the Guano Bat, this species is now formally recognized by mammalogists as the Brazilian Free-tailed Bat. Its common name provides a useful clue for identifying the species—a mouse-like tail extends well beyond the flight membrane connecting the hind legs, a characteristic shared with only two other species of free-tailed bats occurring in Texas. Unlike the other two free-tails, however, the Brazilian Free-tailed Bat has large, rounded ears that are not connected at the midline of the forehead. As with other bats that roost in caves, the pelage of the species is uniformly dark brown.

Estimates vary, but up to 100 million Brazilian Free-tailed Bats migrate to Texas each spring to bear and rear their young. The largest of these maternity colonies, which often exceed ten million bats, occupy limestone caves in Central and West Texas, but smaller clusters dwell in abandoned buildings, hollow trees, cliff crevices, Cliff Swallow nests, and abandoned mines, as well as under bridges and clay roof tiles. To nourish their young (called "pups," one per female each year), nursing females leave each night in search of high-energy foods, leaving their helpless offspring clinging to the roof of the roost. When returning, each female unerringly locates her own pup among the millions of others using a combination of site-specific memory and recognition of its cry and smell.

Brazilian Free-tailed Bats emerge from their retreats at dusk, swarming upward to elevations between 1,000 and 10,000 feet where they disperse to forage on mosquitoes, moths, and other protein-rich flying insects. For some bats, these nightly explorations may extend more than 30 miles from their roosts and cover more than 154 square miles. The bats themselves fly through a gauntlet of predators when exiting and returning to their roosts. Perched on cave walls, Great Plains Ratsnakes strike at passing bats and Raccoons leap to attack from the floor below. In the air, Red-tailed Hawks pluck bats at twilight, as do Great Horned Owls after dark.

Like most insectivorous bats, Brazilian Free-tailed Bats rely on echolocation, a radar-like mechanism, to communicate, navigate, and locate their prey in dark environments. While a bat is flying, its larynx produces high-frequency sounds that are emitted through its open mouth as a series of pulses. After bouncing echo-like off objects (e.g., cave walls, trees, other bats, or insects), the returning pulses collect in the bat's large ears and travel to a highly adapted auditory system. In addition to registering the intensity of the returning echo and the time lapse between the sending and receiving of pulses, the system processes the minute time lag separating each ear's independent reception of the echoes. Together, this information reveals the location, size, and movements of objects ahead. When searching for prey, bats produce pulses known as "clicks" at a low rate (e.g., 10–20 clicks per second), but after an insect is detected, the rate spikes to about 200 clicks per second. Some moths have evolved a means of detecting a feeding bat, and when they hear the ultrasounds, they immediately execute a complex series of evasive maneuvers. Bats hunting near each other sometimes "jam" their neighbor's radar system with their own emission spikes in an effort to steal the same prey. Nevertheless, thanks to echolocation and agile flight, Brazilian Free-tailed Bats remain highly efficient aerial hunters.

Bats consume huge numbers of insect pests, a nightly—and free—service to the Texas agricultural industry. The bats of Bracken Cave alone consume an estimated 250 tons of insects per night, many of which are moths whose larvae severely damage valuable crops. All told, Brazilian Free-tailed Bats foraging unseen in the dark skies across Texas consume approximately four billion insects each night. Consequently, cotton farmers in an eight-county area of south-central Texas realize about $741,000 of extra income per year as the direct result of bats feeding on egg-laden Cotton Bollworm moths, thereby stopping short the production of about five million larvae of the Cotton Bollworm each night. Of all the native animals designated as official symbols of Texas, only the Brazilian Free-tailed Bat provides the state with substantial economic benefits.

THE STATE SMALL MAMMAL OF TEXAS: ARMADILLO (1995)

If Texas had erected a border security fence along the Rio Grande immediately after achieving statehood (1845), the Nine-banded Armadillo would likely not be the official State Small Mammal of Texas. The direct ancestors of the species now celebrated as a state icon evolved about 11 million years ago in South America and gradually moved northward into Central America and Mexico. The first evidence of the species in Mexico dates to the fourth century AD, but its entry into Texas likely occurred about the time Texas joined the Union. In 1849, the gifted artist and naturalist John James Audubon became the first to record the armadillo in Texas on north side of the river in the Lower Rio Grande Valley. After immigrating, the species rapidly colonized most of Texas, reaching the Sabine River to the east and the Pecos River to the west by 1914, and the Red River to the north by the mid-1950s.

Today, the range of the Nine-banded Armadillo extends from eastern New Mexico eastward across the southern tier of states to the Atlantic Ocean and as far north as the southern counties of Nebraska, Illinois, and North Carolina. Populations along the Atlantic Coast likely originated from introductions into Florida in the mid-1800s, but the speed at which the species expanded its range elsewhere remains the subject of much speculation. Undoubtedly, humans facilitated the expansion of armadillos by suppressing prairie fires, creating agricultural developments, eliminating many potential predators, and possibly translocating the armadillos. However, their dispersal is not hindered by small rivers and streams—when encountering water barriers, armadillos simply inflate their intestines and float like corks, or sink to the bottom and walk across the riverbed until reaching the other side. The latter developed as a secondary benefit from an adaptation whose original function enabled armadillos to hold their breath for up to six minutes as they foraged with their heads buried in loose soil.

The Nine-banded Armadillo—the only species of its kind in the United States—is unusual in many respects. Unlike any other living mammal in North America, it has an armor-like shell covering its back, sides, head, tail, and the outer surfaces of its legs. The shell itself consists of nonoverlapping, hardened scales connected by flexible skin that covers a series of dense, bony plates. The hardened carapace caused some early naturalists to mistakenly classify armadillos as turtles. Playfully, Rudyard Kipling (1865–1936) proposed in a fanciful children's story that armadillos were the offspring of a cross between a hedgehog and a tortoise. In addition to its armor, the armadillo's peg-like teeth, which lack enamel, and its reproductive biology differ from those of almost all other mammals.

The mating season in Texas extends from July to August. During this period, females usually mate with only one male, and soon after copulation, a single egg is fertilized. The fertilized egg (now called a zygote) remains in the uterus but undergoes no further development until late fall— or sometimes for two to three years—in a classic example of delayed implantation. Then, when the zygote eventually implants in the uterine lining, cell division forms not one, but two embryos. Shortly thereafter, each of these divides again, with each of the four embryos individually attached by a separate umbilical cord to a common placenta. Since all were derived from the same fertilized egg,

283

the four embryos are genetically identical, and hence all are of the same sex (a unique condition termed polyembryony). After a gestation period lasting about 4.5 months, the mother digs an underground nest and bears her young (contrary to popular lore, armadillos do not lay eggs). The littermates nurse from their mother for about 40 days, emerge and begin foraging in midsummer, and disperse by early fall to live on their own. After reaching sexual maturity at one year of age, a female usually reproduces every year thereafter during a lifespan of 12 to 15 years.

For most of each year, Nine-banded Armadillos remain solitary, and each excavates an extensive burrow system—the territory of some individuals includes up to 12 burrows. Active mainly at night, armadillos forage by sniffing the ground for grubs, beetles, ants, termites, worms, and other foods, which they rapidly dig up and slurp down with their long, sticky tongues. During their wanderings, armadillos stop frequently to sniff the air for indications of danger. When threatened, Nine-banded Armadillos may employ one or more of four defensive behaviors: (1) jump straight up 3–4 feet in the air, startling the attacker; (2) flee with unexpected speed; (3) dive into a burrow; or (4) quickly dig a shallow trench and anchor itself inside, exposing only its durable outer shell. In Texas, armadillos deal primarily with four predators: American Alligators, Bobcats, Coyotes, and the family car. The combined habits of searching highways for road-killed insects and jumping upward when alarmed may well be the major cause of their mortality.

The Nine-banded Armadillo is the only nonhuman vertebrate susceptible to infections of *Mycobacterium leprae,* the bacterium causing leprosy. Armadillos likely acquired the bacterium from humans after Euro-Americans colonized the Western Hemisphere. Although armadillos rarely present signs of infection, the disease may kill infected animals. In Texas, armadillos infected with leprosy generally occur along the Gulf Coast, but even in this area the threat of transmission to humans remains minimal if a few risky situations are simply avoided. Namely, armadillos should not be handled or eaten, despite their reputation as good table fare that "tastes like pork."

In the late 1970s, the Texas legislature quashed the first attempt to establish armadillos as the State Mammal because of their reputation as pests. Then in 1981, Governor Bill Clements used his executive power to declare the armadillo the "Official State Mascot." Debates continued about naming an official State Mammal, but no proposals reached the floor of the Texas House of Representatives again until two competing nominations were submitted in 1994. One group advocated the Texas Longhorn, a breed of cattle imported by early Spanish settlers and widely recognized for their massive horns, which can span 7 feet tip to tip. Another group vigorously endorsed the armadillo. To settle the dispute, the legislature decided to seek the opinion of the state's schoolchildren. When the children's votes were tabulated, support for the two candidates was as evenly divided as the legislature itself. With Solomon-like wisdom, House Concurrent Resolution No. 174 proclaimed the Texas Longhorn the official State Large Mammal of Texas and the armadillo as the official State Small Mammal of Texas. Although it is not specified in the resolution, logic presumes that the Nine-banded Armadillo, the only armadillo occurring in Texas—or in the United States—was the species recognized.

THE STATE NATIVE PEPPER OF TEXAS: CHILTEPIN (1997)

After the Japaleño was named the official State Pepper in 1995, some realized that the popular green pepper known for its spicy bite was not a native plant. To rectify this oversight, in 1997 the Texas legislature designated the only wild native chile in Texas, the fiery Chiltepin, as the State Native Pepper. "Chile," by the way, is correct—"chili" is an Anglicized corruption of the Spanish spelling.

Chiltepins earned the distinction of being the "mother of all peppers" because they may be the oldest species in the genus *Capsicum.* They grow naturally in dry canyons—usually under bushes or trees—from West Texas to southern Arizona, but many domesticated varieties have been cultivated. Although these perennial shrubs have may live for 25 to 30 years and reach heights of nearly 10 feet in frost-free locations, most bushes of the wild type typically grow to about 4 feet and, when cut back by hard freezes, usually produce basal sprouts. Appearing in summer or early fall, each of their tiny white blossoms yields a round, pea-sized green berry that turns red as it ripens.

When chewed, Chiltepins produce a far greater burning sensation than Jalapeños. Scoville heat units (SHUs) measure the "hotness" of peppers with a pungency index reflecting the concentration of capsaicin—the chemical source of the burning sensation. Native Chiltepins register between 100,000 and 265,000 SHUs, depending on the amount of rainfall when the berries develop, whereas Jalapeños, often regarded as too "hot," rate a mere 2,500 to 9,000 SHUs. Some can detect the "heat" from just one ounce of dried Chiltepins— with seeds removed—in 300 gallons of salsa. Fortunately, the searing sensation does not persist for long, described in local lingo as *arrebatado* (rapid).

Wild Chiltepins generally prefer well-drained silty or sandy loams. Many develop under nurse trees from seeds dropped by birds, but they grow equally well in the sun. Birds, because of their weak senses of smell and taste, feast on the berries without suffering ill effects. Northern Mockingbirds, the State Bird of Texas, and Wild Turkeys are especially fond of Chiltepins and serve as key agents for dispersing the seeds. Humans also use the small berries for food as well as medicine. Sun-dried flakes from the berries add a piquant zing to cheese and ice cream and add a zippy component to many recipes. Outlandish as it seems, residents in the Trans-Pecos tout Chiltepins as a treatment for acid indigestion.

A fellow horticulturist presented President Thomas Jefferson with dried fruits and seeds from Chiltepins. The president became so enamored with the distinctive flavor that he considered marketing the peppers. Since then, horticulturists have produced many variants, which led to concerns for preserving the genetic integrity of the native strain. To this end, the USDA Forest Service partnered with Native Seeds/SEARCH to establish the 2,500-acre Wild Chile Botanical Area in Coronado National Forest near Tumacacori, Arizona. The area protects a large, wild Chiltepin population for research and use as a genetic

reserve. All told, the long history, unique taste, and many uses of Chiltepin underscore its recognition as a state icon.

THE STATE NATIVE SHRUB OF TEXAS: TEXAS PURPLE SAGE (2005)

On the same day that Governor George W. Bush signed the resolution naming the Chiltepin as the State Native Pepper of Texas—thereby rectifying an oversight that earlier recognized an alien species, the Jalapeño, as the State Pepper—he also signed a resolution proclaiming the Crepe Myrtle, another introduced species, as the State Shrub of Texas. Such a lapse might be forgiven if the legislature and governor had been unaware of the Crepe Myrtle's foreign origins, yet the proclamation itself boldly stated that the Crepe Myrtle had arrived from China a century earlier.

Several years later, Thomas Adams, a botanist and a member of the Native Plant Society of Texas, began an effort originally intending to replace the Crepe Myrtle with a native shrub better suited as a state icon. After deliberation, Thomas and members of the Native Plant Society proposed Texas Purple Sage as a worthy replacement. However, demoting one state icon in favor of another would involve complicated and perhaps prickly legislative maneuvering. To avoid such a tangle, Representative Dennis Bonnen of Angleton employed a tactic similar to the one that named the Chiltepin the State Native Pepper—he proposed to designate an official *native* shrub. Governor Rick Perry signed the resolution; hence Texas Purple Sage complemented, rather than replaced, Crepe Myrtle on the list of state icons.

Texas Purple Sage, often known as Cenizo in many areas of the state, occurs on rocky, limestone-based soils from northern Mexico through the Edwards Plateau and West Texas and into eastern New Mexico. The species, which evolved in the Chihuahuan Desert, tolerates drought and heat and rarely reaches 5 feet in height or 6 feet in breadth. Sometimes called "Barometer Bush" because it usually flowers following periods of high humidity or soil moisture, Texas Purple Sage offers a stunning display of lavender blooms set against silvery to grayish-green leaves. The flowers attract several species of hummingbirds as well as numerous butterflies, moths, and bees, and Northern Mockingbirds, the State Bird of Texas, often select secluded nesting sites in the dense foliage. Native Americans brewed herbal tea from Cenizo leaves as a remedy for chills and fever.

Although called a "sage," the species actually belongs in the figwort family (Scrophulariaceae) along with common species such as Texas Toadflax and Indian Paintbrush. Some mistake the plant as the species featured in the title of a well-known Zane Grey novel, *Riders of the Purple Sage.* The novel, however, is set in Utah, far beyond the distribution of the Texas Purple Sage, and the plant named in the title more likely refers to a true sage in the genus *Salvia,* which also produces showy purple flowers.

Horticulturists have developed several cultivars of Cenizo suitable for growth in those parts of Texas lying outside its original range. Cultivated varieties serve as windbreaks and as components in landscape designs, especially xeriscapes; today they enhance the beauty of homes, businesses, and highways throughout the state.

THE STATE AMPHIBIAN OF TEXAS: TEXAS TOAD (2009)

The Texas Toad was not the first species proposed and approved by the state legislature as the official State Amphibian. At the suggestion of a school librarian, a class of fourth graders at Danbury Elementary School in Danbury, Texas, undertook a research project designed to acquaint the students with databases and library books. The students discovered that Texas had adopted several plants and animals as official icons, but it had yet to designate a State Amphibian. After further research, the students submitted their suggestion, the Texas Blind Salamander, to state Representative Dennis Bonnen. Impressed with the students' initiative, Rep. Bonnen crafted House Concurrent Resolution No. 30 (HCR 30), which was approved by both the House and Senate. However, Governor Rick Perry refused to sign the resolution, formally explaining his decision as follows: "The official designation of items and objects as much-loved objects of Texas should represent the entire state and not just one region or locality. This resolution designates an amphibian as the official State Amphibian of Texas

that is found in only one Texas county. Such a small area does not adequately represent the State of Texas as a whole. Therefore, I am not signing HCR 30 by Rep. Bonnen."

Although likely dismayed by this turn of events, the students nonetheless pushed forward. Biologists from the Texas Parks and Wildlife Department graciously provided the class with a list of five potential candidates to consider as alternate nominees. Working in five groups—each assigned to research one of the species—the students campaigned for their candidate with speeches, posters, and commercials prior to the class-wide vote for the amphibian the Danbury students would again propose for legislative action. Their selection, the Texas Toad, became the State Amphibian of Texas when Governor Perry signed House Concurrent Resolution No. 118—slightly more than two years after he rejected the first proposal.

In light of the governor's objection to their first submission, the students chose well—the Texas Toad is distributed across most of the state. The species occurs in all but the far northwestern corner of the Panhandle and the forests in the eastern quarter of Texas. Texas Toads thrive in open habitats with cover types as varied as Creosotebush flats, mesquite-dominated pastures, prairies, and open woodlands. The robust amphibian feeds on insects during evening and nighttime hours but escapes heat and desiccation during the day by burrowing in loose soil or hiding in mud cracks or rodent burrows, or under vegetation and other protective cover. During extended dry periods, Texas Toads remain dormant in their refuges.

A few characteristics in combination easily distinguish the Texas Toad from nine other toads of similar appearance. As with the other toads, small wart-like bumps cover its chubby body, but it lacks the middorsal stripe found on most of the other species. The hind feet of the Texas Toad have a pair of sharp-edged black tubercles, the innermost of which is distinctively sickle shaped. The oval parotid glands, located behind the eyes, secrete a toxic substance that deters predators. If humans handle Texas Toads, the irritating secretions easily transfer from unwashed hands to eyes and other moist membranes, but contrary to popular belief, humans cannot get warts from handling these or other toads.

Texas Toads rarely stray far from permanent water, where they breed following heavy rains anytime from April to September. They deposit large clumps of eggs around the base of submerged vegetation, and after hatching, the tadpoles mature in 18 to 60 days, depending on water temperature, resource availability, and the density of tadpoles in the pond. The species usually reaches adult size, 2–3.5 inches, by the following summer.

After learning that their efforts to designate a State Amphibian paid off, one of the students involved with the project said, "It makes me feel like we are going to be put in the history books. So when kids learn about the Texas Toad, we can remember we were the ones that made it our State Amphibian." Clearly, Danbury Elementary School and its students achieved a well-earned place in Texas history.

THE STATE DINOSAUR OF TEXAS: *PALUXYSAURUS JONESI* (2009)

Yogi Berra (1925–2015), the Hall of Fame catcher for the New York Yankees, once declared optimistically, "It ain't over 'til it's over." His celebrated comment related to his team's run for the pennant, but the observation applies equally well to endeavors to identify the State Dinosaur of Texas. In fact, "it still ain't over."

More than 65 million years after a herd of huge dinosaurs rumbled across a mudflat, their fossilized footprints were discovered in a limestone shelf bordering the Paluxy River in Somervell County (see fig. 1.8). Paleontologists who initially examined the tracks attributed the imprints to extremely tall, long-necked and long-tailed sauropods in the genus *Pleurocoelus.* Although hardly ferocious, these herbivores were immense and among the largest of all dinosaurs—a characteristic that no doubt impressed members of the state legislature. To claim a bit of prehistory for the state, *Pleurocoelus* became the official Lone Star State Dinosaur in 1997 when Governor George W. Bush signed Senate Concurrent Resolution No. 57.

However, the scientific name of the dinosaur was already in dispute when the resolution was signed. Some paleontologists mounted a convincing argument that *Astrodon* was the appropriate generic name for the ancient animal and downgraded *Pleurocoelus* to a junior synonym. The issue seemed resolved until 2007, when Peter J. Rose, then a graduate student at Southern Methodist University, examined some dinosaur bones from Hood County, also traversed by the Paluxy River. He identified these as representing a new species, which he named *Paluxysaurus jonesi,* thereby linking the footprints along the Paluxy River with the Jones Ranch where the fossil bones were discovered. To conform with the new taxonomic designation, state Representative Charles Green introduced House Concurrent Resolution No. 19. Accordingly, *Paluxysaurus jonesi* replaced *Pleurocoelus* as the official Texas State Dinosaur on June 16, 2009, when Governor Rick Perry signed the House Resolution.

Under most circumstances, the story would end here, but, just as Yogi declared, it was not over—not yet. In 2012, a pair of paleontologists reexamined the bones from the Jones Ranch and compared these with similar fossils collected from deposits in Oklahoma, Wyoming, and elsewhere in Texas. After careful reanalysis, the scientists concluded that specimens attributed to *Paluxysaurus* were the same as those of *Sauroposeidon,* another huge sauropod. In short, *Sauroposeidon jonesi* emerged as the correct name for the official Texas State Dinosaur, but in order to again update the list, the legislature will have to introduce yet another resolution identifying the state's official dinosaur. Perhaps then, even Yogi would believe "it's over."

THE STATE BISON HERD OF TEXAS: BISON HERD AT CAPROCK CANYONS STATE PARK (2011)

The US Congress designated the American Bison as the National Mammal of the United States in 2016, and the species well fills its role. According to some estimates, as many as 60 million once roamed the North American plains and prairies, the vast herds surely as majestic as any other natural wonder on the continental landscape. Bison remain well known to virtually all Americans, in part because of the infamous story of their tragic fate (chapter 8). By 1889, fewer than 1,000 had survived the relentless slaughter that swept away the herds in the 1870s, and many of these found refuge in rugged ravines in Palo Duro Canyon and its tributary gulches.

When legendary rancher Charles Goodnight (1836–1929; infobox 6.1) moved his cattle into Palo Duro Canyon, where the grassy range required no fencing, his cowboys rescued bison calves orphaned when the main herd was pushed deeper into the canyon. At the urging of his wife, Mary Ann (Molly), Goodnight protected and nurtured this small remnant of the once-great southern bison herd. Goodnight and his successors maintained the Palo Duro herd until 1996, when the owners of the JA Ranch donated the animals to the state of Texas for release at Caprock Canyons State Park, a 15,314-acre area of eroded plains carved into the eastern edge of the Llano Estacado.

Genetic tests of the Caprock Canyon herd revealed two matters of interest. First, the animals represented a genetic strain unlike American Bison elsewhere, and they were not contaminated with latent diseases (e.g., brucellosis). Second, the tests warned that continual inbreeding had significantly reduced the herd's genetic diversity to the point where, unless revitalized with "new blood," it would face extinction within decades from various defects, not the least of which was impaired reproduction. To meet this challenge, three bulls from the ranch of media mogul Ted Turner in New Mexico were added to the breeding population at the state park. Today, the herd has grown to nearly 100 animals that thrive on the park's grasslands, much to the delight of visitors armed with cameras instead of rifles. Because large blocks of native grassland no longer remain on the Llano, the conservation efforts at Caprock Canyons State Park stand alone as a successful restoration program for American Bison in Texas.

The Goodnight herd represents one of five foundation herds of American Bison in the United States that contributed to the majority of bison surviving elsewhere in the nation. In 2011, the descendants of the small herd established at Caprock Canyons State Park were formally designated the Texas State Bison Herd, a group celebrated for their strong genetic link to the ghosts of those once thundering across the Southern Plains.

THE STATE SALTWATER FISH OF TEXAS: RED DRUM (2011)

On a wind-rippled surface of a shallow Texas bay, a sudden splash draws attention to a seagrass meadow where, with luck, a quick glance reveals a fish tail marked with a distinct dark spot. Seasoned fishermen and experienced naturalists recognize this scenario as the "tailing" behavior of Red Drum, which characteristically forage with their heads down while breaking the surface with their tails.

The fish's common name stems from its color and the drumming noises the males make during the spawning season. The color of the adults varies individually from reddish bronze, the most common hue, to either a blackish coppery tone or nearly silver. These variations seem related to habitat: dark copper-colored fish usually occur in muddy-bottomed, murky bays, whereas silvery forms more often appear in surf zones or areas with sandy bottoms. But all bear at least one, and sometimes two or more, black spots on either side near the base of the tail. Older fish tend to retain one large spot after the excess spots fade away. Because the tail spots resemble eyes, the markings suggest a form of mimicry that might direct the attack of a predator toward the tail, a less vulnerable part of the body.

Red Drum spawn offshore, where they usually congregate near passes cutting through barrier islands or at the mouths of rivers. Spawning takes place during an eight- to nine-week period from mid-August to mid-October. Each night during the spawning season, males gather in large schools and attract females with drumming noises produced by vibrating a muscle in their swim bladders. The males will spawn every night, but the females visit these areas about once every two to seven days, and each lays 200,000 to 3 million eggs per event, or about 20–40 million eggs per season.

Fertilized eggs contain minute oil droplets that provide both nourishment and flotation for the developing larvae. Tidal flows carry many eggs to bays and lagoons, where they hatch as tiny larvae within 24 hours; these grow rapidly in the warm water, becoming fry that reach about 11 inches and one pound in the first year. For the first three years of life, young Red Drum, known locally as "Rat Reds," remain in bays and lagoons, feeding on small

Blue Crabs, shrimp, and small fishes. After growing to 22–24 inches and 6–8 pounds, the fish migrate to nearshore waters in the Gulf of Mexico where they remain for the rest of their lives. Most anglers refer to Red Drum larger than 30 inches as "Bull Reds" or simply "Redfish." Some Red Drum live 40 years or longer and grow continuously. The Texas record for a Red Drum is 59.5 pounds, but some grow much larger—a Bull Red caught along the North Carolina coast weighed 94 pounds.

During the 1950s, photographs of proud anglers holding stringers of 20–100 Rat Reds lined the walls of bait and tackle shops along the Texas coast. Enlarged and framed, some of these images now decorate popular seafood restaurants, invoking memories of a long-gone era. The popularity of Red Drum increased when a famous New Orleans chef developed his now-renowned dish—Cajun-style blackened Redfish. Increased pressure in the 1980s from both recreational and commercial fishing initiated a steep decline in the Red Drum population, which led the states along the Gulf Coast to impose strict size and creel limits. Additionally, Texas Governor George W. Bush later protected Red Drum from commercial harvest and sale. Because of these restrictions, and bolstered by the release of young fish raised in a saltwater hatchery, the Redfish population in Texas steadily recovered.

Although highly regarded as a game fish, the Red Drum was not quickly adopted as the official State Saltwater Fish of Texas—the Atlantic Tarpon loomed as a contender for the title. As early as 1903, Charles F. Holder advocated that Texas name the tarpon, popularly called the "Silver King," a state symbol because he believed that the "splendid fish" would attract visitors to Texas. The issue simmered until 2009, when, at the request of tarpon fishermen, Representative Brandon Creighton submitted a resolution naming the Atlantic Tarpon the official saltwater fish of Texas. Despite significant support, however, the proposal died in the House Committee on Culture, Recreation and Tourism.

During the 2011 session of the Texas legislature, Representative Dennis Bonnen proposed that the Red Drum become the official State Saltwater Fish of Texas. The "comeback" of the species— an environmental success story—highlighted the reasons for celebrating the Red Drum, and Governor Rick Perry promptly signed the House Concurrent Resolution. North Carolina also recognizes the Red Drum (known there as "Channel Bass") as that state's saltwater fish.

THE STATE SEA TURTLE OF TEXAS: KEMP'S RIDLEY SEA TURTLE (2013)

Texas schoolchildren figured prominently in two successful campaigns to honor native species— the Nine-banded Armadillo and Texas Toad—as state symbols. Perhaps with that history in mind, a fourth-grade science teacher, Katie Blaser, at Oppe Elementary School in Galveston engaged her after-school environmental group in a quest to name the Kemp's Ridley Sea Turtle the official Texas State Sea Turtle. Known as the "Green Team," the students first researched the turtle's natural history and then met with their local state representative, Craig Eiland, to determine how to present their proposal to the legislature. Duly impressed, Rep. Eiland agreed to help.

Two months after Rep. Eiland filed House Concurrent Resolution No. 31 on behalf of the Green Team, 20 members of the team spoke to the House Committee on Culture, Recreation and Tourism. Four well-prepared students—Blaine Heffernan, Violet Schubert, Nicholas Smecca, and Isabella Walser—answered the committee's questions. Soon thereafter, the resolution naming Kemp's Ridley as the State Sea Turtle of Texas was passed unanimously in both the House and Senate and was signed by Governor Rick Perry.

The smallest of the five sea turtles that regularly occur in the Gulf of Mexico, the Kemp's Ridley alone has an almost circular, olive-green to gray carapace (upper shell), which reaches lengths of 28 inches. Adults, which usually weigh between 75 and 100 pounds, rarely leave the Gulf of Mexico, where they typically remain in nearshore waters less than 165 feet deep. Little is known of their food habits, but they apparently prefer to hunt for Blue Crabs and other bottom-dwelling organisms instead of foraging at the surface. However, they sometimes feed in the drift lines where floating marine vegetation, jellyfish, and debris collect.

At the beginning of the nesting season, female

Kemp's Ridley Sea Turtles gather in the waters near a nesting beach, where they mate with waiting males several weeks before nesting. In events called *arribazones,* throngs of gravid females emerge en masse during daylight hours to nest along a narrow stretch of beach. Each female crawls to a dune area well above the high tide zone, where she digs a nest and deposits and covers about 100 eggs before returning to the sea about 45 minutes later. Females nest at intervals of one to three years, but some produce as many as four clutches during a single nesting season. The soft, leathery eggs hatch after 45–55 days of incubation, during which the temperature in the nest determines the gender of the offspring—cooler temperatures produce males and warmer temperatures, primarily females.

After hatching, the young dig out of the nest and hustle to the surf to evade land predators. Even after reaching the open waters of the Gulf, however, the small turtles must run an additional gauntlet of marine predators. The hatchlings drift for months in open water before returning to bays and lagoons along the coast, where they feed in seagrass meadows on small crabs, snails, bivalves, and fish. Kemp's Ridleys reach sexual maturity at 10–20 years of age, and when ready to mate, most females return to the same beach where they hatched. The hatchlings imprint on some environmental cue, perhaps the smell, chemical composition, or magnetic location of the beach where their nests were located.

Kemp's Ridleys nest on beaches in Tamaulipas, Mexico, as well as near Veracruz, Matagorda Island, and Padre Island National Seashore, but the largest arribazón occurs on an isolated beach near Rancho Nuevo in Tamaulipas. A film made there in 1947 shows about 40,000 females arriving to nest in a single day, but unfortunately, this population later declined precipitously. These losses occurred between 1947 and the early 1970s, probably resulting from intensive harvests of eggs for human food and from turtles drowning in the nets of shrimp trawlers. More recently, large numbers of turtles died after ingesting plastic debris.

Although difficult to enforce, laws enacted by both Mexico and the United States now protect nesting areas, curb egg harvests, and reduce dumping in the Gulf. Drownings lessened after 1990, when the National Marine Fisheries Service (NMFS) required that all trawling nets be equipped with turtle excluder devices. Both nations list the species as endangered and cooperate in recovery efforts and research. One project, designed to study turtle movements, outfitted females leaving the nesting beaches at Rancho Nuevo with radio transmitters. Eggs were also shipped from Mexico to supplement nests at Padre Island National Seashore as part of a "head-starting" program. Hatchlings emerging from these nests were collected and reared in an NMFS laboratory in Galveston and then released in the Gulf of Mexico in anticipation of increasing the nesting population returning to Texas. In any event, the combined efforts of a dedicated teacher, an energetic group of fourth graders, and a responsive legislator played an important role in the conservation of the State Sea Turtle of Texas.

APPENDIX B
SCIENTIFIC NAMES

This appendix lists the scientific names of organisms mentioned in the text. Alternate common names, and scientific names no longer in use but occurring in older sources, appear in brackets; some accounts include informative notations. An asterisk (*) denotes an extinct species, whereas a superscript E ([E]) indicates an introduced (exotic) species.

BACTERIA, ALGAE, FUNGI, AND LICHENS

Algae, photosynthetic unicellular or multicellular organisms lacking distinct cell and tissue types such as xylem and phloem [singular: alga]

Botulism, *Clostridium botulinum* [bacterium producing a toxin that causes illness]

Brown algae, class Phaeophyceae [large, multicellular, brownish seaweeds]

Brown tide, see Texas Brown Tide

Cladocera, tiny freshwater crustaceans commonly called waterfleas

Cyanobacteria, phylum Cyanobacteria [photosynthetic bacteria; singular: cyanobacterium]

Desmid, unicellular green algae with symmetrical shapes

Diatom, algae possessing cell walls composed of silicon

Foraminifera, single-celled protozoa with a "shell" called a test

Fowl Cholera, *Pasteurella multocida* [a bacterium]

Fungi, kingdom Fungi [organisms that acquire materials needed for life by dissolving molecules, usually from decaying organisms; singular: fungus]

Fusiform Rust, *Cronartium fusiforme*

Green algae, division Chlorophyta [nonflowering marine seaweeds or freshwater macrophytes with chloroplasts]

Lichen, many species [composite organisms—part algae or cyanobacteria, part fungi]

Microzooplankton, tiny, one-celled aquatic organisms capable of movement

Phytoplankton, one-celled, floating photosynthetic aquatic organisms

Red algae, division Rhodophyta [seaweeds containing red, yellow, or violet pigments]

Red Heart Fungus, *Phellinus pini*

Red Tide, *Karenia brevis*

Sylvatic Plague, *Yersinia pestis* [a bacterium]

Texas Brown Tide, *Aureoumbra lagunensis*

GREEN PLANTS

Acacia, *Acacia* spp.

Agarita, *Mahonia trifoliata*

Agave, *Agave* spp.

Alkali Sacaton, *Sporobolus airoides*

Alligator Juniper, *Juniperus deppeana*

Allthorn, *Koeberlinia spinosa*

Amelia's Sand Verbena, *Abronia ameliae*

American Beautyberry, *Callicarpa americana*

American Beech, *Fagus grandifolia*

American Elm, *Ulmus americana*

American Holly, *Ilex opaca* [Christmas Holly]

American Hornbeam, *Carpinus caroliniana*

American Sweetgum, *Liquidambar styraciflua*

American Sycamore, *Platanus americanus*

American Tarbush, *Flourensia cernua*

American White Water Lily, *Nymphaea odorata*

Anacua, *Ehretia anacua*

Annual Sunflower, *Helianthus annuus* [Common Sunflower]

Arkansas Dogshade, *Limnosciadium pinnatum*

Arkansas Meadow-rue, *Thalictrum arkansanum*

Arrowhead, *Sagittaria longiloba*

Ashe Juniper, *Juniperus ashei*

Aster, *Aster* spp.

Azalea, *Rhododendron* spp.

Baldcypress, *Taxodium distichum* [Bald Cypress, Bald-Cypress]

Ball Moss, *Tillandsia recurvata* [member of the pineapple family; not a moss]

Baretta, *Helietta parvifolia* [native citrus tree restricted to Starr County]

Barnyard Grass, *Echinochloa crus-galli* [Wild Millet]

Basin Bellflower, *Campanula reverchonii*

Beach Tea, *Croton punctatus*

Beaked Panicgrass, *Panicum anceps*

Beak rushes, *Rhynchospora* spp.

Beechdrops, *Epifagus virginiana*

Berlandier Ash, *Fraxinus berlandieriana* [Rio Grande Ash, Mexican Ash]

Big Bluestem, *Andropogon gerardii*

Big Floating Heart, *Nymphoides aquatica*

Big Red Sage, *Salvia pentstemonoides*

Bigtooth Maple, *Acer grandidentatum*

Bitter Panicum, *Panicum amarum*

Blackbrush, *Acacia rigidula* [Blackbrush Acacia, Chaparro Prieto; *Vachellia rigidula*]

Black Grama, *Bouteloua eriopoda*

Blackgum, *Nyssa sylvatica* [Black Gum]

Black Hickory, *Carya texana* [Texas Hickory]

Black Highbush Blueberry, *Vaccinium fuscatum* [*V. arkansanum*]

Blackjack Oak, *Quercus marilandica*

Black Lace Cactus, *Echinocereus reichenbachii* var. *albertii*

Black Mangrove, *Avicennia germinans*

Black Oak, *Quercus velutina*

Blacksamson, *Echinacea angustifolia*

Black Walnut, *Juglans nigra*

Black Willow, *Salix nigra*

Bladderwort, *Utricularia* spp.

Blood Milkwort, *Polygala sanguinea*

Bloodroot, *Sanguinaria canadensis*

Bluebonnet, *Lupinus* spp.

Blue Grama, *Bouteloua gracilis*

Bluejack Oak, *Quercus incana* [Sandjack Oak]

Blue Oak, *Quercus douglasii*

Bog Hemp, *Boehmeria cylindrica* [Smallspike False Nettle]

Brasil, *Condalia hookeri* [Bluewood, Bluewood Condalia]

Broom Dalea, *Psorothamnus scoparius* [*Dalea scoparia*]

Broomsedge Bluestem, *Andropogon virginicus*

Broomweed, *Amphiachyris* spp. and *Gutierrezia* spp. [small, yellow-flowered asters]

Brownseed Paspalum, *Paspalum plicatulum*

Buffalobur Nightshade, *Solanum rostratum*

Buffalograss, *Buchlöe dactyloides*

Buffelgrass, *Pennisetum ciliare*

Bulrush, *Schoenoplectus* spp. [*Scirpus* spp.]

Bur Oak, *Quercus macrocarpa*

Bush Muhly, *Muhlenbergia porteri*

Butterwort, *Pinguicula* spp.

Buttonbush Flatsedge, *Cyperus cephalanthus*

Buttonwood, *Conocarpus erectus*

Camphor Daisy, *Rayjacksonia phyllocephala* [*Machaeranthera phyllocephala*]

Camphor Weed, *Heterotheca subaxillaris*

Candelilla, *Euphorbia antisyphilitica* [Candelaria, Wax Plant, Wax Weed]

Cane Bluestem, *Bothriochloa barbinodis* [*Andropogon barbinodis*]

Canyon Mock Orange, *Philadelphus ernestii*

Carolina Ash, *Fraxinus caroliniana*

Carolina Laurel Cherry, *Prunus caroliniana*

Carolina Wolfberry, *Lycium carolinianum*

Catclaw Acacia, *Acacia greggii* [an alternate name, Long-flowered Acacia, distinguishes this plant

with recurved thorns from Catclaw Mimosa, which has puffball-like flowers]

Catclaw Mimosa, *Mimosa aculeaticarpa* [Wait-a-Bit, because its recurved thorns catch and hold]

Cattail, *Typha* spp.

Cedar Elm, *Ulmus crassifolia*

Cedar Sage, *Salvia roemeriana*

Cenizo, *Leucophyllum frutescens* [Texas Purple Sage, Texas Silverleaf, Texas Sage]

Chapotillo, *Amyris texana* [Texas Torchwood]

Chatterbox Orchid, *Epipactis gigantea*

Cherrybark Oak, *Quercus pagoda*

Chiltepin, *Capsicum annuum* var. *aviculare* [Chilipiquin, Chile Tepin]

Chinese Tallow, *Triadica sebifera* [E] [*Sapium sebiferum*]

Chittamwood, *Bumelia lanuginosa*

Christmas Fern, *Polystichum acrostichoides*

Cinnamon Fern, *Osmunda cinnamomea*

Climbing Hempweed, *Mikania scandens* [Climbing Hempvine]

Clover Grass, *Halophila engelmannii* [Peanut Grass, Star Grass]

Coastal Gayfeather, *Liatris bracteata*

Coastal Pepperbush, *Clethra alnifolia*

Coast Redwood, *Sequoia sempervirens*

Colima, *Zanthoxylum fagara* [Lime Pricklyash]

Colorado Pinyon Pine, *Pinus edulis*

Columbine, *Aquilegia* spp.

Common Beggar-Tick, *Bidens frondosa*

Common Buttonbush, *Cephalanthus occidentalis*

Common Persimmon, *Diospyros virginiana*

Common Reed, *Phragmites australis* [River Cane]

Common Shooting Star, *Dodecatheon meadia*

Compassplant, *Silphium laciniatum*

Coneflower, *Echinacea atrorubens*

Coralberry, *Symphoricarpos orbiculatus*

Coral Root orchids, *Hexalectris* spp. [*H. spicata, H. nitida, H. warnockii,* and *H. grandiflora* occur in Dallas County]

Coreopsis, *Coreopsis* spp. [several species of brilliant, multiflowered yellow daisies]

Corkwood, *Leitneria pilosa pilosa*

Cottonwood, *Populus* spp. [see Eastern Cottonwood]

Coyotillo, *Karwinskia humboldtiana*

Cranefly Orchid, *Tipularia discolor*

Creosotebush, *Larrea tridentata* [Creosote Bush]

Crested Coral Root, *Hexalectris spicata*

Crucita, *Chromolaena odorata* [Blue Mistflower, Jack in the Bush; *Eupatorium odoratum*]

Curly Mesquite, *Hilaria belangeri* [Curlymesquite]

Deciduous Holly, *Ilex decidua*

Deeprooted Sedge, *Cyperus entrerianus* [E]

Desert Olive, *Forestiera pubescens* [Stretchberry]

Desert Tobacco, *Nicotiana obtusifolia*

Desert Willow, *Chilopsis linearis*

Dewberry, *Rubus* spp.

Douglas-Fir, *Pseudotsuga menziesii* [because this tree is not a true fir, the common name is often hyphenated]

Duckweed, *Lemna minor*

Dwarf Palmetto, *Sabal minor*

Eastern Cottonwood, *Populus deltoides*

Eastern Gamagrass, *Tripsacum dactyloides*

Eastern Prickly Pear, *Opuntia humifusa*

Eastern Redbud, *Cercis canadensis*

Eastern Redcedar, *Juniperus virginiana*

Eastern Swamp Privet, *Forestiera acuminata*

Elegant Waterlily, *Nymphaea elegans*

Elm, *Ulmus* spp.

Erect Colubrine, *Colubrina stricta*

Escarpment Black Cherry, *Prunus serotina* var. *eximia*

Farkleberry, *Vaccinium arboreum*

Fendler's Bladderpod, *Lesquerella fendleri*

Finger Poppy-Mallow, *Callirhoe pedata*

Firewheel, *Gaillardia pulchella* [Indian Blanket]

Florida Maple, *Acer floridanum*

Flowering Dogwood, *Cornus florida*

Four-angle Sedge, *Carex tetrastachya*

Fourwing Saltbush, *Atriplex canescens*

Fox Grape, *Vitis vulpina*

Fragrant Ladies' Tresses, *Spiranthes odorata* [Swamp Tresses]

Gambel Oak, *Quercus gambelii*

Giant Ball Moss, *Tillandsia baileyi* [a member of the pineapple family; not a moss]

Giant Coral Root, *Hexalectris grandiflora*

Giant Ladies' Tresses, *Spiranthes praecox*

Giant Reed, *Arundo donax*

Giant Salvinia, *Salvinia molesta* [E]

Glasswort, *Salicornia* spp. [*S. utahensis* occurs in western Texas; *S. bigelovii* and *S. virginica* occur along the Texas coast]

Glen Rose Yucca, *Yucca necopina*

Globe Mallow, *Sphaeralcea incana*

Goatfoot Morning-Glory, *Ipomoea pes-caprae* [Railroad Vine]

Golden Smoke, *Corydalis aurea*

Golden Wave, *Coreopsis basalis*

Grama grass, *Bouteloua* spp.

Granjeno, *Celtis ehrenbergiana* [Spiny Hackberry; *C. pallida*]

Grapefruit, *Citrus paradisi*

Gray Oak, *Quercus grisea*

Green Ash, *Fraxinus pennsylvanica*

Greenbrier, *Smilax* spp.

Green Milkweed, *Asclepias viridis*

Gregg's Wild Buckwheat, *Eriogonum greggii*

Ground Cherry, *Physalis pruinosa*

Guajillo, *Senegalia berlandieri*

[Berlandier Acacia; *Acacia berlandieri*]

Guayacan, *Guaiacum angustifolium* [Texas Lignum-vitae, Soapbush; *Porlieria angustifolia*]

Guinea Grass, *Megathyrsus maximus* [E]

Gulf Coast Yucca, *Yucca louisianensis* [Beargrass Yucca]

Gulf Cordgrass, *Sporobolus spartinus* [*Spartina spartinae*]

Gulfdune Paspalum, *Paspalum monostachyum*

Gulf Muhly, *Muhlenbergia filipes*

Gypsum Broomscale, *Lepidospartum burgessii* [Burgess Broomshrub]

Gypsum Grama, *Bouteloua breviseta*

Hackberry, *Celtis* spp. [see Sugar Hackberry]

Hairy Crinklewort, *Tiquilia hispidissima* [*Coldenia hispidissima*]

Hairy Grama, *Bouteloua hirsuta*

Hairy Rattleweed, *Baptisia arachnifera*

Hairy Tridens, *Erioneuron pilosum*

Harvestbells, *Gentiana saponaria*

Hazel Alder, *Alnus serrulata*

Hickory, *Carya* spp.

Honey Mesquite, *Prosopis glandulosa*

Hooked Buttercup, *Ranunculus recurvatus*

Hornwort, *Ceratophyllum* spp.

Horsetail, *Equisetum* spp.

Houston Meadow-rue, *Thalictrum texanum*

Huisache, *Acacia farnesiana* [Sweet Acacia, Needlebush; *A. smallii, Vachellia farnesiana, Mimosa farnesiana*]

Hydrilla, *Hydrilla verticillata* [E] [Waterweed]

Indian Blanket, *Gaillardia aestivalis*

Indiangrass, *Sorghastrum nutans* [Yellow Indiangrass]

Jack-in-the-Pulpit, *Arisaema triphyllum*

Joe-Pye Weed, *Eutrochium fistulosum* [*Eupatorium fistulosum, Eupatoriadelphus fistulosus*]

Johnson Grass, *Sorghum halepense* [E]

Jointed Flat Sedge, *Cyperus articulatus*

Junco, *Adolphia infesta*

Juniper, *Juniperus* spp.

King Ranch Bluestem, *Bothriochloa ischaemum* var. *songarica* [E]

Kleberg Bluestem, *Dichanthium annulatum* [E]

Lanceleaf Loosestrife, *Lysimachia lanceolata*

Langtry Rainbow, *Echinocereus pectinatus* var. *wenigeri* [a local form of hedgehog cactus]

Largeflower Flameflower, *Phemeranthus calycinus* [Rock-Pink; *Talinum calycinum*]

Large-fruited Sand-Verbena, *Abronia macrocarpa*

Large Gallberry, *Ilex coriacea*

Leatherplant, *Jatropha cuneata*

Lechuguilla, *Agave lechuguilla*

Lemon Beebalm, *Monarda citriodora*

Lime Pricklyash, see Colima

Limoncillo, *Esenbeckia berlandieri* [Berlandier's Jopoy, Jopoy; *E. runyonii*]

Lindheimer's Beebalm, *Monarda lindheimeri*

Little Bluestem, *Schizachyrium scoparium* var. *scoparium* [*Andropogon scoparius*]

Little Walnut, *Juglans microcarpa*

Live Oak, see Plateau Live Oak

Loblolly Pine, *Pinus taeda* [Carolina Pine, Oldfield Pine]

Longleaf Pine, *Pinus palustris* [Yellow Pine, Heart Pine, Longstraw Pine]

Long-leaf Wild Buckwheat, *Eriogonum longifolium*

Long-spike Tridens, *Tridens strictus*

Lotebush, *Ziziphus obtusifolia*

Lotus, *Nelumbo lutea*

Lovegrass Tridens, *Tridens eragrostoides*

Macartney Rose, *Rosa bracteata* [E] [Cherokee Rose]

Maidencane, *Panicum hemitomon*

Maidenhair Fern, *Adiantum capillus-veneris*

Manatee Grass, *Cymodocea filiforme* [*Syringodium filiforme*]

Marijuana, *Cannabis sativa*

Marshmillet, *Zizaniopsis miliacea*

Maximilian Sunflower, *Helianthus maximiliani*

Mayapple, *Podophyllum peltatum* [May Apple]

Meadow Beauty, *Rhexia mariana*

Meadow Flax, *Linum pratense*

Mead's Sedge, *Carex meadii*

Mesquite, *Prosopis* spp.

Mexican Buckeye, *Ungnadia speciosa* [Texas Buckeye, Canyon Buckeye]

Mexican Flame, *Anisacanthus quadrifidus* var. *wrightii* [Hummingbird Bush, Wright's Desert Honeysuckle, Flame Acanthus]

Mexican Hat, *Ratibida columnifera*

Mexican Palmetto, *Sabal texana* [Sabal Palm, Cabbage Palmetto]

Mexican Pinyon Pine, *Pinus cembroides*

Mexican Plum, *Prunus mexicana*

Milkweed, *Asclepias* spp.

Mockernut Hickory, *Carya tomentosa*

Mock Orange, *Philadelphus* spp.

Montezuma Baldcypress, *Taxodium mucronatum*

Mormon Tea, *Ephedra trifurca*

Mountain Mahogany, *Cercocarpus montanus*

Mountain Torchwood, *Amyris madrensis* [Shrub Amyris]

Multiflowered False Rhodesgrass, *Trichloris pluriflora* [Showy Chloris; *Chloris pluriflora*]

Mustang Grape, *Vitis mustangensis*

Navasota Ladies' Tresses, *Spiranthes parksii*

Netleaf Hackberry, *Celtis laevigata* var. *reticulata*

Nutgrass, *Cyperus* spp. [Nutsedge]

Nuttall's Stonecrop, *Sedum nuttallianum*

Oak, *Quercus* spp.

Ocotillo, *Fouquieria splendens*

Old Plainsman, *Hymenopappus scabiosaeus*

Onion Blanketflower, *Gaillardia multiceps*

Orange, *Citrus sinensis* [Sweet Orange; *C. aurantium* is a bitter orange]

Osage Orange, *Maclura pomifera* [Bois d'Arc because Native Americans valued its wood for making their bows]

Overcup Oak, *Quercus lyrata*

Oyamel Fir, *Abies religiosa*

Pale Pitcher Plant, *Sarracenia alata* [Yellow Trumpet]

Pale Prairie Coneflower, *Echinacea pallida*

Pale Yucca, *Yucca pallida*

Palo Verde, *Parkinsonia* spp. [see Retama]

Pan American Balsamscale, *Elionurus tripsacoides*

Panic grass, *Panicum* spp.

Papershell Pinyon, *Pinus remota*

Parry Agave, *Agave parryi* [one of many species known as "Century Plant"]

Pawpaw, *Asimina triloba*

Pecan, *Carya illinoinensis*

Peyote, *Lophophora williamsii*

Pink Evening Primrose, *Oenothera speciosa* [Showy Evening Primrose]

Pink Pappusgrass, *Pappophorum bicolor*

Pinyon Pine, *Pinus edulis* [spelled "Piñon" in some sources to reflect Spanish orthography]

Pipewort, *Eriocaulon* spp.

Plains Bristlegrass, *Setaria macrostachya*

Plains Cottonwood, *Populus deltoides monilifera*

Plains Nipple Cactus, *Coryphantha missouriensis*

Plateau Live Oak, *Quercus fusiformis* [*Q. virginiana* var. *fusiformis* is Texas Live Oak]

Plateau Milkvine, *Matelea edwardsensis*

Poison Ivy, *Toxicodendron radicans*

Ponderosa Pine, *Pinus ponderosa*

Pondweed, *Potamogeton* spp.

Possumhaw, *Ilex decidua*

Post Oak, *Quercus stellata*

Prairie-Bishop, *Bifora americana*

Prairie Bluet, *Stenaria nigricans* [Baby's Breath, Diamondflower]

Prairie Coneflower, *Ratibida columnifera*

Prairie Sphagnum, *Sphagnum palustre* [Peat Moss, Sphagnum Moss; a bryophyte]

Prairie Thistle, *Cirsium canescens*

Prickly pear, *Opuntia* spp.

Purple Cliffbreak Fern, *Pellaea atropurpurea*

Purple Threeawn, *Aristida purpurea*

Purple Woolly Loco, *Astragalus mollissimus*

Quaking Aspen, *Populus tremuloides* [Trembling Aspen]

Quinine Bush, *Allenrolfea occidentalis* [Pickleweed]

Ragweed, *Ambrosia psilostachya*

Rattlesnake Flower, *Brazoria truncata*

Rattleweed, *Astragalus* spp.

Red Bay, *Persea borbonia* [Sweet Bay, Laurel Tree]

Redberry Juniper, *Juniperus pinchotii*

Red Grama, *Bouteloua trifida*

Red Mangrove, *Rhizophora mangle*

Red Maple, *Acer rubrum*

Red Sprangletop, *Leptochloa panicea*

Rescuegrass, *Bromus unioloides*

Resurrection Plant, see Siempre Viva

Retama, *Parkinsonia aculeata* [Mexican Palo Verde; sometimes placed in genus *Genista*]

Reverchon's Palafox, *Palafoxia reverchonii*

Rice, *Oryza sativa*

Rio Grande Greenthread, *Thelesperma nuecense*

River Birch, *Betula nigra*

Rock Quillwort, *Isoetes lithophila*

Rocky Mountain Juniper, *Juniperus scopulorum*

Rose Pogonia, *Pogonia ophioglossoides* [Snake-mouth Orchid]

Rough-leaved Dogwood, *Cornus drummondii*

Runyon's Water-Willow, *Justicia pacifica*

Rushes, *Juncus* spp.

Russian Olive, *Elaeagnus angustifolia* [E]

Russian Thistle, *Salsola kali* [E] [Tumbleweed; *S. kali* may represents up to five species]

Sago Pondweed, *Stuckenia pectinata* [*Potamogeton pectinatus*]

Saltcedar, *Tamarix ramosissima* [E] [an invasive species]

Salt-flat Grass, *Monanthochloë littoralis* [Shoregrass]

Saltgrass, *Distichlis spicata*

Saltmeadow Cordgrass, *Sporobolus pumilus* [Marshhay Cordgrass; *Spartina patens*]

Saltwort, *Batis maritima*

Sand Bluestem, *Andropogon hallii* [closely related to Big Bluestem]

Sandjack Oak, *Quercus incana* [Bluejack Oak]

Sand Post Oak, *Quercus margaretta* [may be a variety of Post Oak, *Q. stellata* var. *margaretta*]

Sand Sage, *Artemisia filifolia*

Sandyland Bluebonnet, *Lupinus subcarnosus*

Sassafrass, *Sassafras albidum*

Saw Palmetto, *Serenoa repens*

Screwbean, *Prosopis pubescens*

Scrub Oak, *Quercus sinuata* var. *breviloba* [Bigelow's Oak, Shin Oak]

Sea Blite, *Suaeda linearis*

Seacoast Bluestem, *Schizachyrium scoparium* var. *littorale*

Sea Lavender, *Limonium carolinianum*

Seaoats, *Uniola paniculata* [Beachgrass]

Sea Oxeye, *Borrichia frutescens* [Sea Ox-eye Daisy]

Sea Purslane, *Sesuvium portulacastrum*

Seashore Dropseed, *Sporobolus virginicus*

Seaside Heliotrope, *Heliotropium curassavicum*

Sedges, *Carex* spp.

Seep Muhly, *Muhlenbergia reverchonii*

Seepwillow Baccharis, *Baccharis salicifolia*

Shagbark Hickory, *Carya ovata*

Sharpscale Spikerush, *Eleocharis acutisquamata*

Shinnery Oak, *Quercus havardii* [Havard Oak, Shin Oak, Sand Shinnery Oak]

Shoalgrass, *Halodule beaudettei* [*H. wrightii, Diplanthera wrightii*]

Shoregrass, *Monanthochloë littoralis*

Shortleaf Pine, *Pinus echinata* [Shortleaf Pine, Yellow Pine, Shortstraw Pine]

Sideoats Grama, *Bouteloua curtipendula* [two varieties occur in Texas: var. *curtipendula* with long rhizomes, and var. *caespitosa* lacking rhizomes]

Siempre Viva, *Selaginella lepidophylla* [Resurrection Plant, Rose of Jericho]

Silktassel, *Garrya wrightii*

Silverbells, *Styrax americanus*

Silver Bluestem, *Bothriochloa laguroides* [*Andropogon saccharoides*]

Silverleaf Oak, *Quercus hypoleucoides* [Whiteleaf Oak]

Silveus' Dropseed, *Sporobolus silveanus*

Skunkbush Sumac, *Rhus trilobata*

Slender Bluestem, *Schizachyrium tenerum* [*Andropogon tener*]

Slender Passionflower, *Passiflora filipes* [Yellow Passion Flower Vine]

Slim Tridens, *Tridens muticus*

Small Panicgrass, *Panicum oligosanthes*

Small-toothed Sedge, *Carex microdonta*

Smartweed, *Polygonum* spp. [Knotweed]

Smooth Cordgrass, *Sporobolus alterniflorus* [*Spartina alterniflora*]

Snake Eyes, *Phaulothamnus spinescens* [Devilqueen]

Soaptree Yucca, *Yucca elata*

Solomon's Seal, *Polygonatum biflorum*

Sotol, *Dasylirion texanum* [Texas Sotol, Green Sotol]

Southern Cattail, *Typha domingensis*

Southern Dewberry, *Rubus trivialis*

Southern Lady's Slipper, *Cypripedium kentuckiense*

Southern Magnolia, *Magnolia grandiflora*

Southern Red Oak, *Quercus falcata*

Southern Shield Fern, *Thelypteris kunthii*

Southern Wild Rice, *Zizaniopsis miliacea*

Southwestern White Pine, *Pinus strobiformis* [Southwestern Pine]

Spanish Moss, *Tillandsia usneoides* [a member of the pineapple family; not a moss]

Sphagnum Moss, see Prairie Sphagnum

Spiderwort, *Tradescantia* spp.

Spikemoss, *Selaginella apoda*

Spikerush, *Eleocharis* spp.

Spiny-fruited Prickly Pear, *Opuntia spinosibacca*

Spotted Coral Root, *Corallorhiza maculata*

Spring Ladies' Tresses, *Spiranthes vernalis*

Star Cactus, *Astrophytum asterias*

Stinging Cevallia, *Cevallia sinuata* [Stingleaf, Stinging Serpent]

Strawberry Cactus, *Echinocereus stramineus*

Sugar Hackberry, *Celtis laevigata* [Granjeno, Sugarberry, Hackberry]

Sundew, *Drosera* spp.

Sunflowers, *Helianthus* spp.

Swamp Blackgum, *Nyssa sylvatica*

Swamp Chestnut Oak, *Quercus michauxii*

Swamp Palmetto, *Sabal palmetto* [Swamp Cabbage, Sabal Palm]

Swamp Titi, *Cyrilla racemiflora*

Sweet Bay, *Magnolia virginiana*

Switchgrass, *Panicum virgatum*

Swollen Bladderwort, *Utricularia inflata*

Tall Dropseed, *Sporobolus asper* [*Sporobolus compositus*]

Tasajillo, *Cylindropuntia leptocaulis* [Pencil Cactus, Jumping Cactus; *Opuntia leptocaulis*]

Tepeguaje, *Leucaena pulverulenta* [Great Leadtree]

Texas Ash, *Fraxinus texensis*

Texas Bluebonnet, *Lupinus texensis*

Texas Bullnettle, *Cnidoscolus texanus* [*Jatropha texana*]

Texas Doveweed, *Croton texensis*

Texas Ebony, *Ebenopsis ebano* [*Pithecellobium flexicaule*]

Texas Grama, *Bouteloua rigidiseta*

Texas Lantana, *Lantana urticoides* [Calico Bush; *L. horrida, L. hispida*]

Texas Madrone, *Arbutus xalapensis*

Texas Mallow, *Callirhoe scabriuscula*

Texas Mountain Laurel, *Sophora secundiflora*

Texas Mulberry, *Morus microphylla*

Texas Paintbrush, *Castilleja indivisa* [Indian Paintbrush]

Texas Paloverde, *Parkinsonia texana* [Border Paloverde]

Texas Persimmon, *Diospyros texana*

Texas Poppy-Mallow, *Callirhoe scabriuscula* [an endangered species]

Texas Prickly Pear, *Opuntia lindheimeri* [Lindheimer's Prickly Pear]

Texas Purple Sage, see Cenizo

Texas Redbud, *Cercis canadensis* var. *texensis*

Texas Red Oak, *Quercus buckleyi* [Texas Oak]

Texas Sandmint, *Rhododon ciliatus*

Texas Snowbells, *Styrax platanifolius texanus*

Texas Wild Olive, *Cordia boissieri* [Anacahuita, Mexican Olive]

Texas Wild Rice, *Zizania texana*

Texas Wintergrass, *Nassella leucotricha* [Speargrass; *Stipa leucotricha*]

Threeawn grass, *Aristida* spp.

Three Birds Orchid, *Triphora trianthophoros*

Tickseed, *Coreopsis* spp.

Tobosagrass, *Pleuraphis mutica* [*Hilaria mutica*]

Tree Cholla, *Cylindropuntia imbricata* [*Opuntia imbricata*]

Trillium, *Trillium* spp.

Tropical Threefold, *Trixis inula* [*T. radialis*]

Trout Lily, *Erythronium rostratum*

Trumpet Creeper, *Campsis radicans*

Tufted Rockmat, *Petrophytum caespitosum*

Tule, *Schoenoplectus californicus*

Tumbleweed, see Russian Thistle

Tupelo, *Nyssa* spp.

Turtle Grass, *Thalassia testudinum*

Twig Rush, *Cladium mariscoides*

Virginia Creeper, *Parthenocissus quinquefolia*

Warnock's Ragwort, *Senecio warnockii*

Water Elm, *Planera aquatica* [Planetree]

Water Fern, *Marsilea* spp. [often resembles a four-leaf clover]

Water Hickory, *Carya aquatica*

Water Hyacinth, *Eichhornia crassipes* E

Water Oak, *Quercus nigra*

Water Paspalum, *Paspalum repens*

Water Stargrass, *Heteranthera dubia* [Grassleaf Mudplantain]

Water Tupelo, *Nyssa aquatica*

Wax Myrtle, *Myrica cerifera* [Southern Bayberry]

Weeping Lovegrass, *Eragrostis curvula* E

Whitebrush, *Aloysia gratissima* [Beebrush, Beebush]

White Mangrove, *Laguncularia racemosa*

White Oak, *Quercus alba*

White Rosinweed, *Silphium albiflorum*

White Spikerush, *Eleocharis albida*

White-tipped Sedge, *Rhynchospora colorata*

White Tridens, *Tridens albescens*

White Trout Lily, *Erythronium albidum* [Adder's Tongue, White Dog-tooth Violet]

White Water Lily, *Nymphaea odorata*

Widgeon Grass, *Ruppia maritima* [Widgeongrass]

Widow's Tear, *Commelina erecta* [Erect Dayflower]

Wild Blue Indigo, *Baptisia australis*

Wild plum, *Prunus* spp.

Willdenow's Sedge, *Carex willdenowii*

Willow, *Salix* spp. [four species occur in the Trans-Pecos: Goodding Willow, *S. gooddingii,* is common along the Rio Grande; see also Black Willow]

Willow Oak, *Quercus phellos*

Winecup, *Callirhoe digitata*

Winged Elm, *Ulmus alata*

Witch Hazel, *Hamamelis virginiana*

Woolly Rose-Mallow, *Hibiscus lasiocarpos*

Yaupon, *Ilex vomitoria* [Yaupon Holly]

Yellow-eyed grass, *Xyris* spp.

Yellow-fringed Orchid, *Platanthera ciliaris*

Yellow Rock Nettle, *Eucnide bartonioides* [Yellow Stingbush]

Yucca, *Yucca* spp.

INVERTEBRATE ANIMALS

American Burying Beetle, *Nicrophorus americanus*

Amphipods, order Amphipoda [small swimming crustaceans]

Barnacles, infraclass Cirripedia [shelled marine crustaceans that attach permanently to a surface]

Bay Scallop, *Argopecten irradians amplicostatus*

Beautiful Caecum, *Caecum pulchellum*

Bee Assassin, *Apiomerus spissipes*

Bivalve, see Brachiopods

Black Widow Spider, *Latrodectus mactans*

Blue Crab, *Callinectes sapidus*

Boll Weevil, *Anthonomus grandis*

Brachiopods, phylum Brachiopoda [animals such as clams, mussels, and oysters that have upper and lower valves (shells) encasing their body]

Broken-rays Mussel, *Lampsilis reeveiana*

Brown Shrimp, *Farfantepenaeus aztecus* [*Penaeus aztecus*]

Bumblebee, *Bombus* spp.

By-the-Wind Sailor, *Velella velella*

Cactus Moth, *Cactoblastis cactorum* [E]

Carpenter bee, *Xylocopa* spp.

Cave crickets, superfamily Rhaphidophoroidea

Chigger, family Trombiculidae [tiny ectoparasitic mites; *Trombicula alfreddugesi* is the most common in North America]

Cladocerans, order Cladocera [small crustaceans known as "water fleas"; most are freshwater species, many in the genus *Daphnia*]

Cochineal, *Dactylopius coccus*

Comanche Harvester Ant, *Pogonomyrmex comanche*

Copepods, subclass Copepoda [tiny saltwater and freshwater crustaceans]

Cotton Bollworm Moth, *Helicoverpa zea*

Crayfish, freshwater crustaceans known as crawdads, crawfish, or mudbugs

Crustacean, subphylum Crustacea [a large group of arthropods that includes crabs, crayfish, shrimp, lobsters, isopods, and barnacles]

Daddy longlegs, order Opiliones [Harvestmen; long-legged spiderlike arachnids]

Desert Cicada, *Diceroprocta apache*

Desert Termite, *Gnathamitermes tubiformans*

Devil's Toenail, *Texigryphaea mucronata** [an extinct oyster]

Dung Beetle, family Scarabaeidae

Dwarf Surf Clam, *Mulinia lateralis*

Eastern Oyster, *Crassostrea virginica* [American, Atlantic, or Virginia Oyster]

Echinoderms, phylum Echinodermata [a group including starfish, sea stars, and sea urchins]

Face Fly, *Musca autumnalis*

Fairy shrimp, order Anostraca, order Cladocera, phylum Arthropoda [tiny shrimplike crustaceans]

False Spike, *Quadrula mitchelli*

Fiddler Crab, *Uca subcylindrica*

Flatworm, *Planaria* spp.

Flesh fly, family Sarcophagidae

Gastropods, class Gastropoda [snails and slugs]

Ghost Crab, *Ocypode quadrata*

Giant Red-headed Centipede, *Scolopendra heros*

Giant Red Velvet Mite, *Dinothrombium pandorae* [Angelitos]

Giant Regal Moth, *Citheronia regalis*

Grass Cerith, *Bittiolum varium*

Guava Skipper, *Phocides polybius*

Gulf Stone Crab, *Menippe adina*

Hagen's Sphinx Moth, *Ceratomia hageni*

Harvester ants, *Pogonomyrmex* spp. [Red Harvester Ant, *P. barbatus,* is a common prey of the Texas Horned Lizard]

Hays' Rocksnail, *Stramonita canaliculata* [*Thais haemastoma haysae*]

Honey Bee, *Apis mellifera* [E] [all Honey Bees in North America were introduced]

Horn Fly, *Haematobia irritans*

Isopods, order Isopoda [crustaceans with seven pairs of legs]

Lightning Whelk, *Busycon perversum* [*B. p. pulleyi* is the Texas State Shell; alternate genus: *Sinistrofulgur*]

Limpet, order Gastropoda [marine snails with a cone-shaped shell]

Luna Moth, *Actias luna*

Malachite, *Siproeta stelenes* [a green-winged butterfly]

Marsh Periwinkle, *Littoraria irrorata*

Mexican Bluewing Butterfly, *Myscelia ethusa*

Midges, family Chironomidae [their wormlike larvae often indicate water quality]

Mole Crab, *Emerita benedicti*

Monarch Butterfly, *Danaus plexippus*
Muddy Cerith, *Cerithium lutosum*
Oyster Mosquito, *Boonea impressa*
Parasitic Eyeworm, *Oxyspirura petrowi*
Parkhill Prairie Crayfish, *Procambarus steigmani* [may be a subspecies of the Regal Burrowing Crayfish]
Pecos Assiminea Snail, *Assiminea pecos* [an endangered species]
Pinacate Beetle, *Eleodes obscurus* [Stink Beetle]
Pink Shrimp, *Farfantepenaeus duorarum* [*Penaeus duorarum*]
Pipe Organ Mud Dauber, *Trypoxylon politum* [a wasp in the family Crabronidae]
Pitcher Plant Mosquito, *Wyeomyia smithii*
Pitcher Plant Sarcophagid, *Fletcherimyia fletcheri* [*Blaesoxipha fletcheri*]
Plicate Hornshell, *Cerithidea pliculosa*
Polychaete, class Polychaeta [segmented marine worms with fleshy protrusions on each body segment]
Pseudoscorpion, order Pseudoscorpiones [false scorpions; small arachnids equipped with pincers but lacking a scorpion-like tail]
Ram's Horn Oyster, *Ilmatogrya arietina**
Red Imported Fire Ant, *Solenopsis invicta* E [this harmful invasive species should not be confused with native fire ants in the same genus]
Regal Burrowing Crayfish, *Procambarus regalis*
Rotifer, phylum Rotifera [near-microscopic, aquatic "wheel animals"]
Screwworm, *Cochliomyia hominivorax*
Sea anemone, order Actiniaria [stationary, tube-shaped animals with a flowerlike array of tentacles]
Serpulid worm, family Serpulidae [stationary, tube-building polychaetes]
Smooth Duck Clam, *Anatina anatina*
Spongefly, family Sisyridae [Spongillaflies; the aquatic larvae of six species parasitize freshwater sponges in Texas]
Swallow Bug, *Oeciacus vicarius*
Swallow Tick, *Ornithodoros concanensis*
Tampico Pearlymussel, *Cyrtonaias tampicoensis*
Tarantula, *Aphonopelma* spp. [large ground-dwelling spiders]
Tarantula Hawk, *Pepsis grossa* [*Pepsis thisbe* also occurs in western Texas]
Texas Brown Tarantula, *Aphonopelma hentzi*
Texas Cave Shrimp, *Palaemonetes antrorum*
Texas Fatmucket, *Lampsilis bracteata*
Texas Leaf Cutter Ant, *Atta texana*
Texas River Crayfish, *Orconectes texanus* [*Buannulifictus texanus*]
Tiger beetle, subfamily Cicindelinae, family Carabidae [predatory ground beetles capable of aerodynamic flight to capture prey]
Tropical Bont Tick, *Amblyomma variegatum* E
Tumblebug, see Dung Beetle
Variable Coquina, *Donax variabilis*
Variable Oakleaf Caterpillar, *Lochmaeus manteo*
Viceroy, *Limenitis archippus* [Viceroy Butterfly]
Vinegaroon, *Mastigoproctus giganteus* [Giant Whip-Scorpion]
White Shrimp, *Litopenaeus setiferus* [*Penaeus setiferus*]
Yellow crab spider, family Thomisidae [any of 2,100 crab spiders]
Yucca moth, *Tegeticula* spp. [each species of yucca is presumably pollinated by a single species of yucca moth]
Zebra Heliconian [butterfly], *Heliconius charithonia*

FISHES

American Paddlefish, *Polyodon spathula*
Arkansas River Shiner, *Notropis girardi*
Atlantic Tarpon, *Megalops atlanticus*
Bay Anchovy, *Anchoa mitchelli*
Black Crappie, *Pomoxis nigromaculatus*
Black Drum, *Pogonias cromis*
Comanche Springs Pupfish, *Cyprinodon elegans*
Dusky Pipefish, *Syngnathus floridae*
Fountain Darter, *Etheostoma fonticola* [an endangered species]
Golden Topminnow, *Fundulus chrysotus*
Green Sunfish, *Lepomis cyanellus*
Guadalupe Bass, *Micropterus treculii*
Gulf Menhaden, *Brevoortia patronus*
Largemouth Bass, *Micropterus salmoides*
Leon Springs Pupfish, *Cyprinodon bovinus*
Lined Seahorse, *Hippocampus erectus*
Longnose Gar, *Lepisosteus osseus*
Orangethroat Darter, *Etheostoma spectabile*
Pecos Gambusia, *Gambusia nobilis*
Plains Minnow, *Hybognathus placitus*
Rainbow Trout, *Oncorhynchus mykiss* [*Salmo gairdneri*]
Rainwater Killifish, *Lucania parva*
Redbreast Sunfish, *Lepomis auritus*
Red Drum, *Sciaenops ocellatus* [Redfish, Red, Spottail, Channel Bass]
Red Lionfish, *Pterois volitans* E [an invasive species]
Rio Grande Perch, *Herichthys cyanoguttatus*
Rio Grande Silvery Minnow, *Hybognathus amarus* [an endangered species]

San Marcos Gambusia, *Gambusia georgei*

Sheepshead, *Archosargus probatocephalus*

Smallmouth Bass, *Micropterus dolomieu*

Spotted Bass, *Micropterus punctulatus*

Spotted Seatrout, *Cynoscion nebulosus*

Striped Mullet, *Mugil cephalus*

Tidewater Silversides, *Menidia peninsulae*

Warmouth, *Lepomis gulosus*

Western Mosquitofish, *Gambusia affinis*

Yellowbelly Sunfish, *Lepomis auritus*

AMPHIBIANS

Barred Tiger Salamander, *Ambystoma mavortium* [*A. tigrinum mavortium*]

Barton Springs Salamander, *Eurycea sosorum*

Black-spotted Newt, *Notophthalmus meridionalis*

Blanchard's Cricket Frog, *Acris blanchardi blanchardi* [*A. crepitans blanchardi*]

Bullfrog, *Lithobates catesbeiana* [*Rana catesbeiana*]

Cane Toad, *Rhinella marina* [Marine Toad, Giant Toad; *Bufo marinus*]

Canyon Treefrog, *Hyla arenicolor*

Cliff Chirping Frog, *Eleutherodactylus marnockii*

Couch's Spadefoot, *Scaphiopus couchii*

Dwarf Salamander, *Eurycea quadridigitata*

Eastern Narrow-mouthed Toad, *Gastrophryne carolinensis*

Great Plains Narrow-mouthed Toad, *Gastrophryne olivacea* [Western Narrow-mouthed Toad]

Great Plains Toad, *Anaxyrus cognatus* [*Bufo cognatus*]

Green Frog, *Lithobates clamitans* [*Rana clamitans*]

Green Treefrog, *Hyla cinerea*

Gulf Coast Toad, *Incilius nebulifer* [*Bufo valliceps*]

Houston Toad, *Anaxyrus houstonensis* [*Bufo houstonensis*]

Hurter's Spadefoot, *Scaphiopus hurteri*

Marbled Salamander, *Ambystoma opacum*

Mexican Burrowing Toad, *Rhinophrynus dorsalis*

Mexican White-lipped Frog, *Leptodactylus fragilis*

New Mexico Spadefoot, *Spea multiplicata* [*Scaphiopus multiplicatus*]

Pickerel Frog, *Lithobates palustris* [*Rana palustris*]

Plains Leopard Frog, *Lithobates blairi* [*Rana pipiens*]

Plains Spadefoot, *Spea bombifrons* [*Scaphiopus bombifrons*]

Red-spotted Toad, *Anaxyrus punctatus* [*Bufo punctatus*]

Rio Grande Lesser Siren, *Siren intermedia texana*

Sheep Frog, *Hypopachus variolosus*

Spotted Chorus Frog, *Pseudacris clarkii*

Spring Peeper, *Pseudacris crucifer*

Strecker's Chorus Frog, *Pseudacris streckeri*

Texas Blind Salamander, *Eurycea rathbuni* [*Typhlomolge rathbuni*]

Texas Toad, *Anaxyrus speciosus* [*Bufo speciosus*]

Western Slimy Salamander, *Plethodon albagula* [*P. glutinosus albagula*]

Woodhouse's Toad, *Anaxyrus woodhousii* [*Bufo woodhousii*]

REPTILES

Alligator Snapping Turtle, *Macrochelys temminckii*

American Alligator, *Alligator mississippiensis*

Banded Gecko, see Desert Banded Gecko and Texas Banded Gecko

Black-tailed Rattlesnake, *Crotalus molossus*

Blotched Watersnake, *Nerodia erythrogaster transversa*

Blue Spiny Lizard, *Sceloporus cyanogenys*

Brazos River Watersnake, *Nerodia harteri harteri*

Broad-banded Copperhead, *Agkistrodon contortrix laticinctus*

Bullsnake, *Pituophis catenifer sayi* [*Pituophis melanoleucus sayi*]

Common Side-blotched Lizard, *Uta stansburiana*

Cottonmouth, *Agkistrodon piscivorus* [Water Moccasin]

Desert Banded Gecko, *Coleonyx variegatus*

Desert Massasauga, *Sistrurus catenatus edwardsii*

Dunes Sagebrush Lizard, *Sceloporus arenicolus*

Eastern Box Turtle, *Terrapene carolina*

Eastern Collared Lizard, *Crotaphytus collaris*

Eastern Mud Turtle, *Kinosternon subrubrum*

Flat-headed Snake, *Tantilla gracilis*

Glossy Crayfish Snake, *Regina rigida*

Greater Short-horned Lizard, *Phrynosoma hernandesi*

Great Plains Skink, *Plestiodon obsoletus* [*Eumeces obsoletus*]

Green Sea Turtle, *Chelonia mydas*

Harter's Watersnake, *Nerodia harteri*

Keeled Earless Lizard, *Holbrookia propinqua*

Kemp's Ridley Sea Turtle, *Lepidochelys kempii*

Little-striped Whiptail, *Aspidoscelis inornata* [*Cnemidophorus inornatus*]

Long-nosed Snake, *Rhinocheilus lecontei*

Louisiana Pinesnake, *Pituophis ruthveni*

Mexican Spiny-tailed Iguana, *Ctenosaura pectinata*

Mojave Rattlesnake, *Crotalus scutulatus*

Ornate Box Turtle, *Terrapene ornata*

Phytosaur, *Leptosuchus crosbiensis* * [extinct semiaquatic reptile resembling a modern-day crocodile]

Plesiosaur, *Plesiosaurus* spp. * [extinct aquatic reptile with large finlike appendages instead of legs and feet]

Prairie Lizard, *Sceloporus consobrinus* [*S. undulatus consobrinus*]

Prairie Skink, *Plestiodon septentrionalis* [*Eumeces septentrionalis*]

Ratsnake, *Pantherophis* spp. [*Elaphe* spp.]

Red-striped Ribbonsnake, *Thamnophis proximus rubrilineatus*

Regal Black-striped Snake, *Coniophanes imperialis*

Reticulate Collared Lizard, *Crotaphytus reticulatus*

Rock Rattlesnake, *Crotalus lepidus*

Rose-bellied Lizard, *Sceloporus variabilis*

Round-tailed Horned Lizard, *Phrynosoma modestum*

Rusty Lizard, see Texas Spiny Lizard

Side-blotched Lizard, *Uta stansburiana*

Southern Copperhead, *Agkistrodon contortrix contortrix*

Speckled Racer, *Drymobius margaritiferus*

Spiny Soft-shelled Turtle, *Apalone spinifera*

Tamaulipan Hook-nosed Snake, *Ficimia streckeri*

Texas Alligator Lizard, *Gerrhonotus infernalis*

Texas Banded Gecko, *Coleonyx brevis*

Texas Greater Earless Lizard, *Cophosaurus texanus texanus*

Texas Horned Lizard, *Phrynosoma cornutum* [a state-listed endangered species]

Texas Indigo Snake, *Drymarchon melanurus erebennus*

Texas Map Turtle, *Graptemys versa*

Texas Ratsnake, *Pantherophis guttata* [*Elaphe guttata*]

Texas Spiny Lizard, *Sceloporus olivaceus* [Rusty Lizard]

Texas Spiny Softshell, *Apalone spinifera emoryi*

Texas Tortoise, *Gopherus berlandieri* [Berlandier's Tortoise]

Trans-Pecos Black-headed Snake, *Tantilla cucullata* [Big Bend Black-headed Snake; its distributional range extends onto the western Edwards Plateau]

Western Coachwhip, *Coluber flagellum testaceus* [*Masticophis f. testaceus*]

Western Diamond-backed Rattlesnake, *Crotalus atrox*

Western Massasauga, *Sistrurus catenatus tergeminus*

BIRDS

Altimira Oriole, *Icterus gularis*

American Avocet, *Recurvirostra americana*

American Coot, *Fulica americana*

American Crow, *Corvus brachyrhynchos*

American Kestrel, *Falco sparverius*

American Oystercatcher, *Haematopus palliatus*

American Robin, *Turdus migratorius*

American Wigeon, *Anas americana* [*Mareca americana*]

Aplomado Falcon, *Falco femoralis*

Ash-throated Flycatcher, *Myiarchus cinerascens*

Attwater's Prairie-Chicken, *Tympanuchus cupido attwateri* [an endangered subspecies of the Greater Prairie-Chicken]

Audubon's Oriole, *Icterus graduacauda*

Bald Eagle, *Haliaeetus leucocephalus*

Baltimore Oriole, *Icterus galbula*

Band-tailed Pigeon, *Columba fasciata*

Barn Owl, *Tyto alba*

Barred Owl, *Strix varia*

Black-bellied Plover, *Pluvialis squatarola*

Black-bellied Whistling-Duck, *Dendrocygna autumnalis* [Black-bellied Tree Duck]

Black-capped Vireo, *Vireo atricapilla* [*Vireo atricapillus*]

Black-chinned Hummingbird, *Archilochus alexandri*

Black-crested Titmouse, *Baeolophus atricristatus* [*Parus atricristatus*]

Black-necked Stilt, *Himantopus mexicanus*

Black Skimmer, *Rynchops niger*

Black-throated Sparrow, *Amphispiza bilineata*

Black Vulture, *Coragyps atratus*

Blue-gray Gnatcatcher, *Polioptila caerulea*

Blue Jay, *Cyanocitta cristata*

Blue-winged Teal, *Anas discors*

Brown-crested Flycatcher, *Myiarchus tyrannulus* [Wied's Crested Flycatcher]

Brown-headed Cowbird, *Molothrus ater*

Brown-headed Nuthatch, *Sitta pusilla*

Brown Jay, *Cyanocorax morio*

Brown Pelican, *Pelecanus occidentalis*

Buff-bellied Hummingbird, *Amazilia yucatanensis*

Bufflehead, *Bucephala albeola*

Bullock's Oriole, *Icterus bullockii*

Burrowing Owl, *Athene cunicularia* [*Speotyto cunicularia*]

Cactus Wren, *Campylorhynchus brunneicapillus*

California Condor, *Gymnogyps californianus*

Canada Goose, *Branta canadensis*

Canyon Wren, *Catherpes mexicanus*

Carolina Chickadee, *Poecile carolinensis* [*Parus carolinensis*]

Carolina Wren, *Thryothorus ludovicianus*

Cassin's Sparrow, *Aimophila cassinii*

Cattle Egret, *Bubulcus ibis*

Cave Swallow, *Petrochelidon fulva* [*Hirundo fulva*]

Cedar Waxwing, *Bombycilla cedrorum*

Chihuahuan Raven, *Corvus*

cryptoleucus [White-necked Raven]

Cinnamon Teal, *Anas cyanoptera*

Cliff Swallow, *Petrochelidon pyrrhonota*

Colima Warbler, *Vermivora crissalis*

Common Black-Hawk, *Buteogallus anthracinus*

Common Gallinule, *Gallinula galeata*

Common Pauraque, *Nyctidromus albicollis*

Couch's Kingbird, *Tyrannus couchii*

Crested Caracara, *Caracara cheriway* ["Mexican Eagle"]

Curve-billed Thrasher, *Toxostoma curvirostre*

Dickcissel, *Spiza americana* [called Black-throated Bunting by Audubon]

Eared Grebe, *Podiceps nigricollis*

Eastern Bluebird, *Sialia sialis*

Eastern Meadowlark, *Sturnella magna*

Eastern Phoebe, *Sayornis phoebe*

Eastern Tufted Titmouse, *Baeolophus bicolor bicolor* [*Parus b. bicolor*]

Elf Owl, *Micrathene whitneyi*

Emu, *Dromaius novaehollandiae* E

Ferruginous Hawk, *Buteo regalis*

Ferruginous Pygmy-Owl, *Glaucidium brasilianum*

Fish Crow, *Corvus ossifragus*

Fulvous Whistling-Duck, *Dendrocygna bicolor* [Fulvous Tree Duck]

Golden-cheeked Warbler, *Setophaga chrysoparia* [*Dendroica chrysoparia*]

Golden Eagle, *Aquila chrysaetos*

Golden-fronted Woodpecker, *Melanerpes aurifrons*

Grasshopper Sparrow, *Ammodramus savannarum*

Gray-crowned Yellowthroat, *Geothlypis trichas*

Gray Hawk, *Buteo nitidus*

Great Blue Heron, *Ardea herodias*

Great Crested Flycatcher, *Myiarchus crinitus*

Great Egret, *Ardea alba* [*Casmerodius albus*]

Greater Rhea, *Rhea americana* E [American Rhea]

Greater Roadrunner, *Geococcyx californianus*

Great Horned Owl, *Bubo virginianus*

Great Kiskadee, *Pitangus sulphuratus*

Green Jay, *Cyanocorax yncas*

Green Kingfisher, *Chloroceryle americana*

Green Parakeet, *Aratinga holochlora*

Green-winged Teal, *Anas crecca* [*Anas carolinensis*]

Groove-billed Ani, *Crotophaga sulcirostris*

Gull-billed Tern, *Sterna nilotica*

Harris's Hawk, *Parabuteo unicinctus*

Heath Hen, *Tympanuchus cupido cupido* * [an extinct subspecies of the Greater Prairie-Chicken]

Hooded Oriole, *Icterus cucullatus*

Hook-billed Kite, *Chondrohierax uncinatus*

Horned Lark, *Eremophila alpestris*

House Wren, *Troglodytes aedon*

Hudsonian Godwit, *Limosa haemastica*

Indigo Bunting, *Passerina cyanea*

Interior Least Tern, *Sterna antillarum athalassos* [an endangered subspecies that nests on sandbars of Texas rivers; *S. a. antillarum* breeds on the Gulf Coast]

Ivory-billed Woodpecker, *Campephilus principalis* * [presumed extinct in North America, but subspecies may persist in Cuba]

Killdeer, *Charadrius vociferus*

Laughing Gull, *Larus atricilla*

Least Grebe, *Tachybaptus dominicus*

Least Sandpiper, *Calidris minutilla*

Lesser Goldfinch, *Carduelis psaltria*

Lesser Prairie-Chicken, *Tympanuchus pallidicinctus*

Lesser Scaup, *Aythya affinis*

Lesser Snow Goose, *Anser caerulescens* [*Chen caerulescens*]

Little Blue Heron, *Egretta caerulea*

Loggerhead Shrike, *Lanius ludovicianus*

Long-billed Curlew, *Numenius americanus*

Long-billed Dowitcher, *Limnodromus scolopaceus*

Mallard, *Anas platyrhynchos*

Masked Duck, *Nomonyx dominicus* [*Oxyura dominica*]

Mexican Jay, *Aphelocoma ultramarina*

Mississippi Kite, *Ictinia mississippiensis*

Montezuma Quail, *Cyrtonyx montezumae*

Mountain Plover, *Eupoda montana*

Mourning Dove, *Zenaida macroura*

Neotropic Cormorant, *Phalacrocorax brasilianus*

Northern Beardless Tyrannulet, *Camptostoma imberbe*

Northern Bobwhite, *Colinus virginianus*

Northern Cardinal, *Cardinalis cardinalis*

Northern Flicker, *Colaptes auratus*

Northern Harrier, *Circus cyaneus* [Marsh Hawk]

Northern Mockingbird, *Mimus polyglottos*

Northern Pintail, *Anas acuta*

Orchard Oriole, *Icterus spurius*

Osprey, *Pandion haliaetus*

Painted Bunting, *Passerina ciris*

Painted Redstart, *Myioborus pictus*

Passenger Pigeon, *Ectopistes migratorius* *

Peregrine Falcon, *Falco peregrinus*

Phainopepla, *Phainopepla nitens*

Pileated Woodpecker, *Dryocopus pileatus*

Piping Plover, *Charadrius melodus*

Plain Chachalaca, *Ortalis vetula*

Purple Gallinule, *Porphyrio martinica*

Ravens, *Corvus* spp.

Red-bellied Woodpecker, *Melanerpes carolinus*

Red-billed Pigeon, *Columba flavirostris*

Red-breasted Merganser, *Mergus serrator*

Red-cockaded Woodpecker, *Picoides borealis* [*Leuconotopicus borealis*]
Red-crowned Parrot, *Amazona viridigenalis*
Reddish Egret, *Egretta rufescens*
Red-eyed Vireo, *Vireo olivaceus*
Redhead, *Aythya americana*
Red-headed Woodpecker, *Melanerpes erythrocephalus*
Red-tailed Hawk, *Buteo jamaicensis*
Ringed Kingfisher, *Ceryle torquata*
Ring-necked Pheasant, *Phasianus colchicus* [E]
Rock Wren, *Salpinctes obsoletus*
Rose-throated Becard, *Pachyramphus aglaiae*
Royal Tern, *Sterna maxima*
Ruby-throated Hummingbird, *Archilochus colubris*
Ruddy Duck, *Oxyura jamaicensis*
Ruddy Turnstone, *Arenaria interpres*
Sanderling, *Calidris alba*
Sandhill Crane, *Antigone* (*Grus*) *canadensis*
Scissor-tailed Flycatcher, *Tyrannus forficatus*
Screech Owl, *Otus asio*
Short-eared Owl, *Asio flammeus*
Snow Goose, *Anser caerulescens* [*Chen caerulescens* in some classifications]
Snowy Egret, *Egretta thula*
Snowy Plover, *Charadrius nivosus* [*Charadrius alexandrinus*]
Summer Tanager, *Piranga rubra*
Swainson's Hawk, *Buteo swainsoni*
Swallow-tailed Kite, *Elanoides forficatus*
Tricolored Heron, *Egretta tricolor* [Louisiana Heron]
Tropical Parula, *Setophaga pitiayumi* [*Parula pitiayumi*]
Tufted Titmouse, *Baeolophus bicolor* [*Parus bicolor*]
Turkey Vulture, *Cathartes aura*
Western Kingbird, *Tyrannus verticalis*
Western Meadowlark, *Sturnella neglecta*
White-collared Seedeater, *Sporophila torqueola*

White-eyed Vireo, *Vireo griseus*
White-fronted Goose, *Anser albifrons*
White Pelican, *Pelecanus erythrorhynchos*
White-tailed Hawk, *Buteo albicaudatus*
White-tipped Dove, *Leptotila verreauxi* [White-fronted Dove]
White-winged Dove, *Zenaida asiatica*
Whooping Crane, *Grus americana*
Wild Turkey, *Meleagris gallopavo*
Willet, *Catoptrophorus semipalmatus*
Wilson's Phalarope, *Phalaropus tricolor*
Wood Duck, *Aix sponsa*
Wood warblers, family Parulidae [e.g., Yellow Warbler]
Yellow-breasted Chat, *Icteria virens*
Yellow Warbler, *Stetophaga petechia* [*Dendroica petechia*]
Zone-tailed Hawk, *Buteo albonotatus*

MAMMALS
African Elephant, *Loxodonta africana* or *L. cyclotis*
African Lion, *Panthera leo* [*Felis leo*]
American Badger, *Taxidea taxus*
American Beaver, *Castor canadensis*
American Bison, *Bos bison* [incorrectly known as "Buffalo"; *Bison bison*]
American Cheetah, *Miracinonyx trumani* *
American Mastodon, *Mammut americanum* *
American Mink, *Vison vison* [*Mustela vison*]
Arctic Fox, *Alopex lagopus*
Armadillo, see Nine-banded Armadillo
Atlantic Bottle-nosed Dolphin, *Tursiops truncatus*
Attwater's Pocket Gopher, *Geomys attwateri*
Axis Deer, *Axis axis* [E] [Chital Deer, Chital]
Baird's Pocket Gopher, *Geomys breviceps* [*G. bursarius breviceps*]
Banner-tailed Kangaroo Rat, *Dipodomys spectabilis*

Big Brown Bat, *Eptesicus fuscus*
Bison, see American Bison
Black Bear, *Ursus americanus*
Blackbuck Antelope, *Antilope cervicapra* [E]
Black-footed Ferret, *Mustela nigripes*
Black Rhinoceros, *Diceros bicornis* [E]
Black-tailed Jackrabbit, *Lepus californicus*
Black-tailed Prairie Dog, *Cynomys ludovicianus*
Bobcat, *Lynx rufus* [*Felis rufus*]
Brazilian Free-tailed Bat, *Tadarida brasiliensis* [Mexican Free-tailed Bat]
Buffalo, see American Bison
Cattle, *Bos taurus*
Clouded Leopard, *Neofelis nebulosa* [*Felis macrocelis, F. mamota*]
Collared Peccary, *Pecari tajacu* [Javelina; *Tayassu tajacu, Pecari angulatus*]
Columbian Mammoth, *Mammuthus columbi* *
Common Muskrat, *Ondatra zibethicus*
Cougar, see Mountain Lion
Coyote, *Canis latrans*
Davis Mountains Cottontail, *Sylvilagus floridanus robustus* [*S. robustus* in some references]
Deer Mouse, *Peromyscus maniculatus*
Desert Bighorn Sheep, *Ovis canadensis*
Desert Cottontail *Sylvilagus audubonii*
Domestic Cat, *Felis catus*
Eastern Cottontail, *Sylvilagus floridanus*
Eastern Fox Squirrel, *Sciurus niger*
Eastern Gray Squirrel, *Sciurus carolinensis*
Eastern Harvest Mouse, *Reithrodontomys humulis*
Eastern Mole, *Scalopus aquaticus*
Eastern Red Bat, *Lasiurus borealis*
Eastern White-throated Woodrat, *Neotoma leucodon* [*N. albigula*]
Eastern Woodrat, *Neotoma floridana*
Evening Bat, *Nycticeius humeralis*

Fallow Deer, *Dama dama* [E]

Feral Pig, *Sus scrofa* [E]

Fox Squirrel, see Eastern Fox Squirrel

Fulvous Harvest Mouse, *Reithrodontomys fulvescens*

Giant Bison, *Bison latifrons* * [Longhorn Bison; horns spanned 84 inches, tip to tip]

Golden Mouse, *Ochrotomys nuttalli*

Gray-footed Chipmunk, *Neotamias canipes* [*Tamias canipes*]

Gray Fox, *Urocyon cinereoargenteus*

Gray Squirrel, see Eastern Gray Squirrel

Gray Wolf, *Canis lupus*

Ground Sloth, *Nothrotheriops* spp. *

Gulf Coast Kangaroo Rat, *Dipodomys compactus*

Harlan's Ground Sloth, *Paramylodon harlani* *

Hispid Cotton Rat, *Sigmodon hispidus*

Hispid Pocket Mouse, *Chaetodipus hispidus*

Hoary Bat, *Lasiurus cinereus*

Jaguar, *Panthera onca*

Jefferson's Ground Sloth, *Megalonyx jeffersonii* *

Kangaroo rat, *Dipodomys* spp.

Kit Fox, *Vulpes macrotis* [*V. velox macrotis*]

Louisiana Black Bear, *Ursus americanus luteolus* [a subspecies of American Black Bear]

Mammoth, *Mammuthus* spp. *

Maritime Pocket Gopher, *Geomys personatus maritimus*

Merriam's Pocket Mouse, *Perognathus merriami*

Mexican Ground Squirrel, *Ictidomys mexicanus* [*Spermophilus mexicanus*]

Mexican Long-nosed Bat, *Leptonycteris nivalis*

Mexican Spiny Pocket Mouse, *Liomys irroratus*

Mountain Lion, *Puma concolor* [Cougar, Puma, Panther; *Felis concolor*]

Mule Deer, *Odocoileus hemionus*

Musk Ox, *Ovibos moschatus* [Muskox, Musk-Ox; now restricted to Arctic habitats]

Nilgai, *Boselaphus tragocamelus* [E]

Nine-banded Armadillo, *Dasypus novemcinctus*

North American Porcupine, *Erethizon dorsatum*

Northern Grasshopper Mouse, *Onychomys leucogaster*

Northern Pygmy Mouse, *Baiomys taylori*

Northern River Otter, *Lontra canadensis* [*Lutra canadensis*]

Nutria, *Myocastor coypus* [E] [Coypu]

Ocelot, *Leopardus pardalis* [*Felis pardalis*]

Ord's Kangaroo Rat, *Dipodomys ordii*

Pallid Bat, *Antrozous pallidus*

Palo Duro Mouse, *Peromyscus truei comanche* [endemic to the upper Palo Duro Canyon]

Piñon Mouse, *Peromyscus truei*

Plains Pocket Gopher, *Geomys bursarius*

Plains Pocket Mouse, *Perognathus flavescens*

Plains Zebra, *Equus quagga* [E]

Pocket gopher, *Geomys* spp.

Pocket mouse, *Perognathus* spp. and *Chaetodipus* spp.

Porcupine, see North American Porcupine

Pronghorn, *Antilocapra americana* [popularly called "Antelope," but not related to the true antelope of Africa]

Raccoon, *Procyon lotor*

Rafinesque's Big-eared Bat, *Corynorhinus rafinesquii macrotis*

Razorback, see Feral Pig

Red Fox, *Vulpes fulva*

Red Wolf, *Canis rufus* [*C. niger*]

Reticulated Giraffe, *Giraffa camelopardalis reticulata* [E]

Ringtail, *Bassariscus astutus*

Rock Squirrel, *Otospermophilus variegatus* [*Spermophilus variegatus*]

Rocky Mountain Elk, *Cervus canadensis nelsoni* [Wapiti; *C. elaphus*]

Saber-toothed cat, *Smilodon* spp. *

Seminole Bat, *Lasiurus seminolus*

Short-faced bear, *Arctodus* spp. *

Sika Deer, *Cervus nippon* [E]

Silky Pocket Mouse, *Perognathus flavus*

Sloth, see Ground Sloth, Harlan's Ground Sloth, and Jefferson's Ground Sloth

Southern Flying Squirrel, *Glaucomys volans*

Southern Plains Woodrat, *Neotoma micropus*

Southern Yellow Bat, *Lasiurus ega*

Striped Skunk, *Mephitis mephitis*

Swamp Rabbit, *Sylvilagus aquaticus*

Swift Fox, *Vulpes velox* [represented by two subspecies, *V. v. velox* of grasslands and *V. v. macrotis* (Kit Fox) of deserts]

Texas Pocket Gopher, *Geomys personatus*

Western Perimyotis, *Perimyotis hesperus* [Western Pipistrelle; *Pipistrellus hesperus*]

White-ankled Mouse, *Peromyscus pectoralis*

White-backed Hog-nosed Skunk, *Conepatus leuconotus*

White-footed Mouse, *Peromyscus leucopus*

White-nosed Coati, *Nasua narica*

White-tailed Deer, *Odocoileus virginianus*

Woodrat, *Neotoma* spp.

Woolly Mammoth, *Mammuthus primigenius* *

Yellow-faced Pocket Gopher, *Cratogeomys castanops*

Yellow-nosed Cotton Rat, *Sigmodon ochrognathus*

GLOSSARY

Allee effect: Decline in either reproduction or survival that results when population density becomes too low to maintain normal communication or social behaviors.

Allelopathy: The production of toxic chemicals by one plant to inhibit the germination or survival of other plants.

Allopatric distribution: Refers to the separate distributions of two or more related species (i.e., their geographical ranges do not overlap). For comparison, see **Sympatric distribution.**

Altricial: For birds, a description indicating hatchlings that have little or no plumage and closed eyes; they remain in the nest and rely entirely on their parents for food and warmth until they fledge (e.g., songbirds). Also applies to some mammals born in an undeveloped state (e.g., opossums). For comparison, see **Precocial.**

Aposematism: Color patterns that provide a warning to a potential predator. The striking black-and-white pattern of skunks and the red, yellow, and black of coral snakes are examples.

Artesian: An adjective used to describe groundwater under positive pressure. An artesian aquifer, when penetrated by a well, may be under enough pressure from overlying rocks to push water upward without the use of mechanical means.

Asexual reproduction: One of several forms of reproduction in which gametes are not exchanged to create new individuals. Examples include sprouts arising from stolons and rhizomes; also budding and cloning.

Bajada: An alluvial fan formed by the deposition of down-washed sediments at the base of a mountain.

Bay: A body of water occupying a shoreline indentation that opens to an ocean or lake. An estuary fed by a river can also be termed a "bay." Embayment is a synonym. See also **Estuary.**

Benthos: The community of organisms found on the bottom of a lake, bay, ocean, or other body of water, including those found within bottom sediments. Benthic is the adjective form.

Bioindicator: A species or community whose presence indicates certain habitat conditions (e.g., soil, salinity, overgrazing).

Biomass: The dry weight of organisms per unit area— a measure often used to gauge productivity of an area. Vegetation contains more biomass per unit area than the herbivores feeding on that vegetation in the same area.

Biospeleology: The scientific study of cave life.

Bolson: A desert valley entirely surrounded by hills or mountains that allow no outlet for rainwater. Typically, a salt pan forms at the center of the bolson where water collects before evaporating.

Butte: A small, isolated hill with relatively steep sides and a flat top. Mesas are larger than buttes, and plateaus are tablelands occupying huge expanses of land.

Carrying capacity: The ability of a habitat to support an animal population without degrading the resources (e.g., overgrazing); often varies from year to year depending on factors such as rainfall. Measured in terms of biomass or animal numbers per unit area.

Coevolution: The adaptations of one species in response to changes in another species. Examples occur when two species have a close, interdependent ecological relationship, as with predators and prey or insects and the flowers they pollinate.

Competitive exclusion: The concept that competition between two species for exactly the same resource will result in the elimination (i.e., exclusion) of one of the species from the habitat.

Convergent evolution: Independent development of similar physical characteristics or behavioral features in different lineages of organisms as adaptations to similar environmental conditions. For example, kangaroos and kangaroo rats have elongated tails and hind legs to facilitate jumping and balance.

Coppice dunes: Small dunes located on the backshore, typically behind a wind-blocking obstacle. These dunes may or may not be stabilized by vegetation.

Coprophagy: The consumption of feces.

Coprophilous: Growing or living on fecal matter.

Crepuscular: A term describing animal activity primarily during twilight (i.e., dawn or dusk).

Cryptobiosis: A living state in which metabolic activity is temporarily suspended or undetectable.

Cryptogam: A plant or plantlike organism that reproduces by spores rather than seeds. Algae, fungi, ferns, mosses, and liverworts are examples.

Cryptogamic soils: Extremely dry, often sandy soils covered with a thin layer of cryptogams that form a biological soil crust. See **Cryptogam.**

Cuesta: A ridge characterized by a cliff or steep slope on one side and a gentle ascent on the other. Most cuestas form when geological forces tilt underlying strata.

Cultch: Any solid substrate, such as oyster shells or rocks, positioned as an attachment site for oyster spat, a developmental stage of the Eastern Oyster.

Desertification: A particular type of land degradation in which an area becomes increasingly arid and assumes the environmental characteristics of a desert.

Detritivore: An organism that consumes detritus—the decomposing parts of dead plants and animals.

Diapause: A period of arrested development or dormancy requiring little metabolic investment and minimal expenditure of nutrient reserves. The condition is usually a means of enduring harsh environmental conditions, such as cold or drought.

Dichromatism: The occurrence of two different colors or color patterns in a single species of animal. Most often associated with sexual differences.

Discontinuous distribution: A geographic arrangement in which a species occurs in two or more isolated areas having no connection.

Doline: A sinkhole; a naturally occurring pit draining descending in karst areas. See **Karst.**

Ecotone: Area of transition between two communities with representative species of both types. Most change gradually, but a few are sharply delineated ("knife-edged" ecotones). May apply to regions as large as the Cross Timbers and Prairies or to those as small as field edges.

Ecotype: A locally adapted variant within a species that results from selection pressures peculiar to the site (e.g., drought-prone locations). Because these physical and/or physiological traits arise genetically, ecotypes retain their uniqueness even if moved to environments lacking similar limitations (e.g., more moisture).

Ectoparasite: A parasite that lives on the exterior of its host. For comparison, see **Endoparasite.**

Edaphic: A descriptive term referring to soils. Texture, chemistry, fertility, and drainage represent edaphic conditions relevant to natural history (e.g., vegetational development and the occurrence of burrowing animals).

Endemic: Term indicating that a species or higher taxon is limited to only a certain region.

Endoparasite: A parasite that lives within an internal body organ or tissue of its host.

Eolian: Denotes materials deposited by winds, especially soils carried to locations far removed from where the soils originally developed. "Aeolian" is an alternate spelling.

Ephemeral: A description for short-lived plants that germinate only when conditions are ripe for survival rather than during a predictable season. The term is also used to describe ponds and other habitats that exist for only short periods.

Epiphytes: Plants growing on the trunks and limbs of trees, but not as parasites. Instead, they survive on nutrients gained from rainwater, air, and in some cases debris. Usually harmless, although their growth may become heavy enough to break branches. Spanish Moss is a widely known example, but others include some species of orchids, ferns, and even cacti.

Epizootic: A severe outbreak of a disease in which large numbers of animals die in a short time. Equivalent of an epidemic in humans.

Estivation: A state of dormancy characterized by lowered metabolic and breathing rates during periods of extreme heat or drought. Hibernation, a similar condition, occurs in response to prolonged cold. The term is sometimes spelled "aestivation."

Estuary: A partially enclosed body of water fed by one or more rivers or streams and having an open connection to the ocean.

Exotic: A species living outside its native distributional range after being introduced there by deliberate or unintentional human activity.

Extinction: Refers to a species no longer existing anywhere. Often misused in situations where "extirpation" (locally extinct) is the correct terminology. For comparison, see **Extirpation.**

Extinction vortex: A threat to small populations from a series of adversities, any one of which by itself might not cause extinction, but which in tandem push a species too far. In simple terms, a harmful event increases the vulnerability of a small population to each successive event.

Extirpation: The absence of a species from part of its former range. For comparison, see **Extinction.**

Fecundity rate: The rate at which new individuals are added to the population as a result of reproductive success.

Food chain: The pathway of food and energy through an ecosystem, typically beginning with plants, followed by herbivores, and ending with predators. Most ecosystems have several food chains that together represent a food web.

Fossorial: A term describing organisms that spend most of their life underground. Examples include earthworms, moles, and pocket gophers.

Gilgai: A small, water-retaining depression in an area of clay soils. Gilgai are initiated when clay soils dry and crack during dry periods in a region with distinct wet and dry seasons. Dry, wind-borne sediments fill the crack and then swell when moistened, forcing the walls of the crack apart to create a depression. Gilgai can occur on areas of high ground.

Graminoid: Collective term for grasses and grasslike plants (e.g., sedges).

Guano: The excrement of bats or seabirds.

Guanophile: An animal adapted to living in or on guano deposits.

Gyre: A pattern of rotating currents within an ocean basin generated by the Coriolis effect.

Halophyte: A plant adapted to living in salty environments.

Head-starting: A conservation technique designed to supplement wild populations of endangered species by increasing survival rates of younger, more vulnerable life stages (e.g., eggs or larvae such as tadpoles). Typically accomplished by collecting and hatching and/or hand-rearing young individuals until they reach a certain age or size that will increase their chances for survival after they are returned to their original habitat. In some cases, the released animals originate from a captive breeding population. Because of potential genetic differences between populations, animals should be returned to the same locale from which their stock originated.

Helictite: A contorted speleothem found in some limestone caves. Portions of these deposits seem to defy gravity because the axis varies from vertical during stages of growth.

High-energy beach: Any sandy seashore influenced by large, powerful waves from an open ocean. For contrast, see **Low-energy beach.**

Homing behavior: A phenomenon in which migratory birds return to breed at the same location where they hatched the previous year; long-lived species thereafter continue returning to nest at the same site. In ducks, only the females demonstrate this fidelity, but in other groups, both sexes return (e.g., geese). Also applies to some mammals (e.g., bats).

Hurricane: A massive cyclonic storm with sustained surface winds exceeding 74 miles per hour. Most hurricanes also possess a cloud-free "eye" in the center of circulation.

Hydroperiod: The length of time water inundates a site; usually characterized in relative terms (i.e., long or short). Locations with long hydroperiods develop wetland vegetation (cattails and rushes), whereas those inundated less often form swales typically characterized by sedges and spikerushes. Distinctive soils also develop at sites with a long history of regular inundation. In tidal marshes, hydroperiod is based on the length of inundation during a 24-hour period.

Inbreeding depression: The undesired effects of repeated matings among individuals in small and/or isolated populations, which thereby reduce genetic variation and increase abnormalities in their offspring; an influx of animals from other populations counteracts its occurrence.

Indicator species: A species that is so characteristic of a certain type of environmental condition that its presence reflects that condition. For example, mosses indicate acidic soil and Lechuguilla is an indicator species of the Chihuahuan Desert.

Inflorescence: The cluster or other arrangement of flowers on a stem-like structure.

Inquilines: Organisms inhabiting water enclosed on plant structures. See **Phytotelmata.**

Intertidal zone: Portion of a shoreline or exposed hard surface submerged during high tides and exposed during low tides. In marine environments, the intertidal zone is also known as the littoral zone.

Introgression: Loss of genetic identity by a species because of the incorporation of genes from one species into the gene pool of another. Also known as "genetic swamping."

Isohyet: A map line connecting points with equal amounts of annual rainfall.

Karst: A landscape formed from limestone, dolomite, or gypsum—soluble rocks deposited by ancient seas—and characterized by cracks or sinkholes leading to underground caves or aquifers.

Keystone species: A species whose presence exerts a type of control on the structure and species composition of the community in which it lives. The existence of a keystone species is usually not discovered until the species is reduced or eliminated.

Lagoon: Any shallow body of water separated from a much larger body of water by a barrier island or reef.

Littoral zone: Shallow-water zone where plants such as cattails flourish along with a wealth of invertebrates and small fishes. Because sunlight can penetrate to the bottom, productivity at these sites exceeds that in deeper water.

Longshore current: A current flowing parallel to the beach, resulting from waves breaking at an angle to the shoreline.

Low-energy beach: Sandy seashores typically lining bays, lagoons, or other bodies of water where wave action is limited and rarely causes erosion. For contrast, see **High-energy beach.**

Macrofauna: Marine or freshwater invertebrates too large to pass through a sieve with 0.5 mm mesh. For comparison, see **Meiofauna.**

Mast: Collective term for woody fruits such as the litter of acorns, walnuts, or pecans beneath trees.

Meiofauna: Marine or freshwater invertebrates small enough to pass through a sieve with 0.5–1 mm mesh, but not through a sieve with 30–45 μm mesh. For comparison, see **Macrofauna.**

Mesic: A relative term used to characterize conditions—typically soil moisture—that are less extreme than elsewhere in the immediate area. Community composition and structure accordingly change as conditions become more mesic (e.g., in dry regions, stands of deciduous trees often characterize valley floors, which are mesic sites compared to the surrounding hills where range vegetation develops).

Mesopredators: Predators (e.g., foxes and Raccoons) in the middle levels of a food chain. When apex predators (e.g., wolves) disappear, mesopredator populations typically increase and disturb previous predator-prey relationships. Under similar circumstances, the same pattern prevails in aquatic systems.

Microhabitat: A constrained site where the environmental conditions differ enough from those in the surrounding habitat to provide suitable conditions for certain organisms (e.g., the moist soil beneath a decaying log provides a microhabitat for salamanders and certain beetles).

Midden: An ancient deposit of refuse, such as shells, bones, or seeds, that offers clues about past environments or civilizations.

Monadnock: An isolated hill, knob, or mountain standing amid lower environs as the survivor in an area in which erosion reduced the surrounding elevation.

Mutualism: A type of symbiotic relationship in which two species derive benefits from their close association.

Nekton: Small, swimming aquatic organisms that move independently of water currents.

Neoteny: A developmental condition, often in amphibians, in which sexually mature adults retain some or all of their juvenile or larval features (e.g., adult frogs of some species resemble tadpoles).

Neotropical migrants: Birds that breed in North America and overwinter in Central or South America and the Caribbean Islands. Mostly songbirds (e.g., wood warblers) but also species from other groups (e.g., Purple Martin, Common Nighthawk, and Swainson's Hawk). Some fly nonstop across large expanses of water, whereas others follow coastlines or mountain ranges.

Nutrient pulses: Short-term, often sudden, influxes of nutrients into a community or ecosystem. Examples include seasonal floods and decaying carcasses of Pacific Salmon or American Bison.

Oophagy: The practice of eating eggs.

Parapatric: Describes the occupation of areas that either partially overlap or have a barrier between them.

Periglacial: Refers to environments affected by glaciers, but not themselves buried under ice (e.g., areas influenced by the cool climate immediately south of the ice sheets once covering much of North America); also, the prevailing conditions (e.g., cool, wet climate).

Periphyton: The velvety coating of detritus and microorganisms that accumulates on surfaces of submerged plants and objects in marine or freshwater habitats. Periphyton is harvested for food by many aquatic organisms.

Petroglyph: A type of rock engraving produced by removing part of a (usually darker) surface. This differs from a pictograph, which is produced by adding a pigment to a rock surface to create an image.

Phagocytosis: Process by which a cell membrane surrounds a particle on the cell surface and eventually forms a vacuole to take the object into the cell.

Phreatophyte: A plant with roots deep enough to obtain water from the permanent ground supply. Some phreatophytes release copious amounts of water vapor into the air, whereas others outcompete surrounding vegetation where water is limited.

Phytotelmata: Small pools of water held in vegetative structures such as tree cavities and leaf axils, and in pitcher plants. These sometimes support inquilines—species unique to these habitats. Singular: phytotelma. See **Inquilines.**

Pluvial: A term relating to rain.

A pluvial lake is one whose only water source is rainfall.

Precocial: For birds, a description indicating that nestlings, although remaining with one or both parents, hatch open eyed and feathered, feed themselves, and quickly leave their nests (e.g., ducklings). Also applies to some newborn mammals (e.g., fawns). For comparison, see **Altricial.**

Prehensile: Capable of grasping; usually used in reference to a tail, tongue, or snout.

Primary productivity: The formation of plant tissues, including the energy-rich compounds stored within as a product of photosynthesis; usually estimated by measuring the biomass of new growth per unit area. See **Biomass.**

Pyrophyte: Plants adapted to withstand fires. Pyrophytic is the adjective form.

Resource partitioning: Any change in behavior or other specialization that lessens competition between species for the same resource.

Ruderal: Term indicating locations or organisms, usually plants, subject to recurring disturbance (e.g., roadsides and their vegetation).

Seed bank: The residual source of seeds that persist in the soil until conditions favoring germination return. At a given location, a seed bank may contain several species, each with its own environmental requirements for germination and growth.

Sexual selection: A type of natural selection that favors the development of traits such as size, color patterns, male antlers, and so forth that enhance the probability of mating success.

Sky islands: Isolated mountains tall enough to develop communities greatly dissimilar from those in the surrounding area and well separated from mountains elsewhere (e.g., basin-and-range landscapes); often sites with endemic taxa and/or relict populations of more widely occurring species stranded at the end of the Pleistocene. Commonly high-elevation forests surrounded by lowland desert.

Speleothem: A secondary mineral formation typically found in a limestone cave. See **Helictite, Stalactite,** and **Stalagmite.**

Stalactite: A type of rock formation (a speleothem) that hangs from the ceiling of a cave.

Stalagmite: A type of rock formation (a speleothem) that arises from the floor of a cave as the result of mineral-laden water dripping from the cave ceiling.

Succulent: In botany, a plant having thick, fleshy parts adapted to retain water.

Supratidal zone: The area above the high tide line that is often wetted by wave splash or spay, but not submerged. Also known as the supralittoral, splash, or spray zone.

Swale: Any natural low area that remains marshy or moist for a substantial period. A bioswale is a similar depression formed by human activities.

Sympatric distribution: Refers to the overlapping geographical ranges of two or more species, usually in the context of closely related taxa. Compare with Allopatric distribution.

Taxon: Any of the units in the hierarchy of classification (species, genus, family, order, class, and phylum are each a taxon). Plural: taxa.

Territory: A defended area typically involved with breeding activities. The site itself is seldom of biological significance and represents only a possession held by dominance (e.g., an area surrounding a fence post on which a male bobwhite utters its familiar call). Territorial behavior typically has two functions: defense against other males and a means of attracting females.

Transpiration: The evaporative loss of water through the stems, leaves, and flowers of plants.

Troglobite: An animal obligated to live in a cave because it cannot survive elsewhere. Most troglobitic species lack pigments and functional eyes.

Troglophile: An animal with adaptations that allow it to survive equally well in caves or aboveground.

Trogloxene: An animal that inhabits caves while roosting or hibernating but exits to feed and carry out other activities (e.g., Brazilian Free-tailed Bat).

Trophic cascade: Collapse of community structure occurring when the population of a significant species in a food chain (i.e., an apex predator) is reduced or eliminated.

Tropical depression: A low-pressure area in the tropics; that is, an area between 10° and 20° north or south latitude that generates strong thunderstorms.

Tropical storm: A storm originally generated in a tropical zone that develops sustained winds between 39 and 74 miles per hour.

Urohydrosis: The habit of some birds (e.g., Turkey Vultures) of defecating on the legs to allow evaporative cooling of the fluids to reduce the bird's body temperature.

Viviparous: Giving birth to mobile or nonshelled young.

Xeric: A relative term describing dry habitat conditions

Xerophyte: A plant adapted to living in dry environments.

READINGS AND REFERENCES

The most significant references consulted during the preparation of each chapter, as well as a few suggestions for additional readings of related interest, are presented for each chapter in a format that mirrors the headings and subheadings in the chapter. References consulted for multiple chapters are listed below and are not repeated elsewhere. Similarly, references used to construct the text for more than one section of a single chapter are listed under the chapter title and are not repeated thereafter. Some references are annotated to provide information about content or significance.

Ammerman, L.K., C.L. Hice, and D.J. Schmidly. 2012. Bats of Texas. Texas A&M University Press, College Station. 305 pp.

Bezanson, D. 2000. Natural vegetation types of Texas and their representation in conservation areas. Unpublished MS thesis, University of Texas Press, Austin. 215 pp.

Blair, F. 1950. The biotic provinces of Texas. Texas Journal of Science 2:93–117.

Bomar, G.W. 1995. Texas weather. 2nd rev. ed. University of Texas Press, Austin. 287 pp.

Brady, N.C., and R.R. Weil. 2007. The nature and properties of soils. 14th ed. Prentice Hall, Upper Saddle River, NJ. 980 pp.

Carter, W.T. 1931. The soils of Texas. Texas Agricultural Experiment Station Bulletin 431:1–190.

Correll, D.S., and M.C. Johnston. 1970. The manual of the vascular plants of Texas. Texas Research Foundation, Renner. 1881 pp.

Diggs, G.M., Jr., B.L. Lipscomb, and R.J. O'Kennon. 1999. Shinners & Mahler's illustrated flora of north central Texas. Center for Environmental Studies, Austin College, Sherman, TX, and Botanical Research Institute of Texas, Fort Worth. 1626 pp.

Diggs, G.M., Jr., B.L. Lipscomb, M.D. Reed, and R.J. O'Kennon. 2006. Illustrated flora of East Texas. 3 vols. Botanical Research Institute of Texas, Fort Worth. 1594 pp.

Godfrey, C., G.S. McKee, and H. Oats. 1973. General soils map of Texas. Texas Agricultural Experiment Station Miscellaneous Publication MP-1304.

Gould, F.W. 1969. Texas plants: A checklist and ecological summary. Revised. Texas Agricultural Experiment Station, Texas A&M University, College Station. 121 pp. [Maps and descriptions of the state's major vegetational areas provide the framework for this book.]

Graves, J. 2002. Texas rivers. Texas Parks and Wildlife Press, Austin. 144 pp.

Griffith, G., S. Bryce, J. Omernik, and A. Rogers. 2007. Ecoregions of Texas. Texas Commission on Environmental Quality, Austin. 125 pp.

Huser, V. 2000. Rivers of Texas. Texas A&M University Press, College Station. 264 pp.

Larkin, T.J., and G.W. Bomar. 1983. Climatic atlas of Texas. Texas Department of Water Resources LP-192:1–151.

Lockwood, M.W., and B. Freeman. 2014. The TOS handbook of Texas birds. 2nd ed. Texas A&M University Press, College Station. 403 pp.

Oberholser, H.C. 1974. The bird life of Texas. Edited by E. G. Kincaid Jr. 2 vols. University of Texas Press, Austin. 1069 pp.

Schmidly, D.J. 2002. Texas natural history: A century of change. Texas Tech University Press, Lubbock. 534 pp.

Schmidly, D.J., and R.D. Bradley. 2016. The mammals of Texas. 7th ed. University of Texas Press, Austin. 720 pp.

Sellards, E.H., W.S. Adkins, and F.B. Plummer. 1932. The geology of Texas. University of Texas Bulletin 3232(1):1–1007.

Spearing, D. 1991. Roadside geology of Texas. Mountain Press, Missoula, MT. 418 pp.

Werler, J.E., and J.R. Dixon. 2000. Texas snakes: Identification, distribution, and natural history. University of Texas Press, Austin. 437 pp.

CHAPTER 1
Early Natural History in Texas

Bragg, A.N. 1961. Strecker—Naturalist and man. Bios 32:177–181.

Brown, L., H. Jackson, and J. Brown. 1972. Burrowing behavior of the chorus frog, *Pseudacris streckeri*. Herpetologica 25:325–328.

Burke, H.R. Pioneer Texas naturalists: A contribution to the history of Texas natural history from its beginning to 1940. 111 pp. [An unpublished manuscript provided by the author.]

Casto, S.D. 2001. Texas ornithology during the 1880s. Texas Birds 3(2):14–19.

Evans, H.E. 1997. The natural history of the Long Expedition to the Rocky Mountains, 1819–1920. Oxford University Press, New York. 268 pp.

Flores, D.L., ed. 1984. Jefferson and southwestern exploration: The Freeman and Custis accounts of the Red River Expedition of 1806. University of Oklahoma Press, Norman. 386 pp.

Geiser, S.W. 1948. Naturalists of

the frontier. 2nd ed. Southern Methodist University Press, Dallas, TX. 296 pp. [Source for Audubon's quotes.]

Goetzmann, W.H. 1959. Army exploration in the American West, 1803–1863. Yale University Press, New Haven, CT. 509 pp.

———. 1966. Exploration and empire: The explorer and the scientist in winning the American West. Knopf, New York. 656 pp.

Goyle, M.A. 1991. A life among the Texas flora: Ferdinand Lindheimer's letters to George Engelmann. Texas A&M University Press, College Station. 236 pp.

Lawson, R.M. 2012. Frontier naturalist: Jean Louis Berlandier and the exploration of northern Mexico. University of New Mexico Press, Albuquerque. 262 pp.

Peterson, R.T. 1963. A field guide to the birds of Texas and adjacent states. Houghton Mifflin, Boston. 304 pp. [The original edition, cited here, was later revised.]

Romer, A.S. 1927. Notes on the Permo-Carboniferous reptile *Dimetrodon*. Journal of Geology 35:673–689.

Schuler, E.W. 1917. Dinosaur tracks in the Glen Rose limestone near Glen Rose Texas. American Journal of Science 44:294–298.

Establishing Ecological Boundaries

Daubenmire, R.F. 1938. Merriam's life zones of North America. Quarterly Review of Biology 13:327–332.

Dice, L.R. 1943. The biotic provinces of North America. University of Michigan Press, Ann Arbor. 78 pp.

Edwards, R.J., G. Longley, R. Moss, et al. 1989. A classification of Texas aquatic communities with special consideration toward the conservation of endangered and threatened taxa. Texas Journal of Science 41:231–240.

Evans, F.C. 1978. Lee Raymond Dice (1887–1977). Journal of Mammalogy 59:635–644.

Evans, J.A. 1955. Use and misuse of the biotic province concept. American Naturalist 89 (844):21–28.

Omernik, J.M. 2004. Perspectives on the nature and definition of ecological regions. Environmental Management 34 (Supplement 1): s27–s38.

Sterling, K.B. 1977. The last of the naturalists: The career of C. Hart Merriam. Arno Press, New York. 472 pp.

Telfair, R.C., II. 2009. Vegetation of Texas: Concept and commentary. Journal of the Botanical Research Institute of Texas 3:395–399.

The State of Natural History

Enderson, J.H., and D.D. Berger. 1970. Pesticides: Eggshell thinness and lowered production of young in prairie falcons. BioScience 20: 355–356.

Grant, P.R. 2000. What does it mean to be a naturalist at the end of the twentieth century? American Naturalist 155:1–12.

Pyle, R.M. 2001. The rise and fall of natural history. Orion (Autumn):16–23.

Schmidly, D.J. 2005. What it means to be a naturalist and the future of natural history at American universities. Journal of Mammalogy 86:449–456.

Weigl, P.D. 2009. The natural history conundrum revisited: Mammalogy begins at home. Journal of Mammalogy 90:265–269.

Wheeler, Q.D., P.H. Raven, and E.O. Wilson. 2004. Taxonomy: Impediment or expedient? Science 303:285.

Wilcove, D.S., and T. Eisner. 2000. The impending extinction of natural history. Chronicle of Higher Education 47(3):B24.

Wilson, E.O. 1994. Naturalist. Island Press, Washington, DC. 380 pp.

Infobox 1.1. Vernon O. Bailey (1864–1942): Field Naturalist of the Old School

Bailey, V. 1905. Biological survey of Texas. Fauna of North America No. 25. US Department of Agriculture, Washington, DC. 222 pp.

———. 1919. A new subspecies of beaver from North Dakota. Journal of Mammalogy 1:31–32.

Kofalk, H. 1989. No woman tenderfoot: Frances Merriam Bailey, pioneer naturalist. Texas A&M University Press, College Station. 225 pp.

Zahniser, H. 1942. Vernon Orlando Bailey, 1864–1942. Science 96:6–7.

Infobox 1.2. *The Bird Life of Texas:* The Life's Work of Harry C. Oberholser (1870–1963)

Aldrich, J.W. 1968. In memoriam: Harry Church Oberholser. Auk 85: 25–29.

Casto, S.D. 2001. Oberholser's bibliography of Texas birds. Bulletin of the Texas Ornithological Society 34:24–25.

———. 2014. Harry Church Oberholser and *The Bird Life of Texas.* Bulletin of the Texas Ornithological Society 45:30–44.

Casto, S.D., and H.R. Burke. 2007. Louis Agassiz Fuertes and the biological survey of Trans-Pecos Texas. Bulletin of the Texas Ornithological Society 40:49–61.

Chapman, F.M. 1928. In memoriam: Louis Agassiz Fuertes, 1874–1927. Auk 45:1–26.

Morony, J.J. 1976. The bird life of Texas (review). Auk 93:393–396.

Infobox 1.3. W. Frank Blair (1912–1985): Herpetologist and Evolutionary Biologist

Blair, W.F. 1960.The rusty lizard: A population study. University of Texas Press, Austin. 185 pp.

———. 1972. Evolution of the genus *Bufo*. University of Texas Press, Austin. 459 pp.

Blair, W.F., A.P. Blair, P. Broadkorb, et al. 1957. Vertebrates of the United States. McGraw-Hill, New York. 819 pp. [A second edition appeared in 1968.]

Dice, L.R. 1943. The biotic provinces of North America. University of Michigan Press, Ann Arbor. 78 pp.

Hubbs, C. 1985. William Franklin Blair. Copeia 1985:529–531.

Infobox 1.4. Frank W. Gould (1913–1981): Botanist

Gould, F.W. 1968. Grass systematics. McGraw-Hill, New York. 32 pp.

———. 1969. Texas plants: A checklist and ecological summary. Rev. ed. Texas Agricultural Experiment Station, Texas A&M University, College Station. 121 pp.

Gould, L. 1981. Frank Walton Gould, 1913–1981. Annals of the Missouri Botanical Garden 68:1. [Obituary by Gould's wife, Lucile.]

Hatch, S.L. 1981. Frank Walton Gould. Taxon 30:733.

CHAPTER 2

Braun, E.L. 1950. Deciduous forests of eastern North America. Blackburn Press, Caldwell, NJ. 596 pp. [Reprint of first edition.]

Chapman, H.H. 1952. The place of fire in the ecology of pines. Bartonia 26:39–44.

Leipnik, M.R., I. Perry, P.S. Shrestha, and R. Evans. 1997. Geographical information systems based analysis of current and historical vegetation composition in East Texas. Texas Journal of Science 49 (Supplement):13–20.

Schafale, M.P., and P.A. Harcombe. 1983. Presettlement vegetation of Hardin County, Texas. American Midland Naturalist 109:355–366.

Schmidly, D.J. 1983. Texas mammals east of the Balcones Fault Zone. Texas A&M University Press, College Station. 400 pp.

Vines, R.A. 1977. Trees of East Texas. University of Texas Press, Austin. 556 pp.

Walker, L.C., and B.P. Oswald. 1999. The southern forest: Geography, ecology, and silviculture. CRC Press, Boca Raton, FL. 352 pp.

Major Biotic Associations

Allen, J.A. 1977. Restoration of bottomland hardwoods and the issue of woody species diversity. Restoration Ecology 5:125–134.

Bridges, E.L., and S.L. Orzell. 1989. Longleaf Pine communities of the West Gulf Coastal Plain. Natural Areas Journal 9:246–263.

Chambless, L.F., and E.S. Nixon. 1975. Woody vegetation-soil relations in a bottomland forest in East Texas. Texas Journal of Science 41: 105–122.

Chapman, H.H. 1932. Is the Longleaf type a climax? Ecology 13:328–334.

———. 1944. Fire and pines. American Forests 50:62–64, 91–93.

Cotton, M.H., R.R. Hicks Jr., and R.H. Flake. 1975. Morphological variability among Loblolly and Shortleaf Pines of East Texas with reference to natural hybridization. Castanea 40:309–319.

Jose, S., E.J. Jokela, D. Miller, and D.L. Miller. 2006. The Longleaf Pine ecosystem: Ecology, silviculture, and restoration. Springer, New York. 438 pp.

MacRoberts, M.H., and B.R. MacRoberts. 2001. Bog communities of the West Gulf Coastal Plain. Botany and Ecology 1:1–151.

Marks, P.L., and P.A. Harcombe. 1981. Forest vegetation of the Big Thicket, southeast Texas. Ecological Monographs 51:287–305.

McLeod, C.A. 1971. The Big Thicket forest of East Texas. Texas Journal of Science 23:221–233.

Nixon, E.S., L.F. Chambless, and J.L. Malloy. 1973. Woody vegetation of a palmetto (*Sabal minor* (Jacq.) Pers.) area in East Texas. Texas Journal of Science 24:535–541.

Nixon, E.S., and J.A. Raines. 1976. Woody creekside vegetation of Nacogdoches County, Texas. Texas Journal of Science 27:443–452.

Nixon, E.S., R.L. Willet, and P.W. Cox. 1977. Woody vegetation of a virgin forest in an eastern Texas river bottom. Castanea 42:227–236.

Phillips, T.C., S.B. Walker, B.R. MacRoberts, and M.H. MacRoberts. 2007. Vascular flora of a Longleaf Pine upland in Sabine County, Texas. Phytologia 89:317–338.

Platt, W.J., G.W. Evans, and S.L. Rathbun. 1988. The population dynamics of a long-lived conifer (*Pinus palustris*). American Naturalist 131:491–525.

Schneider, R.L., and R.R. Sharitz. 1988. Hydrochory and regeneration in a Bald Cypress-Water Tupelo swamp forest. Ecology 69:1055–1063.

Schultz, R.P. 1997. Loblolly Pine: The ecology and culture of Loblolly Pine. Agriculture Handbook 713. USDA Forest Service, Washington, DC. 492 pp.

Scott, V.E., K.E. Evans, D.R. Patton, and C.P. Stone. 1977. Cavity-nesting birds of North American forests. Agricultural Handbook 511. US Department of Agriculture, Washington, DC. 112 pp.

South, D.B., and E.R. Buckner. 2003. The decline of Southern Yellow

Pine timberland. Journal of Forestry 101:30–35.

Tauer, C.G., J.F. Stewart, R.E. Will, et al. 2012. Hybridization leads to loss of genetic integrity in Shortleaf Pine: Unexpected consequences of pine management and fire suppression. Journal of Forestry 110:216–224.

Varner, J.M., and J.S. Kush. 2004. Remnant old-growth Longleaf Pine (*Pinus palustris* Mill.) savannas and forests of the southeastern USA: Status and threats. Natural Areas Journal 24: 141–149.

Walker, L.C., and H.V. Wiant Jr. 1966. Silviculture of Shortleaf Pine. Stephen F. Austin State College Bulletin No. 9. Nacogdoches, TX. 60 pp.

The Sawdust Empire

Butler, C.B. 1997. Treasures of Longleaf Pines: Naval stores. 2nd ed. Tarkel Publishing, Shalimar, FL. 270 pp.

Gerland, J.K. 1966. Sawdust City: Beaumont, Texas, on the eve of the Petroleum Age. Texas Gulf Historical and Biographical Record 32 (November):20–47.

———. 2004. Steam in the pines: A history of the Texas State Railroad. East Texas Historical Association, Nacogdoches. 66 pp.

Grissino-Mayer, H.D., H.C. Blount, and A.C. Miller. 2001. Tree-ring dating and the ethnohistory of the naval stores industry in southern Georgia. Tree Ring Research 57:3–13.

Endangered Species

Fredrickson, R.J., and P.W. Hedrick. 2006. Dynamics of hybridization and introgression in Red Wolves and Coyotes. Conservation Biology 20:1272–1283.

Hines, J.G., L.M. Hardy, D.C.

Rudolph, and S.J. Burgdorf. 2006. Movement patterns and habitat selection by native and repatriated Louisiana Pine Snakes (*Pituophis ruthveni*): Implications for conservation. Herpetological Natural History 9:103–116.

Jackson, J.A. 2006. In search of the Ivory-billed Woodpecker. Smithsonian Books, Washington, DC. 322 pp.

LeBreton, G.T.O., F.W.H. Beamish, and S.R. McKinley, eds. 2006. Sturgeons and paddlefish of North America. Kluwer Academic Publishers, Dordrecht, Netherlands. 324 pp.

MacFarlane, R.W. 1992. A stillness in the pines: The ecology of the Red-cockaded Woodpecker. W.W. Norton, New York. 270 pp.

McCarley, H. 1972. The taxonomic status of wild *Canis* (Canidae) in the south central United States. Southwestern Naturalist 7:227–235.

Nowak, R.M. 2002. The original status of wolves in North America. Southeastern Naturalist 1:95–130.

Rudolph, D.C., and S. Burgdorf. 1997. Timber Rattlesnake and Louisiana Pine Snake of the west Gulf Coastal Plain: Hypotheses of decline. Texas Journal of Science 49:111–122.

Rudolph, D.C., R.N. Conner, and J. Turner. 1990. Competition for Red-cockaded Woodpecker roost and nest cavities: Effects of resin age and entrance diameter. Wilson Bulletin 102:23–36.

Tanner, J.T. 1942. The Ivory-billed Woodpecker. National Audubon Society, New York. 144 pp.

US Fish and Wildlife Service. 2007. Red Wolf (*Canis rufus*) 5-year status review: Summary and evaluation. US Fish and Wildlife Service, Manteo, NC. 58 pp.

Wilkens, L.A., and M.H. Hofmann. 2007. The paddlefish rostrum as

an electrosensory organ: A novel adaptation for plankton feeding. BioScience 57:399–407.

Highlights

THE BIG THICKET

Ajilvsgi, G. 1979. Wildflowers of the Big Thicket: East Texas and western Louisiana. Texas A&M University Press, College Station. 360 pp.

Bonney, L.G. 2011. The Big Thicket guidebook: Exploring the backroads and history of southeast Texas. University of North Texas Press, Denton. 848 pp.

Callicott, J.B., M. Acevendo, P. Gunter, et al. 2006. Biocomplexity in the Big Thicket. Ethics, Place and Environment 9:21–45.

Eisner, T. 1973. The Big Thicket National Park. Science 179(4073):525. [A plea by an influential ecologist to save the biological diversity by creating a national park.]

Gunter, P.Y.A. 1993. The Big Thicket: An ecological reevaluation. University of North Texas Press, Denton. 243 pp.

Marks, P.L., and P.A. Harcombe. 1981. Forest vegetation of the Big Thicket, southeast Texas. Ecological Monographs 51:287–305.

McLeod, C.A. 1967. The Big Thicket of East Texas. Sam Houston Press, Huntsville, TX. 33 pp.

Watson, G.E. 2006. Big Thicket plant ecology: An introduction. 3rd ed. University of North Texas Press, Denton. 136 pp.

CADDO LAKE

Flores, D.L., ed. 1984. Jefferson and southwestern exploration: The Freeman and Custis accounts of the Red River Expedition of 1806. University of Oklahoma Press, Norman. 386 pp.

Kirk, P. 2015. Secrets of the bayou. Texas Parks and Wildlife Magazine 73(6):38–43.

Nelson, W.A. 1924. Reelfoot—An earthquake lake. National Geographic 45:94–114.

Parent, L. 2008. Official guide to Texas state parks and historic sites. Rev. ed. University of Texas Press, Austin. 188 pp.

Shira, A.F. 1913. The mussel fisheries of Caddo Lake and the Cypress and Sulphur Rivers of Texas and Louisiana. US Bureau of Fisheries Economic Circular 6:1–10.

Smith, F.T. 2000. The Caddo Indians: Tribes at the convergence of empires, 1542–1854. Texas A&M University Press, College Station. 229 pp.

SAVAGE SAVANNAS—CARNIVOROUS PLANTS

Adlassnig, W., M. Peroutka, and T. Lendl. 2011. Traps of carnivorous pitcher plants as a habitat: Composition of the fluid, biodiversity, and mutualistic activities. Annals of Botany 107:181–194.

Cresswell, J.E. 1991. Capture rates and composition of insect prey of the pitcher plant *Sarracenia purpurea*. American Midland Naturalist 125:1–9.

Folkerts, D.R. 1999. Pitcher plants in wetlands of the southern United States: Arthropod associates. Pp. 247–275 *in* Invertebrates in freshwater wetlands of North America: Ecology and management (D.P. Batzer, R.B. Rader, and S.A. Wissinger, eds.). John Wiley and Sons, New York. 1120 pp.

Folkerts, G.W. 1982. The Gulf Coast pitcher plant bogs. American Scientist 70:260–267.

Koopman, M.M., D.M. Fusellier, S. Hird, and B.C. Carstens. 2010. The carnivorous Pale Pitcher Plant harbors diverse, distinct, and time-dependent bacterial communities. Applied and Environmental Microbiology 76:1851–1860.

Conservation and Management

Cozine, J.J., Jr. 2004. Saving the Big Thicket: From exploration to preservation, 1685–2003. University of North Texas Press, Denton. 312 pp.

Jose, S., E.J. Jokela, D. Miller, and D.L. Miller. 2006. The Longleaf Pine ecosystem: Ecology, silviculture, and restoration. Springer, New York. 438 pp.

McMahon, C.K., D.J. Tomczak, and R.M. Jeffers. 1998. Longleaf Pine ecosystem restoration: The role of the USDA Forest Service. Longleaf Alliance Report 3:20–31.

Infobox 2.1. "Turpentiners" and the Naval Stores Industry

Perry, P. 1968. The naval stores industry in the Old South. Journal of Southern History 34:509–526.

Silvester, P. 2009. The story of Boogie Woogie: A left hand like God. 2nd ed. Scarecrow Press, Lanham, MD. 438 pp.

Walker, L.C. 1991. The southern forest: A chronicle. University of Texas Press, Austin. 336 pp.

Infobox 2.2. Squirrels on the Move

Audubon, J.J., and J. Bachman. 1989. Audubon's quadrupeds of North America. Wellfleet Press, Secaucus, NJ. 440 pp.

Dillon, R. 1988. Meriwether Lewis: A biography. Western Tanager Press, Santa Cruz, CA. 364 pp. [See p. 64 for quotes.]

Flyger, V. 1969. The 1968 squirrel migration in the eastern United States. Contribution No. 379. Natural Resources Institute, University of Maryland, College Park. 9 pp.

Schorger, A.W. 1947. An emigration of squirrels in Wisconsin. Journal of Mammalogy 28:401–403.

Seton, E.T. 1920. Migrations of Gray Squirrels (*Sciurus carolinensis*). Journal of Mammalogy 1:53–58.

Infobox 2.3. Geraldine Ellis Watson (1925–2012): Champion of the Big Thicket

Austin American-Statesman. 2012. Obituary—Geraldine Ellis Watson. April 14.

Dembling, S. 2012. Big Thicket Preserve in southeast Texas owes its survival in part to one naturalist. Dallas Morning News, March 31.

Houston Chronicle. 2012. Obituary—Geraldine Watson. April 15.

Todd, D., and D. Weisman, eds. 2010. Big timber and the Big Thicket—Geraldine Watson. Pp. 23–26 *in* The Texas legacy project: Stories of courage and conservation. Texas A&M University Press, College Station. 278 pp.

CHAPTER 3

Olmsted, F.L. 1857. A journey through Texas: Or, a saddle-trip on the southwestern frontier. Dix, Edwards, New York. [See 2004 reprint, University of Nebraska Press, Lincoln. 39 pp.]

Stahle, D.W., and J.G. Hehr. 1984. Dendroclimatic relationships of Post Oak across a precipitation gradient in the southcentral United States. Annals of the Association of American Geographers 74:561–573.

Regional Overview

MacRoberts, M.H., and B.R. MacRoberts. 2004. The Post Oak Savanna ecoregion: A floristic assessment of its uniqueness. Sida 31:399–407.

Singhurst, J.R., J.C. Cathy, D. Prochaska, et al. 2003. The

vascular flora of Gus Engeling Wildlife Management Area, Anderson County, Texas. Southeastern Naturalist 2:347–368.

Stahle, D.W.1996. Tree rings and ancient forest relics. Arnoldia 56:2–10.

Some Interesting Communities

LOST PINES

Al-Rabag'ah, M., and C.G. Williams. 2004. An ancient bottleneck in the Lost Pines of Central Texas. Molecular Ecology 13:1075–1084.

Williams, C.G. 2012. Replanting the (really) Lost Pines. Forest History Today (Spring):12–15.

XERIC SANDYLANDS

Belnap, J., and O.L. Lange, eds. 2001. Biological soil crusts: Structure, function, and management. Springer-Verlag Ecological Studies 150. Berlin, Germany. 506 pp.

MacRoberts, B.R., M.H. MacRoberts, and J.C. Cathey. 2002. Floristics of xeric sandylands in the Post Oak Savanna region of East Texas. Sida 20:373–386.

MacRoberts, M.H., B.R. MacRoberts, B.A. Sorrie, and R.E. Evans. 2002. Endemism in the West Gulf Coastal Plain: Importance of xeric habitats. Sida 20:767–780.

BOGS

Bridges, E.L., and S.L. Orzell. 1989. Additions and noteworthy vascular plant collections from Texas and Louisiana, with historical, ecological and geological notes. Phytologia 66:12–69.

Bryant, V.M., Jr. 1977. A 16,000-year pollen record of vegetational changes in Central Texas. Palynology 1:143–156.

Kral, R. 1955. A floristic comparison of two hillside bog localities in northeastern Texas. Field and Laboratory 23:47–69.

MacRoberts, B.R., and M.H. MacRoberts. 1998. Floristics of muck bogs in east central Texas. Phytologia 85:61–73.

Rowell, C.M., Jr. 1949. A preliminary report on the floral composition of a *Sphagnum* bog in Robertson County, Texas. Texas Journal of Science 1:50–53.

Highlights

SOILS AND FIRE FREQUENCY

Diggs, G.M., Jr., and P.C. Schulze. 2003. Soil-dependent fire frequency: A new hypothesis for the distribution of prairies and oak woodlands/savannas in north central and east Texas. Sida 20: 1139–1153.

LENNOX WOODS, A TREASURE OF THE RED RIVER COUNTRY

Bartlett, R.C. 1995. Saving the best of Texas: A partnership approach to conservation. University of Texas Press, Austin. 221 pp.

Sanders, R.W. 1994. Vegetational survey: Lennox Woods Preserve, Red River County, Texas. Unpublished report. The Nature Conservancy of Texas, San Antonio, and Botanical Research Institute of Texas, Fort Worth. 37 pp.

Trousdale, A.W., and D.C. Beckett. 2005. Characteristics of tree roosts of Rafinesque's Big-eared Bat (*Corynorhinus rafinesquii*) in southeastern Mississippi. American Midland Naturalist 154: 442–449.

AMERICAN BURYING BEETLES, NATURE'S UNDERTAKERS

Anderson, R.S. 1982. On the decreasing abundance of *Nicrophorus americanus* Olivier (Coleoptera: Silphidae) in eastern North America. Coleopterists Bulletin 36:362–365.

Creighton, J.C., C.C. Vaughn, and B.R. Chapman. 1993. Habitat preference of the American Burying Beetle (*Nicrophorus americanus*) in Oklahoma. Southwestern Naturalist 38:275–277.

Kozol, A.J., M.P. Scott, and J.F.A. Traniello. 1988. The American Burying Beetle, *Nicrophorus americanus:* Studies on the natural history of a declining species. Psyche 95:167–195.

Scott, M.P. 1998. The ecology and behavior of burying beetles. Annual Review of Entomology 43: 595–618.

NAVASOTA LADIES' TRESSES

Hammons, J.R., F.E. Smeins, and W.E. Rogers. 2010. Transplant methods for the endangered orchid *Spiranthes parksii* Correll. North American Native Orchid Journal 16:38–46.

Wonkka, C.L., W.E. Rogers, F.E. Smeins, et al. 2012. Biology, ecology, and conservation of Navasota Ladies' Tresses (*Spiranthes parksii* Correll), an endangered terrestrial orchid of Texas. Native Plants Journal 13: 236–243.

HEAVENLY CONTACT— OF A SORT

Alvarez, L.W., W. Alvarez, F. Asaro, and H.V. Michel. 1980. Extraterrestrial cause for the Cretaceous-Tertiary extinction. Science 208:1095–1103.

Alvarez, W. 1997. *T. rex* and the crater of doom. Vintage Books, New York. 185 pp.

Ganapathy, R., S. Gartner, and M.-J. Jiang. 1981. Iridium anomaly at the Cretaceous-Tertiary boundary in Texas. Earth and Planetary Science Letters 54:393–396.

Geller, G. 2007. The Chicxulub impact and K-T mass extinctions in Texas. Bulletin of the South

Texas Geological Society 47(9): 2-26.

Hart, M.B., T.E. Yancey, A.D. Leighton, et al. 2012. The Cretaceous-Paleogene boundary on the Brazos River, Texas: New stratigraphic sections and revised interpretations. Gulf Coast Association of Geological Societies 1:69-80.

Conservation and Management

Abbott, J.C., and T.D. Hibbitts. 2011. *Cordulegaster sarracenia,* n. sp. (Odonata: Cordulegastridae) from east Texas and western Louisiana, with a key to adult Cordulegastridae of the New World. Zootaxa 2899:60-68.

Lomolino, M.V., and J.C. Creighton. 1996. Habitat selection, breeding success and conservation of the endangered American Burying Beetle, *Nicrophorus americanus.* Biological Conservation 77:235-241.

Prugh, L.R., C.J. Stoner, C.W. Epps, et al. 2009. The rise of mesopredators. BioScience 59: 779-791.

Rasmussen, H.N., and F.N. Rasmussen. 2007. Trophic relationships in orchid mycorrhiza—Diversity and implications for conservation. Lankesteriana 7:334-341.

Sikes, D.S., and R.J. Raithel. 2002. A review of hypotheses of decline of the endangered American Burying Beetle (Silphidae: *Nicrophorus americanus* Olivier). Journal of Insect Conservation 6:103-113.

Stahle, D.W., and M.K. Cleaveland. 1988. Texas drought history reconstructed and analyzed from 1698 to 1980. Journal of Climate 1:59-74.

Telfair, R.C., II. 1988. Conservation of the Catfish Creek Ecosystem: A National Natural Landmark in eastern Texas. Texas Journal of Science 40:11-23.

Timmons, J.B., B. Alldredge, W.E. Rogers, and J.C. Cathey. 2012. Feral Hogs negatively affect native plant communities. AgriLife Extension, Texas A&M System Publication SP-467. 9 pp.

Trumbo, S.T., and P.L. Bloch. 2000. Habitat fragmentation and burying beetle abundance and success. Journal of Insect Conservation 4:245-252.

US Fish and Wildlife Service. 2009. Navasota Ladies' Tresses (*Spiranthes parksii*) 5-year review: Summary and evaluation. Ecological Services Field Office, Austin, TX. 65 pp.

Chapter 3 infobox. Houston Toads and Their Troubles

Brown, L.E. 1971. Natural hybridization and trend toward extinction in some relict Texas toad populations. Southwestern Naturalist 16:185-199.

Dodd, C.K., Jr., and R.A. Seigel. 1991. Relocation, repatriation, and translocation of amphibians and reptiles: Are they conservation strategies that work? Herpetologica 47:336-360.

Duarte, A., D.J. Brown, and M.R.J. Forstner. 2011. Estimating abundance of the endangered Houston Toad on a primary recovery site. Journal of Fish and Wildlife Management 2:207-215.

Gaston, M., A. Fugi, F.W. Weckerly, and M.R.J. Forstner. 2010. Potential component Allee effects and their impact on wetland management in the conservation of endangered anuarans. PLoS ONE 5:e10102.

Hillis, D.M., A.M. Hillis, and R.F. Martin. 1984. Reproductive ecology and hybridization of the endangered Houston Toad (*Bufo houstonesis*). Journal of Herpetology 18:56-72.

Najvar, P. 2011. Houston Toad (*Bufo houstonensis*) 5-year review: Summary and evaluation. US Fish and Wildlife Service Field Office, Austin, TX. 22 pp.

CHAPTER 4

Sharpless, M.R., and J.C. Yelderman Jr., eds. 1993. The Blackland Prairie: Land, history, and culture. Baylor University Program for Regional Studies, Waco, TX. 369 pp.

White, M. 2006. Prairie time: A Blackland Prairie portrait. Texas A&M University Press, College Station. 251 pp.

Vegetation and Soils

Diamond, D.D., and F.E. Smeins. 1985. Composition, classification, and species response patterns of remnant tallgrass prairies in Texas. American Midland Naturalist 113:294-308.

———. 1993. The native plant communities of the Blackland Prairie. Pp. 66-81 *in* The Texas Blackland Prairie: Land, history, and culture (M.R. Sharpless and J.C. Yelderman Jr., eds.). Baylor University Program for Regional Studies, Waco, TX. 369 pp.

Kennemer, G.W. 1987. A quantitative analysis of the vegetation on the Dallas County White Rock Escarpment. Sida, Botanical Miscellany 1:1-10.

Springer, V.G. 1957. A new genus and species of elopid fish (*Laminospondylus transversus*) from the Upper Cretaceous of Texas. Copeia 1957:135-140.

Highlights

GILGAI AND MIMA MOUNDS

Berg, A.W. 1990. Formation of Mima mounds: A seismic hypothesis. Geology 18:281-285.

Collins, O.B., F.E. Smeins, and D.H. Riskind. 1975. Plant communities of the Blackland Prairie of Texas. Pp. 75-88 *in* Prairie: A multiple view (M.K. Wali, ed.). University of North Dakota Press, Grand Forks. 433 pp.

Cox, G.W. 1984. Mounds of mystery. Natural History 93(6):36-45. [See also Ecology 65:1397-1405 (1984)].

Gabet, E.J., J.T. Perron, and D.L. Johnson. 2014. Biotic origin for Mima mounds supported by numerical modeling. Geomorphology 206:58-66.

Hobbs, H.H., Jr. 1991. *Procambarus (Girardiella) steigmani,* a new crayfish (Decapoda, Cambaridae) from a long-grass prairie in northeastern Texas. Proceedings of the Biological Society of Washington 104:309-316.

Miller, D.L., and F.E. Smeins. 1988. Vegetation pattern within a remnant San Antonio Prairie as influenced by soil and microrelief variation. Paper 01.10 *in* Proceedings of the Tenth North American Prairie Conference, Denton, TX (A. Davis and G. Stanford, eds.). [Unnumbered pages.]

DOGWOOD CANYON—
A WOODLAND TREASURE IN
THE BLACKLAND PRAIRIES
Brown-Marsden, M., and A.B. Collins. 2006. Range expansion of *Hexalectris grandiflora* (Orchidaceae) in Texas. Sida 22: 1239-1244.

Collins, A.B., J.E. Varnum, and M. Brown-Marsden. 2005. Soil and ecological features of *Hexalectris* (Orchidaceae) sites. Sida 21:1879-1891.

Engel, V.S. 1987. Saprophytic orchids of Dallas. American Orchid Society Bulletin 56:831-835.

Fischer, R.C., A. Richter, F. Hadacek, and V. Mayer. 2008. Chemical differences between seeds and elaiosomes indicate an adaptation to nutritional needs of ants. Oecologia 55:539-547.

Fokuhl, G., J. Heinze, and P. Poschlod. 2012. Myrmecochory by small ants—Beneficial effects through elaiosome nutrition and seed dispersal. Acta Oecologica 38:71-76.

Gilliam, F.S. 2007. The ecological significance of the herbaceous layer in temperate forest ecosystems. BioScience 57:845-858.

Handel, S.N., S.B. Fisch, and G.E. Schatz. 1981. Ants disperse a majority of herbs in a mesic forest community in New York State. Bulletin of the Torrey Botanical Club 108:430-437.

Muller, R.N., and F.H. Bormann. 1976. Role of *Erythronium americanum* Ker. in energy flow and nutrient dynamics of a northern hardwood forest ecosystem. Science 193: 1126-1128.

Stiles, E.W. 1984. A fruit for all seasons. Natural History 93(8):43-53.

Tessier, J.T., and D.J. Raynal. 2003. Vernal nitrogen and phosphorus retention by forest understory vegetation and microbes. Plant and Soil 256:443-453.

MAMMALS OF THE
BLACKLAND PRAIRIES
Schmidly, D.J., D.L. Scarbrough, and M.A. Horner. 1993. Wildlife diversity in the Blackland Prairies. Pp. 82-95 *in* The Blackland Prairie: Land, history, and culture (M.R. Sharpless and J.C. Yelderman Jr., eds.). Baylor University Program for Regional Studies, Waco, TX. 369 pp. [See especially as a source concerning the region's mammalian fauna.]

Strecker, J.K. 1927. The trade in deer skins in early Texas. Journal of Mammalogy 8:106-110.

Wilkins, K.T. 1995. The rodent community and associated vegetation in a tallgrass Blackland Prairie in Texas. Texas Journal of Science 47:243-262.

MR. DEERE AND HIS PLOW—
AN EARTH-TURNING EVENT
Dahlstrom, N., and J. Dahlstrom. 2005. The John Deere story: A biography of plowmakers John and Charles Deere. Northern Illinois University Press, DeKalb. 224 pp.

May, J.M. 1851. Breaking prairie. Transactions of the Wisconsin State Agricultural Society 1:243-246.

Conservation and Management
Herkert, J.R., D.L. Reinking, D.A. Wiedenfeld, et al. 2003. Effects of prairie fragmentation on the nest success of breeding birds in the midcontinental United States. Conservation Biology 17:587-594.

Johnson, S.K., and N.K. Johnson. 2008. Texas crawdads. Crawdad Club Designs, College Station, TX. 160 pp.

Morris, J.R., and K.L. Steigman. 1993. Effects of polygyne fire ant invasion on native ants of a Blackland Prairie in Texas. Southwestern Naturalist 38:136-140.

Richardson, C.W. 1993. Disappearing land: Erosion in the Blacklands. Pp. 262-270 *in* The Texas Blackland Prairie: Land, history, and culture (M.R. Sharpless and J.C. Yelderman Jr., eds.). Baylor University Program for Regional Studies, Waco, TX. 369 pp.

Riskind, D.H., and O.B. Collins. 1975. The Blackland Prairie of Texas: Conservation of representative climax communities. Pp. 361-373 *in* Prairie: A multiple view (M.K.

Wali, ed.). University of North Dakota Press, Grand Forks. 433 pp.

Chapter 4 infobox.
Of Sloths and Seeds

Alroy, J. 2001. A multispecies overkill simulation of the end-Pleistocene megafaunal mass extinction. Science 292:1893–1896.

Anonymous. 2015. Dining with the giant sloth. Early American Life 46(5):52–59.

Barlow, C. 2000. Ghosts of evolution: Nonsensical fruits, missing partners, and other ecological anachronisms. Basic Books, New York. 291 pp.

———. 2001. Anachronistic fruits and the ghosts who haunt them. Arnoldia 61(2):14–21.

Bronaugh, W. 2010. The trees that miss the mammoths. American Forests (Winter):38–43.

Dayton, L. 2001. Mass extinctions pinned on ice age hunters. Science 292:1819.

Grayson, D.K., and D.J. Meltzer. 2003. A requiem for North American overkill. Journal of Archaeological Science 30:585–593.

Hatch, P.J. 2003. "Public treasures": Thomas Jefferson and the garden plants of Lewis and Clark. Twinleaf Journal Online. Center for Historic Plants, Monticello, Charlottesville, VA.

Janzen, D.H., and P.S. Martin. 1982. Neotropical anachronisms: The fruits the gomphotheres ate. Science 215:19–27.

Schambach, F.F. 2000. Spiroan traders, the Sanders Site, and the Plains Interaction Sphere. Plains Anthropologist 45:7–33.

Weniger, D. 1996. Catalpa (*Catalpa bignonioides,* Bignoniaceae) and Bois d'Arc (*Maclura pomifera,* Moraceae) in early Texas records. Sida 17:231–242.

CHAPTER 5

Irving, W. 1835. A tour of the prairies. A. and W. Galignani, Paris. 199 pp. [Modern reprints available.]

Francasviglia, R.V. 2000. The cast iron forest: A natural and cultural history of the North American Cross Timbers. University of Texas Press, Austin. 276 pp.

Vegetation

Dyksterhuis, E.J. 1946. The vegetation of the Fort Worth Prairie. Ecological Monographs 16:1–29.

———. 1948. The vegetation of the Western Cross Timbers. Ecological Monographs 18:325–376.

Green, N.S., and K.T. Wilkins. 2014. Habitat associations of the rodent community in a Grand Prairie preserve. Southwestern Naturalist 59:349–355.

Hoagland, B.W., I.H. Butler, F.L. Johnson, and S. Glenn. 1999. The Cross Timbers. Pp. 231–245 *in* Savannas, barrens, and rock outcrop plant communities of North America (R.C. Anderson, J.S. Fralish, and J.M. Baskin, eds.). Cambridge University Press, New York. 470 pp.

Rykiel, E.J., Jr., and T.L. Cook. 1986. Hardwood-cedar clusters in the Post Oak Savanna of Texas. Southwestern Naturalist 31:73–78.

Stahle, D.W., and J.G. Hehr. 1984. Dendroclimatic relationships of Post Oak across a precipitation gradient in the southeastern United States. Annals of the Association of American Geographers 74:561–573.

Highlights
CUESTAS

Greene, H.W., and G.V. Oliver Jr. 1965. Notes on the natural history of the Western Massasauga. Herpetologica 21:225–228.

Langkilde, T. 2009. Invasive fire ants alter behavior and morphology of native lizards. Ecology 90:208–217.

Patten, T.J., J.D. Fawcett, and D.D. Fogell. 2009. Natural history of the Western Massasauga in Nebraska. Journal of Kansas Herpetology 30:15–20.

AN ENDEMIC SNAKE ON THE BRAZOS

Densmore, L.D., III, F.L. Rose, and S.J. Kain. 1992. Mitochondrial DNA evolution and speciation in water snakes (*Nerodia*), with special attention to *N. harteri.* Herpetologica 4:60–68.

McBride, D.L. 2009. Distribution and status of the Brazos Water Snake (*Nerodia harteri harteri*). MS thesis, Tarleton State University, Stephenville, TX. 80 pp.

Porter, S.T. 1969. An ecological survey of the herpetofauna of Palo Pinto County, Texas. MS thesis, North Texas State University, Denton. 50 pp.

Rodriquez, D., M.R J. Forstner, D.L. McBride, et al. 2012 Low genetic diversity and evidence of population structure among subspecies of *Nerodia harteri,* a threatened water snake endemic to Texas. Conservation Genetics 13:977–986.

Scott, N.J., Jr., T.C. Maxwell, O.K. Thornton Jr., et al. 1989. Distribution, habitat, and future of Harter's Water Snake, *Nerodia harteri,* in Texas. Journal of Herpetology 23:373–389.

BLACK-CAPPED VIREOS AND BROOD PARASITES

Graber, J.W. 1961. Distribution, habitat requirements, and life history of the Black-capped Vireo (*Vireo atricapilla*). Ecological Monographs 31:313–336.

Grzybowski, J.A. 1995. Black-capped Vireo (*Vireo atricapillus*). The birds of North America No. 181 (A. Poole and F. Gill, eds.). Academy of Natural Sciences, Philadelphia, and American Ornithologists' Union, Washington, DC. 24 pp.

Smith, J.N.M., and S.I. Rothstein. 2000. Brown-headed Cowbirds as a model system for studies of behavior, ecology, evolution, and conservation biology. Pp. 1–9 *in* Ecology and management of cowbirds and their hosts: Studies in the conservation of North American passerine birds (J.N.M. Smith, T.L. Cook, S.I. Rothstein, et al., eds.). University of Texas Press, Austin. 388 pp.

A LAND TROD BY GIANTS

Falkingham, P.L., K.T. Bates, and J.O. Farlow. 2014. Historical photogrammetry: Bird's Paluxy River dinosaur chase sequence digitally reconstructed as it was prior to excavation 70 years ago. PLoS ONE 9(4):e93247. doi:10.1371/journal.pone.0093247.

Farlow, J.O. 1987. A guide to the Lower Cretaceous dinosaur footprints and tracksites of Paluxy River Valley, Somervell County, Texas. South Central Geological Society of America, Baylor University, Waco, TX. 50 pp.

Shuler, E.W. 1917. Dinosaur tracks in the Glen Rose Limestone near Glen Rose, Texas. American Journal of Science 44:294–298.

Thomas, D.A., and J.O. Farlow. 1997. Tracking a dinosaur attack. Scientific American 277(6):74–79.

FOSSIL OYSTERS AND ANCIENT SEAS

Flatt, C.D. 1976. Origin and significance of the oyster banks in the Walnut Clay Formation, Central Texas. Geological Studies Bulletin 30. Baylor University, Waco, TX. 47 pp.

LaBarbera, M. 1981. The ecology of Mesozoic *Gryphaea, Exogyra,* and *Ilymatogyra* (Bivalva: Mollusca) in a modern ocean. Paleobiology 7:510–526.

PLEISTOCENE MAMMOTHS

Bishop, A.L. 1977. Flood potential of the Bosque Basin. Baylor Geological Studies Bulletin 30. Baylor University, Waco, TX. 36 pp.

Bongino, J.D. 2007. Late quaternary history of the Waco mammoth site: Environmental reconstruction and interpreting the cause of death. MS thesis, Baylor University, Waco, TX. 136 pp.

Hoppe, K.A. 2004. Late Pleistocene mammoth herd structure, migration patterns, and Clovis hunting strategies inferred from isotopic analysis of multiple death assemblages. Paleobiology 30: 129–145.

COMANCHE HARVESTER ANTS

MacMahon, J.A., J.F. Mull, and T.O. Crist. 2000. Harvester ants (*Pogonomyrmex* spp.): Their community and ecosystem influences. Annual Review of Ecology and Systematics 31:265–291.

Mayo, A.B. 2015. No place like home: The Comanche Harvester Ant in the Cross Timbers. Post Oak & Prairie Journal 1(1):5–8.

Sanders, D., and F.J.F. Van Veer. 2011. Ecosystem engineering and predation: The multi-trophic impact on two ant species. Journal of Animal Ecology 80: 569–576.

Conservation and Management

Blackburn, W.H., R.W. Knight, and J.C. Schuster. 1982. Saltcedar influence on sedimentation in the Brazos River. Journal of Soil and Water Conservation 37:298–301.

Cimprich, D.A., and R.M. Kostecke. 2006. Distribution of the Black-capped Vireo at Fort Hood, Texas. Southwestern Naturalist 51: 99–102.

Hayden, T.J., D.J. Tazik, R.H. Melton, and J.D. Cornelius. 2000. Cowbird control program at Fort Hood, Texas: Lessons for mitigation of cowbird parasitism on a landscape scale. Pp. 357–370 *in* Ecology and management of cowbirds and their hosts: Studies in the conservation of North American passerine birds (J.N.M. Smith, T.L. Cook, S.J. Rothstein, et al., eds.). University of Texas Press, Austin. 388 pp.

Rossi, J., and R. Rossi. 1999. A population survey of the Brazos Water Snake, *Nerodia harteri harteri,* and other water snakes on the Brazos River, Texas, with notes on a captive breeding and release program. Bulletin of the Chicago Herpetological Society 34:251–253.

Infobox 5.1. Lloyd H. Shinners: Systematic Botanist (1918–1971)

Correll, D.S. 1971. Lloyd Herbert Shinners—A portrait. Brittonia 23: 101–104.

Ginsburg, R. 2002. Lloyd Herbert Shinners: By himself. Botanical Research Institute of Texas, Fort Worth. 183 pp.

Mahler, W.F. 1971. Lloyd Herbert Shinners, 1918–1971. Sida 4:228–231.

Rowell, C.M., Jr. 1972. Lloyd Herbert Shinners, 1918–1971, in memoriam. Texas Journal of Science 24:266–271.

Shinners, L.H. 1958. Spring flora of the Dallas-Fort Worth area, Texas. Self-published. 514 pp. [W. F. Mahler edited a second edition, published in 1972 by Prestige Press, Fort Worth, TX.]

Infobox 5.2. Passenger Pigeon: Gone Forever

Blockstein, D.E. 2002. Passenger Pigeon, *Ectopistes migratorius*. The birds of North America No. 611 (A. Poole and F. Gill, eds.). Academy of Natural Sciences, Philadelphia, and American Ornithologists' Union, Washington, DC. 28 pp.

Blockstein, D.E., and H.B. Tordoff. 1985. Gone forever: A contemporary look at the extinction of the Passenger Pigeon. American Birds 39:845–851.

Casto, S.D. 2001. Additional records of the Passenger Pigeon in Texas. Bulletin of the Texas Ornithological Society 34:5–16.

Fulton, T.L., S.M. Wagner, C. Fisher, and B. Shapiro. 2012. Nuclear DNA from extinct Passenger Pigeon (*Ectopistes migratorius*) confirms single origin of New World pigeons. Annals of Anatomy 194:52–57.

Greenburg, J. 2014. A feathered river across the sky: The Passenger Pigeon's flight to extinction. Bloomburg, New York. 289 pp.

Schorger, A.W. 1955. The Passenger Pigeon: Its natural history and extinction. University of Wisconsin Press, Madison. 424 pp.

CHAPTER 6

Flores, D.L. 2010. Caprock canyonlands: Journeys into the heart of the southern plains. 20th anniversary ed. Texas A&M University Press, College Station. 232 pp.

Philips, S. 1942. Big Spring: The casual biography of a prairie town. Prentice-Hall, New York. 231 pp.

Weaver, J.E., and F.W. Albertson. 1956. Grasslands of the Great Plains. Johnson Publishing, Lincoln, NE. 273 pp.

Wright, H.A. 1972. Effect of fire on southern mixed prairie grasses. Journal of Range Management 27:417–419.

Distinctive Subregions

Archer, S. 1995. Tree-grass dynamics in a *Prosopis*-thornscrub savanna parkland: Reconstructing the past and predicting the future. Ecoscience 2:83–99.

Bonner, T.H., and G.R. Wilde. 2000. Changes in the Canadian River fish assemblage associated with reservoir construction. Journal of Freshwater Ecology 15:189–198.

Brown, J.R., and S. Archer. 1989. Woody plant invasion of grasslands: Establishment of Honey Mesquite (*Prosopis glandulosa* var. *glandulosa*) on sites differing in herbaceous biomass and grazing history. Oecologia 80:19–26.

Case, E.C. 1922. New reptiles and stegocephalians from the Upper Triassic of western Texas. Carnegie Institution of Washington Publication 321:1–84.

Guy, F.D., ed. 2001. The story of Palo Duro Canyon. Texas Tech University Press, Lubbock. 226 pp.

Haley, J.E. 1981. Charles Goodnight: Cowman and plainsman. University of Oklahoma Press, Norman. 504 pp.

Hammer, R.L. 2015. Rolling Plains rarity: Dwarf Broomspurge is found in only two Texas counties. Texas Parks and Wildlife Magazine (August/September):18.

Harris, D. 1971. Recent plant invasions in the arid and semi-arid Southwest of the United States. Pp. 459–481 *in* Man's impact on the environment (T. Detwyler, ed.). McGraw-Hill, New York. 731 pp.

Jones, J.O., and T.F. Hentz. 1988. Permian strata of north-central Texas. Pp. 309–316 *in* Centennial field guide, vol. 4 (O.T. Hayward, ed.). South-Central Section of the Geological Society of America, Boulder, CO. 468 pp.

Kramp, B.A., R.J. Ansley, and T.R. Tunnell. 1998. Survival of mesquite seedlings emerging from cattle and wildlife feces in a semi-arid grassland. Southwestern Naturalist 43:300–312.

Neuenschwander, L.F., H.A. Wright, and S.C. Bunting. 1978. The effect of fire on a Tobosagrass-mesquite community in the Rolling Plains of Texas. Southwestern Naturalist 23:315–338.

Rahjen, F.E. 1998. The Texas Panhandle frontier. Rev. ed. Texas Tech University Press, Lubbock. 288 pp.

Seyffert, K.D. 2001. Birds of the Texas Panhandle. Texas A&M University Press, College Station. 501 pp.

Ward, P.E. 1963. Geology and ground-water features of salt springs, seeps and plains in the Arkansas and Red River basins of western Oklahoma and adjacent parts of Kansas and Texas. US Geological Survey Report 63–132. Washington, DC. 71 pp.

Weishampel, D.B. 2004. Dinosaurian distribution. Pp. 63–140 *in* The Dinosauria, 2nd ed. (D.B. Weishampel, P. Dodson, and O. Halszka, eds.). University of California Press, Berkeley. 880 pp.

Rolling Plains Wildlife

NORTHERN BOBWHITE

Brennan, L.A., ed. 2007. Texas quails: Ecology and management. Texas A&M University Press, College Station. 512 pp.

Brennan, L.A., F. Hernandez, and D. Williford. 2014. Northern Bobwhite (*Colinus virginianus*). The birds of North America online (A. Poole, ed.). Cornell Laboratory of Ornithology, Ithaca, NY. http://

bna.birds.cornell.edu/bma/species/397, doi:10.2173/bna.397 (accessed August 15, 2015).

Dunham, N.R., S.T. Peper, C.E. Baxter, and R.J. Kendall. 2014. The Parasitic Eyeworm *Oxyspirura petrowi* as a possible cause of decline in the threatened Lesser Prairie Chicken (*Tympanuchus pallidicinctus*). PLoS ONE 9(9):e10824.

Guthery, F.S. 2000. On bobwhites. Texas A&M University Press, College Station. 224 pp.

Hernández, F., S.E. Henke, N.J. Silvy, and D. Rollins. 2003. The use of prickly pear cactus as nesting cover by Northern Bobwhites. Journal of Wildlife Management 67:417–423.

Hernández, F., and M.J. Peterson. 2007. Northern Bobwhite ecology and life history. Pp. 40–64 *in* Texas quails: Ecology and management (L.A. Brennan, ed.). Texas A&M University Press, College Station. 512 pp.

Jackson, A.S. 1969. A handbook for Bobwhite Quail management in West Texas Rolling Plains. Texas Parks and Wildlife Bulletin 48. Texas Parks and Wildlife Department, Austin. 77 pp.

Larson, J.A., T.E. Fulbright, L.A. Brennan, et al. 2010. Texas Bobwhites: A guide to their foods and habitat management. University of Texas Press, Austin. 294 pp.

Ransom, D., Jr., R.R. Lopez, G.G. Schulz, and J.S. Wagner. 2008. Northern Bobwhite habitat selection in relation to brush management in the Rolling Plains of Texas. Western North American Naturalist 68:186–193.

Renwald, J.D., H.A. Wright, and J.T. Flinders. 1978. Effect of prescribed fire on Bobwhite Quail habitat in the Rolling Plains of Texas.

Journal of Range Management 31:65–69.

Rosene, W. 1969. The Bobwhite Quail: Its life and management. Rutgers University Press, New Brunswick, NJ. 418 pp.

Slater, S.C., D. Rollins, R.C. Dowler, and C.B. Scott. 2001. *Opuntia:* A prickly paradigm for quail management in west-central Texas. Wildlife Society Bulletin 29:713–719.

Tolleson, D.R., W.E. Pinchak, D. Rollins, and L.J. Hunt. 1995. Feral Hogs in the Rolling Plains of Texas: Perspectives, problems, and potential. Pp. 124–128 *in* Great Plains wildlife damage control workshop proceedings (R.E. Masters and J.G. Huggins, eds.). Noble Foundation, Ardmore, OK.

Villareal, S.M., A.M. Fedynich, L.A. Brennan, and D. Rollins. 2012. Parasitic Eyeworm (*Oxyspirura petrowi*) in Northern Bobwhite from the Rolling Plains of Texas, 2007–2011. Proceedings of the National Quail Symposium 7:241–243.

TEXAS BROWN TARANTULA

Baerg, W.J. 1958. The tarantula. University Press of Kansas, Lawrence. 88 pp.

Comstock, J.H. 1975. The spider book. Cornell University Press, Ithaca, NY. 879 pp.

Janowski-Bell, M.E. 2001. Movement of the male Brown Tarantula *Aphonopelma hentzi* (Araneae: Theraphosidae) using radio telemetry. Journal of Arachnology 27:503–512.

Marcy, R.B. 1854. A report on the exploration of the Red River of Louisiana in the year 1852. US Government Printing Office, Washington, DC. 310 pp.

Passmore, L. 1936. Tarantula and Tarantula Hawk. Nature Magazine 27:155–159.

Punzo, F. 2005. Studies on the natural history, ecology, and behavior of *Pepsis cerberus* and *P. mexicana* (Hymenoptera: Pompilidae) from Big Bend National Park, Texas. Journal of the New York Entomological Society 113:84–95.

Punzo, F., and B. Garman. 1989. Effects of encounter experience on the hunting behavior of the Spider Wasp, *Pepsis formosa* (Say) (Hymenoptera: Pompilidae). Southwestern Naturalist 34:513–518.

Shillington, C., and P. Verrell. 1997. Sexual strategies of a North American "tarantula" (Araneae: Theraphosidae). Ethology 103:588–598.

INTERIOR LEAST TERN

Crawford, M., M. Simpson, and S. Schneider. 2003. Designing an island habitat for the Interior Least Tern. Biosystems and Agricultural Engineering Department, Oklahoma State University, Stillwater. 55 pp.

Thompson, B.S., J.A. Jackson, J. Burger, et al. 1997. Least tern (*Sterna antiellarum*). The birds of North America online (A. Poole, ed.). Cornell Laboratory of Ornithology, Ithaca, NY. https://birdsna.org/Species-Account/bna/species/leater1/introduction, doi:10.2173/bna.290 (accessed August 4, 2015).

US Fish and Wildlife Service. 1990. Recovery plan for the interior population of the Least Tern (*Sterna antillarum*). US Fish and Wildlife Service, Twin Cities, MN. 90 pp.

Whitman, P.I. 1988. Biology and conservation of the endangered Interior Least Tern: A literature review. US Fish and Wildlife Service Biological Report 88:1–22.

MOUNTAIN BOOMERS

Blair, W.F., and A.P. Blair. 1941. Food habits of the Collared Lizard in northeastern Oklahoma. American Midland Naturalist 26: 230–232.

Husak, J.F., and S.F. Fox. 2003. Spatial organization and the dear enemy phenomenon in adult female Collared Lizards, *Crotaphytus collaris*. Journal of Herpetology 37:211–215.

———. 2006. Field use of maximum sprint speed by Collard Lizards (*Crotaphytus collaris*): Compensation and sexual selection. Evolution 60:1888–1895.

Husak, J.F., J.M. Macedonia, S.F. Fox, and R.C. Sauceda. 2006. Predation cost of conspicuous male coloration in Collared Lizards (*Crotaphytus collaris*): An experimental test using clay-covered model lizards. Ethology 112:572–580.

Husak, J.F., and J.K. McCoy. 2000. Diet composition of the Collared Lizard (*Crotaphytus collaris*) in west-central Texas. Texas Journal of Science 52:93–100.

Peterson, C.C., and J.F. Husak. 2006. Locomotor performance and sexual selection: Individual variation in sprint speed of Collared Lizards (*Crotaphytus collaris*). Copeia 2006:216–224.

BLACK-TAILED JACKRABBIT

Dunn, J.P., J.A. Chapman, and R.E. Marsh. 1982. Jackrabbits: *Lepus californicus* and allies. Pp. 124–145 *in* Wild mammals of North America: Biology, management, and economics (J.A. Chapman and G.A. Feldhamer, eds.). Johns Hopkins University Press, Baltimore, MD. 1147 pp.

Hansen, R.M., and J.T. Flinders. 1969. Food habits of North American hares. Range Science Department Science Series 1. Colorado State University, Fort Collins. 17 pp.

Lechleitner, R.R. 1957. Reingestion in the Black-tailed Jackrabbit. Journal of Mammalogy 38:481–485.

Pontrelli, M.J. 1968. Mating behavior of the Black-tailed Jackrabbit. Journal of Mammalogy 49:785–786.

Highlights

PRICKLY PLANTS AND PACKRATS

Bement, R.E. 1969. Plains Pricklypear: Relation to grazing intensity and Blue Grama yield on central Great Plains. Journal of Range Management 21:83–86.

Brown, J.R., and S. Archer. 1989. Woody plant invasion of grasslands: Establishment of Honey Mesquite (*Prosopis glandulosa* var. *glandulosa*) on sites differing in herbaceous biomass and grazing history. Oecologia 80:19–26.

Bunting, S.C., H.A. Wright, and L.F. Neuenschwander. 1980. Long-term effects of fire on cactus in the southern mixed prairie of Texas. Journal of Range Management 33:85–88.

Hernández, F., S.E. Henke, N.J. Silvy, and D. Rollins. 2003. Effects of prickly pear control on survival and nest success of Northern Bobwhite in Texas. Wildlife Society Bulletin 31:521–527.

Kingsley, K., and M. Kurzius. 1978. The hospitable rat and the free-loaders: Desert creatures move in on the packrat to share food, keep cool. Defenders 53:196–201.

Kramp, B.A., R.J. Ansley, and T.R. Tunnell. 1998. Survival of mesquite seedlings emerging from cattle and wildlife feces in a semi-arid grassland. Southwestern Naturalist 43:300–312.

Lee, A.K. 1963. The adaptations to arid environments in woodrats of the genus *Neotoma*. University of California Publications in Zoology 64:57–96.

Marcy, R.B. 1854. A report on the exploration of the Red River of Louisiana in the year 1852. US Government Printing Office, Washington, DC. 310 pp.

Perez, J.C., W.C. Haws, and C.H. Hatch. 1978. Resistance of woodrats (*Neotoma micropus*) to *Crotalus atrox* venom. Toxicon 16: 198–200.

Perez, J.C., S. Pichyangkul, and V.E. Garcia. 1979. The resistance of three species of warm-blooded animals to Western Diamond-backed Rattlesnake (*Crotalus atrox*) venom. Toxicon 17:601–607.

Thies, K.M., M.L. Thies, and W. Caire. 1996. House construction by the Southern Plains Woodrat (*Neotoma micropus*) in western Oklahoma. Southwestern Naturalist 41:116–122.

Timmons, F.L. 1941. The dissemination of pricklypear seed by jackrabbits. Journal of the American Society of Agronomy 34:513–520.

Ueckert, D.N. 1997. Pricklypear ecology. Pp. 33–42 *in* Brush sculptors: Symposium proceedings (D. Rollins, D.N. Ueckert, and C.G. Brown, eds.). Texas Agricultural Extension Service, San Angelo. 150 pp.

Wright, H.A., S.C. Bunting, and L.F. Neuenschwander. 1976. Effect of fire on Honey Mesquite. Journal of Range Management 29:467–471.

TUMBLEBUGS: NATURE'S CLEANUP CREW

Blume, R.R. 1985. A check-list, distributional record, and annotated bibliography of the insects associated with bovine droppings on pasture in America north of Mexico. Southwestern Entomologist (Supplement) 9:1–55.

Bowling, G.A. 1942. The introduction of cattle into colonial North America. Journal of Dairy Science 25:129–154.

Byrne, M., M. Dacke, P. Nordström, et al. 2003. Visual cues used by ball-rolling dung beetles for orientation. Journal of Comparative Physiology A. Neuroethology, Sensory, Neural, Behavioral Physiology 189:411–418.

Dacke, M., E. Baird, M. Byrne, et al. 2013. Dung beetles use the Milky Way for orientation. Current Biology 23:298–300.

Dacke, M., D.E. Nilsson, C.H. Scholz, et al. 2003. Animal behavior: Insect orientation to polarized moonlight. Nature 424:33.

Hornaday, W.T. 2002. The extermination of the American Bison. Smithsonian Institution Press, Washington, DC. 548 pp. [This is a reprint of the original work published in 1889.]

Isenberg, A.C. 2000. The destruction of the bison: An environmental history, 1750–1920. Cambridge University Press, New York. 220 pp.

Mohr, C.O. 1943. Cattle droppings as ecological units. Ecological Monographs 13:275–298.

Plumb, G.E., and J.L. Doud. 1994. Foraging ecology of bison and cattle. Rangelands 16:107–109.

Richardson, P.Q., and R.H. Richardson. 2000. Dung beetles improve the soil community (Texas/Oklahoma). Ecological Restoration 18:116–117.

Riley, E.G., and C.S. Wolfe. 2003. An annotated checklist of the Scarabaeoidea of Texas (Coleoptera). Southwestern Entomologist Supplement 26:1–37.

Soper, J.D. 1941. History, range, and home life of the Northern Bison. Ecological Monographs 11:347–412.

Tiberg, K., and K.D. Floate. 2011. Where went the dung-breeding insects of the American bison? Canadian Entomologist 143:470–478.

CLIFF DWELLERS

Chapman, B.R., and J.E. George. 1991. The effects of ectoparasites on Cliff Swallow growth and survival. Pp. 69–92 in Bird-parasite interactions: Ecology, evolution, and behavior (J.E. Loye and M. Zuk, eds.). Oxford University Press, New York. 406 pp.

Cook. B. 1972. Hosts of *Argas cooleyi* and *Ornithodoros concanensis* (Acarina: Argasidae) in a cliff-face habitat. Journal of Medical Entomology 9:315–317.

———. 1973. The effects of repeated desiccation and rehydration on lipid and haemoglobin concentrations in fasting *Ornithodoros concanensis* (Acarina: Argasidae). Comparative Biochemistry and Physiology Part A 44:1141–1148.

Howell, F.G., and B.R. Chapman. 1976. Acarines associated with Cliff Swallow communities in northwest Texas. Southwestern Naturalist 21:275–280.

Larson, D.W., U. Matthes, and P.E. Kelly. 2000. Cliff ecology: Pattern and process in cliff ecosystems. Cambridge University Press, Cambridge, UK. 340 pp.

Merola, M. 1995. Observations of the nesting and breeding behavior of the Rock Wren. Condor 97:585–587.

Myers, L.E. 1928. The American Swallow Bug, *Oeciacus vicarius* Horvath (Hemiptera: Cimicidae). Parasitology 20:159–172.

Webb, J.P., Jr., J.E. George, and B. Cook. 1976. Sound as a host-detection cue for the soft tick *Ornithodoros concanensis*. Nature 265:443–444.

SHIMMERING LAKES AND DUST DEVILS

Hallett, J., and T. Hoffer. 1971. Dust devil systems. Weather 26:247–250.

Ludlum, D.M. 1997. National Audubon Society field guide to North American weather. Alfred A. Knopf, New York. 656 pp.

Lynch, D.K., and W. Livingston. 2001. Color and light in nature. Cambridge University Press, Cambridge, UK. 58 pp.

Minnaert, M. 1954. The nature of light and colour in the open air. Dover, New York. 416 pp. [Translated from the 1948 edition produced by G. Bell and Sons.]

Conservation and Management

Ansley, R.J., W.E. Pinchak, and D.N. Uechert. 1995. Changes in Redberry Juniper distribution in northwest Texas (1948 to 1982). Rangelands 17:49–53.

Archer, S. 1995. Tree-grass dynamics in a *Prosopis*-thornscrub savanna parkland: Reconstructing the past and predicting the future. Ecoscience 2:83–99.

Brown, J.R., and S. Archer. 1989. Woody plant invasion of grasslands: Establishment of Honey Mesquite (*Prosopis glandulosa* var. *glandulosa*) on sites differing in herbaceous biomass and grazing history. Oecologia 80:19–26.

Dickson, J.G. 1992. The Wild Turkey: Biology and management. Stackpole Books, Mechanicsburg, PA. 480 pp.

Everitt, J.H., C. Yang, B.J. Racher, et al. 2001. Remote sensing of Redberry Juniper in the Texas

Rolling Plains. Journal of Range Management 54:254–259.

Gilbert, A.A., D.L. Marshall, and S.A. Gilbert. 2000. Competition between native *Populus deltoides* and invasive *Tamarix ramosissima* and the implications of reestablishing flooding disturbance. Conservation Biology 14:1744–1754.

Harris, D. 1971. Recent plant invasions in the arid and semi-arid Southwest of the United States. Pp. 459–481 *in* Man's impact on the environment (T. Detwyler, ed.). McGraw-Hill, New York. 731 pp.

Leschper, L. 2004. Rolling Plains rios. Texas Parks and Wildlife Magazine (February).

Neuenschwander, L.F., H.A. Wright, and S.C. Bunting. 1978. The effect of fire on a Tobosagrass-mesquite community in the Rolling Plains of Texas. Southwestern Naturalist 23:315–338.

Sher, A.A., and D.L. Marshall. 2003. Competition between native and exotic floodplain tree species across water regimes and soil textures. American Journal of Botany 90:413–422.

Taylor, J.P., and K.C. McDaniel. 1998. Restoration of Saltcedar (*Tamarix* sp.)-infested floodplains on the Bosque del Apache National Wildlife Refuge. Weed Technology 12:345–352.

Wright, H.A. 1969. Effect of spring burning on Tobosagrass. Journal of Range Management 22:425–427.

Infobox 6.1. Charles Goodnight (1836–1929): "Father of the Texas Panhandle"

Hagen, W.T. 2011. Charles Goodnight: Father of the Texas Panhandle. University of Oklahoma Press, Norman. 168 pp.

Haley, J.E. 1981. Charles Goodnight: Cowman and plainsman. University of Oklahoma Press, Norman. 504 pp.

McCoy, D.A. 1987. Texas ranchmen. Eakin Press, Austin, TX. 177 pp.

Infobox 6.2. *Seymouria*: A "Missing Link"?

Broli, F. 1904a. Permische stegocephalen und reptilian aus Texas. Paleontographica 51:1–120.

————. 1904b. Stammreptilien. Anatomischer Anzeiger 25:577–587.

Carroll, R.L. 1995. Problems of the phylogenic analysis of Paleozoic choanates. Bulletin du Muséum National d'Histoire Naturelle de Paris 17:389–445.

Laurin, M. 1996. A redescription of the cranial anatomy of *Seymouria baylorensis,* the best known seymouriamorph (Vertebrata: Seymouriamorpha). PaleoBios 17:1–16.

Olson, E.C. 1979. *Seymouria grandis* n. sp. (Batrachosauria: Amphibia) from the Middle Clear Fork (Permian) of Oklahoma and Texas. Journal of Paleontology 53:720–728.

Vaughn, P.P. 1966. *Seymouria* from the Lower Permian of southeastern Utah, and possible sexual dimorphism in that genus. Journal of Paleontology 40:603–612.

Watson, D.M.S. 1918. On *Seymouria,* the most primitive known reptile. Proceedings of the Zoological Society of London 1918:267–301.

Infobox 6.3. Rattlesnake Roundups: Inglorious Festivals

Adams, C.E., and J.K. Thomas. 2008. Texas rattlesnake roundups. Texas A&M University Press, College Station. 128 pp.

Adams, C.E., J.K. Thomas, K.J. Strnadel, and S.L. Jester. 1994. Texas rattlesnake roundups: Implications of unregulated commercial use of wildlife. Wildlife Society Bulletin 22:324–330.

Arena, P.C., C. Warwick, and D. Duvall. 1995. Rattlesnake roundups. Pp. 313–324 *in* Wildlife and recreationists: Coexistence through management and research (R.L. Knight and P. Kerlinger, eds.). Island Press, New York. 389 pp.

Campbell, J.D., D.R. Formanowicz Jr., and E.D. Brodie Jr. 1989. Potential impact of rattlesnake roundups on natural populations. Texas Journal of Science 41:301–317.

Fitzgerald, L.A., and C.W. Painter. 2000. Rattlesnake commercialization: Long-term trends, issues, and implications for conservation. Wildlife Society Bulletin 28:235–253.

Franke, J. 2000. Rattlesnake roundups: Uncontrolled wildlife exploitation and the rites of spring. Journal of Applied Animal Welfare Science 3:151–160.

Shelton, H. 1981. A history of the Sweetwater Jaycees Rattlesnake Roundup. Pp. 95–234 *in* Rattlesnakes in America (J. Kilmon and H. Shelton, eds.). Shelton Press, Sweetwater, TX.

Warwick, C., C. Steedman, and T. Holford. 1991. Rattlesnake collection drives—Their implications for species and environmental conservation. Oryx 25:39–44.

Wier, J. 1992. The Sweetwater Rattlesnake Round-up: A case study in environmental ethics. Conservation Biology 6:116–127.

CHAPTER 7

Enquist, M. 1987. Wildflowers of the Texas Hill Country. Lone Star Botanical, Austin, TX. 275 pp.

Goetze, J.R. 1998. The mammals of the Edwards Plateau, Texas.

Special Publication No. 41. Museum of Texas Tech University, Lubbock. 263 pp.

Graves, J., and W. Meinzer. 2003. Texas Hill Country. University of Texas Press, Austin. 119 pp. [Source of the chapter epigraph.]

Matthews, W.H. 1951. Some aspects of reef paleontology and lithology in the Edwards Formation of Texas. Texas Journal of Science 3:217–226.

Riskind, D.H., and D.D. Diamond. 1988. An introduction to environments and vegetation. Pp. 1–16 *in* Edwards Plateau vegetation: Plant ecological studies in Central Texas (B.B. Amos and F.R. Gelbach, eds.). Baylor University Press, Waco, TX. 144 pp.

Rose, P.R. 1992. Edwards Group, surface and subsurface, Central Texas. University of Texas, Bureau of Economic Geology Report 72:1–198.

Structure and Climate

GEOCHRONOLOGY AND STRUCTURE

Abbott, P.L., Jr., and C.M. Woodruff, eds. 1986. The Balcones Escarpment: Geology, hydrology, ecology, and social development in Central Texas. Geological Society of America, San Antonio, TX. 200 pp.

Frost, J.G. 1967. Edwards limestone of Central Texas. Pp. 133–157 *in* Comanchean (Lower Cretaceous) stratigraphy and paleontology of Texas (L. Hendricks, ed.). Society of Economic Paleontologists and Mineralogists, Permian Basin Section, Midland, TX. 411 pp.

Hardin, R.W. 1987. The Edwards Aquifer: Underground river of Texas. Guadalupe-Blanco River Authority, Sequin, TX. 63 pp.

Lewis, J.G. 2014. North Harris Geology Hill Country field trip.

Geology Department, Lone Star College–North Harris, Houston, TX. 33 pp. [Unpublished field trip guide.]

Peterson, J.F. 1988. Enchanted Rock State Natural Area: A guidebook to the landforms. Terra Cognita Press, San Antonio, TX. 56 pp.

Rougvie, J.R., W.D. Carlson, P. Copeland, and J.N. Connelly. 1999. Late thermal evolution of Proterozoic rocks in the northeastern Llano Uplift, Central Texas. Precambrian Research 94: 49–72.

Toomey, R.S., III, M.D. Blum, and S. Valastro Jr. 1993. Late Quaternary climates and environments of the Edwards Plateau, Texas. Global and Planetary Change 7:299–320.

CLIMATIC CONDITIONS

Bishop, A.L. 1977. Flood potential of the Bosque Basin. Baylor Geological Studies Bulletin 33:1–36.

Caran, S.C., and V.R. Baker. 1986. Flooding along the Balcones Escarpment. Pp. 1–14 *in* The Balcones Escarpment (C.M. Woodruff Jr. and P.L. Abbott, eds.). Geological Society of America Annual Meeting, San Antonio, TX. 198 pp.

National Weather Service. 1979. The disastrous Texas flash floods of August 1–4, 1978. National Disaster Survey Report 79–1. National Oceanic and Atmospheric Administration, Washington, DC. 153 pp.

Slade, R.M., Jr. 1986. Large rainstorms along the Balcones Escarpment in Central Texas. Pp. 15–19 *in* The Balcones Escarpment (C.M. Woodruff Jr. and P.L. Abbott, eds.). Geological Society of America Annual Meeting, San Antonio, TX. 198 pp.

Biophysiographic Associations

Amos, B.B., and F.R. Gehlbach. 1988. Summary. Pp. 115–119 *in* Edwards Plateau vegetation: Plant ecological studies in Central Texas (B.B. Amos and F.R. Gehlbach, eds.). Baylor University Press, Waco, TX. 144 pp.

McMahan, C.A., R.G. Frye, and K.L. Brown. 1984. The vegetation types of Texas including cropland: An illustrated synopsis to accompany the map. Texas Parks and Wildlife Department, Wildlife Division, Austin. 40 pp.

Weniger, D. 1988. Vegetation before 1860. Pp. 17–24 *in* Edwards Plateau vegetation: Plant ecological studies in Central Texas (B.B. Amos and F.R. Gehlbach, eds.). Baylor University Press, Waco, TX. 144 pp.

BALCONES CANYONLANDS

Baccus, J.T., and M.W. Wallace. 1997. Distribution and habitat affinity of the Swamp Rabbit (*Sylvilagus aquaticus:* Lagomorpha: Leporidae) on the Edwards Plateau of Texas. Occasional Papers of the Museum of Texas Tech University 167:1–13.

Emery, H.P. 1977. Current status of Texas Wild Rice (*Zizania texana* Hitchc.). Southwestern Naturalist 22:393–394.

Gonsoulin, G.J. 1974. A revision of *Styrax* (Styracaceae) in North America, Central America, and the Caribbean. Sida 5:191–258.

Poole, J., and D.E. Bowles. 1999. Habitat characterization of Texas Wild Rice (*Zizania texana* Hitchcock), an endangered aquatic macrophyte from the San Marcos River, TX, USA. Aquatic Conservation: Marine and Freshwater Ecosystems 9:291–302.

EDWARDS PLATEAU WOODLANDS

Cartwright, W.J. 1966. The cedar choppers. Southwestern Historical Quarterly 70:247–255.

Dye, K.L., D.N. Ueckert, and S.G. Whisenant. 1995. Redberry Juniper-herbaceous understory interactions. Journal of Range Management 48:100–107.

Foster, J.H. 1917. The spread of timbered areas in Central Texas. Journal of Forestry 15:442–445.

Garriga, M., A.P. Thurow, T. Thurow, et al. 2015. Commercial value of juniper on the Edwards Plateau. Texas Natural Resources Server, Texas AgriLife Research and Extension Center, College Station. 11 pp. http://texnat.tamu.edu /library/symposia/juniper -ecology-and-management/ (accessed 15 June 2015).

Gehlbach, F.R. 1988. Forests and woodlands of the northeastern Balcones Escarpment. Pp. 57–77 in Edwards Plateau vegetation: Plant ecological studies in central Texas (B.B. Amos and F.R. Gehlbach, eds.). Baylor University Press, Waco, TX. 144 pp.

Griffin, R.C., and B.A. McCarl. 1989. Brushland management for increased water yield in Texas. Water Resources Bulletin 25: 175–186.

Smeins, F.E., and S.D. Fuhlendorf. 2015. Biology and ecology of Ashe Juniper. Texas Natural Resources Server, Texas AgriLife Research and Extension Center, College Station. 16 pp. http://texnat.tamu .edu/library/symposia/juniper -ecology-and-management/ (accessed 15 June 2015).

Taylor, C.A., ed. 1994. Juniper symposium 1994. Technical Report 94-2. Texas A&M University Research Station, Sonora. 80 pp.

Van Auken, O.W., A.L. Ford, and J.L. Allen. 1981. An ecological comparison of upland deciduous and evergreen forests of Central Texas. American Journal of Botany 68:1249–1256.

Van Auken, O.W., A.L. Ford., A. Stein, and A.G. Stein. 1980. Woody vegetation of the upland plant communities in the southern Edwards Plateau. Texas Journal of Science 32:23–34.

LLANO UPLIFT

Allred, L. 2009. Enchanted Rock: A natural and human history. University of Texas Press, Austin. 314 pp.

Kennedy, I. 2010. The history of Enchanted Rock in the Texas Hill Country. Xlibris, Bloomington, IN. 40 pp.

SEMIARID EDWARDS PLATEAU

Brant, J.G., and R.C. Dowler. 2001. The mammals of Devils River State Natural Area, Texas. Occasional Papers of the Museum of Texas Tech University 211:1–32.

Bray, W.L. 1901. The ecological relations of the vegetation of western Texas. Botanical Gazette 32:99–291.

Cottle, H.J. 1931. Studies in the vegetation of southwestern Texas. Ecology 12:105–155.

Dearen, P. 2011. Devils River: Treacherous twin to the Pecos, 1535–1900. Texas Christian University Press, Fort Worth. 224 pp.

Powell, A.M. 1998. Trees and shrubs of the Trans-Pecos and adjacent areas. University of Texas Press, Austin. 498 pp.

Karst, Sinkholes, and Caves

Ford, D.C., and P.D. Williams. 2007. Karst hydrogeology and geomorphology. John Wiley and Sons, Chichester, West Sussex, England. 576 pp.

Hovorka, S., R. Mace, and E. Collins. 1995. Regional distribution of permeability in the Edwards Aquifer. Gulf Coast Association of Geological Societies Transactions 45:259–265.

Veni, G. 2013. Government Canyon State Natural Area: An emerging model for karst management. Proceedings of the 13th Sinkhole Conference, Natural Cave and Karst Research Institute Symposium 2:433–440.

SINKHOLES

Hunt, B.B., B.A. Smith, M.T. Adams, et al. 2013. Cover-collapse sinkhole development in the Cretaceous Edwards Limestone, Central Texas. Proceedings of the 13th Sinkhole Conference, Natural Cave and Karst Research Institute Symposium 2:89–102.

Parent, L. 2008. Official guide to Texas state parks and historic sites. Rev. ed. University of Texas Press, Austin. 200 pp. [Includes Devil's Sinkhole.]

Reddell, J.R., and A.R. Smith. 1965. The caves of Edwards County. Texas Speleological Survey 2(5–6): 19–28.

White, P.J. 1948. The Devil's Sinkhole. Bulletin of the National Speleological Society 10:2–14.

Williams, P. 2004. Dolines. Pp. 305–310 in Encyclopedia of caves and karst science (J. Gunn, ed.). Taylor and Francis, New York. 902 pp.

CAVE FORMATIONS

Elliott, W.R., and G. Veni, eds. 1994. The caves and karst of Texas. 1994 convention guidebook. National Speleological Society, Huntsville, AL. 342 pp.

Fieseler, R., J. Jasek, and M. Jasek. 1978. An introduction to the caves of Texas. National Speleological Society Guidebook 19:93–94.

Kastning, E.H., Jr. 1983. Relict caves as evidence of landscape and aquifer evolution in a deeply dissected carbonate terrain: Southwest Edwards Plateau, Texas. Journal of Hydrology 61: 89–112.

Palmer, A.N. 2009. Cave geology. Cave Books, Cave Research Foundation, Dayton, OH. 454 pp.

Pittman, B. 1999. Texas caves. Texas A&M University Press, College Station. 122 pp.

Reddell, J.A. 1961. The caves of Uvalde County, part I. Texas Speleological Survey 1(3):1–34.

———. 1963. The caves of Uvalde County, part II. Texas Speleological Survey 1(7):1–53.

CAVE FAUNA

Bechler, D.L. 1986. Pheromonal and tactile communication in the subterranean salamander, *Typhlomolge rathbuni*. Proceedings of the Ninth International Congress of Speleology, Barcelona, Spain 2:120–122.

Betke, M., D.E. Hirsch, N.C. Makris, et al. 2008. Thermal imaging reveals significantly smaller Brazilian Free-tailed Bat colonies than previously estimated. Journal of Mammalogy 89:18–24.

Lundelius, E.L., and B.H. Slaughter. 2014. Natural history of Texas caves. Texas Speleological Survey Special Publication, Austin. 174 pp. [Reprint of the 1971 publication.]

Reddell, J.A. 1994. The cave fauna of Texas with special reference to the western Edwards Plateau. Pp. 31–50 *in* The caves and karst of Texas (W.R. Elliott and G. Veni, eds.). National Speleological Society, Huntsville, AL. 252 pp.

Sealander, R.K., and J.K. Baker. 1957. The Cave Swallow in Texas. Condor 59:345–363.

Stejneger, L. 1896. Description of a new genus and species of blind tailed batrachians from the subterranean water of Texas. Proceedings of the US National Museum 18:619–621.

Strenth, N.E. 1976. A review of the systematics and zoogeography of the freshwater species of *Palaemonetes* Heller of North America (Crustacea: Decopoda). Smithsonian Contributions in Zoology 228:1–27.

US Fish and Wildlife Service. 2003. Endangered and threatened wildlife and plants: Designation of critical habitat for seven Bexar County, TX, invertebrate species. Federal Register 68:17155–17231.

Highlights

LAND OF 1,100 SPRINGS

Ashworth, J.B., and J. Hopkins. 1995. The aquifers of Texas. Texas Water Quality Development Board Report 345. Austin. 69 pp.

Bousman, C.B., and D.L. Nickels, eds. 2003. Archaeological testing of the Burleson Homestead at 41HY37 Hays County, Texas. Archaeological Studies Report No. 4. Center for Archaeological Studies, Texas State University, San Marcos.

Brune, G. 1981. Springs of Texas. Branch-Smith, Fort Worth, TX. 566 pp.

Edwards, R.J., G.P. Garrett, and N.L. Allan. 2001. Aquifer-dependent fishes of the Edwards Plateau Region. Pp. 253–268 *in* Aquifers of West Texas (R.E. Mace, W.F. Mullican III, and E.S. Angle, eds.). Texas Water Development Board Report 356. Austin. 263 pp.

Longley, G. 1981. The Edwards Aquifer: Earth's most diverse groundwater ecosystem? International Journal of Speleology 11:123–128.

———. 1986. The biota of the Edwards Aquifer and implications for paleozoogeography. Pp. 51–54 *in* The Balcones Escarpment: Geology, hydrology, ecology and social development in Central Texas (P.L. Abbott and C.M. Woodruff Jr., eds.). Geological Society of America, Boulder, CO. 198 pp.

McKinney, D.C., and D.W. Watkins Jr. 1993. Management of the Edwards Aquifer: A critical assessment. Technical Report CRWR 244. Center for Research in Water Resources, Bureau of Engineering Research, University of Texas at Austin. 94 pp.

Orchard, C.D., and T.N. Campbell. 1954. Evidence of early men from the vicinity of San Antonio, Texas. Texas Journal of Science 6:454–465.

HILL COUNTRY RIVERS

Gregory, S.V., F.J. Swanson, W.A. McKee, and K.W. Cummins. 1991. An ecosystem perspective of riparian zones. BioScience 41: 540–551.

Wagner, M. 2003. Managing riparian habitats for wildlife. Texas Parks and Wildlife Department, Austin. 6 pp.

BRIGHT-FACED SONGSTERS

Kroll, J.C. 1980. Habitat requirements of the Golden-cheeked Warbler: Management implications. Journal of Range Management 33: 60–65.

Pulich, W. 1976. The Golden-cheeked Warbler: A bioecological study. Texas Parks and Wildlife Department, Austin. 172 pp.

Rappole, J.H., D. King, J. Diez, and J.V. Rivera. 2005. Factors affecting population size in Texas' Golden-cheeked Warbler. Endangered Species Update 22:95–103.

US Fish and Wildlife Service. 1992. Golden-cheeked Warbler recovery plan. US Fish and Wildlife Service, Endangered

Species Office, Albuquerque, NM. 88 pp.

NATURE'S WATER FILTERS

Howells, R.G. 2013. Field guide to Texas freshwater mussels. Biostudies, Kerrville, TX. 141 pp.

Howells, R.G., R.W. Neck, and H.D. Murray. 1996. Freshwater mussels of Texas. Texas Parks and Wildlife Department, Austin. 218 pp.

Parmalee, P.W., and A.E. Bogan. 1998. The freshwater mussels of Tennessee. University of Tennessee Press, Knoxville. 328 pp.

LOST MAPLES

Habel, J.C., and T. Assmann, eds. 2010. Relict species: Phylogeography and conservation biology. Springer-Verlag, Berlin. 451 pp.

McGee, B.K., and R.W. Manning. 2000. Mammals of Lost Maples State Natural Area, Texas. Occasional Papers of the Museum of Texas Tech University 198:1–24.

A WORLD WITH OXYGEN

Canfield, D.E. 2014. Oxygen, a four billion year history. Princeton University Press, Princeton, NJ. 196 pp.

Gould, S.J. 1989. Wonderful life: The Burgess Shale and the nature of history. W.W. Norton, New York. 347 pp.

Leis, B., and B.L. Stinchcomb. 2015. Stromatolites: Ancient, beautiful, and earth-altering. Schiffer, Atglen, PA. 176 pp.

Lyons, T.W., C.T. Reinhard, and N.J. Planavsky. 2014. The rise of oxygen in Earth's early ocean and atmosphere. Nature 506:307–315.

Mills, D.B., and D.E. Canfield. 2014. Oxygen and animal evolution: Did a rise of atmospheric oxygen "trigger" the origin of animals? Bioessays 36:1145–1155.

Plutino, T. 2009. Mason's world class stromatolites and thrombolites. Mason County News 134(9):B10, March 4.

EXOTIC SPECIES: THE WORLD'S WILDLIFE IN TEXAS

Armstrong, W.E., and E.L. Young. 2000. White-tailed Deer management in the Texas Hill Country. PWD RP W7000-0828. Texas Parks and Wildlife Department. 53 pp.

Baccus, J.T. 2002. Impacts of game ranching on wildlife management in Texas. Transactions of the North American Wildlife and Natural Resources Conference 67: 276–288.

Bolen, E.G., and W.L. Robinson. 2003. Wildlife ecology and management. 5th ed. Prentice Hall, Upper Saddle River, NJ. 634 pp.

Butler, L.D. 1991. White-tailed Deer hunting leases: Hunter costs and rancher revenues. Rangelands 13: 20–22.

Butler, M.J., A.P. Teaschner, W.B. Ballard, and B.K. McGee. 2005. Commentary: Wildlife ranching in North America—arguments, issues, and perspectives. Wildlife Society Bulletin 33:381–389.

Faas, C.J., and F.W. Weckerly. 2010. Habitat interference by Axis Deer on White-tailed Deer. Journal of Wildlife Management 74:698–706.

Henke, D.E., S. Demarais, and J.A. Pfister. 1988. Digestive capacity and diets of White-tailed Deer and exotic ruminants. Journal of Wildlife Management 52:595–598.

Mungall, C. 2000. Exotics. Pp. 736–764 in Ecology and management of large mammals in North America (S. Demarais and P.R. Krausman, eds.). Prentice Hall, Upper Saddle River, NJ. 778 pp.

Richardson, M.L., and S. Demarais. 1992. Parasites and condition of

coexisting White-tailed Deer and exotic deer in south-central Texas. Journal of Wildlife Diseases 28: 485–489.

Teer, J.G. 1975. Commercial uses of game animals on rangelands of Texas. Journal of Animal Science 40:1000–1008.

The Wildlife Society. 2009. Final position statement: Confinement of wild ungulates within high fences. The Wildlife Society, Bethesda, MD. 2 pp.

Conservation and Management

Allen, C.R., S. Demarais, and R.S. Lutz. 1994. Red Imported Fire Ant impact on wildlife: An overview. Texas Journal of Science 46:51–60.

Burkalova, L.E., A.Y. Karateyev, V.A. Karatayev, et al. 2011. Biogeography and conservation of freshwater mussels (Bivalvia: Unionidae) in Texas: Patterns of diversity and threats. Diversity and Distributions 17:393–407.

Fowler, N.L., and D.W. Dunlap. 1986. Grassland vegetation of the eastern Edwards Plateau. American Midland Naturalist 115: 146–155.

Howells, R.G., R.W. Neck, and H.D. Murray. 1996. Freshwater mussels of Texas. Texas Parks and Wildlife Department, Austin. 218 pp.

Mabe, J.A., and J. Kennedy. 2014. Habitat conditions associated with a reproducing population of the critically endangered freshwater mussel *Quadrula mitchelli* in Central Texas. Southwestern Naturalist 59:297–300.

Pool, W.C. 1975. A historical atlas of Texas. Encino Press, Austin, TX. 190 pp.

Smeins, F.E., and L.B. Merrill. 1988. Long-term change in a semiarid grassland. Pp. 101–104 in Edwards Plateau vegetation: Plant ecological studies in Central Texas

(B.B. Amos and F.R. Gelbach, eds.). Baylor University Press, Waco, TX. 144 pp.

Texas Water Commission. 1989. Ground water quality of Texas—An overview of natural and man-affected conditions. Texas Water Commission Report 89-01:1-197.

Infobox 7.1. Texas Wildflowers: Lady Bird's Legacy

Gillette, M.L. 2012. Lady Bird Johnson: An oral history. Oxford University Press, New York. 416 pp.

Gould, L.L. 1999. Lady Bird Johnson: Our environmental First Lady. University Press of Kansas, Lawrence. 176 pp.

Johnson, L.B., and C.B. Lees. 1998. Wildflowers across America. Abbeville Press, New York. 312 pp.

Paulson, A. 1989. The National Wildflower Research Center handbook. Texas Monthly Press, Austin. 337 pp.

Infobox 7.2. Living Sponges in Texas Rivers?

Anakina, R.P. 2010. Sponges as biological indicators and remedial components of freshwater ecological systems. Biosphere 2:397-408.

Davis, J.R. 1980a. Species composition and diversity of benthic macroinvertebrate populations of the Pecos River, Texas. Southwestern Naturalist 25:241-256.

———. 1980b. Species composition and diversity of benthic macroinvertebrates of lower Devil's River, Texas. Southwestern Naturalist 25:379-384.

Nichols, H.T., and T.H. Bonner. 2014. First record and habitat associations of *Spongilla cenota* (Class Demospongiae) within streams of the Edwards Plateau,

Texas, USA. Southwestern Naturalist 59:467-472.

Porrier, M.A. 1969. Fresh-water sponge hosts of Louisiana and Texas spongilla-flies, with new locality records. American Midland Naturalist 81:573-575.

———. 1972. Additional records of Texas fresh-water sponges (Spongillidae) with the first record of *Radiospongilla cerebellata* (Bowerbank, 1863) from the Western Hemisphere. Southwestern Naturalist 16:434-435.

Reiswig, H., T.M. Frost, and A. Ricciardi. 2010. Porifera. Pp. 91-123 *in* Ecology and classification of North American freshwater invertebrates (J. Thorp and A. Covich, eds.). Elsevier, Oxford, UK. 1021 pp.

Infobox 7.3. Guano, Gunpowder, and Bat Bombs

Campbell, R.B. 2003. Gone to Texas: A history of the Lone Star State. Oxford University Press, New York. 265 pp.

Constantine, D.G. 1957. Color variation and molt in *Tadarida brasiliensis* and *Myotis velifer*. Journal of Mammalogy 38:461-466.

———. 1958. Bleaching of hair pigment of bats by the atmosphere in caves. Journal of Mammalogy 39:513-520.

Couffer, J. 1992. Bat bomb: World War II's other secret weapon. University of Texas Press, Austin. 252 pp.

Davis, R.B., C.F. Herrid II, and H.L. Short. 1962. Mexican Free-tailed Bats in Texas. Ecological Monographs 32:311-346.

Mohr, C.E. 1948. Texas bat caves served in three wars. National Speleological Bulletin 10:89-96.

Infobox 7.4. Clark Hubbs (1921–2008): Ichthyologist and Eminent Naturalist

Holtcamp, W. 2011. Legend, lore and legacy: The fish wrangler. Texas Parks and Wildlife Magazine (April).

Marsh-Matthews, E. 2008. Clark Hubbs: 1921-2008. Southwestern Naturalist 53:539-541.

Martin, F.D., R.J. Edwards, D.A. Hendrickson, and G.P. Garrett. 2008. Obituary: Clark Hubbs, 1921-2008, Ichthyologist. Fisheries 33:302.

Matthews, R. 2008. Dr. Clark Hubbs: March 15, 1921-February 3, 2008. Texas Academy of Science, 11th Annual Meeting Program and Abstracts 2008:14-15.

Infobox 7.5. The Nature of Roy Bedichek (1878–1959)

Bedichek, R. 1947. Adventures with a Texas naturalist. Doubleday, Garden City, NY. 293 pp.

———. 1950. Karankaway Country. Doubleday, Garden City, NY. 290 pp.

Dugger, R., ed. 1967. Three men in Texas: Bedichek, Webb, and Dobie. University of Texas Press, Austin. 285 pp.

Owens, W.A., and L. Grant, eds. 1985. The letters of Roy Bedichek. University of Texas Press, Austin. 542 pp.

Walker, S. 1948. The lively hermit of Friday Mountain. Saturday Evening Post, October 16. Pp. 38-39, 54, 57-58, 61.

CHAPTER 8

Choate, L.L. 1997. The mammals of the Llano Estacado. Special Publication No. 40. Museum of Texas Tech University, Lubbock. 240 pp.

Flores, D. 1990. Caprock Canyonlands: Journeys into the heart of the Southern Plains.

University of Texas Press, Austin. 200 pp.

———. 2001. The natural west. University of Oklahoma Press, Norman. 285 pp.

Seyffert, K.D. 2001. Birds of the Texas Panhandle: Their status, distribution, and history. Texas A&M University Press, College Station. 501 pp.

Wester, D.B. 2007. The Southern High Plains: A history of vegetation since 1540 to the present. Pp. 24–47 *in* Proceedings: Shrubland dynamics—Fire and water (R.E. Sosebee, D.B. Wester, C.M. Britton, et al., eds.). Proceedings RMRS P-47. Rocky Mountain Research Station, USDA Forest Service, Fort Collins, CO. 173 pp.

The Llano Estacado

Dhillion, S.S., and M. Mills. 1999. The Sand Shinnery oak (*Quercus havardii*) communities of the Llano Estacado: History, structure, and restoration. Pp. 262–274 *in* Savannas, barrens, and rock outcrop communities of North America (R.C. Anderson, J.S. Fralish, and J.M. Baskin, eds.). Cambridge University Press, New York. 470 pp.

Flint, R. 2008. No settlement, no conquest: A history of the Coronado entrada. University of New Mexico Press, Albuquerque. 358 pp.

Goetzmann, W.H. 1959. Army exploration in the American West, 1803–1863. Yale University Press, New Haven, CT. 518 pp.

Holliday, V.T. 1991. The geologic record of wind erosion, eolian deposition, and the aridity of the Southern High Plains. Great Plains Research 1:7–15.

Machenberg, M.D. 1984. Geology of Monahans Sandhills State Park. Bureau of Economic Geology,

Guidebook 21. University of Texas, Austin. 39 pp.

Morris, J.M. 1997. El Llano Estacado: Exploration and imagination on the High Plains of Texas and New Mexico, 1836–1860. Texas State Historical Association, Austin. 414 pp.

Pesaturo, R.J., J.K. Jones Jr., R.W. Manning, and C. Jones. 1990. Mammals of the Muleshoe Sandhills in Bailey, Hale, and Lamb Counties, Texas. Occasional Papers of the Museum of Texas Tech University 136:1–32.

Peterson, R., and C.S. Boyd. 1998. Ecology and management of Sand Shinnery communities: A literature review. General Technical Report RMRS-GTR-16. Rocky Mountain Research Station, USDA Forest Service, Fort Collins, CO. 44 pp.

Pettit, R.D. 1986. Sand Shinnery Oak: Control and management. Management Note No. 3. Department of Range and Wildlife Management, Texas Tech University, Lubbock. 5 pp.

Weaver, J.E. 1958. Summary and interpretation of underground development in natural grassland communities. Ecological Monographs 28:55–78.

Prairie Dogs and Bison: Grassland Icons

Bailey, V. 1905. Biological survey of Texas. North American Fauna No. 25. Bureau of Biological Survey, US Department of Agriculture, Washington, DC. 222 pp.

PRAIRIE DOGS AND ASSOCIATES

Baker, B.W., D.J. Augustine, J.A. Sedgwick, and B.C. Lubow. 2013. Ecosystem engineering varies spatially: A test of the vegetation modification paradigm for prairie dogs. Ecography 36:230–239.

Bangert, R.K., and C.N.

Slobodchikoff. 2006. Conservation of prairie dog ecosystem engineering may support arthropod beta and gamma diversity. Journal of Arid Environments 67:100–115.

Barnes, A.M. 1993. A review of plague and its relevance to prairie dog populations and the Black-footed Ferret. Pp. 28–37 *in* Proceedings of the symposium on the management of prairie dog complexes for the reintroduction of the Black-footed Ferret (J.L. Oldemeyer, D.E. Biggins, B.J. Miller, and R. Crete, eds.). US Fish and Wildlife Special Report 13:1–96.

Bonham, C.D., and J.S. Hannan. 1978. Blue Grama and Buffalograss patterns in and near a prairie dog town. Journal of Range Management 31:63–65.

Bonham, C.D., and A. Lerwick. 1976. Vegetation changes induced by prairie dogs on shortgrass range. Journal of Range Management 29:221–225.

Cully, J.F., Jr., S.K. Collinge, R.E. VanNimwegen, et al. 2010. Spatial variation in keystone effects: Small mammal diversity associated with Black-tailed Prairie Dog colonies. Ecography 33:667–677.

Johnsgard, P.A. 2005. Prairie dog empire: A saga of the shortgrass prairie. University of Nebraska Press, Lincoln. 243 pp.

Johnson, T.J., J.F. Cully Jr., S.H. Collinge, et al. 2011. Spread of plague among Black-tailed Prairie Dogs is associated with colony spatial characteristics. Journal of Wildlife Management 75:357–368.

Knowles, C.J., C.J. Stoner, and S.P. Gieb. 1982. Selective use of Black-tailed Prairie Dog towns by Mountain Plovers. Condor 84:71–74.

Kotliar, N.B., B.W. Baker, A.D.

Whicker, and G. Plumb. 1999. A critical review of assumptions about the prairie dog as a keystone species. Environmental Management 24:177–192.

Kretzler, J.E., and J.F. Cully Jr. 2001. Effects of Black-tailed Prairie Dogs on reptiles and amphibians in Kansas shortgrass prairie. Southwestern Naturalist 46:171–177.

Miller, B., R.P. Reading, and S. Forrest. 1996. Prairie night: Black-footed Ferrets and the recovery of an endangered species. Smithsonian Institution Press, Washington, DC. 254 pp.

Singhurst, J.R., J.H. Young, G. Kerouac, and H.A. Whitlaw. 2010. Estimating Black-tailed Prairie Dog (*Cynomys ludovicianus*) distribution in Texas. Texas Journal of Science 62:243–262.

Thacker, P. 2001. A new wind sweeps the plains. Science 292:2427.

VanNimwegen, R.E., J. Kretzer, and J.F. Cully Jr. 2008. Ecosystem engineering by a colonial mammal: How prairie dogs structure rodent communities. Ecology 89:3298–3305.

AMERICAN BISON

Collins, S.L., and G.E. Uno. 1983. The effect of early spring burning on vegetation in Buffalo wallows. Bulletin of the Torrey Botanical Club 110:474–481.

Fuhlendorf, S.D., B.W. Allred, and R.G. Hamilton. 2010. Bison as keystone herbivores on the Great Plains: Can cattle serve as proxy for evolutionary grazing patterns? Working Paper No. 4. American Bison Society, Bozeman, MT. 39 pp.

Haley, J.L. 1976. The Buffalo War: The history of the Red River Indian uprising of 1884. Doubleday, Garden City, NY. 290 pp.

Hornaday, W.T. 2002. The extermination of the American Bison. Smithsonian Institution Scholarly Press, Washington, DC. 240 pp. [Reprint of an 1889 classic.]

Knapp, A.K., J.M. Blair, J.M. Briggs, et al. 1999. The keystone role of Bison in North American tallgrass prairie. BioScience 49:39–50.

Larson, F. 1940. The role of Bison in maintaining the short grass plains. Ecology 21:113–121.

Lott, D.F. 2002. American Bison: A natural history. University of California Press, Berkeley. 264 pp.

Peden, D.G. 1976. Botanical composition of Bison diets on shortgrass plains. American Midland Naturalist 96:225–229.

Peden, D.G., G.M. Van Dyne, R.W. Rice, and R.M. Hansen. 1974. The trophic ecology of *Bison bison* L. on shortgrass plains. Journal of Applied Ecology 11:489–497.

Polley, H.W., and S.L. Collins. 1984. Relationships of vegetation and environment in Buffalo wallows. American Midland Naturalist 112:178–186.

Playas: "Round like Plates"

GEOLOGY, GEOGRAPHY, AND VEGETATION

Bolen, E.G., and D.L. Flores. 1988. Prairie wetlands of West Texas: The history and ecology of playa lakes. Paper 23.02 *in* Proceedings of the Tenth North American Prairie Conference, Denton, TX (A. Davis and G. Stanford, eds.). [Unnumbered pages.]

Bolen, E.G., L.M. Smith, and H.L. Schramm Jr. 1989. Playa lakes: Prairie wetlands of the Southern High Plains. BioScience 39:615–623.

Guthery, F.S., J.M. Pates, and F.A. Stormer. 1982. Characterization of playas of the north-central Llano Estacado in Texas. Transactions of

the North American Wildlife and Natural Resources Conference 47:516–527.

Haukos, D.A., and L.M. Smith. 1994. The importance of playa wetlands to biodiversity of the Southern High Plains. Landscape and Urban Planning 28:83–98.

———. 2004. Plant communities of playa wetlands in the Southern High Plains. Special Publication No. 47. Museum of Texas Tech University, Lubbock. 63 pp.

Oserkamp, W.P., and W.W. Wood. 1987. Playa-lake basins on the Southern High Plains of Texas and New Mexico, part I. Hydrologic, geomorphic, and geologic evidence for their development. Geological Society of America Bulletin 99:215–223.

Simpson, C.D., and E.G. Bolen. 1981. Wildlife assessment of playa lakes. Final report submitted to Bureau of Reclamation, Southwest Region, Amarillo, TX. 159 pp.

Smith, L.M. 2003. Playas of the Great Plains. University of Texas Press, Austin. 257 pp.

WATERFOWL

Baldassarre, G.A., and E.G. Bolen. 1984. Field-feeding ecology of waterfowl wintering on the Southern High Plains of Texas. Journal of Wildlife Management 48:63–71.

Johnson, W.P., L. Baar, R.S. Matlack, and R.B. Barron. 2010. Hatching chronology of ducks using playas in the Southern High Plains of Texas. American Midland Naturalist 163:247–253.

Moore, R.L., and C.D. Simpson. 1981. Disease mortality of waterfowl on Texas playa lakes. Southwestern Naturalist 25:566–568.

Ray, J.D., B.D. Sullivan, and H.W. Miller. 2003. Breeding ducks and their habitats in the High Plains of Texas. Southwestern Naturalist 48:241–248.

Rhodes, M.J. 1979. Redheads breeding in the Texas Panhandle. Southwestern Naturalist 24:691–692.

Rhodes, M.J., and J.D. Garcia. 1981. Characteristics of playa lakes related to summer waterfowl use. Southwestern Naturalist 26:231–235.

SHOREBIRDS

Baldassarre, G.A., and D.H. Fischer. 1984. Food habits of fall migrant shorebirds on the Texas High Plains. Journal of Field Ornithology 55:220–229.

Davis, C.A., and L.M. Smith. 1999. Ecology and management of migrant shorebirds in the playa lakes region of Texas. Wildlife Monographs 140. 45 pp.

GRASSLAND BIRDS

Guthery, F.S., and R.W. Whiteside. 1984. Playas important to pheasants on the Texas High Plains. Wildlife Society Bulletin 12:40–43.

Littlefield, C.D. 1970. A Marsh Hawk roost in Texas. Condor 72:245.

Littlefield, C.D., and D.H. Johnson. 2005. Habitat preferences of migrant and wintering Northern Harriers in northwestern Texas. Southwestern Naturalist 50:448–452.

Schibler, M.D. 1981. Some aspects of the winter ecology of harriers on the Texas High Plains. MS thesis, Texas Tech University, Lubbock. 14 pp.

Whiteside, R.W., and F.S. Guthery. 1983. Ring-necked Pheasant movements, home ranges, and habitat use in West Texas. Journal of Wildlife Management 47:1097–1104.

MAMMALS

Choate, L.L. 1997. The mammals of the Llano Estacado. Special Publication No. 40. Museum of Texas Tech University, Lubbock. 240 pp.

Scribner, K.T., and R.K. Chesser. 1993. Environmental and demographic correlates of spatial and seasonal genetic structure in the Eastern Cottontail (*Sylvilagus floridanus*). Journal of Mammalogy 74:1026–1044.

Scribner, K.T., and L.J. Krysl. 1982. Summer foods of the Audubon's Cottontail (*Sylvilagus auduboni*: Leporidae) on Texas Panhandle playa basins. Southwestern Naturalist 27:460–463.

Scribner, K.T., and R.J. Warren. 1990. Seasonal demography and movements of cottontail rabbits on isolated playa basins. Journal of Wildlife Management 54:403–409.

Sovada, M.A., A.B. Sargeant, and J.W. Grier. 1995. Differential effects of Coyotes and Red Foxes on duck nest success. Journal of Wildlife Management 59:1–9.

Van Den Bussche, R.A., M.J. Hamilton, R.K. Chesser, and K.T. Scribner. 1987. Genetic differentiation among cottontails from isolated playa basins. Genetica 75:153–157.

Whiteside, R.W., and F.S. Guthery. 1981. Coyote use of playas on the Texas High Plains. Prairie Naturalist 13:42–44.

TOADS AND SALAMANDERS

Anderson, A.M., D.A. Haukos, and J.T. Anderson. 1999a. Diet composition of three anurans from the playa wetlands of northwest Texas. Copeia 1999:515–520.

———. 1999b. Habitat use of anurans emerging and breeding in playa wetlands. Wildlife Society Bulletin 27:759–769.

Justis, J.T., M. Sandomir, T. Urquhart, and B.O. Ewan. 1977. Developmental rates of two species of toads from the Desert Southwest. Copeia 1977:592–594.

Rose, F.L., and D. Armentrout. 1976. Adaptive strategies of *Ambystoma tigrinum* Green inhabiting the Llano Estacado of West Texas. Journal of Animal Ecology 45:713–729.

Other Wildlife on the Llano

MISSISSIPPI KITE

Bolen, E.G., and D.L. Flores. 1993. The Mississippi Kite: Portrait of a southern hawk. University of Texas Press, Austin. 128 pp.

LESSER PRAIRIE-CHICKEN

Bell, L.A., S.D. Fuhlendorf, M.A. Patten, et al. 2010. Lesser Prairie-Chicken hen and brood habitat use on Sand Shinnery Oak. Rangeland Ecology and Management 63:478–486.

Crawford, J.A., and E.G. Bolen. 1976a. Effects of land use on Lesser Prairie Chickens in Texas. Journal of Wildlife Management 40:96–104.

———. 1976b. Effects of lek disturbances on Lesser Prairie Chickens. Southwestern Naturalist 21:238–240.

———. 1976c. Fall diet of Lesser Prairie Chickens in West Texas. Condor 78:142–144.

Grisham, B.A., P.K. Borsdorf, C.W. Boal, and K.K. Boydston. 2014. Nesting ecology and nest survival of Lesser Prairie-Chickens on the Southern High Plains of Texas. Journal of Wildlife Management 78:857–866.

Jackson, A.S., and R. DeArment. 1963. The Lesser Prairie Chicken in the Texas Panhandle. Journal of Wildlife Management 27:733–737.

Sullivan, R.M., J.P. Hughes, and J.E. Lionberger. 2000. Review of the historical and present status of the Lesser Prairie-Chicken (*Tympanuchus pallidicinctus*) in Texas. Prairie Naturalist 32: 177–188.

AN ENDEMIC LIZARD

Chan, L.M., L.A. Fitzgerald, and K.R. Zamudio. 2009. The scale of genetic differentiation in the Dunes Sagebrush Lizard (*Sceloporus arenicolus*), an endemic species. Conservation Genetics 10:131–142.

Smolensky, N., and L.A. Fitzgerald. 2011. Population variation in dune-dwelling lizards in response to patch size, patch quality, and oil and gas development. Southwestern Naturalist 56:315–324.

SANDHILL CRANE

Andrei, A.E., L.M. Smith, D.A. Haukos, and J.G. Surles. 2008. Habitat use by migrant shorebirds in saline lakes of the Southern High Plains. Journal of Wildlife Management 72:246–253.

Iverson, C., P.A. Vohs, and T.C. Tacha. 1985. Distribution and abundance of Sandhill Cranes in western Texas. Journal of Wildlife Management 49:250–255.

Reeves, C.C. 1966. Pluvial basins of West Texas. Journal of Geology 74: 269–291.

Rosen, D.J., A.D. Caskey, W.C. Conway, and D.A. Haukos. 2013. Vascular flora of saline lakes in the Southern High Plains of Texas and eastern New Mexico. Journal of the Botanical Research Institute of Texas 7:595–602.

Saalfeld, S.T., W.C. Conway, D.A. Haukos, and W.P. Johnson. 2100. Nest success of Snowy Plovers (*Charadrius nivosus*) in the Southern High Plains. Waterbirds 34:389–399.

Windingstad, R.M., R.J. Cole, P.E. Nelson, et al. 1989. *Fusarium* mycotoxins from peanuts suspected as a cause of Sandhill Crane mortality. Journal of Wildlife Diseases 25:38–46.

Highlights

THE DUST BOWL

Worster, D. 1979. Dust Bowl—The southern plains in the 1930s. Oxford University Press, New York. 277 pp.

A LITTLE FOX

Cutter, W.L. 1958. Food habits of the Swift Fox in northern Texas. Journal of Mammalogy 39:527–532.

Dragoo, J.W., J.R. Choate, T.L. Yates, and T.P. O'Farrell. 1990. Evolutionary and taxonomic relationships among North American arid-land foxes. Journal of Mammalogy 71:318–332.

Henke, S.E., and F.C. Bryant. 1999. Effects of Coyote removal on the faunal community in western Texas. Journal of Wildlife Management 63:1066–1081.

Kamler, J.F., W.B. Ballard, R.L. Gilliland, et al. 2003. Impacts of Coyotes on Swift Foxes in northwest Texas. Journal of Wildlife Management 67:317–323.

CONTACT IN THE PANHANDLE

Dixon, K.L. 1989. Contact zones of avian congeners on the Southern Great Plains. Condor 91:15–22.

Rising, J.D. 1983. The Great Plains hybrid zones. Current Ornithology 1:131–157.

Short, L.L. 1965. Hybridization in flickers (*Colaptes*) of North America. American Museum of Natural History Bulletin 129: 307–428.

Sibley, C.G., and L.L. Short. 1964. Hybridization of the orioles of the Great Plains. Condor 66:130–150.

Smith, J.I. 1987. Evidence of hybridization between Red-bellied and Golden-fronted Woodpeckers. Condor 89:377–386.

Szijj, L.J. 1966. Hybridization and the nature of the isolating mechanisms in sympatric populations of meadowlarks (*Sturnella*) in Ontario. Zeitschrift für Tierpsychologie 23:677–690.

TUMBLING TUMBLEWEEDS

Howard, J.L. 1992. *Salsola kali*. Fire Effects Information System. Rocky Mountain Research Station, USDA Forest Service, Fort Collins, CO. http://www.feis-crs.org/feis/ (accessed January 8, 2015).

Young, J.A. 1991. Tumbleweed. Scientific American 264(3):82–87.

OGALLALA AQUIFER

Reeves, C.C., Jr., and J.A. Reeves. 1996. The Ogallala Aquifer of the Southern High Plains. Estacado Books, Lubbock, TX. 360 pp.

Zartman, R.E., R.H. Ramsey, P.W. Evans, et al. 1994. Playa lakes on the Southern High Plains in Texas. Journal of Soil and Water Conservation 49:299–301.

Conservation and Management

Iverson, C., P.A. Vohs, and T.C. Tacha. 1985. Distribution and abundance of Sandhill Cranes in western Texas. Journal of Wildlife Management 49:250–255.

Luo, H.R., L.M. Smith, B.L. Allen, and D.A. Haukos. 1997. Effects of sedimentation on playa wetland volume. Ecological Applications 7:247–252.

Smith, L.M. 2003. Playas of the Great Plains. University of Texas Press, Austin. 257 pp.

Texas Parks and Wildlife Department. 2004. Texas Black-tailed Prairie Dog conservation and management plan. Texas Parks and Wildlife Department, Austin. 45 pp.

Vermeire, L.T., and D.B. Wester. 2001. Shinnery Oak poisoning of rangeland cattle: Causes, effects, and solutions. Rangelands 23: 19–21.

Infobox 8.1. Canadian River Breaks

Morris, J.M. 1997. El Llano Estacado: Exploration and imagination on the High Plains of Texas and New Mexico, 1536–1860. Texas State Historical Association, Austin. 414 pp.

Roper, B. 1988. How did this river come to be called Canadian? Panhandle-Plains Historical Review 61:17–24.

Seyffert, K.D. 2001. Birds of the Texas Panhandle: Their status, distribution, and history. Texas A&M University Press, College Station. 501 pp.

Infobox 8.2. Captain Randolph B. Marcy (1812–1887): Army Explorer

Barnes, J.K. 2002. Giant Redheaded Centipede. Arthropod Museum Notes No. 13. University of Arkansas Arthropod Museum, Fayetteville.

Flores, D. 1990. Caprock Canyonlands. University of Texas Press, Austin. 200 pp.

Goetzmann, W.H. 1991. Army exploration in the American West. Texas State Historical Association, Austin. 509 pp.

Hollon, W.E. 1955. Beyond the Cross Timbers: Travels of Randolph B. Marcy, 1812–1857. University of Oklahoma Press, Norman. 270 pp.

Marcy, R.B. 1850. A report and map of Lt. Simpson of the route from Fort Smith to Santa Fe; also a report on the same subject from Captain R.B. Marcy. House of Representatives Executive Document 7, No. 45. Washington, DC.

———. 1854. A report on the exploration of the Red River of Louisiana in the year 1852. US Government Printing Office, Washington, DC. 310 pp.

Infobox 8.3. Muleshoe National Wildlife Refuge

Moore, R.L., and C.D. Simpson. 1981. Disease mortality of waterfowl on Texas playa lakes. Southwestern Naturalist 25:566–568.

Quortrup, E.R., F.B. Queen, and L J. Merovka. 1946. An outbreak of pasteurellosis in wild ducks. Journal of the American Veterinary Medical Association 108:94–100.

Infobox 8.4. Lubbock Lake Site

Black, C.C., ed. 1974. History and prehistory of the Lubbock Lake Site. West Texas Museum Association, Lubbock. The Museum Journal 15:1–160.

Grayson, D.K. 1991. Late Pleistocene mammalian extinctions in North America: Taxonomy, chronology, and explanations. Journal of World Prehistory 5:193–231.

Guthrie, R.D. 1970. Bison evolution and zoogeography in North America during the Pleistocene. Quarterly Review of Biology 45: 1–15.

Johnson, E., ed. 1987. Lubbock Lake: Late Quaternary studies on the Southern High Plains. Texas A&M University Press, College Station. 192 pp.

Johnson, E., and V. Holliday. 1980. A Plainview kill/butchering locale on the Llano Estacado—The Lubbock Lake Site. Journal of the Plains Conference 25:89–111.

Martin, P.S., and R.G. Klein, eds. 1984. Quaternary extinctions: A prehistoric revolution. University of Arizona Press, Tucson. 892 pp.

CHAPTER 9

Brown, D.E., ed. 1982. Biotic communities of the American Southwest—United States and Mexico. Desert Plants 4:1–342.

Dodson, C. 2012. A guide to plants of the northern Chihuahuan Desert. University of New Mexico Press, Albuquerque. 194 pp.

Echols, W.H. 1860. Camel expedition through the Big Bend Country. US 36th Congress, Second Session, Senate Executive Document No. 1, US Serial Set 1079:36–50.

Maxwell, R.A. 1979. The big bend of the Rio Grande: A guide to the rocks, geologic history, and settlers of the area of Big Bend National Park. Bureau of Economic Geology Guidebook 7. University of Texas, Austin. 138 pp.

Peterson, J., and B.M. Zimmer. 1998. Birds of the Trans Pecos. University of Texas Press, Austin. 216 pp.

Schmidly, D.J. 1977. The mammals of Trans-Pecos Texas. Texas A&M University Press, College Station. 225 pp.

Tyler, R.C. 1996. The Big Bend: A history of the last Texas frontier. Texas A&M University Press, College Station. 286 pp.

Walton, A.W., and C.D. Henry, eds. 1979. Cenozoic geology of the Trans-Pecos volcanic field of Texas. Bureau of Economic Geology Guidebook 19. University of Texas, Austin. 193 pp.

Wauer, R.H., and C.M. Fleming. 2002. Naturalist's Big Bend: An introduction to the trees and shrubs, wildflowers, cacti, mammals, birds, reptiles and amphibians, fish, and insects. Texas A&M University Press, College Station. 185 pp.

Wood, R.A., J.A.D. Dickson, and B. Kirkland-George. 1994. Turning the Capitan Reef upside down: A new appraisal of the ecology of the Permian Capitan Reef, Guadalupe Mountains, Texas and New Mexico. Palaios 9:422–427.

Yarborough, S.C., and M.A. Powell. 2002. Ferns and fern allies of the Trans-Pecos and adjacent areas. Texas Tech University Press, Lubbock. 116 pp.

A Harsh Environment

Scogin, R.A., and E.H. Elam. 1995. The Sanderson Flood: Crisis in a rural community. Sul Ross State University, Alpine, TX. 109 pp.

Van Devender, T.R. 1990. Late Quaternary vegetation and climate change of the Chihuahuan Desert, United States and Mexico. Pp. 104–133 in Packrat middens: The last 40,000 years of biotic change (J.L. Betancourt, T.R. Van Devender, and P.S. Martin, eds.). University of Arizona Press, Tucson. 469 pp.

Wells, P.V. 1966. Late Pleistocene vegetation and degree of pluvial climatic change in the Chihuahuan Desert. Science 153: 970–975.

Adaptations to Desert Life

PLANT ADAPTATIONS

Darrow, R.A. 1943. Vegetative and floral growth of Fouquieria splendens. Ecology 24:310–322.

Evenari, M. 1949. Germination inhibitors. Botanical Review 15: 153–194.

Handel, S.N., and A.J. Beattie. 1990. Seed dispersal by ants. Scientific American 26:76–83A.

Kemp, P.R. 1989. Seed banks and vegetative processes in deserts. Pp. 257–281 in Ecology of soil seed banks (M.A. Leck, V.T. Parker, and R.L. Simpson, eds.). Academic Press, San Diego, CA. 462 pp.

Killingbeck, K.T. 1990. Leaf production can be decoupled from root activity in the desert shrub Ocotillo (Fouquieria

splendens [Engelm.]). American Midland Naturalist 124:124–129.

Pake, C.E., and D.L. Venable. 1996. Seed banks in desert annuals: Implications for persistence and coexistence in variable environments. Ecology 77:1427–1435.

Reichman, O.J. 1979. Desert granivore foraging and its impact on seed densities and distributions. Ecology 60:1086–1092.

Rice, E.L. 1974. Allelopathy. Academic Press, New York. 353 pp.

Tevis, L. 1958. A population of desert ephemerals germinated by less than one inch of rain. Ecology 39: 688–695.

Waser, N.M. 1979. Pollinator availability as a determinant of flowering time in Ocotillo (Fouquieria splendens). Oecologia 39:107–121.

Yeaton, R.I. 1978. Cyclical relationship between Larrea tridentata and Opuntia leptocaulis. Journal of Ecology 66: 651–656.

ANIMAL ADAPTATIONS

Arad, Z., U. Midtgård, and M.H. Bernstein. 1989. Thermoregulation in Turkey Vultures: Vascular anatomy, arteriovenous heat exchange, and behavior. Condor 91:505–514.

Belk, D., and G.A. Cole. 1975. Adaptational biology of desert temporary-pond inhabitants. Pp. 207–266 in Environmental physiology of desert organisms (N.F. Hadley, ed.). Dowden, Hutchinson and Ross, Stroudsberg, PA. 283 pp.

Bentley, P.J. 1966. Adaptations of amphibians to arid environments. Science 152:619–623.

Dayton, G.H., and L.A. Fitzgerald. 2001. Competition, predation, and the distribution of four desert anurans. Oecologia 129:430–435.

Dayton, G.H., R. Skiles, and L. Dayton. 2007. The frogs and toads of Big Bend National Park. Texas A&M University Press, College Station. 64 pp.

Dayton, G.H., and S.D. Wapo. 2002. Cannibalistic behavior in Scaphiopus couchii: More evidence for larval oophagy. Journal of Herpetology 36:531–532.

Dimmitt, M.A., and R. Ruibal. 1980. Environmental correlates of emergence in spadefoot toads (Scaphiopus). Journal of Herpetology 14:21–29.

Hadley, N.F, M.C. Quinlan, and M.L. Kennedy. 1991. Evaporative cooling in the Desert Cicada: Thermal efficiency and water/metabolic costs. Journal of Experimental Biology 159:269–283.

Mayhew, W.W. 1965. Adaptations of the amphibian, Scaphiopus couchi, to desert conditions. American Midland Naturalist 74: 95–109.

Newman, R.A. 1989. Developmental plasticity of Scaphiopus couchi tadpoles in an unpredictable environment. Ecology 70:1775–1789.

Oberholser, H.C. 1918. The migration of North American birds. 2nd series. IV. The waxwings and Phainopepla. Bird-Lore 20:219–222.

INTERDEPENDENCE OF PLANTS AND POLLINATORS

Danforth, B.N. 1999. Emergence dynamics and bet hedging in a desert bee, Perdita portalis. Proceedings of the Royal Society B: Biological Sciences 266(1432):1985–1994.

Pellmyr, O., J.N. Thompson, J. Brown, and R.G. Harrison. 1996. Evolution of pollination and mutualism in the yucca moth lineage. American Naturalist 148:827–847.

Powell, J.A. 1992. Interrelationships of yuccas and yucca moths. Trends in Ecology and Evolution 7:10–15.

Smith, P.E., S.L. Buchmann, and M.K. O'Rourke. 1993. Evidence for mutualism between a flower-piercing carpenter bee and Ocotillo. Ecological Entomology 18:234–240.

Waser, N.M. 1979. Pollinator availability as a determinant of flowering time in Ocotillo (*Fouquieria splendens*). Oecologia 39:107–121.

Yoder, J.B., C. Smith, I. Pellmyr, and O. Pellmyr. 2010. How to become a yucca moth: Minimal trait evolution to establish the obligate pollination mutualism. Biological Journal of the Linnean Society 100:847–855.

Biophysiographic Associations

Carter, W.T., and V.L. Cory. 1931. Soils of the Trans-Pecos, and some of their vegetative relations. Transactions of the Texas Academy of Science 15:19–37.

Powell, A.M. 1998. Trees and shrubs of the Trans-Pecos and adjacent areas. University of Texas Press, Austin. 498 pp.

———. 2000. Grasses of the Trans-Pecos and adjacent areas. Iron Mountain Press, Marathon, TX. 377 pp.

STOCKTON PLATEAU

Bray, W.L. 1901. The ecological relations of the vegetation of western Texas. Botanical Gazette 32:99–291.

Cottle, H.J. 1931. Studies in the vegetation of southwestern Texas. Ecology 12:105–155.

Elliot, W.R., and G. Veni, eds. The caves and karst of Texas. National

Speleological Society, Huntsville, AL. 252 pp.

Hollander, R.R., C. Jones, J.K. Jones Jr., and R.W. Manning. 1990. Preliminary analysis of the effects of the Pecos River on geographic distribution of small mammals in western Texas. Journal of Big Bend Studies 2:97–107.

Jones, C., and D.A. Parish. 2001. Effects of the Pecos River on the geographic distributions of mammals in western Texas. Occasional Papers of the Museum of Texas Tech University 204:1–11.

Suttkus, R.D., and C. Jones. 2006. Fishes of Independence Creek and Pecos River. Occasional Papers of the Museum of Texas Tech University 248:1–7.

CHIHUAHUAN DESERT

Bray, W.L. 1905. Vegetation of the Sotol country in Texas. Bulletin of the University of Texas Science Series 6:1–24.

Milstead, W.W. 1960. Relict species of the Chihuahuan Desert. Southwestern Naturalist 5:75–88.

Plumb, G.A. 1992. Vegetation classification of Big Bend National Park, Texas. Texas Journal of Science 44:375–387.

Schmidt, R.H., Jr. 1979. A climatic delineation of the "real" Chihuahuan Desert. Journal of Arid Environments 2:243–250.

———. 1986. Chihuahuan climate. Pp.40–63 *in* Second symposium on resources of the Chihuahuan Desert region: United States and Mexico. Chihuahuan Desert Research Institute, Alpine, TX.

DESERT BASINS AND SALT FLATS

Angle, E.S. 2001. Hydrogeology of the Salt Basin. Pp. 232–247 *in* Aquifers of West Texas (R.E. Mace, W.F. Mullican III, and E.S. Angle, eds.). Texas Water Development Board Report 356. Austin. 263 pp.

Denyes, H.A. 1956. Natural terrestrial communities of Brewster County, Texas, with special reference to the distribution of mammals. American Midland Naturalist 55: 289–320.

Milstead, W.W. 1960. Relict species of the Chihuahuan Desert. Southwestern Naturalist 5:75–88.

Toweill, D.E, V. Geist, T. Story, et al. 1999. Return of royalty: Wild sheep of North America. Boone and Crockett Club, Missoula, MT. 224 pp.

Wells, P.V. 1965. Vegetation of the Dead Horse Mountains, Brewster County, Texas. Southwestern Naturalist 10:256–260.

TRANS-PECOS GRASSLANDS

Buechner, H.K. 1950. Life history, ecology, and range use of the Pronghorn Antelope in Trans-Pecos Texas. American Midland Naturalist 43:257–354.

Echols, W.H. 1860. Camel expedition through the Big Bend Country. US 36th Congress, Second Session, Senate Executive Document No. 1, US Serial Set 1079:36–50.

LOW MOUNTAINS AND BAJADAS

Davis, W.B., and W.P. Taylor. 1939. The bighorn sheep of Texas. Journal of Mammalogy 20:440–455.

Frick, W.F., P.A. Heady III, and J.P. Hayes. 2009. Facultative nectar-feeding behavior in a gleaning insectivorous bat, *Antrozous pallidus.* Journal of Mammalogy 90:1157–1164.

Lenhart, P.A., V. Mata-Silva, and J.D. Johnson. 2010. Foods of the Pallid Bat, *Antrozous pallidus* (Chiroptera: Vespertilionidae), in the Chihuahuan Desert of western Texas. Southwestern Naturalist 55:110–115.

Monson, G., and L. Sumner, eds. 1999. The Desert Bighorn:

Its life history, ecology, and management. University of Arizona Press, Tucson. 320 pp.

SKY ISLANDS AND MADREAN PINE-OAK WOODLANDS

Dugelby, B.L., D. Foreman, R. List, et al. 2001. Rewilding the sky islands region of the Southwest. Pp. 65–81 *in* Large mammal restoration ecological and sociological challenges in the 21st century (D. Maehr, R.F. Noss, and J.L. Larkin, eds.). Island Press, Washington, DC. 336 pp.

Harveson, L.A., T.H. Allen, F. Hernandez, et al. 2006. Montezuma Quail ecology and life history. Pp. 23–39 *in* Texas quails: Ecology and management (L.A. Brennan, ed.). Texas A&M University Press, College Station. 512 pp.

Myers, N., R.A. Mittermeier, C.G. Mittermeier, et al. 2000. Biodiversity hotspots for conservation priorities. Nature 403:853–858.

Nall, A.V., L.K. Ammerman, and R.C. Dowler. 2012. Genetic and morphologic variation in the Davis Mountains Cottontail (*Sylvilagus robustus*). Southwestern Naturalist 57:1–7.

Northington, D.K., and T.L. Burgess. 1979. Summary of the vegetative zones of the Guadalupe Mountains National Park, Texas. Pp. 51–57 *in* Biological investigations in the Guadalupe Mountains National Park, Texas. Symposium proceedings, National Park Service Transactions and Proceedings Series No. 4. 442 pp.

Parent, L., and J.N. Patoski. 2001. Texas mountains. University of Texas Press, Austin. 155 pp.

Possingham, H., and K. Wilson. 2005. Turning up the heat on hotspots. Nature 436:919–920.

Warnock, B.H. 1977. Wildflowers of the Davis Mountains and the Marathon Basin, Texas. Sul Ross State University, Alpine, TX. 276 pp.

SPRINGS, CIÉNAGAS, AND TINAJAS

Brannan, D.K., C.R. Brannan, and T.E. Lee Jr. 2003. Reproductive and territorial behavior of Comanche Springs Pupfish (*Cyprinodon elegans*) in San Solomon Spring pool, Balmorhea State Park, Reeves County, Texas. Southwestern Naturalist 48:85–88.

Echelle, A.A., E.W. Carson, A.F. Echelle, et al. 2005. Historical biogeography of the new-world pupfish genus *Cyprinodon* (Teleostei: Cyprinodontidae). Copeia 2005:320–339.

Garrett, G.P., and N.L. Allan. 2003. Aquatic fauna of the northern Chihuahuan Desert: Contributed papers from a special session within the Thirty-Third Annual Symposium of the Desert Fishes Council. Special Publications of the Museum of Texas Tech University 46:1–160.

Hubbs, C. 2014. Differences in spring versus stream fish assemblages. Pp. 376–395 *in* Proceedings of the Sixth Symposium on the Natural Resources of the Chihuahuan Desert Region (C.A. Hoyt and J. Karges, eds.). Chihuahuan Desert Research Institute, Fort Davis, TX. 427 pp.

Kirkland, F. 1940. Pictographs of Indian masks at Hueco Tanks. Bulletin of the Texas Archaeological and Paleontological Society 12:8–29.

Newcomb, W.W., Jr., and F. Kirkland.1967. The rock art of the Texas Indians. University of Texas Press, Austin. 239 pp.

US Fish and Wildlife Service. 1982. Leon Springs Pupfish (*Cyprinodon bovinus*) recovery plan. US Fish and Wildlife Service, Albuquerque, NM. 26 pp.

US Fish and Wildlife Service. 1985. Comanche Springs Pupfish (*Cyprinodon elegans*) recovery plan. US Fish and Wildlife Service, Albuquerque, NM. 25 pp.

RIPARIAN ZONES

Gregory, S.V., F.J. Swanson, W.A. McKee, and K.W. Cummins. 1991. An ecosystem perspective of riparian zones: Focus on links between land and water. BioScience 41:540–551.

Perlichek, K.B., L.A. Harveson, B.J. Warnock, and B. Tarrant. 2009. Habitat characteristics of winter roost sites of Wild Turkeys in Trans-Pecos, Texas. Southwestern Naturalist 54:446–452.

Powell, A.M. 1998. Trees and shrubs of the Trans-Pecos and adjacent areas. University of Texas Press, Austin. 498 pp.

Schade, J.D., E. Marti, J.R. Weller, et al. 2002. Sources of nitrogen to the riparian zone of a desert stream: Implications for riparian vegetation and nitrogen retention. Ecosystems 5:68–79.

Highlights

ROCK OUTCROPS

Northington, D.K., and T.L. Burgess. 1979. Summary of the vegetative zones of the Guadalupe Mountains National Park, Texas. Pp. 51–57 *in* Biological investigations in the Guadalupe Mountains National Park, Texas. Symposium proceedings, National Park Service Transactions and Proceedings Series No. 4.

Potter, R.M., and G.R. Rossman. 1981. Desert varnish: The importance of clay minerals. Science 213: 1245–1247.

RESURRECTION PLANTS

Curtin, L.S.M., and W. Moore. 2003. Healing herbs of the Upper Rio Grande: Traditional medicine of the Southwest. Western Edge Press, Santa Fe, NM. 256 pp.

Eickmeier, W.G. 1983. Photosynthetic recovery of the Resurrection Plant *Selaginella lepidophylla* (Hook. & Grev.) Spring: Effects of prior desiccation rate and mechanisms of desiccation damage. Oecologia 58:115–120.

Lebkuecher, J.G., and W.G. Eickmeier. 1993. Physiological benefits of stem curling for Resurrection Plants in the field. Ecology 74:1073–1080.

DESERT TERMITES

Allen, C.T., D.E. Foster, and D.N. Uechert. 1980. Seasonal food habits of a desert termite, *Gnathamitermes tubiformans,* in West Texas. Environmental Entomology 9:461–466.

Elkins, N.Z., G.V. Sabol, T.J. Ward, and W.G. Whitford. 1986. The influence of subterranean termites on the hydrological characteristics of a Chihuahuan Desert ecosystem. Oecologia 68: 521–528.

Nash, M.S., J.P. Anderson, and W.G. Whitford. 1999. Spatial and temporal variability in relative abundance and foraging behavior of subterranean termites in desertified and relatively intact Chihuahuan Desert ecosystems. Applied Soil Ecology 12:149–157.

Whitford, W.G. 1991. Subterranean termites and long-term productivity of desert rangelands. Sociobiology 19:235–243.

Whitford, W.G., Y. Steinberger, and G. Ettershank. 1982. Contributions of subterranean termites to the "economy" of desert ecosystems. Oecologia 55:298–302.

LAKE CABEZA DE VACA

Axtell, R.W. 1977. Ancient playas and their influence on the recent herpetofauna of the northern Chihuahuan Desert. Pp. 493–512 *in* Transactions of the Symposium on the Biological Resources of the Chihuahuan Desert Region, United States and Mexico (R.H. Wauer and D.H. Riskind, eds.). National Park Service Transactions and Proceedings Series No. 3. National Park Service, Washington, DC.

Manier, M.K. 2004. Geographic variation in the Long-nosed Snake, *Rhinocheilus lecontei* (Colubridae): Beyond the subspecies debate. Biological Journal of the Linnean Society 83: 65–85.

Morafka, D.J. 1977. Biogeographical analysis of the Chihuahuan Desert through its herpetofauna. D.W. Junk, The Hague, Netherlands. 321 pp.

Rosenthal, J., and M.R. J. Forstner. 2004. Effects of a Plio-Pleistocene barrier on Chihuahuan Desert herpetofauna. Pp. 269–282 *in* Proceedings of the Sixth Symposium on the Natural Resources of the Chihuahuan Desert Region (C.A. Hoyt and J. Karges, eds.). Chihuahuan Desert Research Institute, Fort Davis, TX. 427 pp.

Zamudio, K.R., K.B. Jones, and R.H. Ward. 1997. Molecular systematics of short-horned lizards: Biogeography and taxonomy of a widespread species complex. Systematic Biology 46:284–305.

TEXAS BONES AND CALIFORNIA CONDORS

Casto, S.D. 2002. Extinct and extirpated birds of Texas. Bulletin of the Texas Ornithological Society 35:17–32.

Emslie, S.D. 1987. Age and diet of fossil California Condors in Grand Canyon, Arizona. Science 237: 768–770.

Wetmore, A., and H. Friedmann. 1933. The California Condor in Texas. Condor 35:37–38.

THE MARFA LIGHTS

Brueske, J.M. 1989. The Marfa Lights, being a collection of first-hand accounts by people who have seen the lights close-up or in unusual circumstances, and related material. 2nd ed. Ocotillo Enterprises, Marfa, TX. 50 pp.

Bunnell, J. 2003. Night orbs. Lacey Publishing, Benbrook, TX. 320 pp.

Darack, E. 2008. Unlocking the atmospheric secrets of the Marfa Mystery Lights. Weatherwise 61: 36–44.

Stephan, K.D., J. Bunnell, J. Klier, and L. Komala-Noor. 2011. Quantitative intensity and location measurements of an intense long-duration luminous object near Marfa, Texas. Journal of Atmospheric and Solar-Terrestrial Physics 73:1953–1958.

Conservation and Management

Bell, G.P., S. Yanoff, J. Karges, et al. 2004. Conservation blueprint for the Chihuahuan Desert ecoregion. Pp. 1–36 *in* Proceedings of the Sixth Symposium on the Natural Resources of the Chihuahuan Desert Region (C.A. Hoyt and J. Karges, eds.). Chihuahuan Desert Research Institute, Fort Davis, TX.

Hall, G.E. 2002. High and dry: The Texas-New Mexico struggle for the Pecos River. University of New Mexico Press, Albuquerque. 303 pp.

James, T.L., and A. De La Cruz. 1989. *Prymnesium parvus* Carter as a suspect of mass mortalities of fish and shellfish communities in western Texas. Texas Journal of Science 41:429–430.

Rhodes, K., and C. Hubbs. 1992. Recovery of Pecos River fishes from a red tide fish kill. Southwestern Naturalist 37:178–187.

Tyler, R.C. 1996. The Big Bend. Texas A&M University Press, College Station. 286 pp.

Infobox 9.1. *Curanderos:* **The Healers**

Kane, C.W. 2011. Medicinal plants of the American Southwest. Lincoln Town Press, Tucson, AZ. 368 pp.

Smithers, W.D. 1999. Chronicles of the Big Bend. Texas State Historical Association, Austin. 144 pp.

Infobox 9.2. Law West of the Pecos

Davis, J.T. 1985. Legendary Texians. Vol. 2. Eakin Press, Austin, TX. 192 pp.

Eisner, T., J. Meinwald, A. Monro, and R. Ghent. 1961. Defense mechanisms in Arthropods. I. The composition and function of the spray of the Whip Scorpion, *Mastigoproctus giganteus* (Lucas) (Arachnida, Penipalpida). Journal of Insect Physiology 6:272–292.

Punzo, F. 2000. Diel activity patterns and diet of the Giant Whipscorpion *Mastigoproctus giganteus* (Lucas) (Arachnida, Uropygi) in Big Bend National Park (Chihuahuan Desert). Bulletin of the British Arachnological Society 11:385–387.

Robinson, C.M. 2005. American frontier lawmen, 1850–1930. Osprey, Oxford, UK. 72 pp.

Skiles, J. 1996. Judge Roy Bean country. Texas Tech University Press, Lubbock. 204 pp.

Sonnichsen, C.L. 1943. The story of Roy Bean: Law west of the Pecos. Devin-Adair, New York. 207 pp.

Infobox 9.3. Return of the Black Bears

Hellgren, E.C. 1993. Status, distribution, and summer food habits of Black Bears in Big Bend National Park. Southwestern Naturalist 38:77–80.

Hellgren, E.C., D.P. Onorato, and J.R. Skiles Jr. 2005. Dynamics of a Black Bear population within a desert metapopulation. Biological Conservation 122:131–140.

Onorato, D.P., and E.C. Hellgren. 2001. Black Bear at the border: Natural recolonization of the Trans-Pecos. Pp. 245–259 *in* Large mammal restoration: Ecological and sociological considerations in the 21st century (D.S. Maehr, R.F. Noss, and J.L. Larkin, eds.). Island Press, Washington, DC. 336 pp.

Onorato, D.P., E.C. Hellgren, F.S. Mitchell, and J.R. Skiles Jr. 2003. Home range and habitat use of American Black Bears on a desert montane island in Texas. Ursus 14: 120–129.

Onorato, D.P., E.C. Hellgren, R.A. Van Den Bussche, and D.L. Doan-Crider. 2004. Phylogeographic patterns within a metapopulation of Black Bears (*Ursus americanus*) in the American Southwest. Journal of Mammalogy 85:140–147.

Infobox 9.4. *Angelitos:* **The Little Red Angels**

Hartke, T.R., and B. Baer. 2011. The mating biology of termites. Animal Behavior 82:927–936.

Newell, I.M., and L. Tevis Jr. 1960. *Angelothrombium pandorae* n.g., n.sp. (Acari, Trombidiidae), and notes on the biology of the Giant Red Velvet Mites. Annals of the Entomological Society of America 53:293–304.

Tevis, L., Jr., and I.M. Newell. 1962. Studies on the biology and seasonal cycle of the Giant Red Velvet Mite, *Dinothrombium pandorae* (Acari, Trombidiidae). Ecology 43:497–505.

Zhang, Z.-Q. 1998. Biology and ecology of trombidiid mites (Acari: Trombidioidea). Experimental and Applied Acarology 22:139–155.

CHAPTER 10

Clover, E.V. 1937. Vegetational survey of the lower Rio Grande Valley, Texas. Madroño 4:41–66, 77–100.

Everitt, J.H., and D.L. Drawe. 1993. Trees, shrubs, and cacti of South Texas. Texas Tech University Press, Lubbock. 213 pp.

Fulbright, T.E., and F.C. Bryant. 2002. The last great habitat. Texas A&M University–Kingsville, Caesar Kleberg Wildlife Research Unit Special Publication 1:1–32.

Oberholser, H.C. 1945. The outstanding bird life of Texas. Southwest Review 30:377–381.

Taylor, R.B., J. Rutledge, and J.G. Herrera. 1999. A field guide to common South Texas shrubs. Texas Parks and Wildlife Press, Austin. 123 pp.

Vegetative Associations

Archer, S. 1989. Have southern Texas savannahs been converted to woodlands in recent history? American Naturalist 134:545–561.

Archer, S., C. Scifres, C.R. Bassham, and R. Maggio. 1988. Autogenic succession in a subtropical savanna: Conversion of a grassland to a thorn woodland. Ecological Monographs 58:111–127.

Bogush, E.R. 1952. Brush invasion in the Rio Grande Plain of Texas. Texas Journal of Science 8:356–359.

Crosswhite, F.S. 1980. Dry country plants of the South Texas plains. Desert Plants 2:141–179.

Hanselka, C.W. 1980. The

historical role of fire on South Texas rangelands. Pp. 2–18 *in* Prescribed range burning in the Coastal Prairie and eastern Rio Grande Plains of Texas (C.W. Hanselka, ed.). Texas Agricultural Experiment Station Contribution No. TA 16277. College Station. 128 pp.

Hart, C.R. 2008. Brush and weeds of Texas rangelands. Cooperative Extension Service, Texas A&M University Press, College Station. 204 pp.

Inglis, J.N. 1964. A history of vegetation on the Rio Grande Plain. Texas Parks and Wildlife Bulletin No. 45:1–122.

Johnston, M.C. 1963. Past and present grasslands of southern Texas and northeastern Mexico. Ecology 44: 456–466.

NORTHERN SEMIARID THORNSCRUB

Archer, S. 1995. Tree-grass dynamics in a *Prosopis*-thornscrub savanna parkland: Reconstructing the past and predicting the future. Ecoscience 2:83–99.

Griffith, G., S. Bryce, J. Omerlink, and A. Rogers. 2007. Ecoregions of Texas. Project report to Texas Commission on Environmental Quality, Austin. 25 pp.

Hanselka, C.W., and D.E. Kilgore. 1987. The Nueces River Valley: The cradle of the western livestock industry. Rangelands 9:195–198.

Hubbs, C., R.J. Edwards, and G.P. Garrett. 1991. An annotated checklist of the freshwater fishes of Texas, with keys to identification of species. Texas Journal of Science 43:1–56.

Telfair, R.C., II, ed. 1999. Texas wildlife resources and land uses. University of Texas Press, Austin. 404 pp.

TAMAULIPAN MEZQUITAL

Barnes, P.W., and S.R. Archer. 1996. Influence of an overstory tree (*Prosopis glandulosa*) on associated shrubs in a savanna parkland: Implications for patch dynamics. Oecologia 105:493–500.

Collins, K. 1984. Status and management of native South Texas brushlands. US Fish and Wildlife Service, Office of Ecological Services, Corpus Christi, TX. 18 pp.

Drawe, D.L., and I. Higginbotham Jr. 1980. Plant communities of the Zachary Ranch in the South Texas plains. Texas Journal of Science 32:319–332.

Fulbright, T., and F.C. Bryant. 2003. The Wild Horse Desert: Climate and ecology. Pp. 23–34 *in* Ranch management: Integrating cattle, wildlife and range (C.A. Forgason, F.C. Bryant, and P.C. Genho, eds.) King Ranch, Kingsville, TX. 296 pp.

Roth, R.R. 1977. The composition of four bird communities in a South Texas brush-grassland. Condor 79:642–654.

Scifres, C.J., J.L. Mutz, and G.P. Durham. 1976. Range improvement following chaining of South Texas mixed brush. Journal of Range Management 29: 418–421.

THE BORDAS ESCARPMENT

Crosswhite, F.S. 1980. Dry country plants of the South Texas plains. Desert Plants 2:141–179.

Jahrsdoefer, S.E., and D.M. Leslie Jr. 1988. Tamaulipan brushland of the Lower Rio Grande Valley of South Texas: Description, human impacts, and management options. US Fish and Wildlife Service Biological Report 88(36):1–63.

Mild, C. 2004. Stinging Cevallia blooms near Rio Grande City.

Rio Delta Wild (April 10). www .riodeltawild.com (accessed September 20, 2015).

Morgan, G.R., and O.C. Stewart. 1984. Peyote trade in South Texas. Southwestern Historical Quarterly 87:269–296.

Price, W.A. 1963. Sedimentary and quaternary geomorphology of South Texas. Transactions of the Gulf Coast Association of Geological Societies 8:41–75.

Sayre, A.N. 1937. Geology and ground-water resources of Duval County, Texas. US Geological Survey Water-Supply Paper 776. Washington, DC. 104 pp.

Thompson, C.M., R.R. Sanders, and D. Williams. 1972. Soil survey of Starr County, Texas. US Department of Agriculture, Soil Conservation Service, Washington, DC. 87 pp.

RIO GRANDE RIPARIAN FORESTS

Blom, C.W.P.M., G.M. Bögeman, P. Laan, et al. 1990. Adaptations to flooding in plants from river areas. Aquatic Botany 38:29–47.

Brush, T., and A. Cantu. 1998. Changes in the breeding bird community of subtropical evergreen forest in the Lower Rio Grande Valley of Texas, 1970s–1990s. Texas Journal of Science 50:123–132.

Gehlbach, F.R. 1981. Natural history sketches, densities, and biomass of breeding birds in evergreen forests of the Rio Grande, Texas, and Rio Corona, Tamaulipas, Mexico. Texas Journal of Science 39:241–251.

Lonard, R.I., and F.W. Judd. 2002. Riparian vegetation of the lower Rio Grande. Southwestern Naturalist 47:420–432.

Passmore, M.F., and B.C. Thompson. 1981. Responses of three species of kingfishers to fluctuating water levels below Falcon Dam. Bulletin of the Texas Ornithological Society 14:13–17.

Small, M.F., T.H. Bonner, and J.T. Baccus. 2009. Hydrologic alteration of the Lower Rio Grande terminus: A quantitative assessment. River Research and Applications 25:241–252.

LOWER RIO GRANDE VALLEY

Brush, T., and A. Cantu. 1998. Changes in the breeding bird community of subtropical evergreen forest in the Lower Rio Grande Valley of Texas, 1970s–1990s. Texas Journal of Science 50:123–132.

Lonard, R.I. 1993. Guide to grasses of the Lower Rio Grande Valley. University of Texas–Pan American Press, Edinburg. 240 pp.

Lonard, R.I., J.H. Everitt, and F.W. Judd. 1991. Woody plants of the Lower Rio Grande Valley, Texas. Texas Memorial Museum, University of Texas Press, Austin. 179 pp.

Richardson, A. 1990. Plants of southernmost Texas. Gorgas Science Foundation, Brownsville, TX. 298 pp.

———. 1995. Plants of the Rio Grande Delta. University of Texas Press, Austin. 332 pp.

———. 2011. Plants of deep South Texas: A field guide to the woody and flowering species. Texas A&M University Press, College Station. 448 pp.

Shindle, D.B., and M.E. Tewes. 1998. Woody species composition of habitats used by Ocelots (*Leopardus pardalis*) in the Tamaulipan Biotic Province. Southwestern Naturalist 43:273–279.

BOSCAJE DE LA PALMA

Brush, T. 2005. Nesting birds of a tropical frontier: The lower Rio Grande Valley of Texas. Texas A&M University Press, College Station. 245 pp.

Chapman, S.S., and B.R. Chapman. 1990. Bats from the coastal region of southern Texas. Texas Journal of Science 42:13–22.

Chipman, D.E. 1995. Alonso Alvarez de Pineda and the Río de las Palmas: Scholars and the mislocation of a river. Southwestern Historical Quarterly 98:369–385.

Everitt, J.H., F.W. Judd, D.E. Escobar, et al. 1996. Using remote sensing and spatial information technologies to map Sabal Palm in the Lower Rio Grande Valley of Texas. Southwestern Naturalist 41: 218–226.

Lockett, L. 1995. Historical evidence of the native presence of *Sabal mexicana* (Palmae) north of the Lower Rio Grande Valley. Sida 16: 717–719.

Lonard, R.I., and F.W. Judd. 2002. Riparian vegetation of the lower Rio Grande. Southwestern Naturalist 47:420–432.

Distinctive Fauna

COOPERATIVELY HUNTING HAWKS

Bednarz, J.C. 1988. Cooperative hunting in Harris Hawks (*Parabuteo unicinctus*). Science 239:1525–1527.

Brown, J.L. 2014. Helping and communal breeding in birds. Princeton University Press, Princeton, NJ. 384 pp.

Coulson, J.O., and T.D. Coulson. 1995. Group hunting by Harris' Hawks in Texas. Journal of Raptor Research 29:265–267.

Dawson, J.W., and R.W. Mannan. 1991. Dominance hierarchies and helper contributions in Harris's Hawks. Auk 108:480–483.

Dwyer, J.F., and J.C. Bednarz. 2011. Harris' Hawk (*Parabuteo unicinctus*). The birds of North America online (A. Poole, ed.). Cornell Laboratory of Ornithology, Ithaca, NY. http://bna.birds.cornell.edu/bna/species/146, doi:10.2173/bna.146 (accessed August 27, 2015).

Mader, W.J. 1975. Extra adults at Harris Hawk nests. Condor 77: 482–485.

Moore, J.H. 1986. The ornithology of Cheyenne religionists. Plains Anthropologist 31:177–192.

COLLARED PECCARY

Bissonette, J.A. 1982. Social behavior and ecology of the Collared Peccary in Big Bend National Park. National Park Service Science Monograph 16:1–85.

Everitt, J.H., C.L. Gonzalez, M.A. Alaniz, and G.V. Latigo. 1981. Food habits of the Collared Peccary on South Texas rangelands. Journal of Range Management 34:141–144.

Gallagher, J.F., L.W. Varner, and W.E. Grant. 1984. Nutrition of the Collared Peccary in South Texas. Journal of Wildlife Management 48:749–761.

Hellgren, E.C., D.R. Synatzske, P.W. Oldenburg, and F.S. Guthery. 1995. Demography of a Collared Peccary population in South Texas. Journal of Wildlife Management 59:153–163.

Manaster, J. 2006. Javelinas: Collared peccaries of the Southwest. Texas Tech University Press, Lubbock. 100 pp.

Oldenburg, P.W., P.J. Ettestad, W.E. Grant, and E. Davis. 1985 Structure of Collared Peccary herds in South Texas: Spatial and temporal dispersion of herd members. Journal of Mammalogy 66:764–770.

Sowls, L.K. 1984. The peccaries. University of Arizona Press, Tucson. 251 pp.

———. 1997. Javelinas and other peccaries: Their biology,

management and use. 2nd ed. Texas A&M University Press, College Station. 352 pp.

IF IT'S AN INDIGO, LET IT GO

Keegan, H.L. 1944. Indigo snakes feeding upon poisonous snakes. Copeia 1944:59.

Wright, A.H., and A.A. Wright. 1994. Handbook of snakes of the United States and Canada. 2 vols. Comstock, Ithaca, NY. 1142 pp. [Reprint of the original 1957 edition.]

TEXAS TORTOISE,
THE "POSTER CHILD"

Auffenberg, W., and W.G. Weaver. 1969. *Gopherus berlandieri* in southeastern Texas. Bulletin of the Florida State Museum 13: 141–203.

Bury, R.B., and E.L. Smith. 1986. Aspects of the ecology and management of the tortoise *Gopherus berlandieri* at Laguna Atascoca, Texas. Southwestern Naturalist 31:387–394.

Engeman, R.M., M.J. Pipas, and H.T. Smith. 2004. *Gopherus berlandieri* (Texas Tortoise) mortality. Herpetological Review 35:54–55.

Hellgren, E.C., R.T. Kazmaier, D.C. Ruthven III, and D.R. Synatzske. 2000. Variation in tortoise life history: Demography of *Gopherus berlandieri*. Ecology 81:1297–1310.

Judd, F.W., and J.D. McQueen. 1980. Incubation, hatching, and growth of the tortoise, *Gopherus berlandieri*. Journal of Herpetology 14:377–380.

Judd, F.W., and F.L. Rose, 1983. Population structure, density and movements of the Texas Tortoise, *Gopherus berlandieri*. Southwestern Naturalist 28:387–398.

Kazmaier, R.T., E.C. Hellgren, and D.C. Ruthven III. 2001. Habitat selection by the Texas Tortoise in a managed thornscrub ecosystem. Journal of Wildlife Management 65:653–660.

Rose, F.L., and F.W. Judd. 2014. The Texas Tortoise: A natural history. University of Oklahoma Press, Norman. 188 pp.

Highlights

FOR KING AND CURLEWS

Andres, B.A., P.A. Smith, R.I.G. Morrison, et al. 2012. Population estimates of North American shorebirds, 2012. Wader Study Group Bulletin 119:178–194.

Dugger, B.D., and K.M. Dugger. 2002. Long-billed Curlew (*Numenius americanus*). The birds of North America online (A. Poole., ed.). Cornell Laboratory of Ornithology, Ithaca, NY. http://bna.birds.cornell.edu/bna/species/628 (accessed August 28, 2015).

Hawkins, W. 1947. El Sal del Rey. Texas State Historical Association, Austin. 68 pp.

Stanley, T.R., and S.K. Skagen. 2007. Estimating the breeding population of Long-billed Curlew in the US. Journal of Wildlife Management 71:2556–2564.

CHIGGERS

Mullen, G.R., and L.A. Durden, eds. 2002. Medical and veterinary entomology. Academic Press, New York. 597 pp.

Waller, D.E., and H. Proctor. 2013. Mites: Ecology, evolution and behaviour; life at a microscale. Springer, New York. 494 pp.

MORNING JEWELS

Dickerson, A.K., and D.L. Hu. 2014. Mosquitoes actively remove drops deposited by fog and dew. Integrated and Comparative Biology 54:1008–1013.

Edwards, R.L. 1946. Some notes on the life history of the Mexican Ground Squirrel in Texas. Journal of Mammalogy 27:105–121.

Ortega-Jimenez, V., and R. Dudley. 2012. Aerial shaking performance of wet Anna's Hummingbirds. Journal of the Royal Society Interface 9:1093–1099.

Sherbrooke, W.C. 1990. Rain-harvesting in the lizard, *Phrynosoma cornutum:* Behavior and integumental morphology. Journal of Herpetology 24:302–308.

Wisdom, K.M., J.A. Watson, X. Qu, et al. 2013. Self-cleaning of superhydrophobic surfaces by self-propelled jumping condensate. Proceedings of the National Academy of Sciences 110:7992–7997.

Conservation and Management

Jahrsdoefer, S.E., and D.M. Leslie Jr. 1988. Tamaulipan Brushland of the Lower Rio Grande Valley of South Texas: Description, human impacts, and management options. US Fish and Wildlife Service Biological Report 88(36):1–63.

Morgan, M.J. 2015. Border sanctuary: The conservation legacy of the Santa Ana Land Grant. Texas A&M University Press, College Station. 215 pp.

Parvin, B. 1988. Valley under siege. Defenders 63:18–29.

Rappole, J.H., C.E. Russell, J.R. Norwine, and T.E. Fulbright. 1986. Anthropogenic pressures and impacts on marginal, neotropical, semiarid ecosystems: The case of South Texas. Journal of Science of the Total Environment 55:91–99.

Infobox 10.1. Hell in Texas

Felleman, H. 1936. The best loved poems of the American people. Garden City Publishing, Garden City, NY. 670 pp. [Source of the excerpt from "Hell in Texas" by an anonymous poet.]

Infobox 10.2. A Paradise for Birders and Butterfliers

Brush, T. 2005. Nesting birds of a tropical frontier: The Lower Rio Grande Valley of Texas. Texas A&M University Press, College Station. 245 pp.

Teter, D., and D.L. McNeely. 1995. Abundance and diversity of aquatic birds on two South Texas oxbow lakes. Texas Journal of Science 47:62–68.

Woodin, M.C., M.K. Skoruppa, and G.C. Hickman. 2000. Surveys of night birds along the Rio Grande in Webb, County, Texas. Final report. US Fish and Wildlife Service, Corpus Christi, TX. 19 pp.

World Birding Center. http://theworldbirdingcenter.com (accessed October 19, 2015).

CHAPTER 11

Smeins, F.E., D.D. Diamond, and C.W. Hanselka. 1991. Coastal prairie. Pp. 269–290 in Natural grasslands: Introduction and Western Hemisphere. Ecosystems of the world 8A (R.T. Coupland, ed.). Elsevier, Amsterdam. 469 pp.

The Vegetation, Past and Present

THE GRASSLANDS

Diamond, D.D., and F.E. Smeins. 1984. Remnant grassland vegetation and ecological affinities of the upper Coastal Prairie of Texas. Southwestern Naturalist 29:321–334.

———. 1988. Gradient analysis of remnant true and upper Coastal Prairie grasslands of North America. Canadian Journal of Botany 66:2152–2161.

Dietz, R.S. 1945. The small mounds of the Gulf Coastal Plain. Science 102:596–597.

Hatch, S.L., J.L. Schuster, and D.L.

Drawe. 1999. Grasses of the Texas gulf prairies and marshes. Texas A&M University Press, College Station. 355 pp.

Johnson, M.C. 1963. Past and present grasslands of southern Texas and northeastern Mexico. Ecology 44: 456–466.

Rosen, D.J. 2007. The vascular flora of Nash Prairie: A coastal prairie remnant in Brazoria County, Texas. Journal of the Botanical Research Institute of Texas 1:679–692.

———. 2010. The vascular plants of Mowotony Prairie: A small remnant coastal grassland in Brazoria County, Texas. Journal of the Botanical Research Institute of Texas 4:489–495.

Singhurst, J.R., D.J. Rosen, A. Cooper, and W.C. Holmes. 2014. The vascular flora and plant communities of Candy Abshier Wildlife Management Area, Chambers County, Texas, U.S.A. Journal of the Botanical Research Institute of Texas 8:667–677.

US Fish and Wildlife Service. 2009. Black Lace Cactus (*Echinocereus reichenbachii* var. *albertii*) 5-year review: Summary and evaluation. Corpus Christi Ecological Services Field Office, Corpus Christi, TX. 31 pp.

COASTAL SAND PLAIN, THE LLANO MESTEÑO

Bogusch, E.R. 1952. Brush invasion in the Rio Grande Plain of Texas. Texas Journal of Science 4:85–91.

Brown, L.F., J.L. Brewton, T.J. Evans, et al. 1980. Environmental geologic atlas of the Texas Coastal Zone: Brownsville-Harlingen Area. Bureau of Economic Geology, University of Texas, Austin. 140 pp.

Brown, L.F., J.H. McGowen, T.J. Evans, et al. 1977. Environmental geologic atlas of the Texas Coastal

Zone: Kingsville Area. Bureau of Economic Geology, University of Texas, Austin. 131 pp.

Diamond, D.D., and T.E. Fulbright. 1990. Contemporary plant communities of the upland grasslands of the Coastal Sand Plain, Texas. Southwestern Naturalist 35:385–392.

Fulbright, T.E., and F.C. Bryant. 2002. The last great habitat. Texas A&M University-Kingsville, Special Publication of the Caesar Kleberg Wildlife Research Institute 1:1–32. [Source for quote by Sheridan.]

Fulbright, T.E., D.D. Diamond, J. Rappole, and J. Norwine. 1990. The Coastal Sand Plain of southern Texas. Rangelands 12: 337–340.

Huffman, G.G., and W.A. Price. 1949. Clay dune formation near Corpus Christi, Texas. Journal of Sedimentary Petrology 19:118–127.

Judd, F.W. 2002. Tamaulipan Biotic Province. Pp. 38–58 in The Laguna Madre of Texas and Tamaulipas (J.W. Tunnell Jr. and F.W. Judd, eds.). Texas A&M University Press, College Station. 372 pp.

Price, W.A. 1963. Physiochemical and environmental factors in clay dune genesis. Journal of Sedimentary Petrology 33:766–778.

Price, W.A., and L.S. Kornicker. 1961. Marine and lagoonal deposits in a clay dune, Gulf Coast, Texas. Journal of Sedimentary Petrology 31:245–255.

Rappole, J.W., and G.W. Blacklock. 1985. Birds of the Texas Coastal Bend. Texas A&M University Press, College Station. 126 pp.

BOTTOMLAND FORESTS AND OAK MOTTES

Evans, A. 2012. Upland habitats. Pp. 125–139 in The ecology and sociology of the Mission-Aransas Estuary: An estuarine and watershed profile. University of Texas Marine Science Institute, Port Aransas. 183 pp.

Fulbright, T.E., D.G. Hewitt, W.P. Kuvlesky Jr., and T.M. Langschied. 2008. The value of Live Oaks. Wildlife Management Bulletin No. 7. Caesar Kleberg Wildlife Research Institute, Texas A&M University-Kingsville. 7 pp.

Rosen, D.J., R. Carter, and C.T. Bryson. 2006. The recent spread of *Cyperus entrerianus* (Cyperaceae) in the southeastern United States and its invasive potential in bottomland hardwood forests. Southeastern Naturalist 5:333–344.

Rosen, D.J., and W.L. Miller. 2005. The vascular flora of an old-growth Columbia Bottomland forest remnant, Brazoria County, Texas. Texas Journal of Science 57: 223–250.

Wood, V.S. 1981. Live oaking: Southern timber for tall ships. Northeastern University Press, Boston. 206 pp.

UNWANTED WOODLAND

Barrow, W.C., Jr., and I. Renne. 2001. Interactions between migrant landbirds and an invasive exotic plant: The Chinese tallow tree. Flyway Newsletter 8:11. Texas Parks and Wildlife Department, Austin. [Unnumbered pages.]

Bell, M. 1996. Some notes and reflections upon a letter from Benjamin Franklin to Noble Wimberly Jones, October 7, 1772. Ashantilly Press, Darien, GA. 10 pp.

Bruce, K.A., G.N. Cameron, and P.A. Harcombe. 1995. Initiation of a new woodland type on the Texas Coastal Prairie by the Chinese Tallow tree (*Sapium sebiferum* (L.) Roxb.). Bulletin of the Torrey Botanical Club 122:215–225.

Bruce, K.A., G.N. Cameron, P.A.

Harcombe, and G. Jubinsky. 1997. Introduction, impact on native habitats, and management of a woody invader, the Chinese Tallow tree (*Sapium sebiferum* (L.) Roxb.). Natural Areas Journal 17: 255–260.

Colson, W., and A. Fedynich. 2016. Observations of unusual feeding behavior of White-winged Dove on Chinese Tallow. Southwestern Naturalist 61:133–135.

Jubinsky, G., and L.C. Anderson. 1996. The invasive potential of Chinese Tallow-tree (*Sapium sebiferum* Roxb.) in the southeast. Castanea 61:226–231.

Highlights

RICE CULTURE AND GEESE

Ankney, C.D. 1996. An embarrassment of riches: Too many geese. Journal of Wildlife Management 60:217–223.

Bolen, E.G., and M.K. Rylander. 1978. Feeding adaptations in the Lesser Snow Goose *Anser caerulescens.* Southwestern Naturalist 23:158–161.

Cooke, F.G., H. Finney, and R.F. Rockwell. 1976. Assortative mating in Lesser Snow Geese (*Anser caerulescens*). Behavior Genetics 6:127–140.

Flickinger, E.L. 1979. Effects of Aldrin exposure on Snow Geese in Texas rice fields. Journal of Wildlife Management 43:94–101.

Hobaugh, W.C. 1984. Habitat use by Snow Geese wintering in southeast Texas. Journal of Wildlife Management 48:1085–1096.

Hobaugh, W.C., C.D. Stutzenbaker, and E.L. Flickinger. 1989. The rice prairies. Pp. 367–383 *in* Habitat management for migrating waterfowl in North America (L.M. Smith, R.L. Pederson, and R.M. Kaminski, eds.). Texas Tech University Press, Lubbock. 560 pp.

Horn, E.E., and L.L. Glasgow.1964. Rice and waterfowl. Pp. 435–443 *in* Waterfowl tomorrow (J.P. Linduska, ed.). US Fish and Wildlife Service, Washington, DC. 770 pp.

Marty, J.R., J.B. Davis, R.M. Kaminski, et al. 2015. Waste rice and natural seed abundances in rice fields in the Louisiana and Texas Coastal Prairies. Journal of the Southeastern Association of Fish and Wildlife Agencies 2:121–126.

Meanley, B., and A.G. Meanley. 1959. Observations on the Fulvous Tree Duck in Louisiana. Wilson Bulletin 71:33–45.

Parsons, K.C., P. Mineau, and R.B. Renfrew. 2010. Effects of pesticide use in rice fields on birds. Waterbirds 33 (Special Publication 1):193–218.

GOPHERS AND GRASSLAND ECOLOGY

Block, S.B., and E.G. Zimmerman. 1991. Allozymic variation and systematics of Plains Pocket Gophers (*Geomys*) of south-central Texas. Southwestern Naturalist 36:29–36.

Spencer, S.R., G.N. Cameron, B.D. Eshelman, et al. 1985. Influence of pocket gopher mounds on a Texas coastal prairie. Oecologia 66:111–115.

Vleck, D. 1979. The energy cost of burrowing by the pocket gopher *Thomomys bottae*. Physiological Zoology 52:122–136.

Williams, L.R., and G.N. Cameron. 1986. Food habits and dietary preferences of Attwater's Pocket Gopher, *Geomys attwateri.* Journal of Mammalogy 67:489–496.

Williams, L.R., G.N. Cameron, S.R. Spencer, et al. 1986. Experimental analysis of the effects of pocket gopher mounds on the Texas Coastal Prairie. Journal of Mammalogy 67:672–679.

THE COASTAL BEND:
A REALM OF DIVERSITY

Anderson, J.T., and T.C. Tacha. 1999. Habitat use by Masked Ducks along the Gulf Coast of Texas. Wilson Bulletin 111:119–121.

Cortez, J.D., S.E. Henke, D.W. Wiemers, et al. 2015. Distribution and habitat selection by the Maritime Pocket Gopher. Southeastern Naturalist 14:41–56.

Cottam, C., and W.C. Glazener. 1959. Late nesting of water birds in South Texas. Transactions of the North American Wildlife Conference 24:382–395.

Gould, F.W., and T.W. Box. 1965. Grasses of the Texas Coastal Bend. Texas A&M University Press, College Station. 186 pp.

Johnsgard, P.A., and M. Carbonell. 1996. Ruddy Ducks and other stifftails: Their behavior and biology. University of Oklahoma Press, Norman. 291 pp.

Lehman, R.L., R. O'Brien, and T. White. 2005. Plants of the Texas Coastal Bend. Texas A&M University Press, College Station. 352 pp.

Rappole, J.H., T. Fulbright, J. Norwine, and R. Gingham. 1998. The forest-grassland boundary in Texas: Population differentiation without isolation. Pp. 167–177 in Proceedings, Third International Rangeland Congress, New Delhi, India. 374 pp.

Rappole, J.H., S. Glasscock, K. Goldberg, et al. 2011. Range change among New World tropical and subtropical birds. Pp. 151–167 in Tropical vertebrates in a changing world (K.-L. Schuchmann, ed.). Bonner Zoologische Monographien 57. Bonn, Germany. 204 pp.

Whitehouse, E. 1962. Common fall flowers of the Coastal Bend of Texas. Rob and Bessie Welder Wildlife Foundation Contribution 3 Series B. Sinton, TX. 116 pp.

AN IMMIGRANT EGRET

Scifres, C.J. 1975. Fall application of herbicides improved Macartney Rose-infested Coastal Prairie rangelands. Journal of Range Management 28:483–486.

Telfair, R.C., II. 1981. Cattle egrets, inland heronries, and the availability of crayfish. Southwestern Naturalist 26:37–41.

A SCAVENGING FALCON

Actkinson, M.A., W.P. Kuvlesky Jr., C.W. Boal, et al. 2007. Nesting habitat relationships of sympatric Crested Caracaras, Red-tailed Hawks, and White-tailed Hawks in South Texas. Wilson Bulletin of Ornithology 119:570–578.

Dickinson, V.M., and K.A. Arnold. 1996. Breeding biology of the Crested Caracara in South Texas. Wilson Bulletin 108:516–523.

Glazener, W.E. 1964. Notes on the feeding habits of the Caracara in South Texas. Condor 66:162.

Yosef, R., and D. Yosef. 1992. Hunting behavior of Audubon's Crested Caracara. Journal of Raptor Research 26:100–101.

A SQUEALING DUCK

Bolen, E.G. 1967. Nesting boxes for Black-bellied Tree Ducks. Journal of Wildlife Management 31:794–797.

———. 1971. Pair-bond tenure in the Black-bellied Tree Duck. Journal of Wildlife Management 34:385–388.

Bolen, E.G., and M.K. Rylander. 1983. Whistling-ducks: Zoogeography, ecology, and anatomy. Special Publications of the Museum of Texas Tech University 20:1–67.

Delnicki, D.E., and E.G. Bolen. 1975. Natural nest site availability for Black-bellied Whistling-Ducks in South Texas. Southwestern Naturalist 20:371–378.

ATTWATER'S PRAIRIE CHICKENS: NEXT TO GO?

Ballard, W.B., and R.J. Robel. 1974. Reproductive importance of dominant male Greater Prairie Chickens. Auk 91:75–85.

Fagan, W.F., and E.E. Holmes. 2006. Quantifying the extinction vortex. Ecology Letters 9:51–60.

Johnsgard, P.A. 1973. Grouse and quails of North America. University of Nebraska Press, Lincoln. 553 pp.

Lehmann, V.W. 1941. Attwater's Prairie Chicken: Its life history and management. North American Fauna 57. US Fish and Wildlife Service, Washington, DC. 65 pp.

Silvy, N.J., M.A. Griffin, M.A. Lockwood, et al. 1999. Attwater's Prairie Chicken: A lesson in conservation biology research. Pp. 153–162 in The Greater Prairie Chicken: A national look (W.D. Svedarsky, R.H. Hier, and N.J. Silvy, eds.). Minnesota Agricultural Experiment Station Miscellaneous Publication 99–1999. University of Minnesota, Saint Paul. 187 pp.

Silvy, N.J., M.J. Peterson, and R.R. Lopez. 2004. The cause of the decline of pinnated grouse: The Texas example. Wildlife Society Bulletin 32:16–21.

US Fish and Wildlife Service. 2010. Attwater's Prairie Chicken recovery plan. 2nd rev. Regional Office, Albuquerque, NM. 106 pp.

Conservation and Management

Abraham, K.F., R.L. Jefferies, and R.T. Alisauskas. 2005. The dynamics of landscape changes and Snow Geese in mid-continent North America. Global Change Biology 11:841–855.

Barrow, W.C., Jr., L.A. Johnson Randall, M.S. Woodrey, et al. 2005. Coastal forests of the Gulf of Mexico: A description and some thoughts on their conservation. Pp. 450–464 *in* Bird conservation, implementation and integration in the Americas (C.J. Ralph and T.D. Rich, eds.). General Technical Report PSW-GTR-191. USDA Forest Service Pacific Research Station, Albany, CA. 643 pp.

Grace, J.B. 1998. Can prescribed fire save the endangered Coastal Prairie ecosystem from Chinese Tallow invasion? Endangered Species Update 15:70–76.

Morrow, M.E., R.E. Chester, S.E. Lehnen, et al. 2015. Indirect effects of Red Imported Fire Ants on Attwater's Prairie-Chicken brood survival. Journal of Wildlife Management 79:898–906.

Rosen, D.J. 2014. Addendum to the vascular flora of Nash Prairie, Texas, U.S.A. Journal of the Botanical Research Institute of Texas 8:381–382.

Rosen, D.J., D. DeSteven, and M.L. Lange. 2008. Conservation strategies and vegetation characterization in the Columbia Bottomlands, an under-recognized southern floodplain forest formation. Natural Areas Journal 28:74–82.

CHAPTER 11 INFOBOX. ROB AND BESSIE WELDER WILDLIFE FOUNDATION

Drawe, D.L., A.D. Chamrad, and T.W. Box. 1978. Plant communities of the Welder Wildlife Refuge. Contribution No. 5, Series B, revised. Welder Wildlife Foundation, Sinton, TX. 38 pp.

Kahlich, S.N. 2014. Leaders in the making: How the Welder Wildlife Foundation shapes future wildlifers. Wildlife Professional 8(3):70–73.

Rezsutek, M.J., and G.N. Cameron. 2011. Diet selection and plant nutritional quality in Attwater's Pocket Gopher (*Geomys attwateri*). Mammalian Biology 76:428–435.

Schroeder, E., C. Boal, and S.N. Glasscock. 2013. Nestling diets and provisioning rates of sympatric Golden-fronted and Ladder-backed Woodpeckers. Wilson Journal of Ornithology 125:188–192.

Teer, J.G., D.L. Drawe, T.L. Blankenship, et al. 1991. Deer and Coyotes: The Welder experiments. Transactions of the North American Wildlife and Natural Resources Conference 56:550–560.

Young, J.K., W.F. Andelt, P.A. Terletzky, and J.A. Shivik. 2006. A comparison of Coyote ecology after 25 years: 1978 versus 2003. Canadian Journal of Zoology 84:573–582.

CHAPTER 12

Blackburn, J. 2004. The book of Texas bays. Texas A&M University Press, College Station. 290 pp.

Britton, J.C., and B. Morton. 1989. Shore ecology of the Gulf of Mexico. University of Texas Press, Austin. 395 pp.

Lehman, R.L. 2013. Marine plants of the Texas Coast. Texas A&M University Press, College Station. 205 pp.

Tunnell, J.W., Jr., J. Andrews, N.C. Barerra, and R. Moretzohn. 2010. Encyclopedia of Texas seashells: Identification, ecology, distribution, and history. Texas A&M University Press, College Station. 512 pp.

Tunnell, J.W., Jr., and F.W. Judd, eds. 2002. The Laguna Madre of Texas and Tamaulipas. Texas A&M University Press, College Station. 346 pp.

Climates and Gulf Currents

Hamilton, P. 1990. Deep currents in the Gulf of Mexico. Journal of Physical Oceanography 20:1087–1104.

Hamilton, P., G.S. Fargion, and D.C. Biggs. 1999. Loop current eddy paths in the western Gulf of Mexico. Journal of Physical Oceanography 29:1180–1207.

Coastal Ecosystems

Behrens, E.W. 1963. Buried Pleistocene river valleys in Aransas and Baffin Bays, Texas. Publications of the Marine Science Institute, University of Texas 9:7–18.

Keddy, P.A. 2010. Wetland ecology: Principles and conservation. Cambridge University Press, Cambridge, UK. 497 pp.

GULF BEACHES

Brown, A.C., and A. McLachlan. 1990. Ecology of sandy shores. Elsevier, Amsterdam. 328 pp.

Bullard, F.M. 1942. Source of beach and river sands on the Gulf of Mexico of Texas. Geological Society of America Bulletin 53:1021–1044.

BARRIER ISLANDS AND PENINSULAS

Drawe, D.L., K.R. Katner, W.H. McFarland, and D.D. Neher. 1981. Vegetation and soil properties of five habitat types on North Padre Island, Texas. Texas Journal of Science 33:145–157.

Lonard, R.I., and F.W. Judd. 1980. Phytogeography of South Padre Island, Texas. Southwestern Naturalist 23:497–510.

Otvos, E.G., Jr. 1970. Development and migration of barrier islands, northern Gulf of Mexico. Geological Society of America Bulletin 81:241–246.

Smith, E.H. 2002. Barrier islands.

Pp. 127–136 *in* The Laguna Madre of Texas and Tamaulipas (J.W. Tunnell Jr. and F.W. Judd, eds.). Texas A&M University Press, College Station. 346 pp.

COASTAL MARSHES

Adam, P. 1990. Saltmarsh ecology. Cambridge University Press, New York. 476 pp.

Chabreck, R.H. 1988. Coastal marshes: Ecology and management. University of Minnesota Press, Minneapolis. 138 pp.

Hinde, H.P. 1954. The vertical distribution of salt marsh phanerograms in relation to tide levels. Ecological Monographs 24: 209–242.

Hunt, H.E., and R.D. Slack. 1989. Winter diets of Whooping and Sandhill Cranes in South Texas. Journal of Wildlife Management 53:1150–1154.

Texas Department of Water Resources. 1980. Guadalupe estuary: A study of the influence of freshwater inflows. Texas Department of Water Resources, Austin. 386 pp.

Tiner, R. 2013. Tidal wetlands primer: An introduction to their ecology, natural history, status, and conservation. University of Massachusetts Press, Amherst. 560 pp.

Wozniak, J.R., T.M. Swannack, R. Butzler, et al. 2012. River inflow, estuarine salinity, and Carolina Wolfberry fruit abundance: Linking abiotic drivers to Whooping Crane food. Journal of Coastal Conservation 16:345–354.

OPEN BAY BOTTOMS

Armstrong, N.E. 1987. The ecology of open-bay bottoms of Texas: A community profile. US Fish and Wildlife Service Biological Report 85(7.12). Washington, DC. 105 pp.

Withers, K. 2002. Open bay. Pp. 102–113 *in* The Laguna Madre of Texas and Tamaulipas (J.W. Tunnell Jr. and F.W. Judd, eds.). Texas A&M University Press, College Station. 346 pp.

SEAGRASS MEADOWS

Hemminga, M.A., and C.M. Duarte. 2000. Seagrass ecology. Cambridge University Press, New York. 312 pp.

Larkum, A., R.J. Orth, and C. Duarte, eds. 2006. Seagrasses: Biology, ecology and conservation. Springer, Dordrecht, Netherlands. 676 pp.

McMahan, C.A. 1968. Biomass and salinity tolerance of Shoalgrass and Manatee-grass in the lower Laguna Madre. Journal of Wildlife Management 32:501–506.

Mitchell, C.A., T.W. Custer, and P.J. Zwank. 1994. Herbivory on Shoalgrass by wintering Redheads in Texas. Journal of Wildlife Management 58:131–141.

Onuf, C.P. 1996. Biomass patterns in seagrass meadows of the Laguna Madre, Texas. Bulletin of Marine Science 58:404–420.

Withers, K. 2002. Seagrass meadows. Pp. 85–101 *in* The Laguna Madre of Texas and Tamaulipas (J.W. Tunnell Jr. and F.W. Judd, eds.). Texas A&M University Press, College Station. 346 pp.

OYSTER REEFS

Doran, E., Jr. 1965. Shell roads in Texas. Geographical Review 55: 223–240.

Galtsoff, P. 1964. The American Oyster, *Crassostrea virginica* Gmelin. US Fish and Wildlife Service Bulletin 64:1–480.

Plinket, J.T., and M. La Peyre, 2005. Oyster reefs as fish and macroinvertebrate habitat in Barataria Bay, Louisiana. Bulletin of Marine Science 77:155–164.

Tunnell, J.W., Jr., and J.W. Tunnell. 2015. Pioneering archaeology in the Texas Coastal Bend: The Pape-Tunnell collection. Texas A&M University Press, College Station. 358 pp.

LAGUNA MADRE

Breuer, J.P. 1962. An ecological survey of the Lower Laguna Madre of Texas, 1953–1959. Publications of the Institute of Marine Science 8:153–183.

Carroll, J.J. 1927. Down Bird Island way. Wilson Bulletin 39:195–208.

Copeland, B.J. 1967. Environmental characteristics of hypersaline lagoons. Contributions in Marine Science 12:207–218.

Hedgepeth, J.W. 1967. Ecological aspects of the Laguna Madre, a hypersaline estuary. Pp. 408–419 *in* Estuaries (G.H. Lauff, ed.). Publication 83. American Association for the Advancement of Science, Washington, DC. 757 pp.

Weller, M.W. 1964. Distribution and migration of the Redhead. Journal of Wildlife Management 28: 69–103.

WIND-TIDAL FLATS

Brush, T. 1995. Habitat use by wintering shorebirds along the lower Laguna Madre of South Texas. Texas Journal of Science 47: 179–190.

Pulich, W., Jr., and S. Rabalais. 1986. Primary production potential of blue-green algal mats on southern Texas tidal flats. Southwestern Naturalist 31:39–47.

Withers, K. 2002. Wind-tidal flats. Pp. 114–126 *in* The Laguna Madre of Texas and Tamaulipas (J.W. Tunnell Jr. and F.W. Judd, eds.). Texas A&M University Press, College Station. 346 pp.

MANGROVE THICKETS

Banchi, T.S., M.A. Allison, J. Zhao, et al. 2013. Historical reconstruction of mangrove expansion in the Gulf of Mexico: Linking climate change with carbon sequestration in coastal wetlands. Estuarine, Coastal and Shelf Science 119:7–16.

McMillan, C., and C.L. Sherrod. 1986. The chilling tolerance of Black Mangrove, *Avicenna germinans,* from the Gulf of Mexico coast of Texas, Louisiana and Florida. Contributions in Marine Science 29:9–16.

Sherrod, C.L., and C. McMillan 1981. Black Mangrove, *Avicennia germinans,* in Texas: Past and present distribution. Contributions in Marine Science 24:129–140.

———. 1986. The distributional history and ecology of mangrove vegetation along the northern Gulf of Mexico coastal region. Contributions in Marine Science 26:129–140.

MARITIME FORESTS

Buler, J.J., F.R. Moore, and S. Woltmann. 2007. A multi-scale examination of stopover habitat use by birds. Ecology 88:1789–1802.

Gauthreaux, S.A. 1971. A radar and direct visual study of passerine spring migration in southern Louisiana. Auk 88:343–365.

Lowery, G.H. 1945. Trans-Gulf migration of birds and the coastal hiatus. Wilson Bulletin 57:92–121.

Highlights

ROCKS AND DOCKS

Breuer, J.P. 1957. An ecological survey of Baffin and Alazan Bays, Texas. Publications of the Institute of Marine Science, University of Texas 4:134–155.

Connell, J.H. 1972. Community interactions on marine rocky intertidal shores. Annual Review of Ecology and Systematics 31:169–192.

Payne, R.T. 1974. Intertidal community structure. Oecologia 15:93–129.

Raffaelli, D.G., and S.J. Hawkins. 1996. Intertidal ecology. Kluwer Academic Publishers, Dordrecht, Netherlands. 356 pp.

WATERBIRD COLONIES

Buckley, F.G., and P.A. Buckley. 1972. Hexagonal packing of Royal Tern nests. Auk 94:36–43.

Burger, J., and C.G. Beer. 1975. Territoriality in the Laughing Gull (*Larus atricilla*). Behaviour 55:307–320.

Chaney, A.H., B.R. Chapman, J.P. Karges, et al. 1978. Use of dredged material islands by colonial seabirds and wading birds in Texas. US Army Corps of Engineers Dredged Material Research Program Technical Report D-78-8. Vicksburg, MS. 170 pp.

Kharitonov, S.P., and D. Siegel-Causey. 1988. Colony formation in seabirds. Current Ornithology 5:223–272.

Mock, D.W., H. Drummond, and C.H. Stinson. 1990. Avian siblicide. American Scientist 78:438–449.

Ward, P., and A. Zahavi. 1973. The importance of certain assemblages of birds as "Information Centres" for food-finding. Ibis 115:517–534.

AN IMPROBABLE FISH

Foster, S.J., and A.C.J. Vincent. 2004. Life history and ecology of seahorses: Implications for conservation and management. Journal of Fish Biology 65:1–61.

Gemmell, B.J., J. Sheng, and E.J. Buskey. 2013. Morphology of seahorse head hydrodynamically aids in capture of elusive prey. Nature Communications 4. doi:10.1038/ncomms3840 (accessed December 31, 2015).

Teske, P.B., and L.B. Beheregaray. 2009. Evolution of seahorses' upright posture was linked to Oligocene expansion of seagrass habitats. Biology Letters 5:521–523.

Vincent, A.C.J. 1990. A seahorse father makes a good mother. Natural History 12:34–43.

———. 1995. A role for daily greetings in maintaining seahorse pair bonds. Animal Behavior 49:258–260.

Conservation and Management

Chapman, B.R. 1984. Seasonal abundance and habitat-use patterns of coastal bird populations on Padre and Mustang Island barrier beaches (Following the IXTOC I oil spill). FWS/OBS-82/10-78. US Fish and Wildlife Service. 73 pp.

Coplin, L.S., and D.L. Galloway. 1999. Houston-Galveston, Texas: Managing coastal subsidence. Pp. 35–48 *in* Land subsidence in the United States. US Geological Circular 1182. Washington, DC. 177 pp.

Davis, R.A., Jr. 2011. Sea-level change in the Gulf of Mexico. Texas A&M University Press, College Station. 192 pp.

Dunton, K.H., S. Schonberg, S. Herzka, et al. 1998. Characterization of anthropogenic and natural disturbance on vegetated and unvegetated bay bottom habitats in the Corpus Christi National Estuary Program study area. Vol. 2. Assessment of propeller scarring in seagrass beds. Texas Natural Resource Conservation Commission Publication CCBNEP-23. Austin. 65 pp.

Emanuel, K.A. 2005. Increasing destructiveness of tropical cyclones over the past 30 years. Nature 436:686–688.

Paine, J.G., and R.A. Morton. 1986. Historical shoreline changes in Trinity, Galveston, West and East Bays, Texas Gulf Coast. Bureau of Economic Geology Circular 86–3. University of Texas, Austin. 58 pp.

Resource Protection Division. 1999. Seagrass monitoring plan for Texas. Texas Parks and Wildlife Department, Austin. 67 pp.

Titus, J.G., and V.K. Narayanan. 1995. The probability of sea level rise. US Environmental Protection Agency, Washington, DC. 184 pp.

Infobox 12.1. Colored Waters: Red and Brown Tides

Buskey, E.J., B. Wysor, and C. Hyatt. 1998. The role of hypersalinity in the persistence of the "Texas Brown Tide" in the Laguna Madre. Journal of Plankton Research 20: 1553–1563.

Flewelling, L.J., J.P. Naar, J.P. Abbott, et al. 205. Red tides and marine mammal mortalities. Nature 435: 753–756.

Magaña, H.A., C. Contreras, and T.A. Villareal. 2003. A historical assessment of *Karenia brevis* in the western Gulf of Mexico. Harmful Algae 2:163–171.

Onuf, C.P. 1996. Seagrass response to long-term light reduction by brown tide in upper Laguna Madre, Texas. Marine Ecology Progress Series 138:219–231.

Potera, C. 2007. Florida red tide brews up drug lead for cystic fibrosis. Science 316:1561–1562.

Walsh, J.J., J.K. Joliff, B.P. Darrow, et al. 2010. Red tides in the Gulf of Mexico: Where, when, and why? Journal of Geophysical Research 111:1–46.

Infobox 12.2. Edward H. Harte (1922–2011): Newsman and Conservationist

Corpus Christi Caller-Times. 2011. Edward H. Harte: A life well spent in service of others. Editorial, May 20.

Heven, D. 2011. Edward H. Harte, Texas newspaper executive, dies at 88. Obituary, New York Times, May 23.

Meyers, R. 2011. Edward H. Harte, former Caller-Times publisher and a philanthropist, dies at 88. Obituary, Corpus Christi Caller-Times, May 18.

Infobox 12.3. Connie Hagar (1886–1973): The Bird Lady of Rockport

McCracken, K.H. 1986. Connie Hagar: The life history of a Texas birdwatcher. Texas A&M University Press, College Station. 296 pp.

APPENDIX A

Andrews, J. 1993. The Texas Bluebonnet. University of Texas Press, Austin. 64 pp.

Bolotin, N., and C. Laing. 2002. The World's Columbian Exposition: The Chicago World's Fair of 1893. University of Illinois Press, Chicago. 176 pp.

Rosenberg, C.M. 2008. America at the fair: Chicago's 1893 World Columbian Exposition. Arcadia Publishing, Mount Pleasant, SC. 288 pp.

The State Flower of Texas: Bluebonnets (1901, 1971)

Andrews, J. 1993. The Texas Bluebonnet. University of Texas Press, Austin. 64 pp.

Christian, C. 2014. UT decides Longhorns have been pranked, will rip out maroon bluebonnets. Houston Chronicle, April 15.

House Concurrent Resolution No. 44. 62nd Legislature, 1971. Texas Statutes, Title 11, Chapter 3101, Section 3101.008 (recognizes all varieties).

House Concurrent Resolution No. 24. 43rd Legislature, 1933. Regular Session. [Flower Song.]

House Concurrent Resolution No. 242. 71st Legislature, 1989. Regular Session. [Tartan.]

House Concurrent Resolution No. 116. 75th Legislature, 1997. Regular Session. [City, Festival, and Trail.]

Parsons, J.M., and T.D. Davis. 1993. 'Abbott Pink' Bluebonnet (*Lupinus texensis* Hook). HortScience 28:65–66.

Senate Concurrent Resolution No. 10. 43rd Legislature, 1901. Regular Session. [Flower.]

Simmons, M.T. 2005. Bullying the bullies: The selective control of an exotic, invasive annual (*Rapistrum rugosum*) by oversowing with a competitive native species (*Gaillardia pulchella*). Restoration Ecology 13:609–615.

The State Tree of Texas: Pecan (1919)

Hall, G.D. 2000. Pecan food potential in prehistoric North America. Economic Botany 54:103–112.

Manaster, J. 2008. Pecans: The story in a nutshell. Texas Tech University Press, Lubbock. 120 pp.

Senate Concurrent Resolution No. 10. Senate Bill No. 97, 36th Legislature, 1919. Approved by House of Representatives by a viva voce vote, 1919. Texas Statutes, Title 11, Chapter 3101, Section 3101.009. [Tree.]

Senate Concurrent Resolution No. 2. 77th Legislature, 2001. Regular Session. [Health Nut.]

Senate Concurrent Resolution No. 22. 83rd Legislature, 2013. Regular Session. [Pie.]

The State Bird of Texas: Northern Mockingbird (1927)

Casto, S.D. 2002. The mockingbird in early Texas. Texas Birds 4(1):5–9.

Derrickson, K.C., and R. Breitwisch. 1992. Northern Mockingbird. The birds of America No. 7 (A. Poole, P. Stettenheim, and F. Gill, eds). Academy of Natural Sciences, Washington, DC, and American Ornithologists' Union. 26 pp.

Dhondt, A.A., and K.M. Kemink. 2008. Wing-flashing in Northern Mockingbirds: Anti-predator defense? Journal of Ethology 26: 361–365.

Doughty, R.W. 1988. The mockingbird. University of Texas Press, Austin. 80 pp.

Hayslette, S.E. 2003. A test of the foraging function of wing-flashing in Northern Mockingbirds. Southeastern Naturalist 2:93–98.

Levey, D.J., G.A. Londoño, J. Ungvari-Martin, et al. 2009. Urban mockingbirds quickly learn to identify individual humans. Proceedings of the National Academy of Sciences 106:8959–8962.

Senate Concurrent Resolution No. 8. 40th Legislature, 1927. Texas Statutes, Title 1, Chapter 3101, Section 3101.007.

The State Stone of Texas: Petrified Palmwood (1969)

Daniels, F.J., and R.D. Dayvault. 2006. Ancient forests: A closer look at fossil wood. Western Colorado Publishing, Grand Junction. 456 pp.

House Concurrent Resolution No. 12. 61st Legislature, 1969. Regular Session.

McMackin, C.E. 1984. Petrified wood from east to west; some we've liked best. Lapidary-Journal 11: 1582–1588.

The State Gem of Texas: Texas Blue Topaz (1969)

House Concurrent Resolution No. 12. 61st Legislature, 1969. Regular Session. [Gem.]

House Concurrent Resolution No. 97, 65th Legislature, 1977. Regular Session. [Gemstone Cut.]

Schumann, W. 2013. Gemstones of the world. Newly revised 5th ed. Sterling, New York. 320 pp.

The State Grass of Texas: Sideoats Grama (1971)

Hatch, S.L., and J. Pluhar. 1993. Texas range plants. Texas A&M University Press, College Station. 344 pp.

Judd, I. 1974. Plant succession of old fields in the Dust Bowl. Southwestern Naturalist 19:227–239.

Senate Concurrent Resolution No. 31. 62nd Legislature, 1971. Regular Session.

Shaw, R.B. 2012. Guide to Texas grasses. Texas A&M University Press, College Station. 1080 pp.

Shaw, R.B., B.S. Rector, and A.M. Dube. 2011. Distribution of grasses in Texas. Botanical Research Institute of Texas, Fort Worth. 196 pp.

Telfair, R.C., II. 2006. Native plants important to wildlife in East Texas. Pp. 1271–1288 in Illustrated flora of East Texas (G.M. Diggs Jr., B.L. Lipscomb, M.D. Reed, and R.J. O'Kennon, eds.). Vol. 1. Center for Environmental Studies, Austin College, Sherman, TX, and Botanical Research Institute of Texas, Fort Worth. 1594 pp.

The State Shell of Texas: Lightning Whelk (1987)

House Concurrent Resolution No. 75. 70th Legislature, 1987. Regular Session.

Sturm, C.F., T.F. Pearce, and A. Valdes, eds. 2006. The mollusks: A guide to their study, collection, and preservation. Universal Publishers, Boca Raton, FL. 445 pp.

Tunnell, J.W., Jr., J. Andrews, N.C. Barerra, and R. Moretzohn. 2010. Encyclopedia of Texas seashells: Identification, ecology, distribution, and history. Texas A&M University Press, College Station. 512 pp.

Tunnell, J.W., Jr., and J.W. Tunnell. 2015. Pioneering archaeology in the Texas Coastal Bend: The Pape-Tunnell collection. Texas A&M University Press, College Station. 358 pp.

The State Fish of Texas: Guadalupe Bass (1989)

House Concurrent Resolution No. 161. 71st Legislature, 1989. Regular Session.

Hubbs, C. 1954. Corrected distributional records for Texas freshwater fishes. Texas Journal of Science 5:277–291.

Hubbs, C., R.A. Kuehne, and J.C. Ball. 1954. The fishes of the upper Guadalupe River, Texas. Texas Journal of Science 5:216–244.

Hurst, H., G. Bass, and C. Hubbs. 1975. The biology of the Guadalupe, Suwanee, and Redeye Basses. Pp. 47–53 in Black Bass biology and management (R. Stroud and H. Clepper, eds.). Sport Fishing Institute, Washington, DC. 534 pp.

Thomas, C., T.H. Bonner, and B.G. Whiteside. 2007. Freshwater fishes of Texas. Texas A&M University Press, College Station. 220 pp.

Whitmore, D.H., and W. Butler. 1982. Interspecific hybridization of Smallmouth and Guadalupe Bass (Micropterus): Evidence based on biochemical genetic and morphological analyses.

Southwestern Naturalist 27: 99–106.

The State Reptile of Texas: Texas Horned Lizard (1993)

House Concurrent Resolution No. 141. 73rd Legislature, 1993. Regular Session.

Schmidt, P.J., W.C. Sherbrooke, and J.O. Schmidt. 1989. The detoxification of ant (*Pogonomyrmex*) venom by a blood factor in horned lizards (*Phrynosoma*). Copeia 1989:603–607.

Sherbrooke, W.C. 1990. Rain-harvesting in the lizard, *Phrynosoma cornutum:* Behavior and integumental morphology. Journal of Herpetology 24:302–308.

Sherbrooke, W.C., J.R. Mason, and C.A. Jones. 2005. Sensory modality used by Coyotes in responding to antipredator compounds in the blood of Texas Horned Lizards. Southwestern Naturalist 50:216–222.

Sherbrooke, W.C., and G.A. Middledorf III. 2004. Responses of Kit Fox (*Vulpes macrotis*) to antipredator blood-squirting and the blood of Texas Horned Lizards, *Phrynosoma cornutum.* Copeia 2004:652–658.

Whitford, W.G., and M. Bryant. 1979. Behavior of a predator and its prey: The horned lizard (*Phrynosoma cornutum*) and harvester ants (*Pogonomyrmex* spp.). Ecology 60:686–694.

The State Plant of Texas: Prickly Pear Cactus (1995)

Donkin, R.A. 1977. Spanish red: An ethnographical study of cochineal and the *Opuntia* cactus. Transactions of the American Philosophical Society 67:1–84.

House Concurrent Resolution No.

44. 74th Legislature, 1995. Regular Session.

Kane, C.W. 2011. Medicinal plants of the American Southwest. Lincoln Town Press, Tucson, AZ. 368 pp.

Pinkava, D.J. 2003. *Opuntia.* Pp. 123–148 *in* Flora of North America. Vol. 4. Oxford University Press, New York. 584 pp.

Pinkava, D.J., B.D. Parfitt, and J.B. Rebman. 2001. Nomenclatural changes in *Cylindropuntia* and *Opuntia* (Cactaceae) and notes on interspecific hybridization. Journal of the Arizona-Nevada Academy of Science 33:162–163.

Powell, A.M., and J.F. Weedin. 2004. Cacti of the Trans-Pecos and adjacent areas. Texas Tech University Press, Lubbock. 509 pp.

Weniger, D. 1984. Cacti of Texas and neighboring states. University of Texas Press, Austin. 356 pp.

Zimmermann, H.G., V.C. Moran, and J.H. Hoffman. 2000. The renowned Cactus Moth, *Cactoblastis cactorum:* Its natural history and threat to native *Opuntia* floras in Mexico and the United States of America. Diversity and Distributions 6:259–269.

The State Insect of Texas: Monarch Butterfly (1995)

Anderson, J.B., and L.P. Brower. 1996. Freeze-protection of overwintering Monarch Butterflies in Mexico: Critical role of the forest as a blanket and an umbrella. Ecological Entomology 21:107–116.

Brower, L.P. 1995. Understanding and misunderstanding the migration of Monarch Butterfly (Nymphalidae) in North America: 1857–1995. Journal of the Lepidopterists' Society 49: 304–385.

Guerra, P., and S.M. Reppert. 2013. Coldness triggers northward flight

in remigrant Monarch Butterflies. Current Biology 23:419–423.

House Concurrent Resolution No. 94. 74th Legislature, 1995. Regular Session.

Pyle, R.M. 2014. Chasing monarchs: Migrating with the butterflies of passage. Yale University Press, New Haven, CT. 336 pp.

Reppert, S.M., H. Zhu, and R.H. White. 2004. Polarized light helps Monarch Butterflies navigate. Current Biology 14:155–158.

Smith-Rodgers, S. 2016. Maiden of the monarchs. Texas Parks and Wildlife 74(2):54–57. [Recounts discovery of a major wintering area in Mexico by a self-taught naturalist now living in Texas.]

Solensky, M.J., and K.S. Oberholser. 2004. The Monarch Butterfly: Biology and conservation. Comstock, Ithaca, NY. 256 pp.

Texas Parks and Wildlife Department. 2015. Texas monarch and native pollinator conservation plan. Texas Parks and Wildlife Department, Austin. 41 pp.

The State Flying Mammal of Texas: Mexican Free-tailed Bat (1995)

Ammerman, L.G., C.L. Hice, and D.J. Schmidly. 2012. Bats of Texas. Texas A&M University Press, College Station. 305 pp.

Barbour, R.W., and W.H. Davis. 1969. Bats of America. University Press of Kentucky, Lexington. 286 pp.

Cleveland, C.J., M. Betke, P. Frederico, et al. 2006. Economic value of pest control service provided by Brazilian Free-tailed Bats in south-central Texas. Frontiers in Ecology and the Environment 5:238–243.

McWilliams, L.A. 2005. Variation in the diet of the Mexican Free-tailed Bat (*Tadarida brasiliensis mexicana*). Journal of Mammalogy 86:599–605.

Schmidly, D.J., and R.D. Bradley. 2016. The mammals of Texas. 7th ed. University of Texas Press, Austin. 720 pp.

Senate Concurrent Resolution No. 95. 74th Legislature, 1995. Regular Session.

Wilkins, K.T. 1989. *Tadarida brasiliensis.* Mammalian Species 331:1–10.

Williams, T.C., L.C. Ireland, and J.M. Williams. 1973. High altitude flights of the Free-tailed Bat, *Tadarida brasiliensis,* observed with radar. Journal of Mammalogy 54:807–821.

The State Small Mammal of Texas: Armadillo (1995)

Audubon, J.J., and J. Bachman. 1989. Audubon's quadrupeds of North America. Wellfleet Press, Secaucus, NJ. 440 pp. [An unabridged reprint of the original work, The viviparous quadrupeds of North America, published as a three-volume folio in 1846.]

Enders, A.C. 2002. Implantation in the Nine-banded Armadillo: How does a single blastocyst form four embryos? Placenta 32:71–86.

Humphrey, S.R. 1974. Zoogeography of the Nine-banded Armadillo (*Dasypus novemcinctus*) in the United States. BioScience 24: 457–462.

Kipling, R. 2014. Just so stories. Aziloth Books, London. 108 pp. [Reprint of a children's story collection first published in 1902.]

Loughry, W.J., and C.M. McDonough. 2013. The Nine-banded Armadillo: A natural history. University of Oklahoma Press, Norman. 323 pp.

Taulman, J.F., and L.W. Robbins. 1996. Recent range expansion and distributional limits of the Nine-banded Armadillo (*Dasypus novemcinctus*) in the United States. Journal of Biogeography 23:635–648.

The State Native Pepper of Texas: Chiltepin (1997)

Horst, T. 2001. Native Seeds/SEARCH tradition and conservation. Cultural Resource Management 24:23–26.

House Concurrent Resolution No. 82. 75th Legislature, 1997. Regular Session.

Logsdon, J.W. 2011. Pick a pepper: A photographic guide to chile peppers, their history, and uses. CreateSpace, Wolcott, CT. 88 pp.

Scoville, W. 1912. Note on capsicums. Journal of the American Pharmaceutical Association 1:453–454.

The State Native Shrub of Texas: Texas Purple Sage (2005)

House Concurrent Resolution No. 71. 79th Legislature, 2005. Regular Session.

Powell, A.M. 1998. Trees and shrubs of the Trans-Pecos and adjacent areas. University of Texas Press, Austin. 498 pp.

Vines, R.A. 2004. Trees, shrubs and woody vines of the southwest. 5th ed. Blackburn Press, Caldwell, NJ. 1104 pp.

Wrede, J. 2010. Trees, shrubs, and vines of the Texas Hill Country: A field guide. 2nd ed. Texas A&M University Press, College Station. 272 pp.

The State Amphibian of Texas: Texas Toad (2009)

Dixon, J.R. 2013. Amphibians and reptiles of Texas: With keys, taxonomic synopses, bibliography, and distribution maps. 3rd ed. Texas A&M University Press, College Station. 447 pp.

Holm, J.A. 2004. Geographic distribution, *Bufo speciosus* (Texas Toad). Herpetological Review 35: 77.

House Concurrent Resolution No. 30. 80th Legislature, 2007. Regular Session. [Texas Blind Salamander.]

House Concurrent Resolution No. 118. 81st Legislature, 2009. Regular Session. [Texas Toad.]

Malone, J.H. 1993. Natural history notes: *Bufo speciosus* (Texas Toad). Diet. Herpetological Review 30:222–223.

Perry, R. 2009. Proclamation by the governor of the state of Texas. House Journal, 80th Legislature, Regular Session. [See page 6668 for the proclamation rejecting the Texas Blind Salamander.]

Saleh Rauf, D. 2009. 4th-graders have hearts set on naming state amphibian. Houston Chronicle, January 16.

The State Dinosaur of Texas: *Paluxysaurus jonesi* (2009)

D'Emic, M.D., and B.Z. Foreman. 2012. The beginning of sauropod dinosaur hiatus in North America: Insights from the Lower Cretaceous Cloverly Formation of Wyoming. Journal of Vertebrate Paleontology 32:883–902.

House Concurrent Resolution No. 16. 81st Legislature, 2009. Regular Session.

Rose, P.J. 2007. A new titanosauriform sauropod (Dinosauria: Saurischia) from the Early Cretaceous of Central Texas and its phylogenetic relationships. Paleontologia Electronica 10:1–65.

Senate Concurrent Resolution No. 57. 75th Legislature, 1997. Regular Session.

The State Bison Herd of Texas: Bison Herd at Caprock Canyons State Park (2011)

Halbert, N.D., W.E. Grant, and J.N. Derr. 2005. Genetic and demographic consequences of importing animals into a small population: A simulation model of the Texas State Bison Herd (USA). Ecological Modeling 181: 263–276.

Halbert, N.D., T. Raudsepp, B.P. Chowdhary, and J.N. Derr. 2004. Conservation genetic analysis of the Texas State Bison Herd. Journal of Mammalogy 85:924–931.

Hornaday, W.T. 2002. The extermination of the American Bison. Smithsonian Institution Scholarly Press, Washington, DC. 240 pp. [Reprint of an 1889 classic.]

House Concurrent Resolution No. 86. 82nd Legislature, 2011. Regular Session.

Ienberg, A.C. 2001. The destruction of the bison: An environmental history, 1750–1920. Cambridge University Press, Cambridge, UK. 220 pp.

Lott, D.F., and H.W. Greene. 2002. American Bison: A natural history. University of California Press, Berkeley. 264 pp.

Meinzer, W., R. Roe, and A. Sansom. 2011. Southern Plains Bison: Restoration of the lost Texas herd. Badlands Blue Star Publications, Dallas, TX. 168 pp.

The State Saltwater Fish of Texas: Red Drum (2011)

Beckman, D.W., C.A. Wilson, and A.L. Stanley. 1989. Age and growth of Red Drum, *Scianops ocellatus,* from offshore waters of the northern Gulf of Mexico. Fishery Bulletin 87:17–28.

Davis, J.T. 1990. Red Drum biology and life history. SRAC Publication No. 320. Southern Regional Aquaculture Center, Texas Agricultural Extension Service, College Station. 2 pp.

Hoese, H.D., and R.H. Moore. 1998. Fishes of the Gulf of Mexico: Texas, Louisiana, and adjacent waters. Texas A&M University Press, College Station. 422 pp.

House Concurrent Resolution No. 133. 82nd Legislature, 2011. Regular Session.

Moore, C., Jr. 2003. Texas reds (saltwater strategies). Texas Fish and Game, Houston. 202 pp.

The State Sea Turtle of Texas: Kemp's Ridley Sea Turtle (2013)

House Concurrent Resolution No. 31. 83rd Legislature, 2013. Regular Session.

National Marine Fisheries Service, US Fish and Wildlife Service, and Secretary of Environment and Natural Resources, Mexico. 2011. National recovery plan for the Kemp's Ridley Sea Turtle (*Lepidochelys kempii*). 2nd rev. National Marine Fisheries Service, Silver Spring, MD. 156 pp.

Plotkin, P.T., ed. 2007. Biology and conservation of ridley sea turtles. Johns Hopkins University Press, Baltimore, MD. 356 pp.

Rostal, D.C., D.W. Owens, J.S. Grumbles, et al. 1998. Seasonal reproductive cycle of the Kemp's Ridley Sea Turtle (*Lepidochelys kempii*). General and Comparative Endocrinology 109: 232–243.

Spotila, J.R. 2004. Sea turtles: A complete guide to their biology, behavior, and conservation. Johns Hopkins University Press, Baltimore, MD. 227 pp.

INDEX

Note: Page numbers in *italics* denote illustrations.